Adiabatic Shear Localization

Adiabatic Shear Localization
Frontiers and Advances

Second Edition

Edited by

Bradley Dodd

Institute of Shock Physics
Imperial College
London, UK

Yilong Bai

State Key Laboratory of Non-linear Mechanics (LNM),
Institute of Mechanics
Chinese Academy of Sciences
Beijing, China

ELSEVIER

AMSTERDAM • BOSTON • HEIDELBERG • LONDON • NEW YORK • OXFORD
PARIS • SAN DIEGO • SAN FRANCISCO • SINGAPORE • SYDNEY • TOKYO

Elsevier
32 Jamestown Road, London NW1 7BY
225 Wyman Street, Waltham, MA 02451, USA

First edition 1992
Second edition 2012

Notices

Knowledge and best practice in this field are constantly changing. As new research and experience broaden our understanding, changes in research methods, professional practices, or medical treatment may become necessary.

Practitioners and researchers must always rely on their own experience and knowledge in evaluating and using any information, methods, compounds, or experiments described herein. In using such information or methods they should be mindful of their own safety and the safety of others, including parties for whom they have a professional responsibility.

To the fullest extent of the law, neither the Publisher nor the authors, contributors, or editors, assume any liability for any injury and/or damage to persons or property as a matter of products liability, negligence or otherwise, or from any use or operation of any methods, products, instructions, or ideas contained in the material herein.

British Library Cataloguing-in-Publication Data
A catalogue record for this book is available from the British Library

Library of Congress Cataloging-in-Publication Data
A catalog record for this book is available from the Library of Congress

ISBN: 978-0-08-097553-5

This book has been manufactured using Print On Demand technology. Each copy is produced to order and is limited to black ink. The online version of this book will show colour figures where appropriate.

Cover figure: The photograph on the cover shows shear bands in a tungsten heavy metal specimen caused by impact loading.

Contents

Preface

Since writing our book on *Shear Band Localization* in 1992, we have seen a burgeoning in the number of researchers and published papers on the subject of adiabatic shear bands, which makes our book seem like a beginner's guide!

But we do not mind because that is what it was always intended to be. Also we hoped that it would stimulate interest in the subject in the scientific community as a whole. A plot of the number of papers versus years is shown here—courtesy of Dr. Stephen Walley who is the keeper of the bibliography of papers on shear bands. In 2002, Tim Wright wrote his excellent book entitled *The Mathematics and Physics of Shear Bands.* This very valuable book for researchers was the second book to be published on the subject and dealt specifically with the detailed mathematics of shear bands. The third book that should be mentioned is by Marc A. Meyers: *Dynamic Behavior of Materials.* This third book has two chapters on shear bands.

As we have seen the subject grow in importance over the years, we felt that it was important to attempt to update our original book; however, the subject is now so vast that it is impossible for us to do justice to the entire subject. Therefore, we decided to write to many of the important researchers in the various subfields of adiabatic shear. We were delighted with the response, and there are individual chapters on the mathematics of shear bands, bands appearing in polymers, energetic materials, bulk metal glasses as well as armour and other areas. Bulk metal glasses (BMGs) are unique in how they shear band. All authors are particularly pleased that Elsevier accepted the book to be placed on ScienceDirect as this gives the maximum possible readership to researchers.

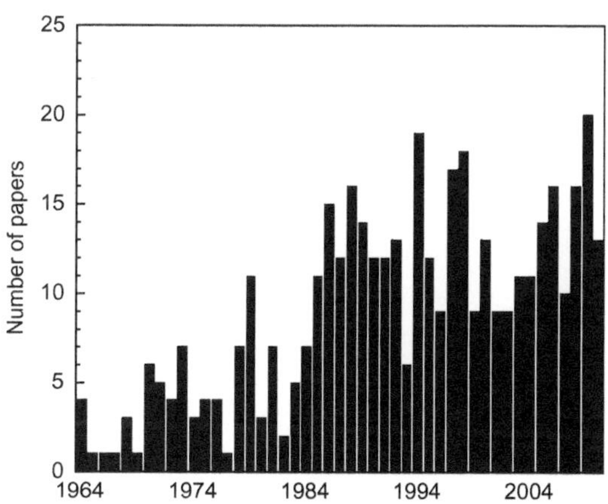

Bradley Dodd
Yilong Bai

Acknowledgements

Scientists and engineers who have inspired us:

J. Dennis Campbell
R. J. Clifton
D. R. Curran
Jaques Duffy
J. L. Ericksen
D. Grady
J. H. Hollomon
James Rice
Henri Tresca
C. A. Truesdell
Alan Turing
Ioannis Vardoulakis
Clarence Zener

List of Contributors

Yilong Bai State Key Laboratory of Non-linear Mechanics (LNM), Institute of Mechanics, Chinese Academy of Sciences, Beijing, China

Romesh C. Batra Department of Engineering Science and Mechanics, Virginia Polytechnic Institute and State University, Blacksburg, VA

Lan Hong Dai State Key Laboratory of Non-linear Mechanics (LNM), Institute of Mechanics, Chinese Academy of Sciences, Beijing, China

Bradley Dodd Institute of Shock Physics, Imperial College, London, UK

Hervé Couque Nexter Munitions, Bourges, France

Lothar W. Meyer Nordmetall GmbH, Adorf/Erzgebirge, Germany

Marc A. Meyers Department of Mechanical and Aerospace Engineering and Department of Nanoengineering, University of California, San Diego, CA

Frank Pursche Nordmetall GmbH, Adorf/Erzgebirge, Germany

Jean Sulem CERMES, Laboratoire Navier, Ecole des Ponts Paris Tech, IFSTTAR, Université Paris-Est, Marne-la-Vallée, France

Gerard M. Swallowe Department of Physics, Loughborough University, Loughborough, Leicestershire, UK

Stephen M. Walley SMF Fracture and Shock Physics Group, Cavendish Laboratory, University of Cambridge, Cambridge, UK

Thomas W. Wright Department of Mechanical Engineering, The Johns Hopkins University, Baltimore, MD

Yongbo Xu Shenyang National Laboratory for Materials Science, Institute of Metal Research, Chinese Academy of Sciences, Shenyang, People's Republic of China

1 Introduction

Bradley Dodd[1] *and Yilong Bai*[2]

[1]Institute of Shock Physics, Imperial College, London, UK, [2]State Key Laboratory of Non-linear Mechanics (LNM), Institute of Mechanics, Chinese Academy of Sciences, Beijing, China

Nomenclature

α metal-forming localization parameter
c specific heat
D diffusion of free volume
f, g the increase rates of morphogen X and Y, respectively, in Turing's formulation
$Le = \kappa/D$ Lewis number
$Pr = \nu/\kappa$ Prandtl number
t time
t^* characteristic time
T temperature
x axis in space
X, Y morphogens in Turing's formulation
$\dot{\gamma}$ shear strain rate
δ_{asb} width of adiabatic shear band
δ_{bl} thickness of boundary layer
δ_{BMG} width of shear bands in bulk metallic glasses (BMG)
Δ spacing between shear bands
κ thermal diffusion
μ, ν diffusion constants of X and Y, respectively, in Turing's formulation
ν viscosity
ρ density
τ shear stress

When metals and other materials are subjected to large plastic deformations, they increase in temperature. This is because a large quantity of the plastic work done is converted into heat, typically about 90% of the work. This was discussed by Joule [1] who derived a constant, now known as Joule's constant, which is also called the mechanical equivalent of heat.

After large plastic deformations, it is possible to cause localization of plastic flow because of the localized heating. This flow localization is often catastrophic, leading to fracture by intense localized shearing.

Adiabatic Shear Localization. DOI: http://dx.doi.org/10.1016/B978-0-08-097781-2.00001-0

This localized shearing was called adiabatic shearing by Zener and Hollomon [2]; the name 'adiabatic' indicates that it is caused by thermal softening, which eventually becomes larger than the rate of strain hardening. The point at which this occurs is called a point of thermal instability.

In reality, often some of the heat generated by plastic flow will be conducted away from the deforming zone; the quantity of this heat will depend on the thermal diffusivity of the deforming material among other variables, such as the local strain rate. In these cases, the phenomenon is often referred to as 'catastrophic thermoplastic shear'. This term was used by Rogers [3] and Bai et al. [4]. 'Catastrophic' is used by these authors to indicate that the shear band often leads to fracture.

1.1 Early Experiments on the Thermoplastic Effect

The first systematic experiments carried out on metals were those of Tresca, who, over an 8-year period, studied a variety of processes on a number of metals. These experiments are described in some detail in the superb book by Bell [5] and also by Johnson [6].

Tresca [7] used a steam hammer for a number of experiments. He wrote that 'the steam hammer, at the same time made a local depression in the bar and lengthened it, also reheating it in the direction of two lines inclined to each other ... so great was the reheating that the metal that was along these lines became luminous and formed a letter X'.

Tresca also coated some metal bars in wax or tallow, and when the hammer impacted the bar, the wax and tallow melted along an X pattern. Figures 1.1–1.3 show some of Tresca's experimental results.

Figure 1.1 shows the plastic deformation of lead stamped with a hammer, and Figure 1.2 shows the lengthening of the bar with successive blows. Figure 1.3 shows the effect of the heat developed with successive blows.

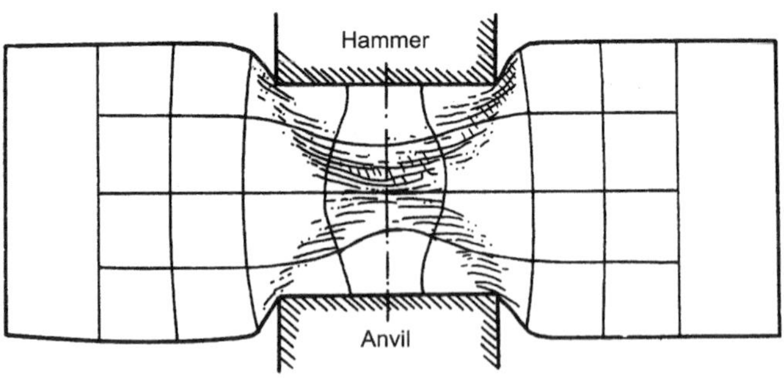

Figure 1.1 Plastic distortion of lead stamped with a hammer.
Source: After Ref. [7].

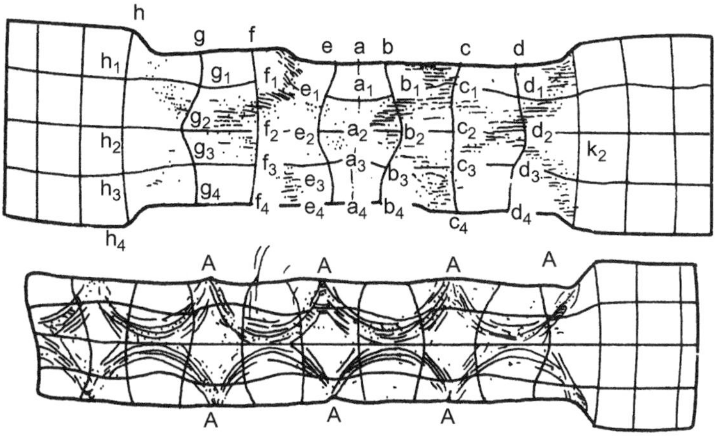

Figure 1.2 As Figure 1.1 but successively stamped after transverse movement a second and a third time [7].

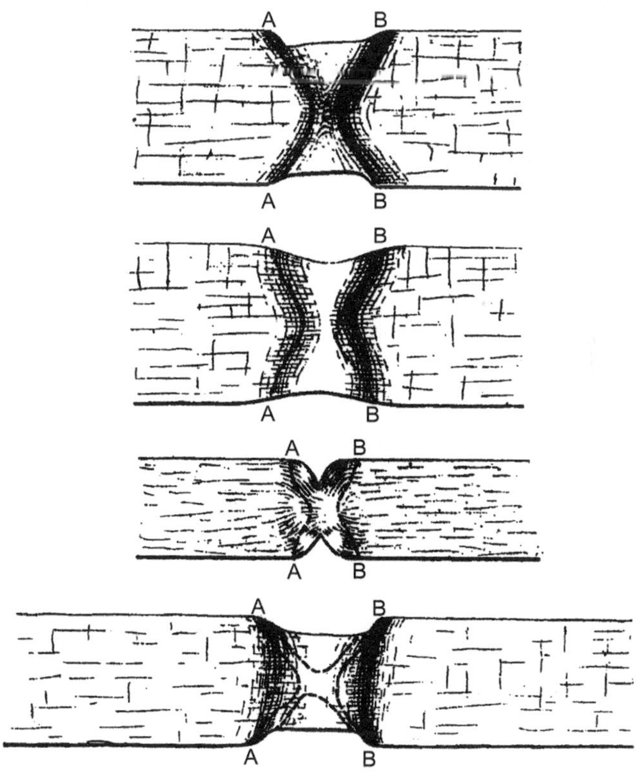

Figure 1.3 Heat-affected zones produced by a hammer blow [7].

Tresca also calculated the efficiency of the process. The efficiencies ranged from 73% to 94%. It took many years before anyone else calculated efficiencies in plastic deformation, and this was done accurately by Taylor and Quinney [8] who obtained results of the order of 90–95% of the plastic work done is converted into heat.

1.2 Fracture and Damage Related to Adiabatic Shear

Adiabatic shear bands can act as precursors to brittle or ductile fracture. Figure 1.4 illustrates planar micro-cracks and adiabatic shear bands near the plugged region of a steel armour. A few visible cracks appear near the periphery of the orifice, but the majority of cracks are within shear bands.

From his experiments on thin-walled tubular specimens, Giovanola [9] found that shear bands nucleate at several locations, coalesce and finally form some well-defined steps. Similar steps have been found by Woodward and co-workers [10]. A rare example of an adiabatic shear band with an associated crack is shown in martensitic steel in Figure 1.5, which was obtained by Dormeval [11]. Backman and Finnegan [12] show several types of cracks in Figure 1.6.

Here there are cracks associated with the shear bands, and some shear bands are confined to them.

In steels, shear bands are often recognized by their white colour. These bands are called transformed bands. From experiments, these white bands were shown to

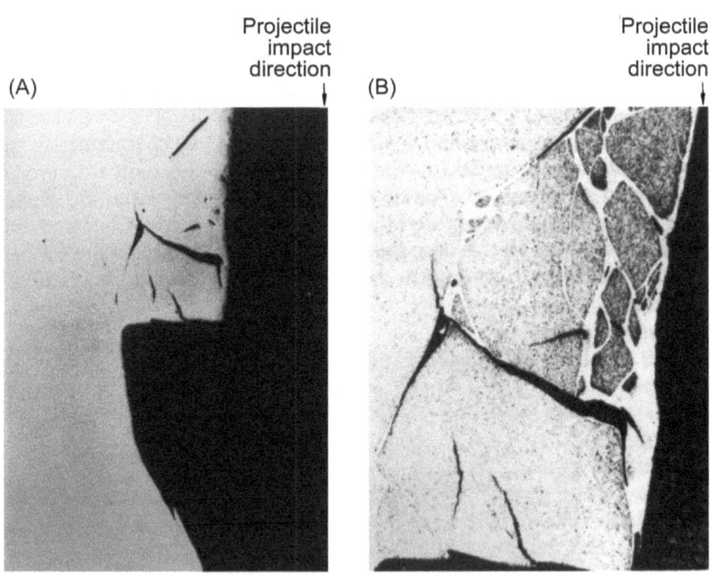

Figure 1.4 Planar micro-cracks and adiabatic shear zones near the plugged region in steel armour: (A) polished section through projectile hole; (B) etched close-up view of (A). *Source*: After Shockey, D.A. Poulter Lab.Tech. Report 004-85, SRI, Int. California.

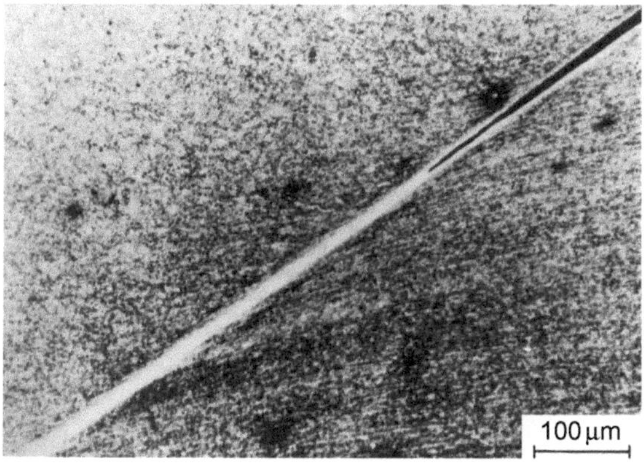

Figure 1.5 Adiabatic shear band and an associated crack in a martensitic steel deformed in a dynamic compression test. *Source*: From Ref. [11]. Society of Manufacturing Engineers, Dearborn, MI.

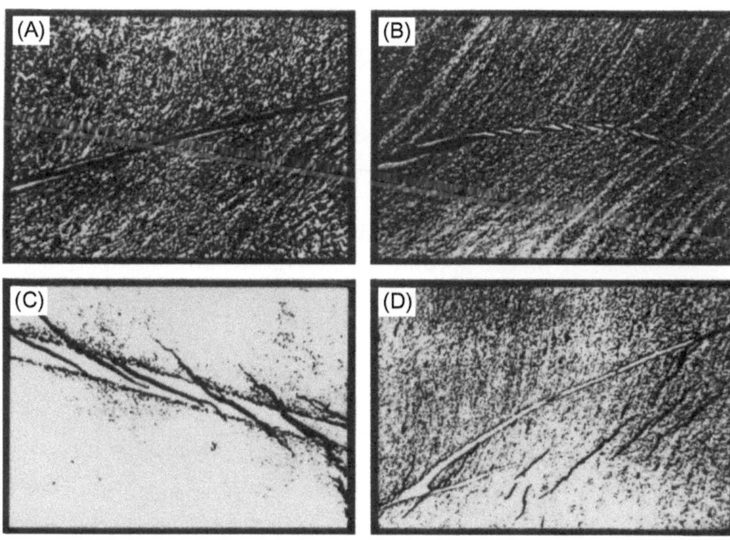

Figure 1.6 Several types of brittle fracture of transformed bands. *Source*: Sketches based on photographs after Ref. [12].

be very fine martensite. Forming martensite in a carbon steel requires a temperature rise in excess of 900°C and a rapid quench. Although the kinetics for the formation of martensite is significantly different at these high strain rates, it has been observed on numerous occasions. For metals that do not form martensites, the shear bands are usually dark and are called deformed bands.

Figure 1.7 is a rather famous photograph of a shear band with voids in it from a report by Irwin [13]. There are a number of nearly equi-axed voids in the U−2Mo sample. Importantly, the void shape shows that they most probably formed after the shearing had taken place.

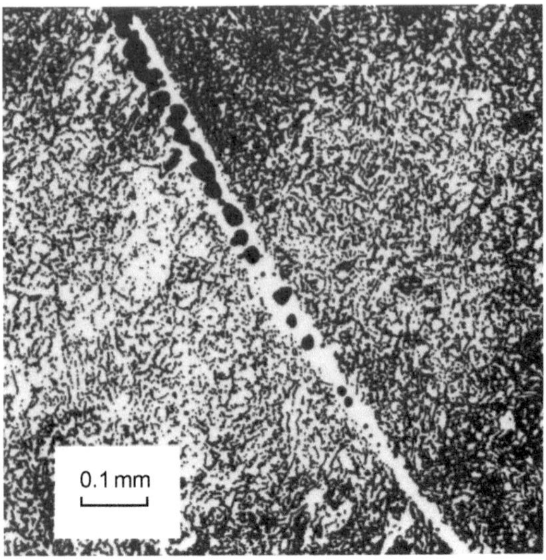

Figure 1.7 Section of a fractured shear band in a U–2Mo alloy.
Source: After Ref. [13].

1.3 Evolution of Adiabatic Shear Bands

In a series of careful dynamic experiments on thin-walled HY-100 steel tubes using a torsional Kolsky bar (split-Hopkinson bar), Marchand and Duffy [14] showed that plastic deformation of the specimens undergoes three distinct stages. Up to 25% strain, the deformation is homogeneous. In the second stage, the deformation

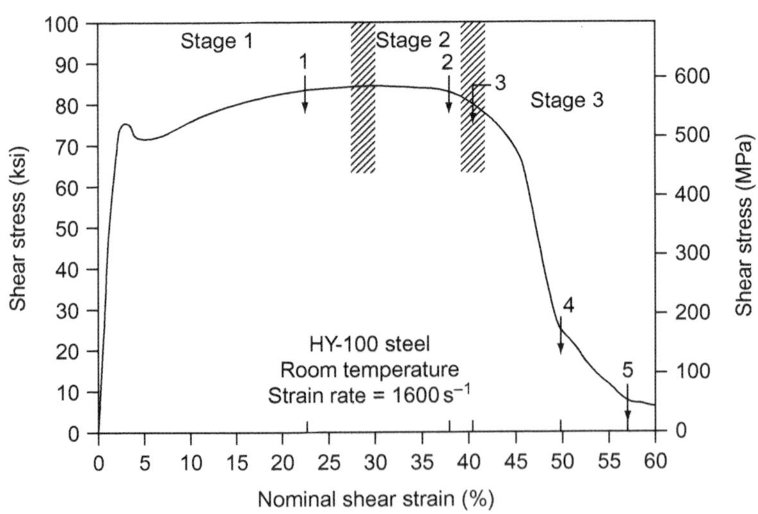

Figure 1.8 A typical stress–strain curve showing the three stages of plastic deformation shown in Figure 1.9.
Source: After Ref. [15].

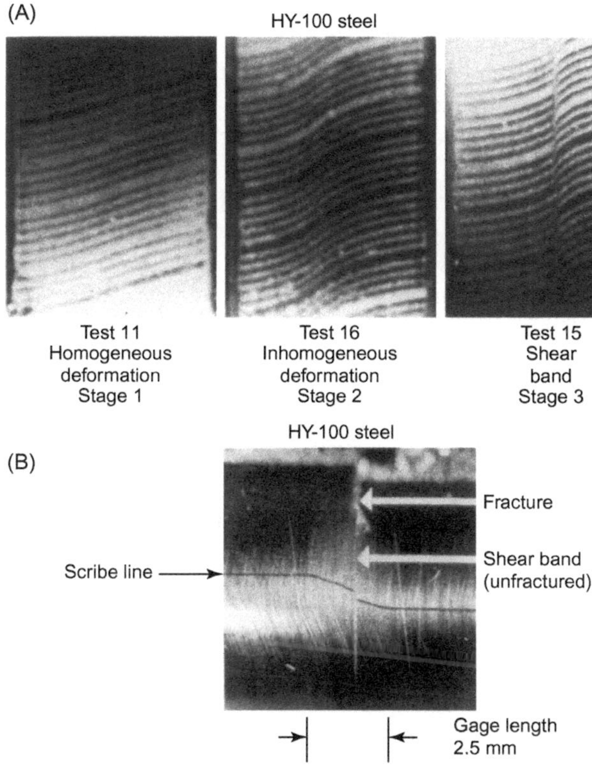

Figure 1.9 (A) Grid patterns shown in the three separate stages at a strain rate of 1600 s^{-1}. (B) A partial fracture shown at a local shear strain of 500%.
Source: After Ref. [15].

is inhomogeneous. Finally, in the third stage, a narrow shear band is formed, which encircles the specimen. Figures 1.8 and 1.9 show the development of the shear band and a partial fracture.

Bai and co-workers [4] carried out an analytical study of the evolution of shear bands. (a) A shear band has a width of 10–100 μm, (b) the strain rate in the shear band is typically $10^2\,s^{-1}$ and (c) the shear band has a distinct structure. In steels, shear bands sometimes manifest themselves as martensitic bands, and, in other cases, they can result in localized melting and re-solidification.

1.4 Metal Shaping and Shear Bands

1.4.1 Metal Machining

Metal cutting or machining is normally carried out at a high speed. Because the process of machining is predominantly a shear process (this is particularly so in orthogonal machining), it is not surprising that in the intense shear bands, there may be adiabatic shear bands that lead to fracture. One might expect the tool to reach a steady-state temperature caused by the fresh metal that is continuously passing through the deformation and cutting zone.

In 1964, Recht [16] carried out some cutting experiments on titanium. He recognized the importance of adiabatic shear banding in the process and even predicted accurately a shear zone temperature of 650°C at a cutting speed of 43 m/min.

Dao and Shockey [17] showed clearly adiabatic shear bands and discontinuous chips in steel and aluminium. They used an infrared camera to measure the temperatures. Peak temperatures of 180°C for the steel and 100°C for the aluminium were found. These figures were corrected because the measured areas were larger than the shear band width, and they were corrected to 500°C and 120°C.

Figure 1.10 shows continuous chips with inhomogeneous shear, according to Shaw [18].

Much more detail is given on machining and adiabatic shear in the books by Dodd and Bai [19] and Bai and Dodd [20]. Also there is a review of adiabatic shear in machining by Childs [21].

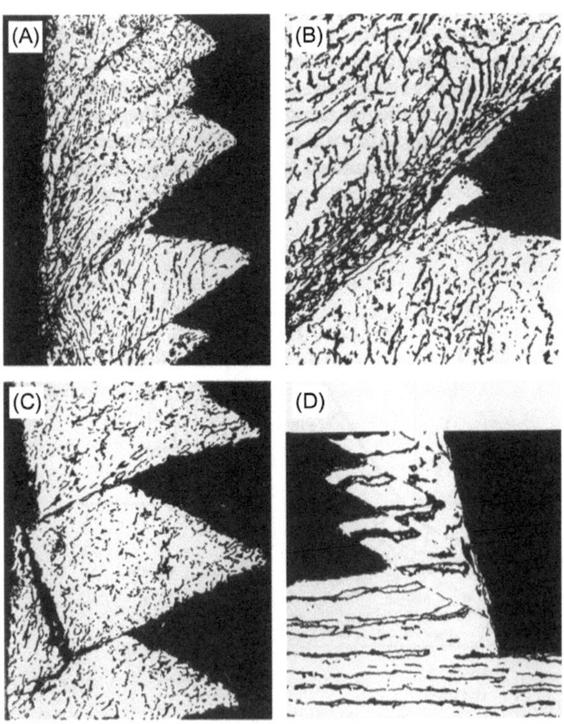

Figure 1.10 Continuous chips with inhomogenous shear: (A) Ti cut at a high speed of 53 m/min, adiabatic shear; (B) enlargement of (A); (C) Ti cut at a low speed of 25 mm/min with periodic fracture, gross sliding and rewelding; (D) 60–40 cold-rolled brass cut with a high-speed tool having an angle of −15°, cutting speed 0.075 m/min. Undeformed chip thickness of 0.16 mm. *Source*: Sketches based on a figure 1.13 in Ref. [18].

1.4.2 Metal Forming

From the early work of Tresca in the nineteenth century, it is well known that high temperatures can be produced by large plastic deformation of metals; forging is an obvious example. Much of the plastic deformation occurs by intense shear. Some of these shear bands become adiabatic shear bands.

There are many metal-forming processes, as described by Lange [22]. Processes can be grouped according to the predominant active stress state. The processes are (a) compressive, (b) combined compressive and tensile, (c) tensile, (d) bending and (e) shearing.

Examples of compressive forming processes are open- and closed-die forging and rolling. Processes that use compressive and tensile forging stresses are flange forming, deep drawing, spinning and bulging. Tensile-forming methods include stretch forming, and expanding and shearing processes include blanking. If shear bands do occur in any of these processes, then we should expect them to occur in (a) compressive, (b) combined compressive and tensile and (c) shear processes.

The first attempt to provide a generalized approach to metal-forming processes was made by Backofen [23], Lange [22] and Poehlandt [24]. This generalized system is shown in Figure 1.11.

Area 1 is the work zone in which the workpiece undergoes plastic deformation; area 2 involves the properties of the material being shaped; area 3 involves the properties of the formed product; area 4 is the contact zone between the tool and the workpiece and includes friction and lubrication and area 5 is the tool itself. High friction in area 4 can lead to shear localization.

One process of interest is blanking shown in Figure 1.12 [25]. In blanking, cropping and punching a part of the workpiece is removed. In most of these processes, the removed parts are often parallel to the direction of travel. The different stages are shown in the diagram.

A number of processes are isothermal, i.e. the tools are at the same temperature as the workpiece. This is certainly so for room-temperature processes and is sometimes so for elevated-temperature processes. Semiatin and Lahoti [26] were able to derive a flow localization parameter, α. This parameter can be used to estimate the tendency of a material to form catastrophic strain concentrations. It has been shown that when this parameter is greater than or equal to 5, flow localization will occur. The isothermal test can be carried out at an elevated temperature, and Figure 1.13 shows workability maps for the plane-strain sidepressing of Ti−6242Si.

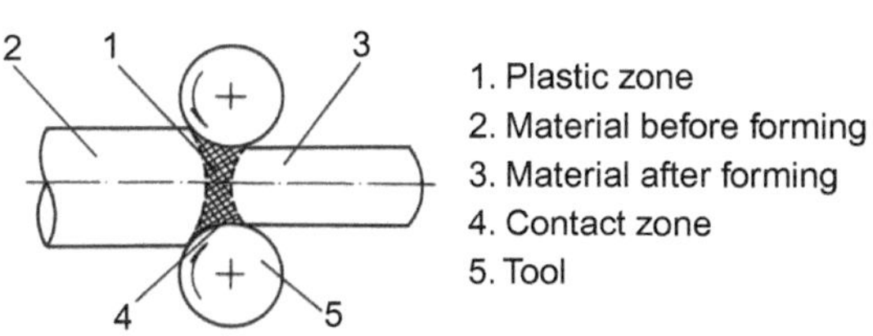

Figure 1.11 A metal-forming system.
Source: After Ref. [24]. Courtesy of Springer-Verlag, Berlin.

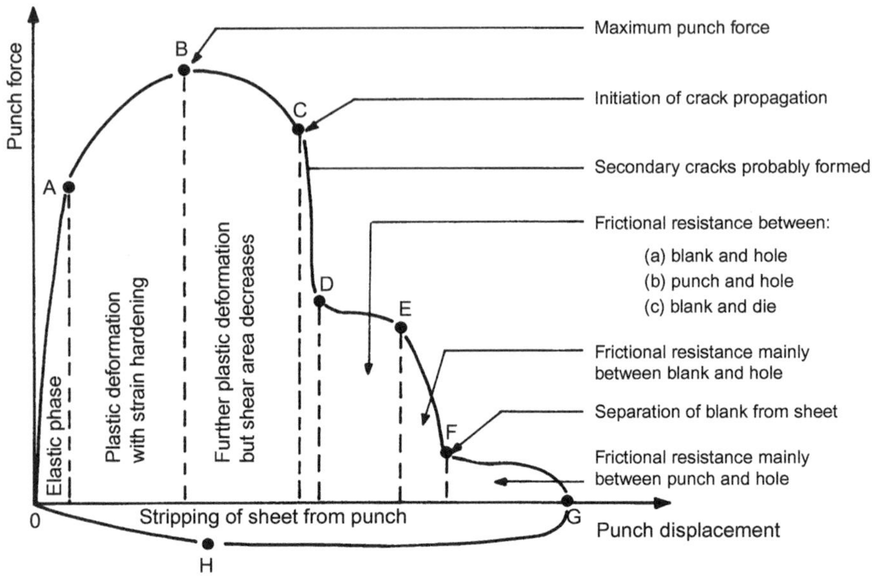

Figure 1.12 Schematic representation of punch force versus punch displacement. *Source*: After Ref. [25].

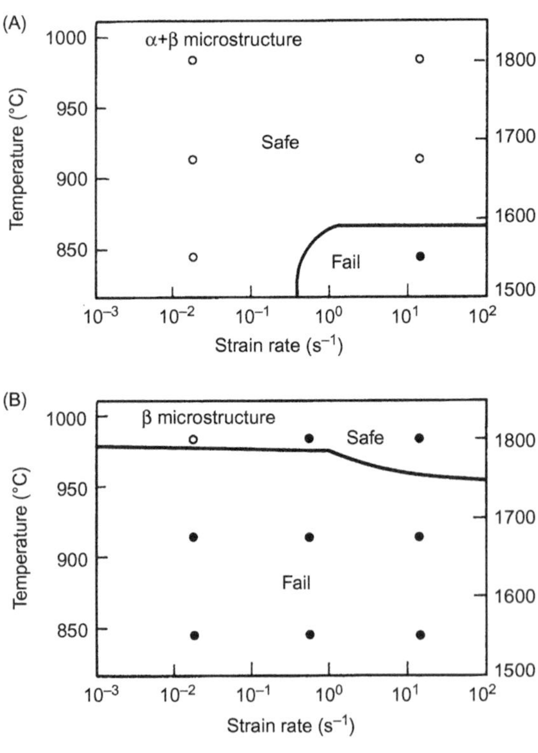

Figure 1.13 Workability maps for the occurrence of shear bands in isothermal sidepressing of Ti–6242Si. *Source*: After Ref. [26].

1.5 Examples of Adiabatic Shear Bands

Figure 1.14 shows a photomicrograph of a shear band formed in a tungsten alloy tested in a split-Hopkinson bar. The two diagrams on the right show the geometry of the specimen. As the shear band grows farther into the specimen, it becomes more diffused. Figure 1.15 shows an adiabatic shear band in a dumbbell tungsten specimen. The morphology of the band at areas a, b and c in A are shown at higher magnification in B, C and D, respectively [27].

Odeshi and Bassim [28] investigated the high strain-rate failure of a high-strength, low-alloy, quench-hardened and tempered AISI 4340 steel. Three sets of specimens were used all were to 843°C, oil quenched and then tempered for 1 h at 205°C, 315°C and 425°C before impact testing. Figure 1.16 shows transformed

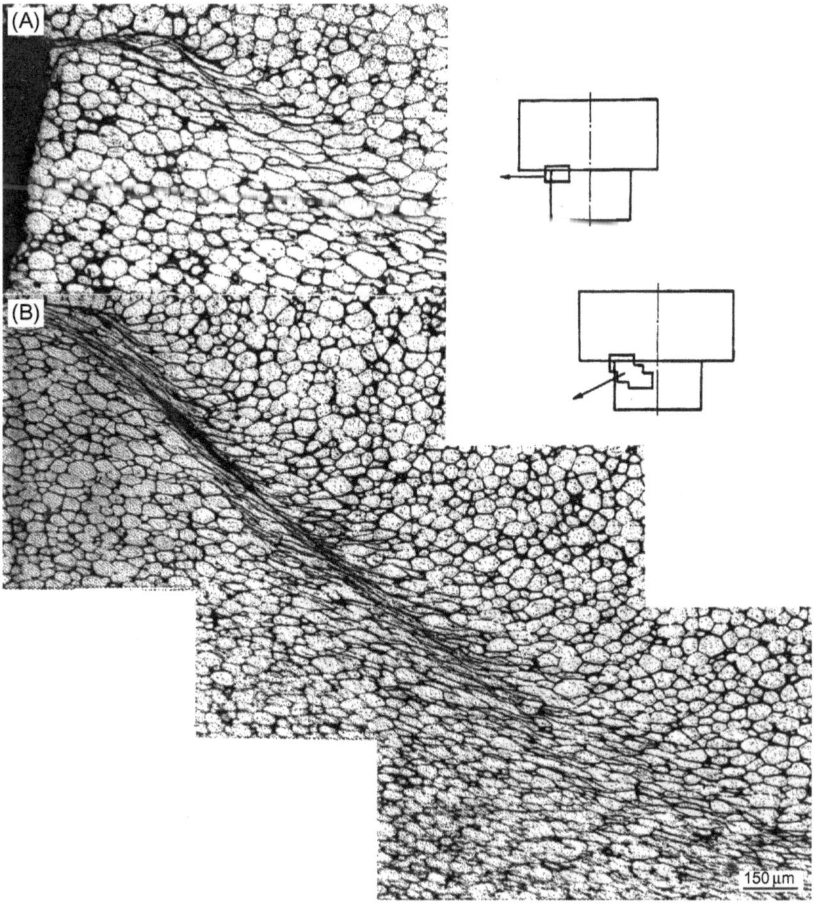

Figure 1.14 Micrographs showing shear bands formed in the areas indicated in tungsten. *Source*: After Ref. [27].

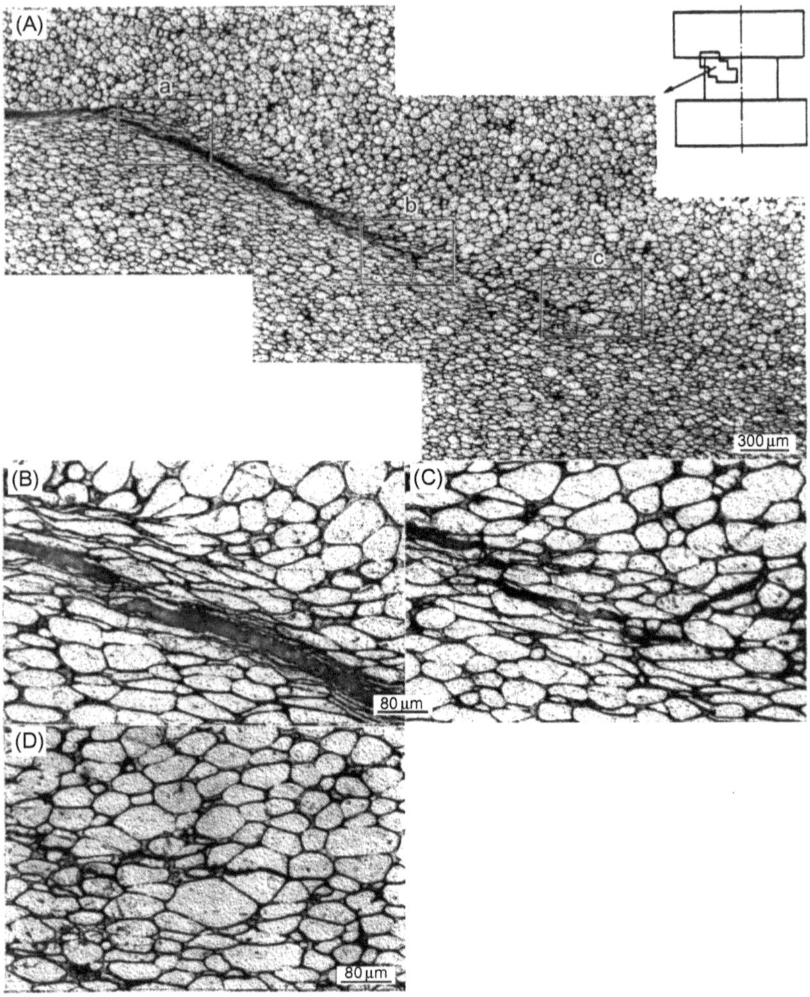

Figure 1.15 Grain shape change in a dumbbell tungsten alloy specimen under impact: (A) low magnification and (B), (C) and (D) high-magnification images corresponding to (a), (b) and (c) in (A).
Source: After Ref. [27].

bands in the steel that have been quench-hardened and tempered at (A) 315°C, (B) 425°C and (C) shows cracking along a shear band.

Figure 1.17 shows a schematic diagram of the shape of the geometry of adiabatic shear banding and the fracture path in steel that was quench-hardened and tempered at 315°C and 425°C. Figure 1.18 shows a schematic diagram of the adiabatic shear band geometry and fracture path in the steel that was quench-hardened and tempered at 205°C. Two collinear shells were observed for the two higher

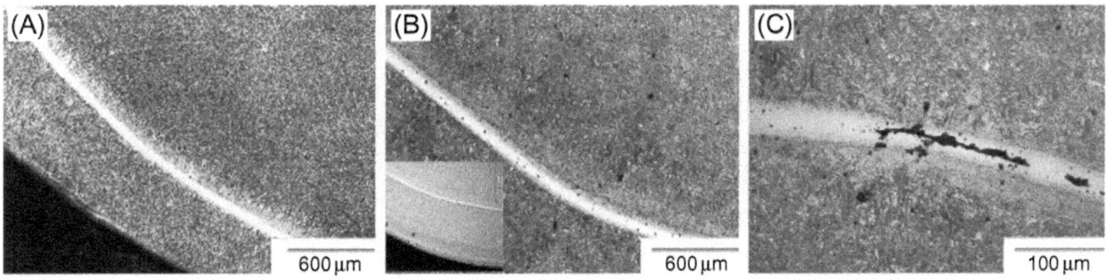

Figure 1.16 Transformed shear bands in AISI 4340 steel specimens quench-hardened and tempered at (A) 315°C, (B) 425°C and (C) cracking along a shear band [28].

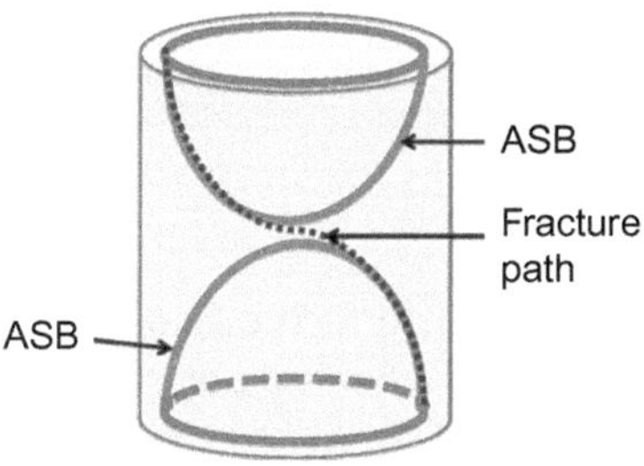

Figure 1.17 A sketch of the geometry of adiabatic shear bands and the fracture path in an AISI steel cylinder that was quench-hardened at 315°C; a similar result was obtained at 425°C [28].

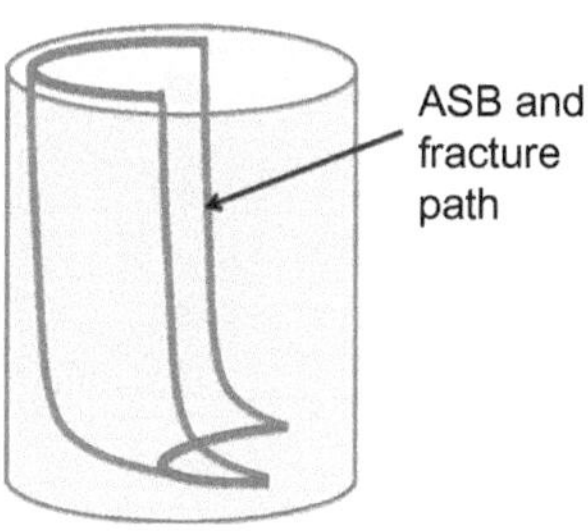

Figure 1.18 A sketch of the geometry of adiabatic shear bands and fracture path in an AISI 4340 steel specimen tempered at 205°C [28].

tempering temperatures (Figure 1.17), and for specimens that had been tempered at the lowest temperature, they cracked rapidly and fragmented.

The evolution of adiabatic shear bands in AM60B magnesium alloy impacted with a projectile 500 m/s was studied by Zhen et al. [29]. Figure 1.19A shows a shear band, and Figure 1.19B shows it at a higher magnification.

Zhou et al. [30] investigated tungsten heavy metal (WHA) penetrators processed by hot hydrostatic extrusion and hot torsion (HE + HT). Under uni-axial dynamic compression, HE + HT specimens show well-defined shear bands as shown in Figure 1.20.

Figure 1.21 shows a macrograph as well as a series of micrographs of an HE + HA WHA remnant. Cracking along the adiabatic shear band as well as shear bands in other areas can be observed.

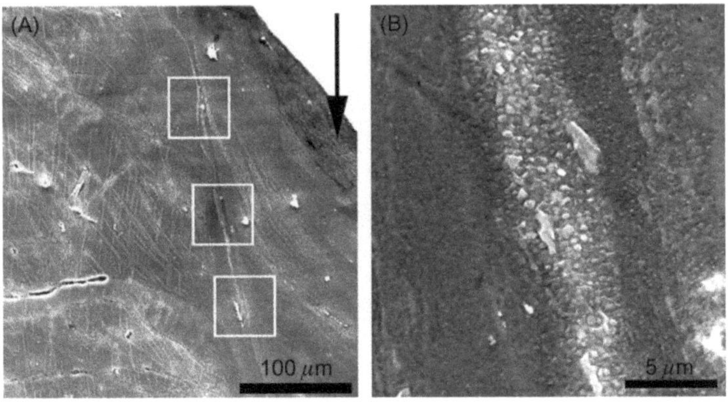

Figure 1.19 (A) SEM images of adiabatic shear band in AM60 Mg alloy target impacted by a projectile at a velocity of 500 m/s and (B) higher magnification [29].

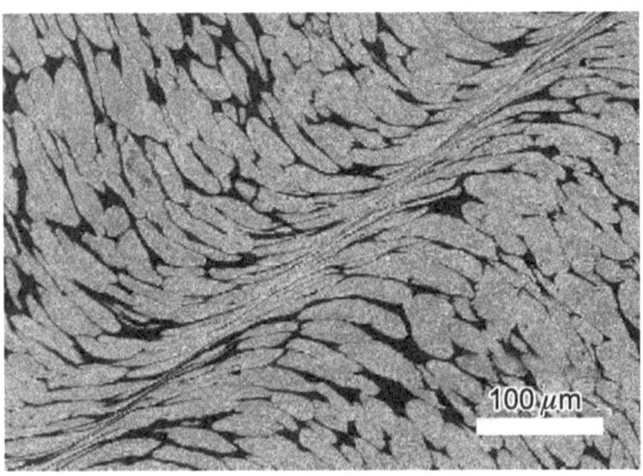

Figure 1.20 SEM micrograph showing a localizing of deformation in an adiabatic shear band in the HE + HT specimens subjected to uni-axial compression [30].

The series of micrographs shown in Figure 1.22 is of a further remnant of the HE + HA WHA penetrator. The arrow shows the penetration direction. Cracking along the adiabatic shear band can be observed as well as shear bands in other areas.

Figure 1.23 shows micrographs of debris from the penetrator. Figure 1.23A shows cracking along the shear band, and Figure 1.23B is an enlarged micrograph of the cracking.

Examples of shear bands can be found in soft metals such as Cu. Yazdani et al. [31] carried out some torsional Hopkinson bar experiments on 99.94% purity Cu with an average grain size of 700 μm. Various tests were carried out at strain rates ranging from 636 to 1166 s^{-1}. Figure 1.24 shows SEM photographs of specimens tested at 886 and 1250 s^{-1}. There is a marked reorientation of the material within the shear bands depending on strain rate.

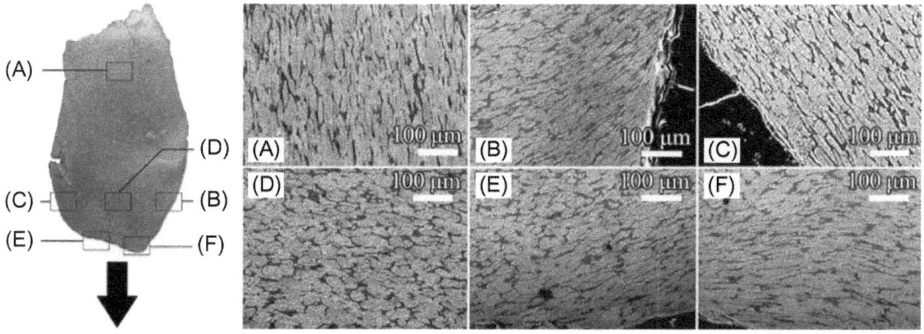

Figure 1.21 Photographic view and micrographs in different regions of the extruded WHA fragment. The arrow depicts the penetration direction. (A) Tungsten grains keep the original fibrous structure indicating a small deformation; (B) shear deformation adjacent to the surface; (C) fibrous shaped tungsten grains have been further elongated; (D) fibrous shaped tungsten grains indicating severe plastic deformation along the penetration direction; (E) and (F) fibrous grains with a fibrous direction perpendicular to the penetration direction, indicating severe plastic deformation [30].

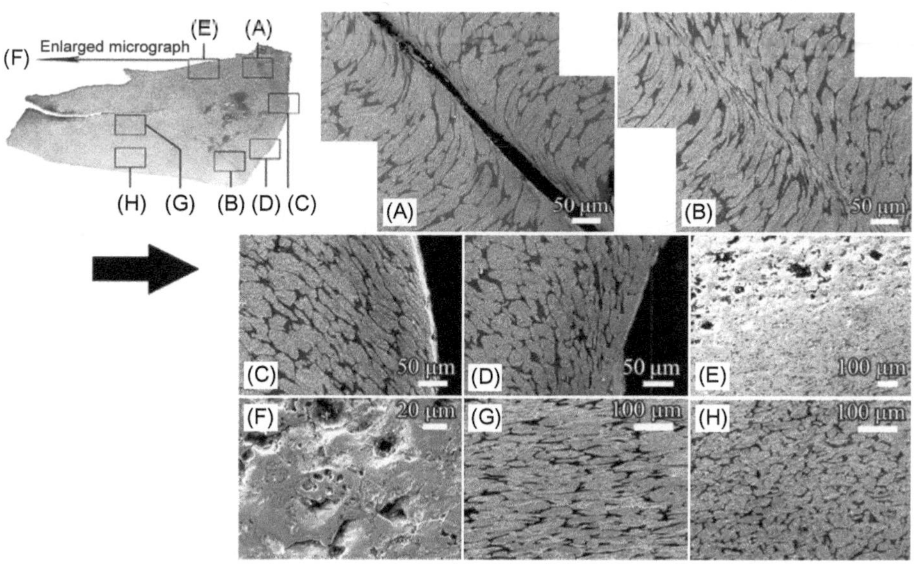

Figure 1.22 Micrographs of different regions of the HE + HT WHA penetrator fragment. The arrow indicates the penetration direction. (A) Cracks formed along the shear band; (B) localized shearing and micro-cracks within the shear band; (C) and (D) localized bands adjacent to the fracture face indicating that the fracture occurred along the shear bands; (E) a large number of voids on one side of the fragment; (F) enlarged micrograph showing that melting occurred in part of the penetrator; (G) and (H) the central region and the edge near the tail of the fragment [30].

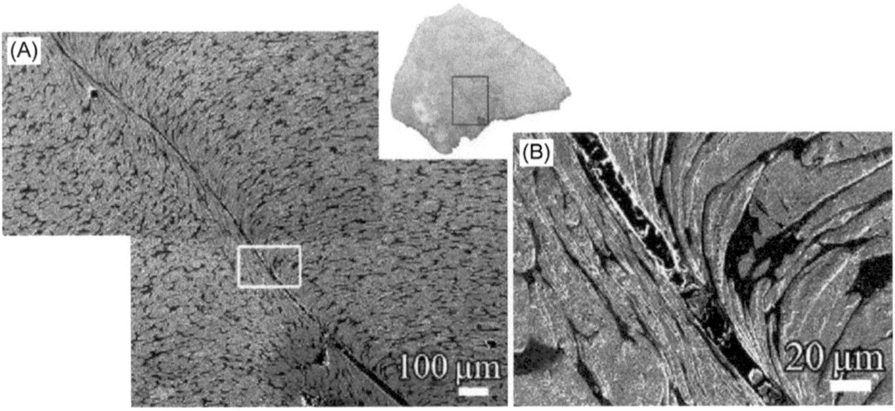

Figure 1.23 Micrographs of debris that has fallen from the HE + HT WHA penetrator: (A) micrograph showing cracking along a shear band; (B) enlarged micrograph showing the detail of the localized shear [30].

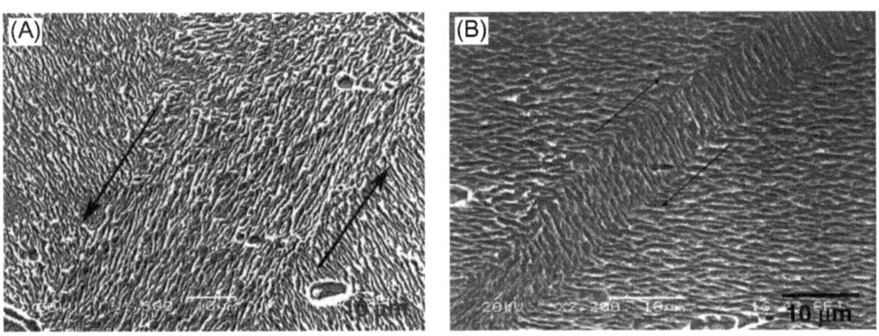

Figure 1.24 (A) SEM micrograph of a sample tested to a 7° angle of twist, strain rate 886 s^{-1} and strain 0.52, showing reorientation of grains within the shear band. (B) SEM micrograph of a sample tested to a 9° angle of twist, strain rate 1250 s^{-1} and strain 0.8, showing a reorientation of grains within the shear band [31].

1.6 The Essence of Localization

It is interesting that the emergence of bands from a uniform field to form a pattern has always inspired investigators. In the mid-twentieth century, Alan Turing (1912–1954) carried out his brilliant work on the Turing pattern. He stated in the beginning of his paper [32] that a system of chemical substances, called morphogens, reacting together and diffusing through a tissue, is adequate to account for the main phenomena of morphogenesis. Such a system, although it may originally be homogeneous, may later develop a pattern or structure due to an instability of the homogeneous equilibrium, which is triggered by random disturbances. Specifically,

Turing proposed the following reaction-diffusion equations with two chemical species are the origin of different biological morphology, for example, the zebra pattern.

For a system consisting of two morphogens, X and Y, suppose that cell r would exchange materials by diffusion with cells $r-1$ and $r+1$, then the equations are

$$\begin{cases} \dfrac{dX_r}{dt} = f(X_r, Y_r) + \mu(X_{r+1} - 2X_r + X_{r-1}) \\ \dfrac{dY_r}{dt} = g(X_r, Y_r) + \nu(Y_{r+1} - 2Y_r + Y_{r-1}) \end{cases} \tag{1.1}$$

where f and g are the rates of increase, and μ and ν are the diffusion constants of X and Y, respectively. After detailed mathematical treatments, Turing restated his biological interpretation of the results as follows: 'After the lapse of a certain period of time form the beginning of instability, a pattern of morphogen concentrations appears'.

Since the 1960s, Prigogine and his group approached the Turing pattern from thermodynamics. They proved that self-organized patterns are possible in open systems [33]. This is called a dissipative structure, which is a non-uniform and steady state, existing only in open systems operating far from thermodynamic equilibrium. They put forward a model – called Brussels oscillators – to justify the existence of the dissipative structure. The theory of 'dissipative structure' unveils the common feature of pattern formation in nature. The Russian–Belgian physical chemist was awarded the Nobel Prize in Chemistry in 1977 for his pioneering work on these structures.

From the preceding discussion of the Turing pattern and dissipative structure, you may note the following conditions for pattern formation:

1. The system (such as Turing's reaction–diffusion system) should be open and far from equilibrium.
2. There must be a certain source in the system (like the increase rates of morphogens).
3. There must be some corresponding diffusion, for which the diffusion constants of two processes should be quite different.

However, both the Turing pattern and Prigogine's dissipative structure are mainly concerned with chemical systems. Actually, in mechanical engineering, similar work was also carried out. Looking back to the well-known boundary layer established by Prandtl in the early twentieth century [34], you may recall the relation between the thickness of the boundary layer δ_{bl} and viscosity ν as $\delta_{bl} \sim \sqrt{\nu t^*}$, where t^* is a characteristic time, $t^* = \frac{\text{sample dimension}}{\text{flow velocity}}$. This means that the width of the boundary layer is proportional to the square root of momentum diffusion ν.

Now let us return to the adiabatic shear bands. A number of authors [20,35] discussed the mechanism and representation of the adiabatic shear bands from the point of view of mechanical diffusions. As a matter of fact, when you look at the narrow shear bands from the outside, you could ignore the thermal diffusion outside the band and adopt the adiabatic assumption. Whereas when you look at a band from the inside, you should note that there is a stationary balance of outward heat diffusion and heat source resulting from plastic work in the band, at late stage

$$\kappa\frac{\partial^2 T}{\partial x^2}+\frac{\beta\tau\dot{\gamma}}{\rho c}=\frac{\partial T}{\partial t}\approx 0 \tag{1.2}$$

where T is temperature, τ is shear stress, $\dot{\gamma}$ is shear strain rate, ρ is density, c is specific heat and κ is the thermal diffusion. Then, the band width could be estimated as:

$$\delta_{\text{asb}}\approx\sqrt{\frac{\rho c\kappa T}{\beta\tau\dot{\gamma}}} \tag{1.3}$$

In principle, this is consistent with Prigogine's idea of dissipative structure.

In fact, the adiabatic shear band is by no means an isolated case of band structures in mechanical engineering. Now, the similarity between various band-like structures in engineering becomes clear. Apart from the width of boundary layer $\delta_{\text{bl}}\sim\sqrt{\nu}$ and the width of the adiabatic shear band $\delta_{\text{asb}}\sim\sqrt{\kappa}\sim(10^1-10^2)$ μm, the width of shear bands in bulk metallic glasses (BMG) $\delta_{\text{BMG}}\sim\sqrt{D}\sim(10^1)$ nm appears to be dependent on mass (void) diffusion D, where D is the diffusion of free volume [35]. So, the three kinds of mechanical diffusion, i.e. momentum, energy and mass, govern three kinds of band-like structures. In addition, the coefficients of the three kinds of diffusion usually present $\nu\gg\kappa\gg D$ in most materials, namely the Prandtl number $Pr=\nu/\kappa$ and the Lewis number $Le=\kappa/D$, leading to different band widths.

More importantly, when we turn to the band pattern with spacing, it will be more interesting. The coupling of two different mechanical diffusions could result in the band pattern with specific band width and spacing, like the Turing pattern. In the case of adiabatic shear banding, the ratio of spacing Δ to width δ of the shear bands is proportional to the square root of the Prandtl number [36–38] as

$$\left(\frac{\Delta}{\delta}\right)_{\text{asb}}\propto\sqrt{\frac{\nu}{\kappa}}=\sqrt{Pr}. \tag{1.4}$$

Hence, this is similar to the Turing pattern, but instead of the diffusions of two morphogens, the pattern of adiabatic shear bands is governed by two kinds of mechanical diffusion, i.e. momentum and energy diffusions, owing to quite different diffusion constants of the two processes, or more specifically very large Prandtl number $Pr=\frac{\nu}{\kappa}\gg 1$.

1.7 Summary

Now, it appears that adiabatic shear localization is a prototype of band-like structures and is merely a small window in the broad area of localization emerging from a uniform field. Perhaps, almost all of the preceding writers in this field, mentioned in the introductory paragraphs of this chapter, did not foresee such wide contexts in the phenomena and powerful influences on the various disciplines in engineering. After considering all other related phenomena of localization in nature and engineering and their importance, this endless frontier in science and technology inspires us to develop more new concepts and techniques to explore novel phenomena, to unveil the mechanisms underlying the phenomena and to benefit various works in engineering and in our society in the future.

References

1. Joule, J. P. (1859), On Some Thermodynamic Properties of Solids, Philos. Trans. R. Soc. London, 140, 91–136.
2. Zener, C. and Holloman, J. H. (1944), Effect of Strain Rate Upon Plastic Flow in Steel, J. Appl. Phys., 15, 22–34.
3. Rogers, H. C. (1979), Adiabatic Plastic Deformation, Ann. Rev. Mater. Sci., 9, 283–311.
4. Bai, Y., Cheng, C. and Yu, S. (1986), On the Evolution of Thermo-Plastic Shear Band, Acta Mech. Sin., 2, 1–7.
5. Bell, J. F. and Truesdell, C. (1973), Mechanics of Solids: Volume 1: The Experimental Foundations of Solid Mechanics (ed. S. Flügge), Springer-Verlag, Berlin.
6. Johnson, W. (1987), Tresca as the Originator of Adiabatic Shear Lines, Int. J. Mech. Sci., 29, 301–320.
7. Tresca, H.-E. (1878), On Further Applications of the Flow of Solids, Proc. Inst. Mech. Eng., 30, 301–345.
8. Farren, W. S. and Taylor, G. I. (1925), The Heat Developed During Plastic Extension of Metals, Proc. R. Soc. London, A107, 422–451.
9. Giovanola, J. H. (1988), Adiabatic Shear Banding Under Pure Shear Loading, I: Direct Observation of Strain Localization and Energy Dissipation Measurements, Mech. Mater., 7, 59–71.
10. Woodward, R. L., Baxter, B. J. and Scarlett, N. V. Y. (1984), Mechanisms of Adiabatic Shear Plugging Failure in High Strength Aluminum and Titanium Alloy. In Mechanical Properties of Materials at High Rates of Strain (ed. J. Harding), 525–532, Institute of Physics Conference Series No. 70, Bristol.
11. Dormeval, R. (1987), The Adiabatic Shear Phenomenon. In Materials at High Strain Rates (ed. T. Z. Blazynski), 47–70, Elsevier, London.
12. Backman, M. E. and Finnegan, S. A. (1973), The Propagation of Adiabatic Shear. In Metallurgical Effects at High Strain Rates (eds. R. W. Rohde, B. M. Butcher, J. R. Holland and C. H. Karnes), 531–543, Plenum Press, New York, NY.
13. Irwin, C. J. (1972), Metallographic Interpretation of Impacted Ogive Penetrators, DREV R-652/72, Canada.
14. Marchand, A. and Duffy, J. (1987), An Experimental Study of the Formation Process of Adiabatic Shear Bands, Brown University Report, Grant: DMR-8316893.
15. Marchand, A. and Duffy, J. (1988), An Experimental Study of the Formation Process of Adiabatic Shear Bands in Structural Steel, J. Mech. Phys. Solids, 36, 261–283.
16. Recht, R. F. (1964), Catastrophic Thermo-Plastic Shear, J. Appl. Mech., 31, 189–193.

17. Dao, K. C. and Shockey, D. A. (1979), A Method for Measuring Shear Band Temperatures, J. Appl. Mech., 50, 8244–8246.
18. Shaw, M. C. (1984), Metal Cutting Principles, Clarendon Press, Oxford.
19. Dodd, B. and Bai, Y. (1987), Ductile Fracture and Ductility, Academic Press, London.
20. Bai, Y. and Dodd, B. (1992), Adiabatic Shear Localization, Pergamon Press, Oxford.
21. Childs, T. H. C. (2011), Adiabatic Shear in Metal Machining, CIRPedia (in press).
22. Lange, K. (editor in chief) (1985), Handbook of Metal Forming, McGraw-Hill, New York.
23. Backofen, W. A. (1972), Deformation Processing, Addison-Wesley, Reading, MA.
24. Poehlandt, K. (1989), Materials Testing for the Metal Forming Industry, Springer-Verlag, Berlin.
25. Johnson, W. and Slater, R. A. C. (1967), A Survey of the Slow and Fast Blanking at Ambient and High Temperatures, Proc. CIRP-ASTME, 825–851.
26. Semiatin, S. L. and Lahoti, G. D. (1981), Deformation and Unstable Flow in Hot Forging of Ti–6Al–4Zr–2Mo–0.1Si, Metall. Trans., A12, 1705–1717.
27. Wei, Z., Li, Y., Li, J. and Hu, S. (2000), Formation Mechanism of Adiabatic Shear Band in Tungsten Heavy Metals, Acta Metall. Sin., 36, 1263–1268.
28. Odeshi, A. G. and Bassim, M. N. (2009), High Strain-Rate Fracture and Failure of a High Strength Low Alloy Steel, Mater. Sci. Eng., A525, 96–101.
29. Zhen, L., Zuo, D. L., Xu, C. Y. and Shao, W. Z. (2010), Microstructure Evolution of Adiabatic Shear Bands in AM60B Magnesium Under Ballistic Impact, Mater. Sci. Eng., A527, 5728–5733.
30. Zhou, X., Shukui, L., Liu, J., Wang, Y. and Wang, X. (2010), Self-Sharpening Behavior During Ballistic Impact of Tungsten Heavy Alloy Rod Penetrators Processed by Hot Extrusion and Hot Torsion, Mater. Sci. Eng., A527, 4881–4886.
31. Yazdani, F., Bassim, M. N. and Odeshi, A. G. (2009), The Formation of Adiabatic Shear Bands in Copper During Torsion at High Strain Rates, Proc. Eng., 1, 225–228.
32. Turing, A. M. (1952), The Chemical Basis of Morphogenesis, Philos. Trans. R. Soc. London, B237, 37–72.
33. Prigogine, I. (1971), Entropy and Dissipative Structure. In Lectures in Statistical Physics (Lecture Notes of Physics, VII) (eds. J. Ehlers, K. Hepp and H. A. Weidenmüller), 1–19, Springer-Verlag, Berlin/Heidelberg/New York.
34. Prandtl, L. (1952), Essentials of Fluid Dynamics, Hafner Publications, New York, NY.
35. Dai, L. H. and Bai, Y. L. (2008), Basic Mechanical Behaviors and Mechanics of Shear Banding in BMGs, Int. J. Impact Eng., 35, 704–716.
36. Grady, D. E. and Kipp, M. E. (1985), The Growth of Inhomogeneous Thermoplastic Shear, J. Phys., 46, 291–298.
37. Bai, Y., Cheng, C. and Ling, Z. (1992), Localization and Pattern of Deformation Thermo-Visco-Plastic Material, J. Mech. Behav. Mater., 4, 19–32.
38. Grady, D. E. (2011), Shear Failure in Brittle Solids, Int. J. Impact Eng., 38, 661–667.

2 Experimental Methods

Lothar W. Meyer and Frank Pursche
Nordmetall GmbH, Adorf/Erzgebirge, Germany

2.1 Introduction

Besides the common failure mechanism based on crack propagation, adiabatic shear failure results from a collapse mechanism, mainly at high deformation rates. This failure incorporates locally extreme high shear strains, but due to the small volume involved, it appears to occur in a macroscopic brittle manner.

This failure behaviour occurs in various technical areas, such as machining, forging, blanking, ballistic (target and penetrator), crash, surface friction and detonation loading. The adiabatic shear-failure behaviour mainly occurs in metallic materials, such as steel, titanium or aluminium, although it can also appear in plastics, rocks or ceramics.

The first part of this chapter deals with the different experimental methods, which are used to determine the propensity of a material to fail under adiabatic shear condition. Several test techniques are based on geometrical discontinuities, such as with hat-shaped, single- or double-edge specimens as well as punch or fracture toughness specimens. On the other hand, some techniques exist in which the failure strain is mainly dependent on material behaviour and is not initially influenced by geometry effects, such as in torsion (when the gauge length is not too small), in mono-axial or several different multi-axial compression tests, the cylinder expansion test or the inclined compression/shear tests. The task of this first part of the chapter is to present, in detail, which test techniques have been developed and are usable to examine the shear propensity of the materials with their advantages and disadvantages. The result of this study shows that no one test procedure can cover all material influences, such as strength or shear strains, and cannot comprise all materials. Hence, the materials have to be discussed separately.

The materials that will be described are metallic materials in the main part from ferrous to non-ferrous engineering alloys, polymers, ceramics and granular or geological materials.

The second part of this chapter discusses the influence of the material properties on the adiabatic shear-failure behaviour for quenched and tempered steels. In the literature, much information can be found, which supports the theory that some material properties influence the occurrence of adiabatic shear-failure behaviour in a positive or negative manner. The determination of the propensity to form shear bands for steels will be done through a special bi-axial dynamic compression/shear

Adiabatic Shear Localization. DOI: http://dx.doi.org/10.1016/B978-0-08-097781-2.00002-2

test in a drop-weight tower. The failure achieved in the test is only material dependent. Furthermore, the test will be concerned, as with the theory of Culver [1], with the competing processes of work hardening and thermal softening. Additionally, we will determine which material properties have a strong influence on the adiabatic shear failure and which properties are insignificant. Further questions include the following: Does a critical value exist for the transition between sheared and non-sheared areas, and is it possible to find a correlation between the material's properties and their adiabatic shear-failure behaviour? Thus, the adiabatic shear-failure behaviour is a process that includes the thermal condition; it is also concerned with dynamic compression behaviour at high temperatures. The question of whether there is any correlation with other properties, such as hardness and tensile strength or the shear capability by hat-shaped specimens for the evaluation of adiabatic failure behaviour, should also be answered.

2.2 Test Methods

This section presents the test techniques that have been developed and are usable to examine the shear propensity of the materials with their advantages and disadvantages. These test methods will be referred to for each material. Because of the different nature and composition of the materials, special tests are necessary to investigate the propensity for adiabatic shear failure. This section mainly focuses on the test technique for metallic materials (ferrous and non-ferrous alloys). Other sections of this chapter are concerning with materials such as ceramics, polymers, and granular and geological materials.

Because of the absence of the Poisson-ratio effect, the complications caused by radial expansion or contraction are eliminated during shear and torsion testing in contrast to uni-axial tension or compression loading, which provides an advantage for these test methods. The problems of necking (tension) and barrelling (compression) do not occur in shear and torsion loading.

To test materials concerning their propensity to adiabatic shear-failure behaviour, a few requirements are necessary to be fulfilled to reach usable results. At first, the test technique should be able to create a sufficient loading velocity to obtain a high strain rate in the material so that the deformation process is mainly adiabatic, and adiabatic shear bands can develop. A second requirement is to reach a high deformation with a localized homogeneously strained area. The third requirement is that necking or barrelling must be avoided. The fourth requirement is to have a well defined and simple stress state in the specimen, and the fifth condition is that the recording of force and strain for shearing during the test must be ensured to assess the adiabatic shear-failure behaviour of the material.

2.2.1 Test Methods for Metallic Materials

Many different experimental methods exist to determine the propensity of a metallic material to fail under adiabatic shear conditions. Several test techniques are based on geometrical discontinuities such as pure torsion, hat-shaped, single-edge or double-edge, punch or fracture toughness specimens. Some techniques exist in which the failure strain is mainly dependent on material behaviour and is not initially influenced by geometrical effects, such as in mono-axial or several different multi-axial compression tests, cylinder expansion or the imploding test, torsion test, compression test and the inclined compression/shear tests or the flyer-plate test.

The most named experimental methods are the torsion test, hat-shaped test, compression test, cylinder expansion test, punch test, compression/shear test and the single-edge or double-edge specimens. In the following sections, each individual test method will be described in detail.

Some practical arrangements, such as projectile penetration, ballistic impact, high-speed machining and indenter and punch tests, are not suitable for systematic studies due to complicated stress states, unrepeatable processes and interactions between different effects of loading.

The Torsion Test

Torsion test techniques include lathe torsion testing machines, a torsional hydraulic actuator, impact torsion techniques and Hopkinson torsion bar systems. These techniques can provide a strain rate range between 10^2 and 10^4 s^{-1}. The specimen used can be made of solid bars or thin-walled tubes. Torsion tests have the following advantages:

- A simple loading condition
- A long loading time
- A uniform stress state
- Good measurement of signals of stress–strain data.

An example for a torsion hydraulic actuator test (Figure 2.1) for the medium velocity range is given by Lindholm and Johnson [2], cited in the ASM Handbook vol. 8 by Lindholm et al. [3], with the determination of shear-stress–shear-strain curves of different materials. The failure strain and the strain-hardening coefficient are taken. The specimen used was a tube form with an outer machined groove. The strain rates reached were about 10 s^{-1}. Through technical changes, a strain rate of up to 300 s^{-1} is possible.

Culver [4] described a torsional impact apparatus to reach higher loading velocities (Figure 2.2). He used a commercial lathe and a drive mechanism. The specimen is rotated in a chuck, and when the rotational speed of the drive unit is established, the loading of the specimen can occur through a triggering rod. An alternative to this arrangement consists of a stationary specimen and a load cell as well as a rotating drive unit, according to Culver [4]. He tested a titanium, a mild steel and an AA 6061-T6 alloy.

Baker and Yew [5] developed a dynamic torsion assembly (Hopkinson principle) to be used on a lathe (Figure 2.3). The test signals were determined from strain gauges on the input and output bars. The test material was a pure copper, which showed a strong strain-rate dependence.

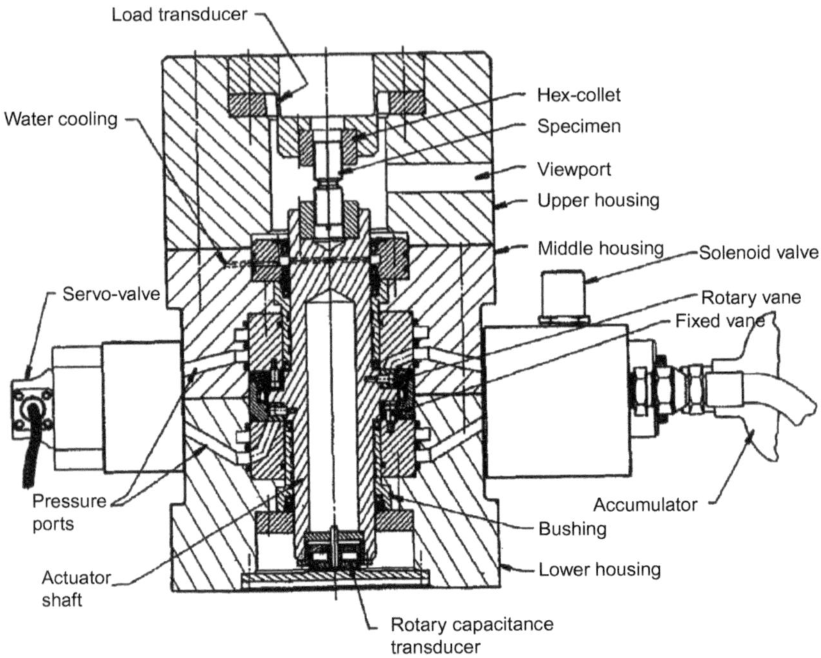

Figure 2.1 Schematic of hydraulic torsion apparatus, according to Lindholm and Johnson [2].

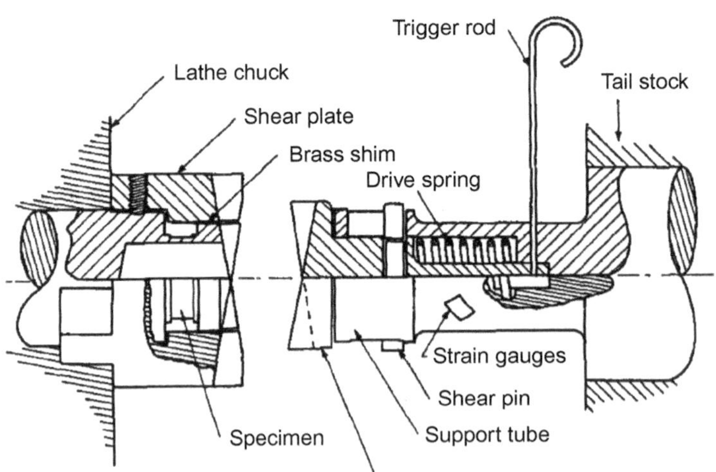

Figure 2.2 Cross section of torsional impact machine, according to Culver [4].

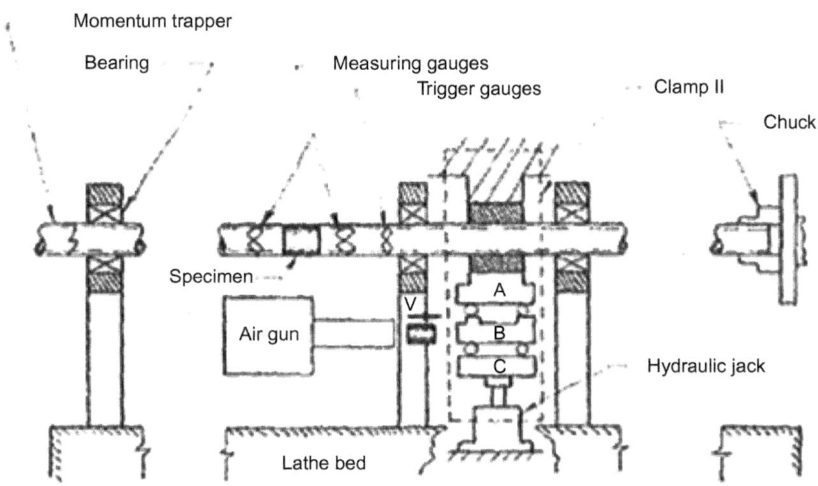

Figure 2.3 Illustration of a dynamic torsional assembly used in a lathe, according to Baker and Yew [5].

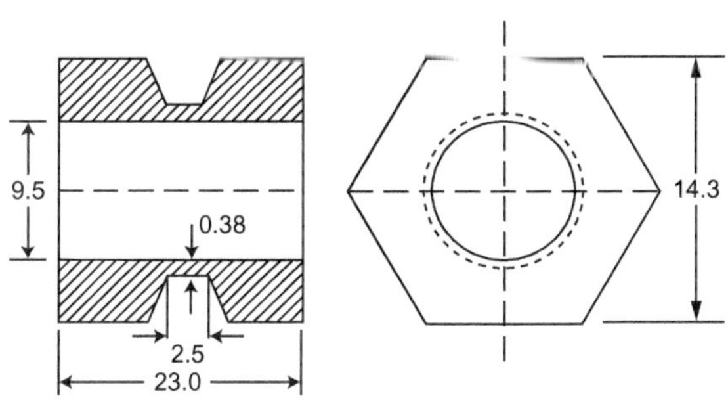

Figure 2.4 Torsional specimen for use in a Hopkinson bar, according to Duffy and Chi [10].

In a bar under torsional loading, a gradient of stress and strain exists from the centre of the bar to the outer surface. To reduce this gradient, hollow tubular specimens are commonly used. The specimen should not be too slender otherwise it might buckle. The thin-walled tubular torsion specimen is a good way to test the shear formability of materials [6–10]. In torsion loading, a simple shear state exists. In a thin-walled tube, a homogeneous state of shear deformation along the gauge length and nearly across the wall thickness is present. A further advantage of the torsion specimen is that very large deformations can be obtained. The most used specimen is tubular with a large reduction of cross section in the gauge length [10], as shown in Figure 2.4. The material is allowed to deform according to its capability to bear the torsion loading. At the tube specimen, there are some disadvantages. The test results can be influenced by buckling or, more likely, a rough surface of the specimen. The determined failure data are not straightforward for engineers. The ratio of gauge length to gauge diameter should be 1 (or higher) when a homogenous stress and strain field is required. But, to reach high strains at high strain rates, a ratio lower than 1 is beneficial.

A specimen geometry that can be simply adapted to the Hopkinson bar is shown by Giovanola [11] and in Figure 2.5. This specimen had an outer diameter of 25 mm and an inner contour that is a hexagon to fix the specimen on the Hopkinson bars. A groove on the inner side was machined in the middle of the length so that a wall thickness of 0.5 mm is generated.

In most cases, the materials will be investigated with a torsion Hopkinson assembly. Campbell et al. [12] investigated the torsional behaviour of copper, titanium and mild steel under static and dynamic conditions. Campbell used a torsion Hopkinson assembly, according to Figure 2.6. He found a larger strain rate dependency of flow stress for titanium in contrast to steel. The influence of strain-rate changes during the test ('jump test') was also conducted for these materials. Titanium and mild steel show a stronger sensitivity to a strain-rate change than copper.

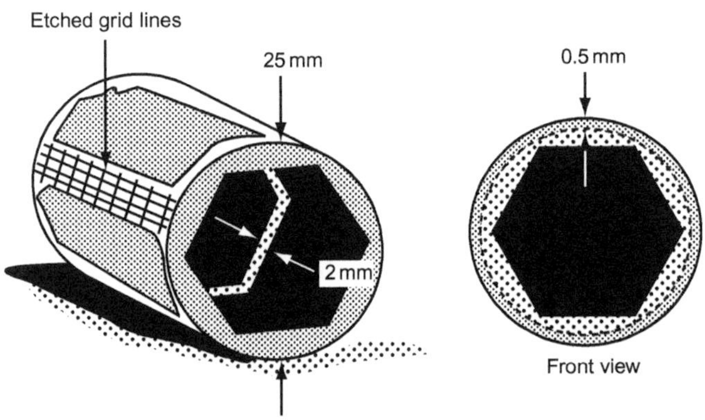

Figure 2.5 Special specimen form for the Hopkinson bar test, according to Giovanola [11].

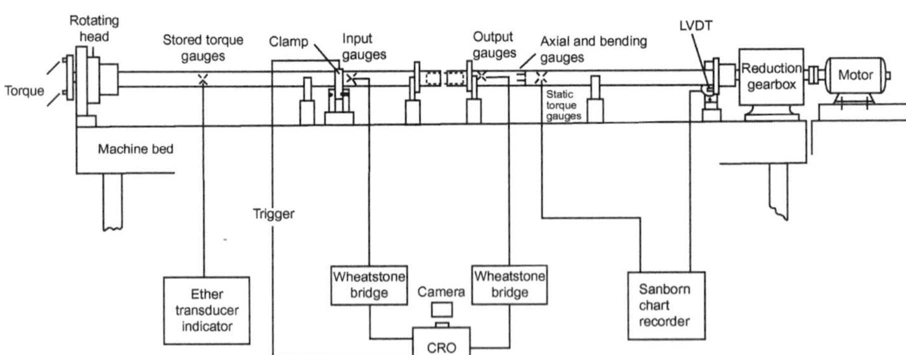

Figure 2.6 Assembly of torsion Hopkinson, according to Campbell et al. [12] and Bai and Dodd [8].

The first use of this technique was by Baker and Yew [5] in a lathe. Duffy et al. [13] and Hartley et al. [14] describe the principle of the torsional Kolsky bar. The torsional Hopkinson bar set-up consists of two bars: the incident bar and the transmission bar. The torque can be stored in the incident bar or can be provided through a flying wheel. Strain rates between 10^2 and 10^3 s^{-1} can be reached.

Marchand and Duffy [15] determined the shear-stress–shear-strain behaviour of a HY100 steel with a torsional Kolsky bar. The same specimen geometry was used by Hwang et al. [16]. They measured the local temperature on the surface and determined the strain distribution during the shearing process, and, respectively, the shear band formation process. Similar investigations with the same test procedure and specimen were done on AISI 4340 steel by Mgbokwere et al. [17].

With this type of torsion specimen, Duffy and Chi [10] and Lee and Lin [18] investigated the dynamic torsion behaviour of Ti–6Al–4V in unaged and overaged (formation of precipitations) conditions, each with two different initial microstructures. The different material behaviour also results in different appearances of the adiabatic shear bands.

Duffy and Chi [10] determined the shear behaviour of a low carbon-rolled steel, a HY100 steel and a AISI 4340 steel in two tempered conditions. The tests were conducted in a torsional Kolsky bar. With increasing defect parameters, the failure shear strain decreases. They observed and documented shear banding evolution. Cho et al. [19] used the same specimen geometry as Duffy and Chi [10] with a Kolsky bar. They investigated the shear behaviour of three steels and observed the microstructure. The AISI 1018 steel showed deformed bands, and the HY100 steel mainly showed deformed bands. However, in the AISI 4340 steel, white etching bands were only found.

Bai et al. [20] determined the shear band development for Ti–6Al–4V alloy by means of torsion Hopkinson experiments with prescribed deformation limits. Thus, an investigation of shear band evolution is possible. To obtain a singular deformation of specimen, the Hopkinson assembly was modified so that the specimen was loaded only with the incident pulse.

Bassim and Panic [21] conducted dynamic torsion tests on a Hopkinson assembly with a similar specimen geometry to Hwang et al. [16]. It was found that the heat treatment and the alloying elements influence the adiabatic shear susceptibility.

Hwang et al. [16] and Hwang et al. [22] conducted torsion tests at ultra-fine-grained low-carbon steel with ECAP-deformation (equal channel angular pressing) in a torsional Kolsky bar. The specimen gauge length was 2.5 mm with an inner diameter of 9.5 mm, and the wall thickness was 0.38 mm. The influence of the ECAP-process and heat treatment on the followed shear-stress behaviour was investigated.

Lee et al. [23] investigated the temperature and strain-rate-dependent behaviour for Ti–6Al–4V alloy in a torsional Hopkinson bar. They used a special cylindrical torsion specimen, which was mounted on the Hopkinson bars (Figure 2.7). The wall thickness in the gauge length is similar to that of Duffy and Chi [10] (Figure 2.4). The fracture behaviour of this alloy is dominated by adiabatic shearing. Similar investigations were carried out by Lee et al. [24] for an Al–Si alloy similar to Figure 2.7.

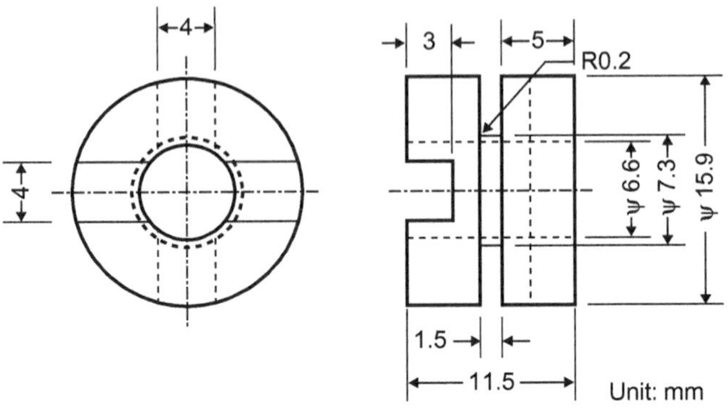

Figure 2.7 Torsion specimen, according to Lee et al. [23].

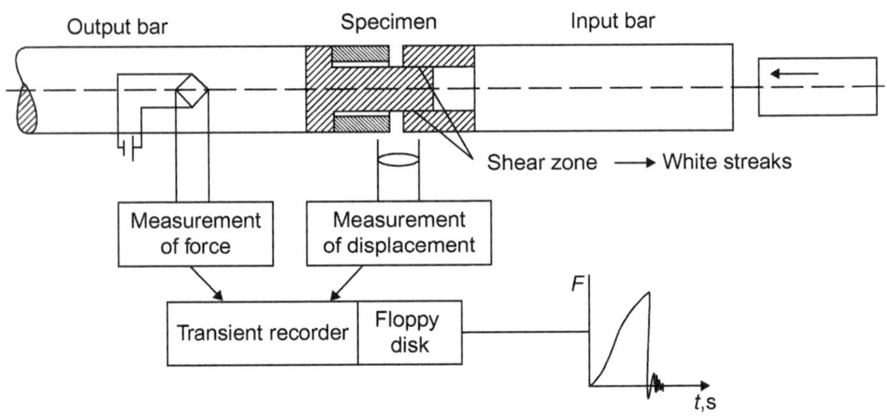

Figure 2.8 Instrumental Hopkinson bar apparatus to produce white bands in a laboratory, according to Hartmann et al. [26].

Latourte et al. [25] conducted dynamic torsion tests in a Hopkinson bar for a TRIP-120 and a HSLA-100 steel. A torsion specimen similar to that of Duffy and Chi [10] was used. The deformation behaviour could be depicted through photographs, and thus the evolution of shear band formation could be observed. The TRIP-120 steel shows a higher susceptibility to adiabatic shear failure than the HSLA-100 steel.

Hat-Shaped Test

The hat-shaped specimen technique is often used in a Hopkinson compression bar (Figure 2.8), which was invented in 1977 by Meyer and Hartmann at the Fraunhofer Institute (IFAM, Bremen, Germany) and was used by Hartmann et al. [26] and cited by Meyer et al. [27]. The shear specimen (hat shape) is located

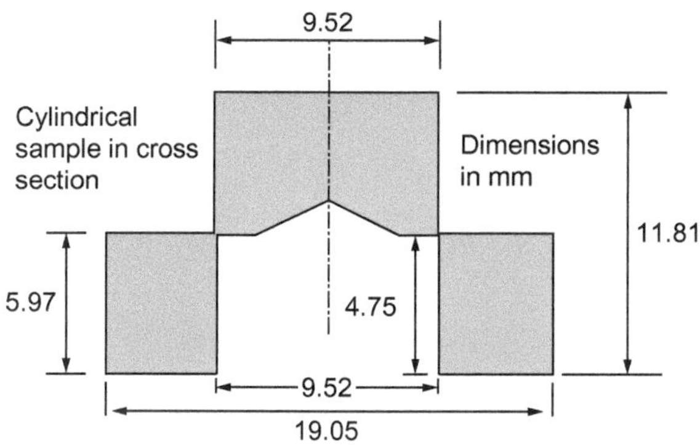

Figure 2.9 Geometry of a hat-shaped specimen, according to Beatty et al. [32].

between the input and output bar. They investigated different steels for mechanical behaviour and their formation of shear bands. The hat specimen is often used by Meyers [28–31] and Beatty et al. [32] (Figure 2.9).

The advantage of the hat-shaped test is that it can be used to produce shear bands even in very ductile materials in a Hopkinson Compression Set-up and that a comparison between small and larger failure strains can be made [33]. Due to the geometry, very high strains at high strain rates can be reached, but the shear-stress measurement is influenced by edge effects. The shear deformability can be measured and assessed. To investigate the microstructure of shear band formation, the displacement of the hat specimen can be limited easily in the axial direction by using stopper rings.

Beatty et al. [32] did experiments on an AISI 4340 steel using the hat specimen to examine the shear band formation of high-strength steels at high loading rates. They examined the influence of austenitization temperature on the energy consumption under shear loading.

Disregarding the hat-shaped geometry [32], where the punch and hole diameter are the same size, there exists in the literature few examinations where the punch diameter is lower than the hole diameter; most investigations are carried out on specimens where the punch diameter is larger than the hole diameter. This configuration leads to an additional compressive load, as a hydrostatic pressure.

A specimen where the punch diameter is lower than the hole diameter is used by Minnaar and Zhou [34] (Figure 2.10). The specimens were loaded in a compression Hopkinson bar and were variously stopped to investigate the shear evolution. The materials examined were HY-80, HY-100, HSLA-80, 4340 VAR and Ti–6Al–4V. The titanium in this study is the hardest material and shows an early failure in contrast to the mild steels. With a similar specimen geometry, as shown in Figure 2.10, Chung et al. [35] performed dynamic tests in a Hopkinson bar for AISI 4340 steel. The shear width is almost the same, but the total diameter is larger.

Pursche [36] and Pursche and Meyer [37] investigated the shear-failure behaviour of different quenched and tempered steels with a hat-shaped specimen where the punch diameter is also lower than the hole diameter (Figure 2.11A). This

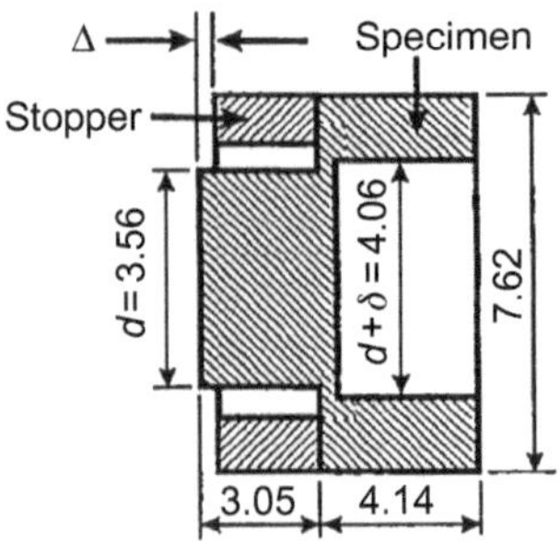

Figure 2.10 Hat-shaped specimen, according to Minnaar and Zhou [34].

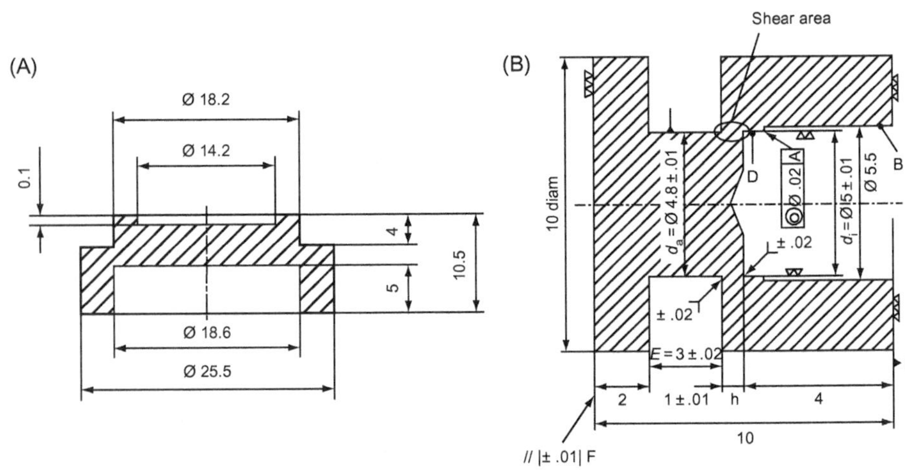

Figure 2.11 Hat-shaped specimen, according to Pursche [36] (A) and Meyer and Krüger [33] (B).

specimen is similar to that of Meyer and Krüger [33] (Figure 2.11B). Pursche found a linear correlation between the shear-failure behaviour for the hat-shaped specimen to the failure strains found under bi-axial compressive shear conditions (Figure 2.94).

By modifying diameters, where the punch diameter is larger than the hole diameter, different compression shear-stress states (hydrostatic pressure) are produced. This stress state hindered the failure, and brittle material can be analysed concerning the susceptibility to adiabatic shearing. Xue [38] used such a hat specimen (Figure 2.12) to investigate the static and dynamic behaviour of 304 stainless steel, titanium and Ti–6Al–4V. For the 304 steel, the influence of grain size was analysed.

Meyers et al. [28] examined the microstructure of sheared materials such as copper, titanium and tantalum. The test methods were hat-shaped specimens loaded in a Hopkinson bar and the thick-walled cylinder test (TWC), which will be explained later.

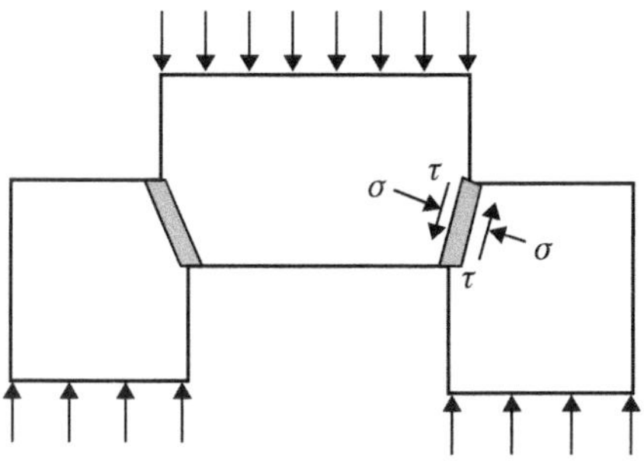

Figure 2.12 Hat-shaped specimen with compression rate, according to Xue [38].

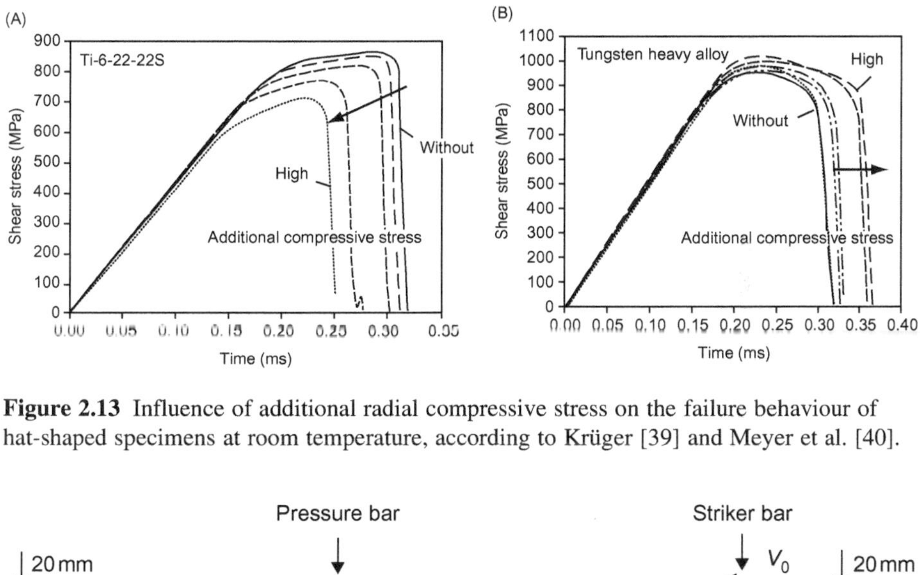

Figure 2.13 Influence of additional radial compressive stress on the failure behaviour of hat-shaped specimens at room temperature, according to Krüger [39] and Meyer et al. [40].

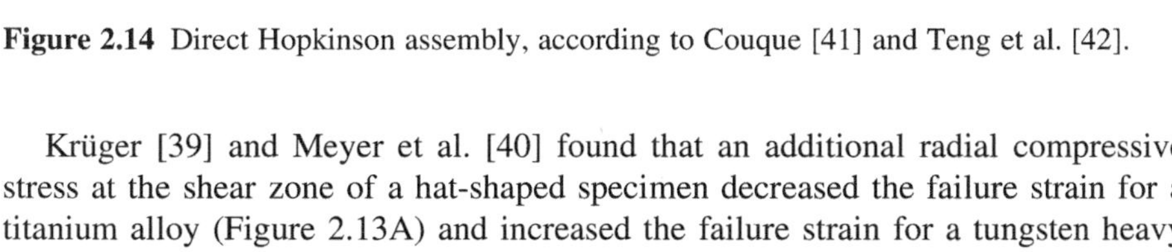

Figure 2.14 Direct Hopkinson assembly, according to Couque [41] and Teng et al. [42].

Krüger [39] and Meyer et al. [40] found that an additional radial compressive stress at the shear zone of a hat-shaped specimen decreased the failure strain for a titanium alloy (Figure 2.13A) and increased the failure strain for a tungsten heavy alloy (Figure 2.13B). The radial compression component was not induced by the specimen geometry itself but by an external cylinder to induce a radial pressure in the shear zone of the specimen.

Couque [41] used a direct Hopkinson bar (Figure 2.14) to investigate the shear-failure behaviour of a tungsten heavy alloy and a steel with different side pressures at a hat specimen (Figure 2.15). Couque found that the length of the adiabatic shear

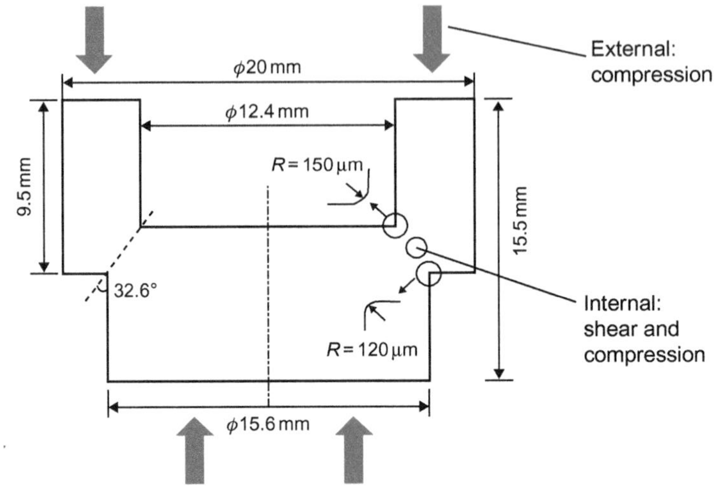

Figure 2.15 Hat-shaped specimen with additional compression stress, according to Couque [41] and Teng et al. [42].

band in the gauge section is strongly influenced by loading velocity. With higher side pressure, the shear-failure strain for tungsten alloy is increased, similar to Krüger's results [39]. Couque explained this behaviour by the decreasing portion of plastic work converted in heat with increasing pressure.

Teng et al. [42] conducted a numerical investigation on a hat-shaped specimen with geometrically induced mutual compression stresses, as shown in Figure 2.15 [41]. The material data from experimental investigations of Couque were used. Teng simulated the formation and propagation of the adiabatic shear bands and the subsequent crack propagation. Teng compared simulated shear band lengths with experimental data. He postulated that the hot spots in a shear band are initiation sites for crack propagation. Thus, the formation of crack within a shear band resembles a linkage of small cracks.

Meyers et al. [43] examined shear bands in AISI 304L stainless steel, which are produced with a hat-shaped specimen in a Kolsky bar and with TWC tests. The shear bands are investigated with Electron Backscatter Diffraction (EBSD) and Transmission Electron Microscope (TEM) techniques.

Clos et al. [44] performed hat-shaped specimen tests with two different geometries: a classical cylindrical specimen and a flat hat-shaped specimen with loading in a Hopkinson bar (Figure 2.16). The flat hat-shaped specimen was designed to reduce edge effects. Furthermore, the flat specimen allows you to measure the temperature rise during the deformation with an infrared detector. Clos et al. varied the compression/shear-stress state and used different deformation limits to examine the development of the shearing process. Clos used steels with 0.15% and 0.45% carbon content and an Inconel 718 alloy. In 0.45% C-steel and in Inconel 718, he found white etching bands (up to 20 μm wide), and in 0.15% C-steel, he only found dark etching bands (about 200 μm). Clos divided the load–displacement behaviour

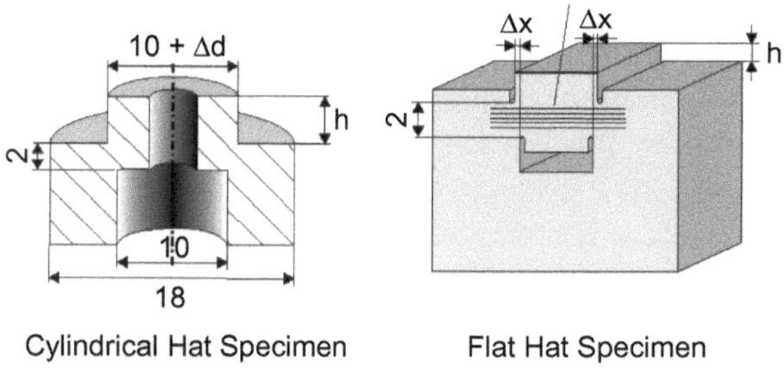

Figure 2.16 Hat specimen as cylindrical and flat geometry, according to Clos et al. [44].

into three stages. Stage I is the pre-localization phase, stage II is the localization phase and stage III is the post-localization phase. The localization was not only created by the instability of the uniformly strained zone but also by edge effects. The failure does not occur simultaneously along the whole shear zone. The shear bands were formed during the post-localization phase by the process of friction. The measured temperatures of 0.45% carbon steel at the beginning of the localization are about 200–250°C, 500°C at the beginning of the post-localization phase and 850°C at the end of the post-localization phase.

Xue and Gray [45] investigated the annealed AISI 316L stainless steel in a Hopkinson bar assembly using hat-shaped specimens with a certain amount of compressive stress. They characterized the shear band formation through deformation steps with various loading times. Comprehensive examinations of the microstructure, including TEM observations, of this steel was done by Xue et al. [46]. Xue et al. [47] performed geometry experiments with pre-shocked materials (304 and 316L) with the same hat-shaped specimen. The pre-shocked material condition shows a higher yield stress and a shorter strain-hardening stage as unshocked material. The failure strain is also lower in contrast to the unshocked state.

Lins et al. [48] also used a hat-shaped specimen with an additional compression stress to investigate a hot-rolled, interstitial-free steel in a Hopkinson assembly with different deformation limits to investigate the microstructural appearances of shear bands.

Lee et al. [49] used a hat-shaped specimen with a slightly larger diameter of the punch as the hole. The materials investigated were C-steels with 0.15%, 0.45% and 1.16% carbon contents. The tests are performed with a Hopkinson bar assembly. Lee examined the formation of adiabatic shear bands. A deformed band for 0.15% C-steel and transformed bands for which medium- and high-carbon content were observed. Lee also investigated the influence of loading velocity on shear band formation.

Xu et al. [29] examined the microstructural aspects of adiabatic shear localization on different materials, such as low-carbon steel, stainless steel, titanium and its alloys, aluminium alloy and others. The materials were tested among other techniques with hat-shaped specimens in a Hopkinson bar assembly.

Peirs et al. [50] conducted dynamic tests on hat-shaped specimens of Ti–6Al–4V with a Hopkinson bar. He examined the influence of the edge radius of the specimen and the shear width on the formation of shear bands. The width generates the homogeneity of the distribution of stress and deformation and of hydrostatic pressure. The radius configuration governs the initiation of adiabatic shear bands. Complementary examinations concerning evolution of shear bands and Finite Elements (FE) calculations were done.

The hat-shaped specimen type was used for titanium Ti–6Al–4V [30], titanium alloy Ti–6Al–4V [51], aluminium alloy [38,52], tantalum [31,53], zirconium [54] and copper [55].

Compression Test

The dynamic compression test method is used frequently to test the susceptibility of materials to the occurrence of adiabatic shear failure. Only a simple specimen geometry is necessary, and the results are straightforward for engineers. Test rigs such as the split-Hopkinson pressure bar (SHPB) and the drop-weight tower are common. With the drop-weight tower (Figure 2.17A), it is possible to reach a strain rate of about $10^3\ s^{-1}$ up to $10^2\ s^{-1}$, and with the Hopkinson bar (Figure 2.18), strain rates of up to $10^4\ s^{-1}$ are possible by use of small specimens (about 1 mm long). A further assembly can be a gas or electromagnetic gun, which can provide a very high loading velocity, strain rates up to $10^6\ s^{-1}$ and high pressure. The disadvantage is that the measuring possibilities are limited.

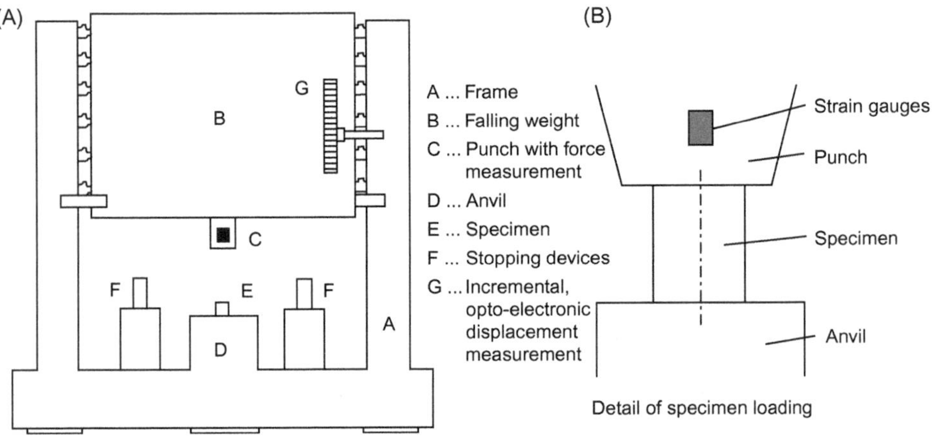

Figure 2.17 Drop-weight tower assembly with loading principle, according to Meyer and Krüger [56].

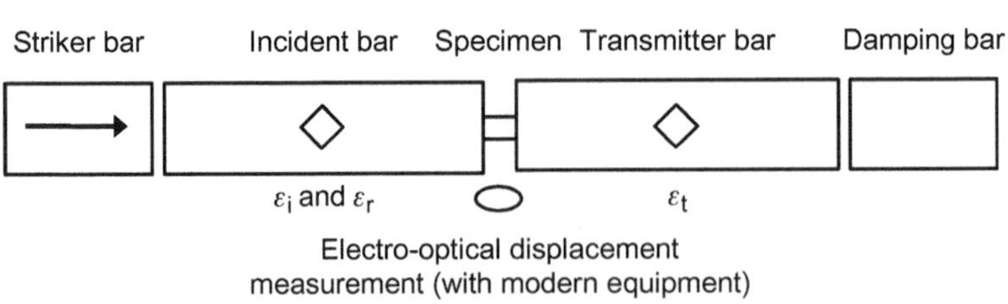

Figure 2.18 Principle of SHPB, according to Kolsky [57].

The drop-weight tower assembly (Figure 2.17) is frequently used to investigate the materials concerning their adiabatic shear propensity. This test principle is described by Meyer and Krüger [56]. With the drop-weight tower, there will be enough energy (because of the falling weight of 600 kg or more) to test high-strength materials or specimens with large dimensions under impact conditions up to a loading velocity of 3 m/s. The measuring of force–time records will be mainly realized on the punch. The measuring of displacement can be done either with a Heidenhain measuring device or with the electro-optical extensometer system of Rudolf. The realization of different temperatures by stopping tests is feasible. With stopped tests, the shear band development can be determined accurately.

The Hopkinson bar (Figure 2.18), used for higher loading velocity, consists of two long bars: the incident bar and the transmission bar. The specimen is positioned between these two bars.

The impulse to the incident bar is given by a short striker bar, which is accelerated by a gas gun or powder gun. All bars have the same diameter. The striker forms a wave-loading impulse to the incident bar. The length of the striker bar determines the duration of the loading impulse. When the loading pulse hits the specimen, a part of the loading wave passes to the transmission bar and is now called a transmission wave. The rest is reflected and is called a reflected wave. A strain gauge, which can record the signals during the test, is glued on the incident and transmission bar. The signal from the transmission bar is used to find the stress-time characteristic, and the signal from the reflected wave (from the incident bar) is used to find the displacement–time characteristic.

Belk and Watson [58] conducted investigations on a tungsten heavy alloy under dynamic compressive loading through a drop-weight tower with compression specimens that had a different ratio of diameter to length. It was found that for slender specimens, the failure is clear and occurs early. Furthermore, it was found that up to a ratio of 1 between the diameter and length of the specimen, the failure behaviour is similar and independent of the specimen length. The failure strain is increasing when the ratio between diameter and length of the specimen exceeds 1.

Sabih et al. [59] investigated the deformability of AISI 1018 steel under cold heading process conditions. In the cold heading process, fracture under tensile stresses can occur, although the materials used are ductile. Sabih and Nemes [60] extended the prior examination to other materials, such as AISI 1008, AISI 1038, AISI 1541, AISI 8640 and a dual-phase (DP) steel. The analysis of the propensity to form adiabatic shear failures of these materials was performed with a drop-weight tower and a special die geometry that has the form of a screw head. The failure behaviour can be divided into two groups: the first group experienced very large cracks, such as the AISI 1008 and AISI 1018, and the second group exhibited short cracks along adiabatic shear bands for the AISI 1038, AISI 1541, AISI 8640 and DP steel.

Examples of failure due to adiabatic shearing in uni-axial compression Hopkinson tests are shown in Figure 2.19A and B. Figure 2.19A is a steel specimen [61], and Figure 2.19B is a broken specimen of AA7039 [9].

Many other authors can be cited concerning cylindrical compression specimens and the use of SHPB to test the material's propensity to adiabatic shear

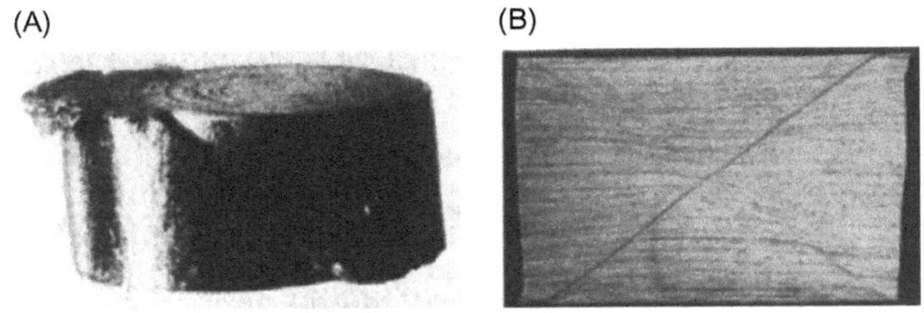

Figure 2.19 Shear band formation under pure compression loading in SHPB: (A) Ansart and Dormeval [61] with steel and (B) Wright [9] with AA7039.

failure: Dormeval [7] on martensitic steels, Meyer [62] on thermomechanically treated steel X41CrMoV51, Odeshi et al. [63] on quenched and tempered steel, Xue et al. [64] on Al-Alloy, Nakkalil et al. [65] on rail steels, Meunier et al. [66] on armour steel, Hanina et al. [67] on Mg alloy and Ti–6Al–4V, Lee and Lin [68] on Ti–6Al–4V, Wei et al. [69], Jinxu et al. [70] and Meyer et al. [71] on tungsten heavy alloy (the latter two had prior pre-deformation), Rittel et al. [72] on Mg alloy, Dai et al. [73], Odeshi et al. [74] and Wu et al. [75] on Al-matrix composites and Lee et al. [76] on Inconel 690 superalloy.

Dormeval [77] conducted dynamic compression tests in an SHPB on different martensitic steels. Depending on the hardness of the material, the shear-failure behaviour is different. With increasing hardness and decreasing strain-hardening coefficient, the failure strain is decreased.

Wang et al. [78] conducted dynamic compression tests in a Hopkinson bar on a titanium alloy and determined the dependence of test temperature on the critical failure strain and critical failure strain rate and combined these on an assessment diagram.

Nguyen [79] conducted compression tests in SHPB on steel and titanium. He found a critical loading velocity below which no adiabatic shearing occurred. Above this critical velocity, a sudden shear failure with a brittle character occurred for steels.

Chen et al. [53] conducted different tests on tantalum in the as-received as well as in pre-shocked conditions at different temperatures. The material was examined under dynamic compression test in a Hopkinson bar, in shear deformation with hat-shaped specimens where the punch diameter was slightly larger than the hole diameter as well as in the explosively driven TWC test. With low temperature and in the pre-shocked condition, the propensity is for adiabatic shear failure.

Xu et al. [64] conducted dynamic compression tests in the Hopkinson bar assembly at room temperature and 77 K for the Al-alloy AA 8090. The alloy was tested in various heat-treatment conditions. It was found deformed as well as transformed (white etching) bands in all material conditions. Their occurrences depend on a critical strain and strain rate.

Wei et al. [69] performed dynamic compression tests on a tungsten heavy metal alloy on a Hopkinson bar. They varied the specimen geometry with a classical cylindrical form, a conical form and a stepped form. With both the latter geometries, the adiabatic shear failure is more probable. In the cylindrical form, no

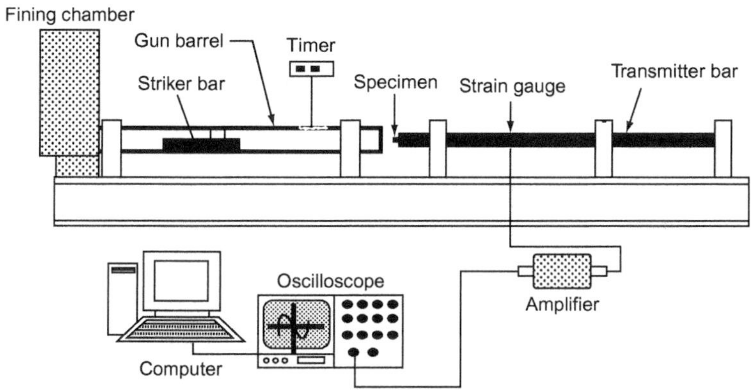

Figure 2.20 Experimental set-up for high strain-rate compression test by direct Hopkinson bar, according to Odeshi et al. [74].

adiabatic shear band was found. The reason can be explained by different stress concentrations between the different specimen geometries.

Dai et al. [73] examined the formation of adiabatic shear bands in AA 2024 matrix composite and its dependence on the particle size of reinforced SiC particles tested in a Hopkinson bar under a compressive load. They found that the onset of adiabatic shear banding strongly depends on the particle size, and the adiabatic shear banding is more readily observed in the composite with smaller particles.

Odeshi et al. [80] investigated the influence of heat treatment on adiabatic shear failure for AISI 4340 steel with a Hopkinson bar. The high-hardness specimen showed white etching band formation, and the highly tempered specimen showed no adiabatic shear failure. The same material was investigated by Odeshi et al. [63] with the influence of loading velocity. With higher loading velocity, the failure strain to fracture is decreased. Odeshi et al. [81] found white etching bands in 300°C and 400°C tempered AISI 4340-steel specimens; in 500°C tempered specimens, there was found a deformed, dark etching band; and in 600°C tempered specimens, only a deformation zone was detected. Odeshi and Bassim [82] conducted dynamic compression tests in a direct Hopkinson pressure bar. They found that DP steel has a higher susceptibility to shear failure than a pure martensitic steel.

Hanina et al. [67] conducted the influence of a confining pressure on the failure behaviour of a compression specimen in a Hopkinson bar for a magnesium and titanium alloy. They found that the failure strain is increased with increasing hydrostatic pressure.

Odeshi et al. [74] examined the propensity of adiabatic shear failure for aluminium 6061 T6 matrix composites reinforced with alumina particles tested in a direct Hopkinson apparatus. The reinforced material conditions show a higher propensity to adiabatic shear failure than the unreinforced condition. A Hopkinson bar was used to investigate the dynamic compression behaviour of the material (Figure 2.20).

Compression/Shear Test

The determination of the propensity of materials for adiabatic shear failure can be examined by a new special bi-axial dynamic compression/shear test. The pure axial loading can be performed with a drop-weight tower (Figure 2.17A) or an SHBP (Figure 2.18). Meyer and Krüger [56] described the test assembly and the use of the drop-weight tower in the ASM Handbook. The drop-weight tower, which is described by Meyer and Krüger [56], has a maximum loading velocity of about 3 m/s and a falling mass of 600 kg. This falling mass can be stopped at desired displacements, at which a visualization of the development of adiabatic shear bands versus applied strains can be observed.

This compression/shear testing with a slightly inclined cylindrical compression specimen (Figure 2.21) in a pure axial compression loading device (Figure 2.17A) was invented by Meyer [83] and used by Meyer and Staskewitsch [84] and Meyer et al. [27]. The advantage of this specimen geometry is to reach a wider differentiation of materials in contrast to plain compression loading. For this reason, the usual compression specimen is inclined a few degrees from the loading axis with the effect of an enhanced multi-axial compression/shear loading.

Meyer and Halle [85] observed the stress distribution in a compression/shear specimen (diameter 6 mm, height 6 mm, inclination 6°) and its dependence on axial deformation. The ratio of shear to normal stress both at the corner and in the centre of the specimen was determined (Figure 2.22). The stress ratio in the centre of the specimen for the entire calculated deformation range increases steadily. In contrast to this, the stress ratio in the corner stays constant at up to 20% axial deformation. From 20% to 30%, the ratio is slightly increased. The point of initiation of the shear band is in the area of the corner, where the ratio of shear to normal stress is nearly constant over a wide range of deformation.

Depending on the desired amount of additional shear stress, the inclination can be varied. This little inclination will induce a certain bi-axial compression/shear state in the specimen.

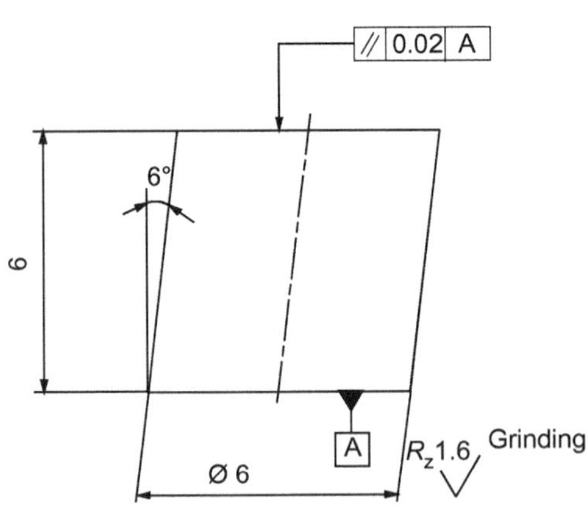

Figure 2.21 Compression/shear specimen, according to Meyer [83] and Krüger [56].

The influence of inclination changes the deformability/compressibility of a tungsten alloy drastically [86], as shown in Figure 2.23. Similar results were found by Meyer et al. [40] for a titanium alloy under a similar rate of strains.

The advantage of this technique is the absence of the influence of a stress enhancement such as a surface flaw or a notch on the failure strain. The failure development is only dependent on the material's properties or propensity to fail. As the result of this test, it can be determined whether or not there is an occurrence of adiabatic shear failure. When shear failure occurs, a corresponding axial failure

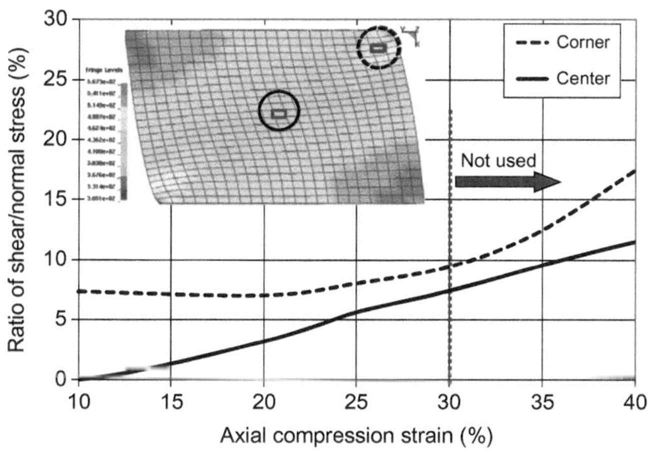

Figure 2.22 Dependence of compression/shear stress ratio in a compression/shear specimen at two different locations, according to Meyer and Halle [85].

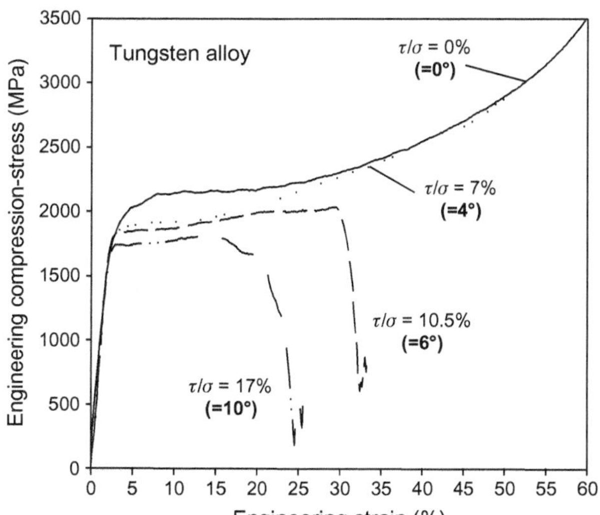

Figure 2.23 Influence of inclination of the compression/shear specimen on failure strain performed in a drop-weight test, according to Meyer et al. [86].

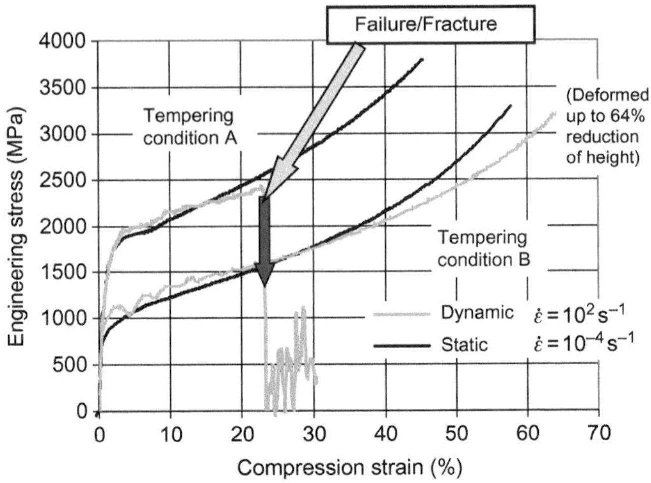

Figure 2.24 Compression stress–strain behaviour of two tempering conditions at bi-axial compression/shear test at static and dynamic loading, according to Pursche [36].

strain can be determined. Therefore, it is possible to give a qualitative and a quantitative result for an evaluation of the propensity of materials for adiabatic shear failure.

Meyer and Staskewitsch [84] investigated the influence of inclination of the compression/shear specimen to their shear-failure behaviour of Ti–6Al–4V for drop-weight tower loading conditions. With increased inclination of the specimen, the ratio of shear to compressive load, the failure strain and the ultimate compressive strength are decreasing. Further investigations were carried out by Meyer et al. [27] for Ti–6Al–4V and a tungsten alloy with various specimen inclinations. The development of shear bands was determined by microscopy and finite element simulations of the compression/shear specimen.

Pursche [36] and Pursche and Meyer [37] examined different quenched and tempered steels with different heat-treatment conditions with a hardness range between 300 and 600 HB. The propensity of adiabatic shear failure of the materials was determined with the bi-axial dynamic compression/shear test. Pursche also compared the static and dynamic behaviour between two different tempering conditions of a quenched and tempered steel (Figure 2.24). Under static loading, both tempering conditions can be deformed to the prescribed deformation limit without failure. When dynamic loading occurs, the harder material condition failed suddenly through adiabatic shear failure. Further results of this investigation will be given in Section 2.3.

A similar behaviour was found by Meyer et al. [40] for an HSLA steel at drop-weight tests between a pure compression specimen (no failure) and a compressions/shear specimen with 10.5% shear-stress loading, where adiabatic shear failure occurred.

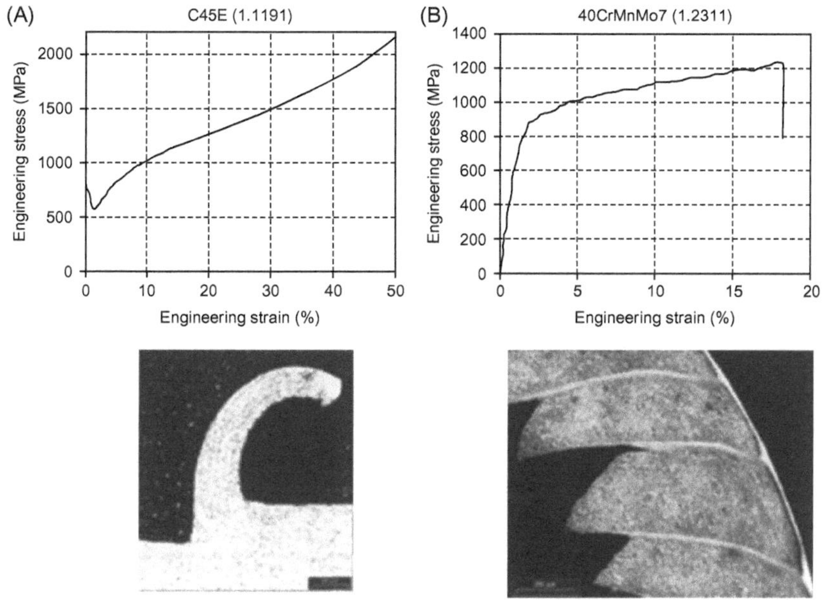

Figure 2.25 Correlation between compression/shear behaviour and chip formation at machining, according to Halle [88].

Investigations of Blümke [87] on an AA 7075 alloy and on a tool steel demonstrated that a good correlation between the results of the dynamic compression/shear test and the chip formation during high-speed cutting does exist. Halle [88] found similar results on a quenched and tempered steel as well as on a 0.45% C-steel (Figure 2.25A) with a clear correlation between the adiabatic shear-failure susceptibility (Figure 2.25B) and the chip formation in cutting. The 0.45% C-steel (Figure 2.25A) shows a continuous chip formation and no adiabatic shear failure up to high axial strains. The alloyed steel (Figure 2.25B) shows a serrated chip formation as well as an early adiabatic shear failure at 15% plastic compression strain.

Meyer et al. [89] found a clear relation between the ballistic performance of titanium alloys and their adiabatic shear properties under dynamic compression/shear loading.

Wei et al. [90] investigated a tungsten heavy metal alloy with differently inclined cylindrical compression/shear specimens. The aim of this investigation was to find the influence of pre-deformation on the susceptibility of adiabatic shear failure tested in a Hopkinson pressure bar. With increasing pre-deformation, the susceptibility is increased.

Cylinder Expansion and Collapse Test

The test arrangement with use of cylindrical specimens is generally called the TWC test. Depending on the loading direction of the explosive charge, they will be

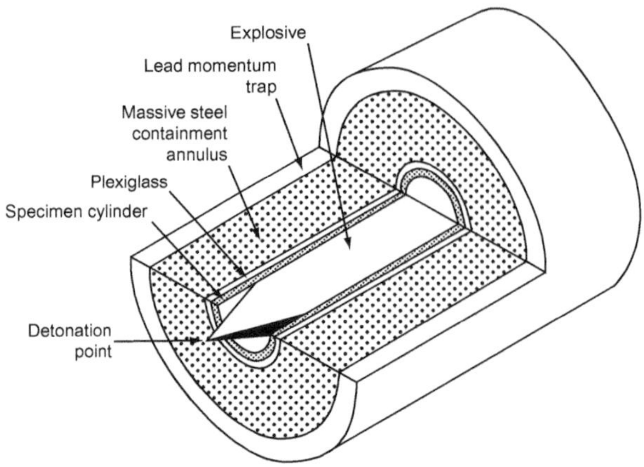

Figure 2.26 Assembly of a cylinder expansion test, according to Shockey and Erlich [91].

divided into either the contained exploding cylinder test (CEC) or the contained imploding cylinder test (CIC).

A schematic test rig for a typical cylinder exploding test (CEC) is described by Shockey and Erlich [91] (Figure 2.26). They investigated the fracture behaviour of 4340 steel at different hardness. The motivation of such configuration was related to former fragmentation problems of round shells or cylinders.

The required energy will be generated through explosive loading. The detonation provides a compressive stress in the radial direction and induces a tensile state in the circumferential direction. In this test, no stress concentrations occur, thus the failure behaviour is only material dependent. The outer lead cylinder is used as a shock wave reducer, whereby the reflection of the compression wave to a tension wave is suppressed. Hence, the additional loading of the investigated material can be avoided. The polymer membrane stops the specimen from coming into contact with the massive steel cylinder. Through variation of the explosive charge, the composition of explosive and the thickness of the polymer cylinder, the strain and strain rate applied to the loaded specimen can be adjusted. This arrangement can provide strain rates up to $10^6\ s^{-1}$ at large plastic deformations.

Wittman et al. [92] examined white etching bands of a 4340 steel produced at a cylinder expansion test (Figure 2.27) to reach further information about the microstructure of shear bands.

Meyers [93] depicted in a simple manner the process of exploding cylinders (Figure 2.28). The shear stress generates shear bands that determine the fragmentation pattern.

Grady and Hightower [94] presented the typical failure behaviour of the specimen in a cylinder expansion test (Figure 2.29). The failure mode of such a cylinder can occur by two fracture modes: tensile fracture or fracture by adiabatic shearing. The dominating fracture mode depends on the material properties and on the existing stress distribution in the cylinder. According to Bai and Dodd [8], the stress and strain distributions are not homogeneous at different axial and radial positions in the test cylinder. Only the middle of the test cylinder can be loaded under a

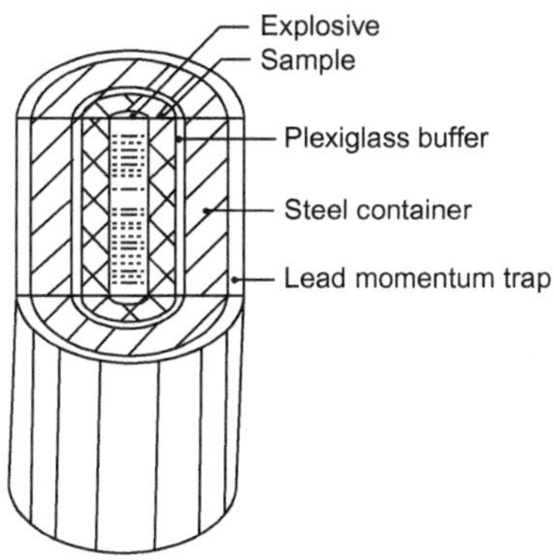

Figure 2.27 Contained fragment configuration used in SRI International, according to Wittman et al. [92].

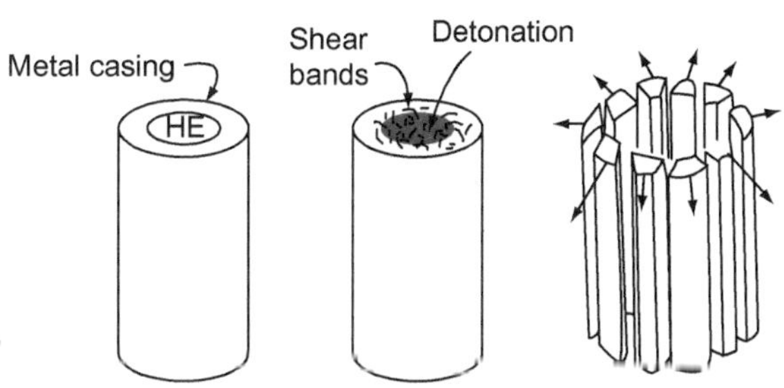

Figure 2.28 Principle of cylinder expansion test driven by explosive, according to Meyers [93].

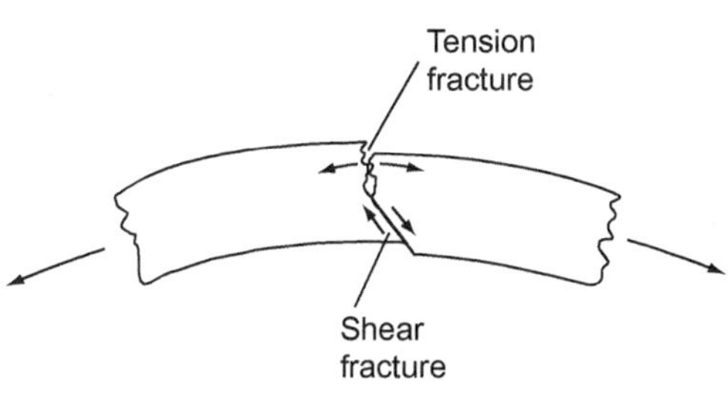

Figure 2.29 Fracture behaviour at a cylinder expansion test, according to Grady and Hightower [94].

uniform state of stress and strain. In the cylinder, a hoop stress will be induced. At the inner surface of the cylinder, large compression stresses shall appear, which induced a large plastic strain. Under this condition, adiabatic shear failure often initiates.

This failure perhaps starts on the outer surface under the influence of tension stresses and propagates to the inner surface by changing the mode into an adiabatic shear band development.

According to Bai and Dodd [8], aluminium and copper fail by a ductile mechanism, and titanium alloys and uranium alloys exhibit brittle failure behaviour at cylinder explosive tests. Steels can exhibit a ductile or brittle behaviour, dependent on heat treatment.

Staker [95] and Staker [96] first used this CEC technique with a quick-stop containment apparatus to determine the critical strain for instability to shear for

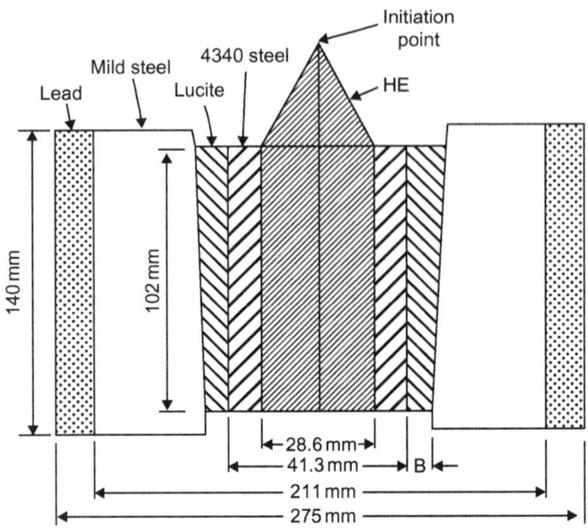

Figure 2.30 Schematic of the containment apparatus, according to Staker [95].

4340 steel in various heat-treated conditions (Figure 2.30). Pak et al. [97] also used this technique for investigation of 4340 steel in a high-hardness condition.

Bai et al. [98] conducted tests to explore the micro-shear fracture characteristic of explosive-driven cylinders for AISI 1045 steel, Ti–6Al–4V and a tungsten alloy. They found deformed as well as transformed bands in AISI 1045 steel and titanium alloy and no adiabatic shear bands in the tungsten alloy.

Sunwoo et al. [99] conducted explosively driven cylindrical expansion tests on AERMET100 steel in the as-received and heat-treated conditions. In the as-received condition, they found deformed shear bands, and in the high-hardness condition, they found transformed shear bands.

Goto et al. [100] conducted explosively driven expansion tests for AERMET100 and 1018 steel in the form of cylinder specimens and ring specimens (Figure 2.31). The purpose was the idea to generate a plane-strain load path with a pure cylinder and a uni-axial stress load path with rings. Many rings were loaded in one test to achieve a better statistical result. The specimen subjected to uni-axial stress state has a higher failure strain than the plane-strain one. The softer material AISI 1018 reaches a higher failure strain than the harder material AERMET100. The results are used to establish a certain fracture model.

The disadvantage of the cylinder explosive test is the presence of a not well-known complex stress distribution, thus the shear bands are statistically distributed, and a defined failure site cannot be predicted. Hence, the explosive test is only useful for qualitative assessment of materials.

As stated earlier, when the loading direction is opposite to the test, it is called CIC test. This test principle was developed by Nesterenko et al. [101] and is described in the ASM Handbook No. 8 by Nesterenko [102] and Nesterenko et al. [103] and cited by Xue [38]. In this CIC test, the external loading is generated by surrounded explosives (Figure 2.32). The specimen is a cylinder that will be

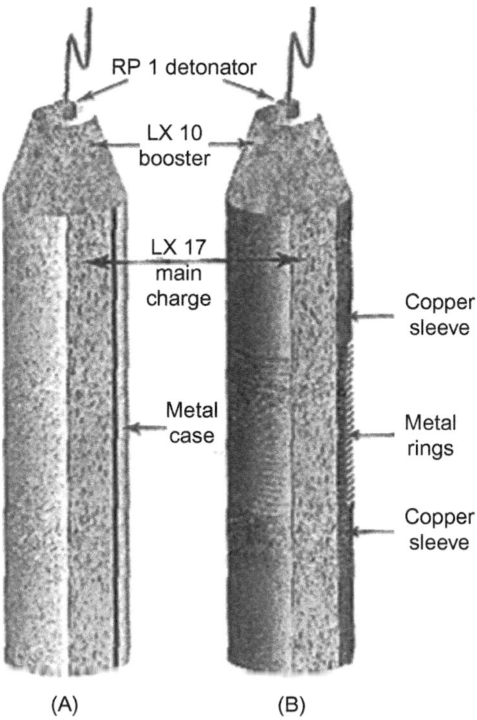

Figure 2.31 Test rig for explosive expansion test of cylinder (A) and rings (B), according to Goto et al. [100].

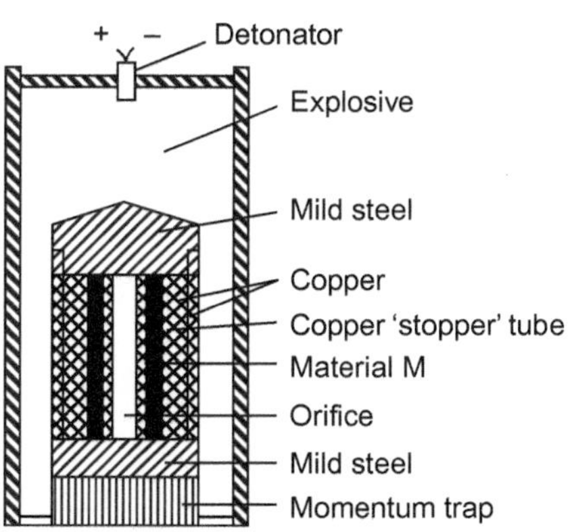

Figure 2.32 Test rig of a CIC test, according to Nesterenko et al. [103].

uniformly loaded to implode by a thick-wall cylinder, which is resistant to show shear failure. The loading state is multi-axial. An advantage of this technique is that a large amount of material will be deformed under pure shear loading before the instability occurs. This principle has no geometrical instability. The loading condition allows a spontaneous shear localization of the specimen. The ratio of cylinder-wall thickness to hole diameter is equal to 1. The TWC test induced compressive stress in the test material, and the shear bands are homogeneously distributed. Through a 'soft' self-recovery system, an investigation of shear band

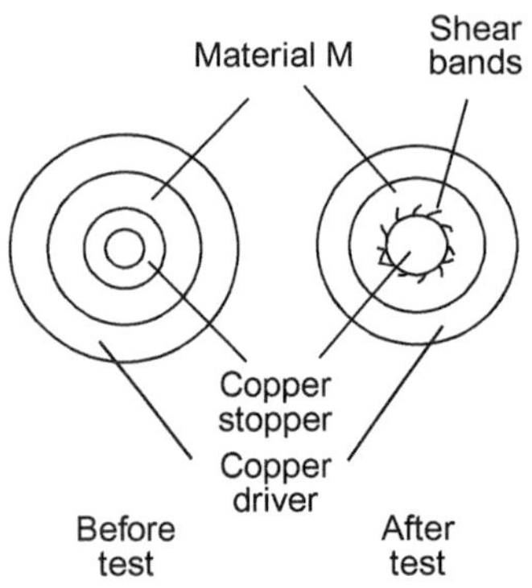

Figure 2.33 TWC test schematically before and after loading, according to Meyers et al. [43].

formation is possible. The explosive expansion test (CEC) is characterized by the action of tensile stresses. Contrary to CECs, the TWC–CIC tests are easier to perform and use lower velocity explosives.

Meyers et al. [43] also examined shear bands in AISI 304L stainless steel, which are produced with TWC tests. The test material is placed as a tube between the outer and the inner hollow copper cylinder (Figure 2.33). After the detonation, the inner cylinder is collapsed, and the shear bands are created at the internal surface of the cylinder and propagate outwards along spiral trajectories. The shear bands are investigated with EBSD and TEM techniques.

Gu and Nesterenko [104] conducted TWC tests to compare the properties of HIPed Ti–6Al–4V alloy in contrast to the baseline material concerning their post-critical behaviour. The Hot Isostatic Pressing (HIPed) condition shows a higher susceptibility to adiabatic shear failure.

Xu [29] examined the microstructural aspects of adiabatic shear localization on different materials, such as low-carbon steel, stainless steel, titanium and its alloys, aluminium alloy and others. The materials were tested with a TWC test.

The TWC test is also used on titanium [103,105], titanium alloy Ti–6Al–4V [106], stainless steel [107] and tantalum [53].

Punch Test

A punch test can be conducted in a Hopkinson bar assembly or in a drop-weight tower. The punch test and the double-shear testing assembly were evolved to reach higher strain rates as in a torsional Kolsky bar.

According to Bai and Dodd [8], a punch test is a simple procedure, where the projectile or punch is designed flat. A schematic assembly for a simple punch test to investigate the propensity of the materials concerning their adiabatic shear-failure behaviour is depicted in Figure 2.34. The loading of the specimen can be done, for example, by a modified Hopkinson bar [108], as shown in Figure 2.35, where the punch is accelerated by a falling weight. Loading velocities up to 23 m/s were reached.

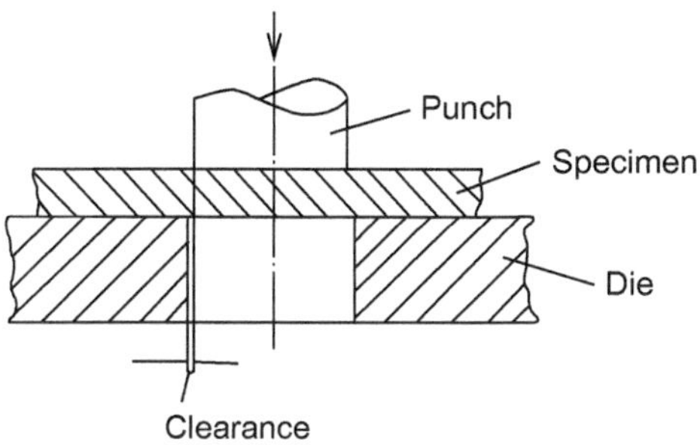

Figure 2.34 Schematic assembly of a punch test, according to Bai and Dodd [8].

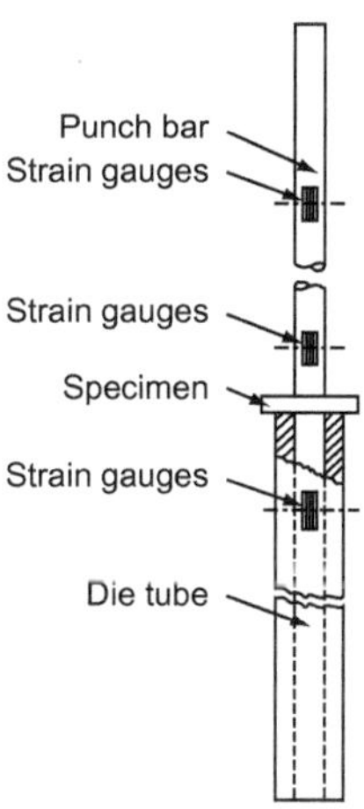

Figure 2.35 Test arrangement for punch loading, according to Dowling et al. [108].

With this punch configuration, a stress concentration will be induced in the material, and the formation of adiabatic shearing can start. The specimen plate will be completely penetrated by the punch. The recorded force–displacement behaviour shows the material property. The distance between punch and die and the radii of the die will influence the gained results [8]. Because of these disadvantages, the punch test is not ideally suited to the study of shear banding.

Moss [109] conducted explosive-driven punch tests on a Ni–Cr steel to investigate the nature of shear bands. The test apparatus is depicted in Figure 2.36. The shear strains obtained were measured by changing the chemical inhomogeneity alignments, which were visible in the deformed shear band. The shear strain level varied across and along the shear band. An average shear strain rate in the shear band was $9 \times 10^5\ s^{-1}$, and the maximum shear strain rate on the fracture surface was $9.4 \times 10^7\ s^{-1}$. It is postulated that thermal softening is not considerable below 570% shear strain.

Zurek [110] conducted punch tests (Figure 2.37) on an AISI 4340 steel in a Hopkinson bar and examined white etching bands. The average strain was 0.5 at average strain rates of $18{,}000\ s^{-1}$. Zurek verified a model of Lindholm and Johnson [2] where the experimental and theoretical values for adiabatic failure

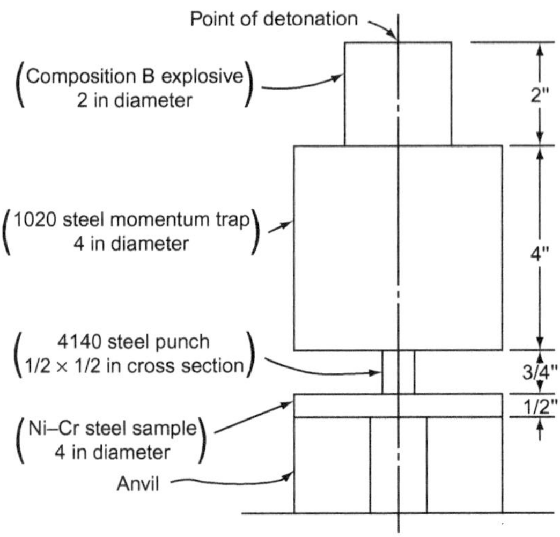

Figure 2.36 Punch assembly, according to Moss [109].

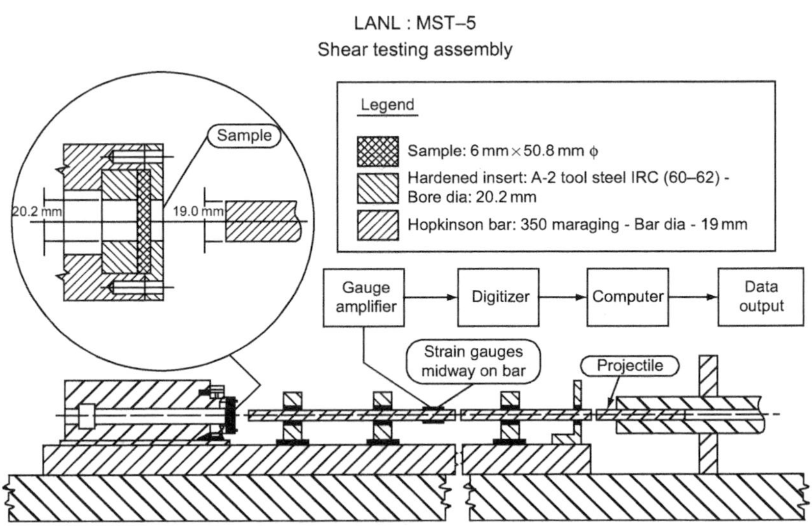

Figure 2.37 Schematic of shear impact test assembly adapted for the Hopkinson pressure bar, according to Zurek [110].

conditions are correlated by means of torsion loading. Zurek found a good correlation between the measured failure data of the punch test and the model of Lindholm and Johnson. The experimental critical strains obtained from the punch test (depending on the material) are between $\gamma_c = 0.2$ and 3.5 and correlate linearly on a log–log scale with the theoretical data.

Roessig and Mason [111] conducted static and dynamic punch tests on an AISI 1018 steel, an AA 6061-T6 alloy and a Ti–6Al–4V titanium alloy. The principle

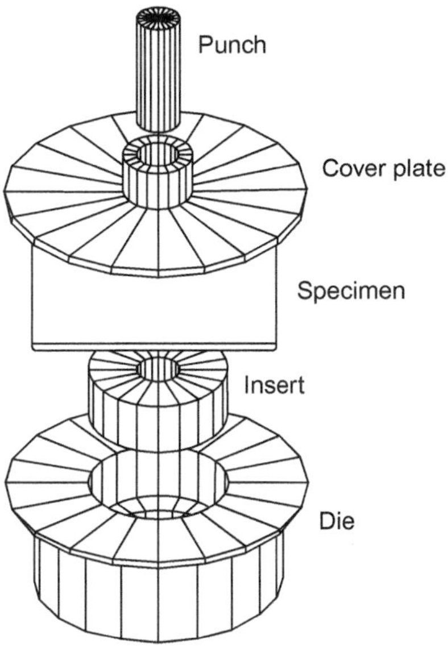

Figure 2.38 Insert die configuration for quasi-static and velocity up to 1 m/s, according to Roessig and Mason [111].

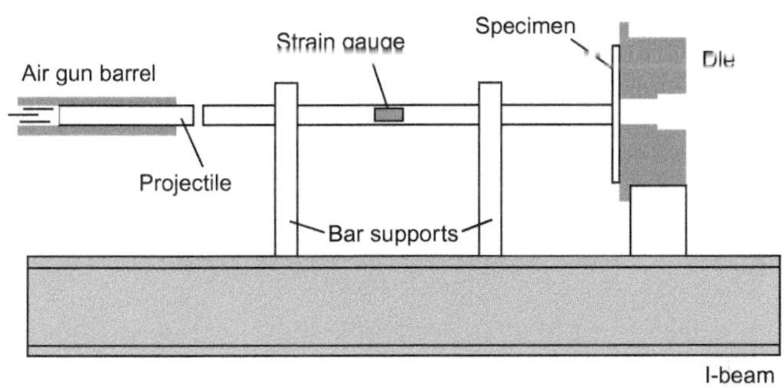

Figure 2.39 Schematic drawing of the punch loading Hopkinson bar apparatus, according to Roessig and Mason [111].

of the punch for quasi-static loading and for velocities up to 1 m/s is shown in Figure 2.38.

The dynamic punch tests were performed in a direct Hopkinson bar assembly, as shown in Figure 2.39.

The clearance between the punch and the die diameter can be varied in a range from 1.6 to 50 mm. For this investigation, a clearance of 1.6 and 3.2 mm were used, compared to a 3 mm plate thickness. The titanium alloy shows the highest susceptibility to adiabatic shear failure. The energy to failure for titanium alloy is decreasing with increasing punch velocity. The aluminium alloy showed no adiabatic shear bands. The influence of the clearance on the energy consumption is

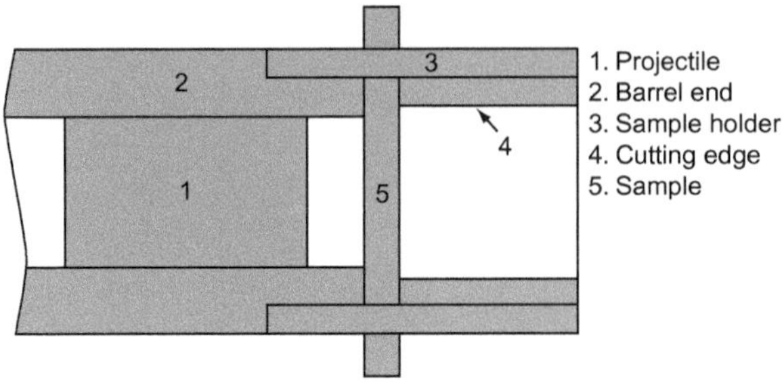

Figure 2.40 Punch assembly for a gas gun, according to Ryttberg [112].

material-failure dependent. The failure strains at the lower clearance are strongly affected by the punching velocity. If adiabatic shear failure occurs, the failure strain is lower at low clearances than at higher clearances. The experimental data show that high-strength, low-strain hardening materials are more susceptible to adiabatic shear localization.

Ryttberg [112] conducted punch tests with a gas gun (Figure 2.40) on 100Cr6 steel at velocities between 38 and 285 m/s. The as-received material (HV 250) was compared to the quenched and tempered condition (HV 310). The conditions in this test are similar to the commercial parting-off process. It was found that the failure by a commercial parting-off process at loading velocities up to 10 m/s and the punch test are the same, which consist of shear fracture and adiabatic shear banding. The energy consumed during the parting-off process is independent of the microstructure or hardness of investigated material conditions. The shear strain at fracture increases with impact speed.

Indentation Test

By means of an indenter test, the punch will be only impressed slightly into the surface of the specimen, and the material will be assessed on the basis of the formation of shear bands. This configuration was invented by Rogers [6]. An example of failure is shown from the work of Rogers [6] with an AISI 1080 steel (Figure 2.42). The indenter test can be performed with a stepped or unstepped projectile. With the stepped projectile, the deformation limit is automatically realized. Chou et al. [113] show such a test rig with a stepped projectile (Figure 2.41). The results are that the shear band length is directly proportional to the flow stress, and the indentation depth is inversely proportional to the back-up hole diameter. For a given depth of penetration, the shear band length increases with a decreasing back-up hole diameter.

Meyer and Manwaring [114] conducted Hopkinson bar tests on extremely hard and thin round blanks of a low-alloy steel and a titanium alloy, and the shear stress–displacement behaviour was measured (Figure 2.43). The flat-plate specimen was punched into the die, and the force was directly measured in the bar. The results led to critical shear strength to initiate adiabatic shear failures, shear deformation and punch velocity of the materials.

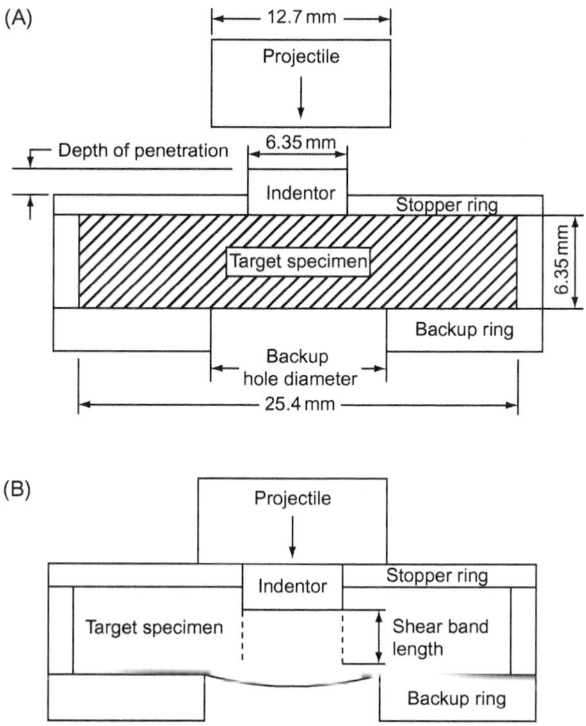

Figure 2.41 Indenter test: (A) before impact and (B) after impact, according to Chou et al. [113].

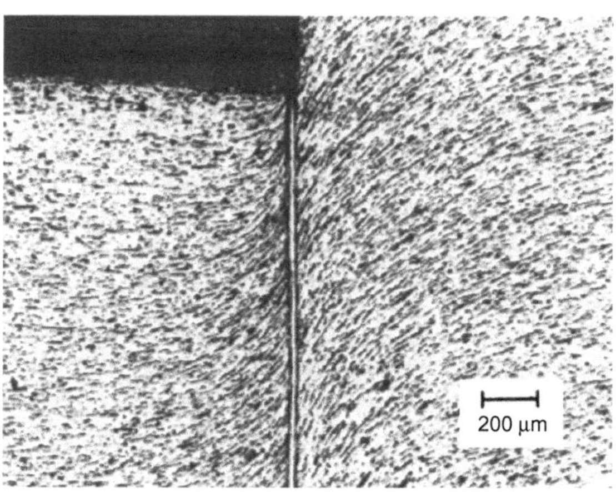

Figure 2.42 Transformation shear band (white etching band) in steel at indenter test, according to Rogers [6].

Bai and Dodd [8] describe results of a 4340 steel where the hardness condition with 20 HRC has not sheared, and the material with high hardness of 52 HRC is sheared in a stepped projectile indenter test.

Irick et al. [115] used a stopped indenter test to investigate pure titanium and pearlitic 4140 steel, as shown in Figure 2.44. The tests were performed at low

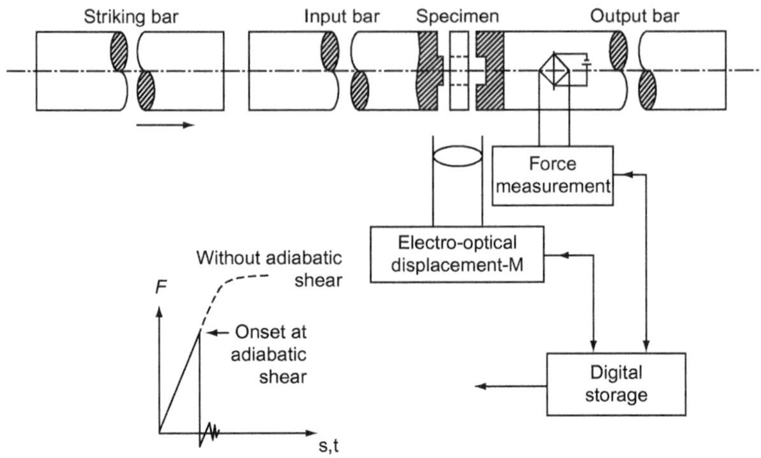

Figure 2.43 Intender test arranged in a Hopkinson bar apparatus, according to Meyer and Manwaring [114].

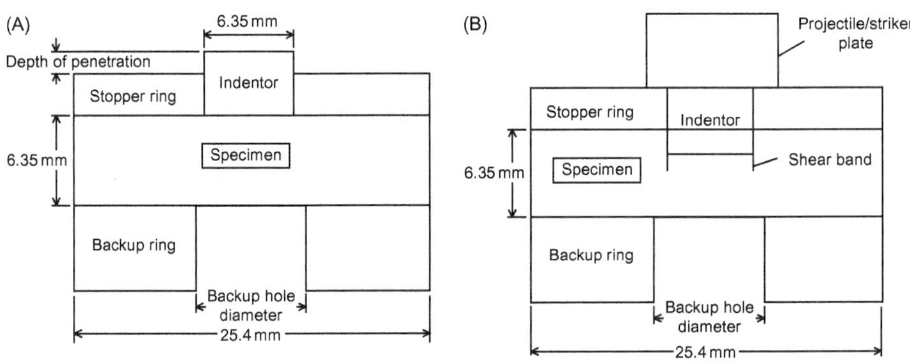

Figure 2.44 Controlled-penetration impact apparatus: (A) before impact and (B) after impact, according to Irick et al. [115].

speeds with a drop-weight tower as well as in a gas gun for high punching velocities. In both materials, they found transformed shear bands, which were examined concerning band width, strain and strain rate.

Mason and Worswick [116] conducted indenter tests in a Hopkinson bar to investigate the susceptibility to adiabatic shear-failure behaviour of annealed ARMCO iron, as-received and shocked REMCO iron as well as a 4340 steel with different tempering temperatures (Figure 2.45). The results indicate that the 4340 steel conditions were susceptible to adiabatic shear failure, and the ARMCO iron and REMCO iron materials were not. The shock-hardened REMCO iron shows a significantly higher flow shear stress and earlier drop in stress in contrast to the as-received REMCO iron. The micrograph indicates a more localized shear

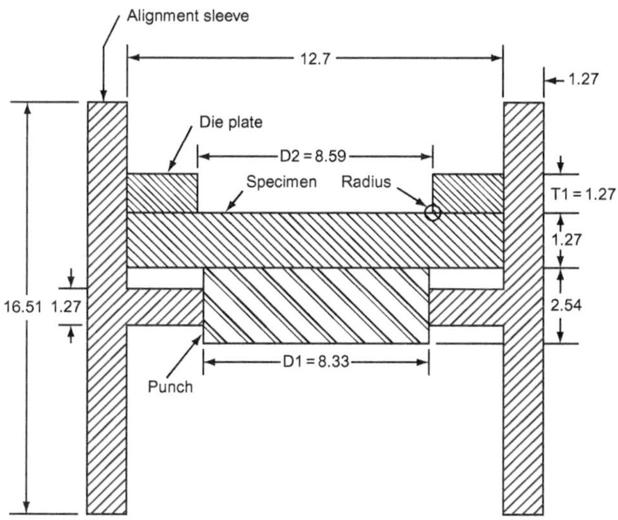

Figure 2.45 Schematic of the die specimen punch assembly, according to Mason and Worswick [116].

deformation as in the as-received condition. For both conditions, a deformation shear band is found. All quenched and tempered 4340 steel conditions show white etching bands on the fracture surface. With a modification of the test assembly, a change in punch configuration is possible.

Double-Shear Specimen

Campbell and Ferguson [117] used a double-shear specimen (Figure 2.46) to investigate the temperature and strain-rate dependence of shear strength of mild steel. The strain-rate sensitivity is decreasing with increasing temperature, except at high strain rates where a damping mechanism by phonon viscosity is named as a reason for the sudden increase in rate sensitivity. The test also attempted to use shear specimens with a smaller slot width (0.25 mm), but the shear zone homogeneity was worse.

Campbell and Ferguson [117] used an SHPB apparatus where the output bar is a tube geometry (Figure 2.47).

Harding and Huddart [118] used an SHPB with a modified double-shear specimen (Figure 2.48) to investigate uranium. The first used specimen (Figure 2.48A) is in accordance with that of Campbell and Ferguson (Figure 2.46). The modified specimen (Figure 2.48B) has a narrower cross section for the shear zone obtained.

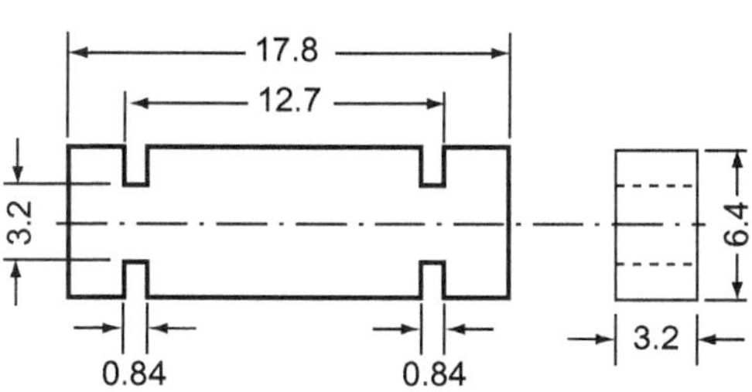

Figure 2.46 Double-shear specimen, according to Campbell and Ferguson [117].

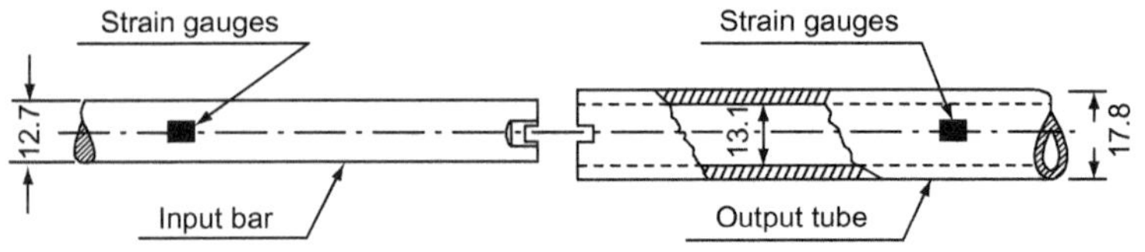

Figure 2.47 SHPB apparatus, according to Campbell and Ferguson [117].

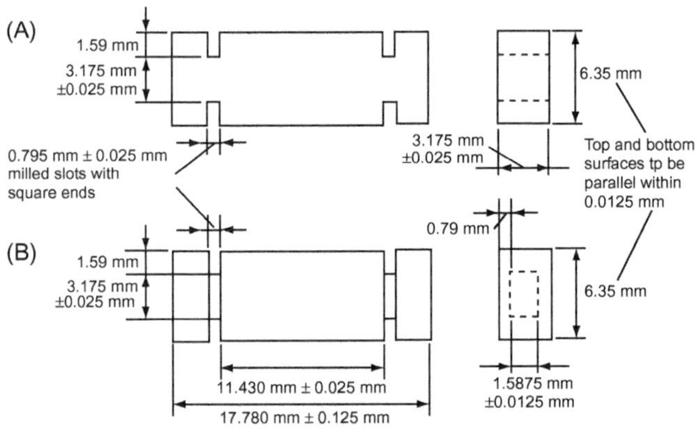

Figure 2.48 Double-shear specimen, according to Harding and Huddart [118].

For this behaviour, the increasing adiabatic condition is possible. The shear behaviour is more accurately expressed with the modified specimen.

Klepaczko [119] used a similar specimen configuration (Figure 2.46) as Campbell and Ferguson [117] and cited in ASM Handbook No. 8 [120], but he extended the velocity up to 200 m/s in a direct Hopkinson set-up (Figure 2.49) with relatively high velocities with a tubular transmitter bar for a low-alloy steel. He postulated a 'critical shear velocity'.

In the same period of time as Campbell and Ferguson [117], Olson et al. [121] conducted tests on high-strength steels with double-shear specimens (Figure 2.50) in an instrumented pendulum-type impact machine. The shear width must be

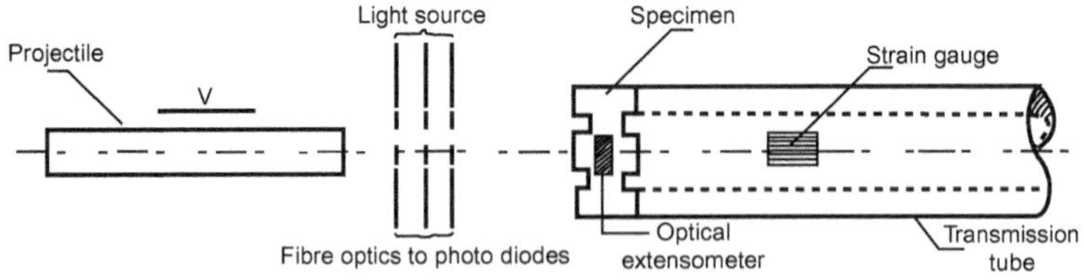

Figure 2.49 Direct Hopkinson assembly for double-shear testing, according to Klepaczko [119].

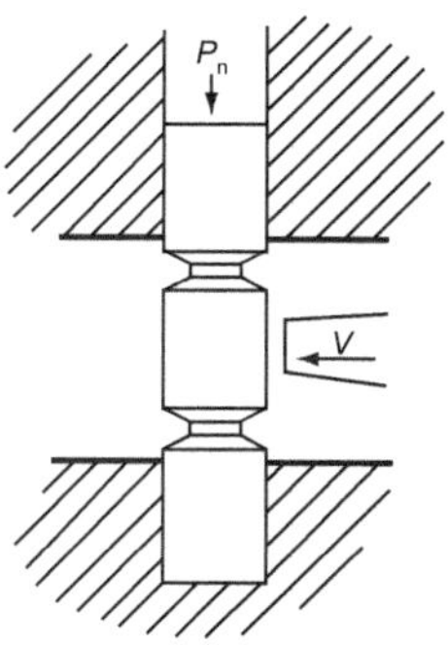

Figure 2.50 Double-shear specimen, according to Olson et al. [121].

narrow to keep the bending influences of the results on a low level. The failure begins at the groove with unknown local stress distributions. At large deformations, a pure shear deformation cannot be ensured.

Because of the uncertainties of stress state in the groove and the inhomogeneous shear strain distribution, the double-notched shear test is not widely used.

Single-Edge and Double-Edge Specimen

Special specimen geometries are the single-edge and double-edge notch specimen. Both geometries can be designed only with a slit or an additional fatigue crack. The loading of the specimen, typically with a projectile impact, produces around the cracks a shear loading similar to mode II in fracture mechanics. The shear band propagation velocity, the shear band propagation and the temperature evolution can be measured. A measuring of stress–strain curves is not possible.

Kalthoff [122] investigated the fracture behaviour of steels with the double-edge specimen (Figure 2.51). A blunt projectile is launched from a gas gun to create an impact next to a pre-crack in the target, which produces shear mode II loading. The deformation pattern at the tip of the propagating shear bands was observed optically. Three types of materials were tested. The epoxy shows failure by tensile cracks. It was found that the fracture behaviour of Maraging steel X2NiCoMo 18-9-5 is dependent on loading velocity. At low striking velocity, only plastic deformation in the notch occurs. With a higher loading velocity, crack propagation starts, and at striking velocities higher than 30 m/s, adiabatic shear failure takes place. This velocity behaviour is influenced by the radius of the slit and the presence of a crack. The limiting velocity for transition to adiabatic shear failure is also dependent on the bluntness of the notch. The aluminium alloy AA 7075 T6 shows adiabatic shear failure down to quasi-static conditions.

Rosakis et al. [123] made impact tests on C300 Maraging steel, HY100 steel and titanium alloy Ti–6Al–4V with single-edge notched specimens (Figure 2.52). Below a critical loading velocity of 20 m/s, common crack propagation appears; above a critical loading velocity, a shear band formation and propagation occurs.

A modification of the single-edge specimen with an additional fatigue crack used by Guduru et al. [124] to investigate the failure behaviour of a

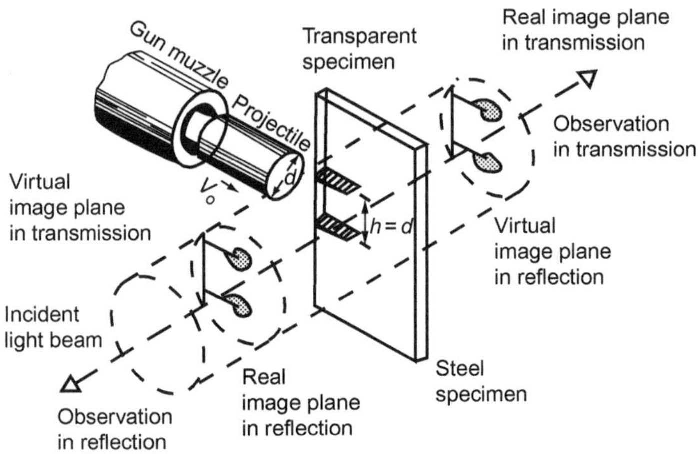

Figure 2.51 Double-edge specimen, according to Kalthoff [122].

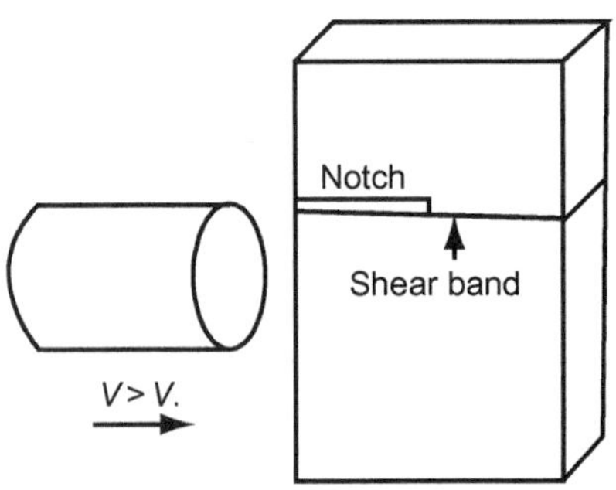

Figure 2.52 Single-edge specimen without fatigue crack, according to Rosakis et al. [123].

C300-Maraging steel versus loading velocity with a gas gun (Figure 2.53). A critical loading velocity of 26 m/s to initiate shear bands were found, which is close to Rosakis's value.

Further Special Tests

Shear-Compression Specimen Test

The shear-compression specimen (SCS), invented by Rittel et al. [125], uses a configuration in which a compression and a shear loading act simultaneously. In a compression specimen, a slit is machined with a 45° inclination (Figure 2.54). The determined failure strains are dependent on specimen geometry and on the material properties, as stated in Rittel et al. [126] and Wang and Rittel [127]. The radii of the notch are important for the results gained. Soft materials are stronger influenced than harder materials. At the manufacturing of this specimen, an additional effort is necessary in contrast to a simple compression specimen. The manufacturing method must be chosen thoughtfully to ensure a high reproducibility of accuracy.

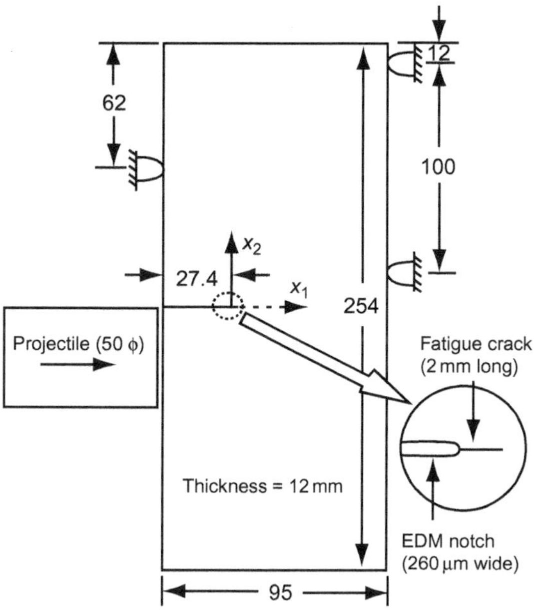

Figure 2.53 Single-edge specimen with fatigue crack, according to Guduru et al. [124].

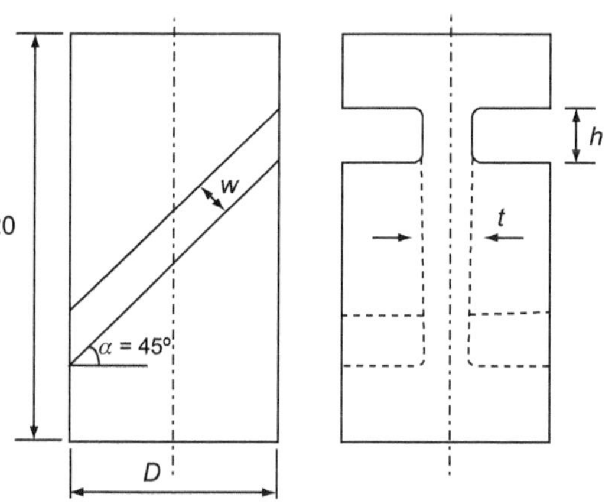

Figure 2.54 SCS specimen (shear-compression specimen), according to Rittel et al. [125].

The measuring procedure is easy to handle. The propensity to adiabatic shear failure of materials is easy to assess. The disadvantage of this specimen geometry is that an influence of slit geometry on the measured failure strain exists, and the shear-stress–shear-strain curves have to be assumed and verified later.

Flyer-Plate Test

Flyer-plate tests are normally used to explore the shock-Hugoniot behaviour under extreme strain rates of 10^5 to 10^7 s^{-1} with plates flying perpendicular to the striking direction [128–131]. Mescheryakov and Atroshenko [130] found circular zones of shear localization on pure copper, alloyed copper and Cr–Ni steel. When the velocity is reduced and the plates are inclined to the flight direction, the

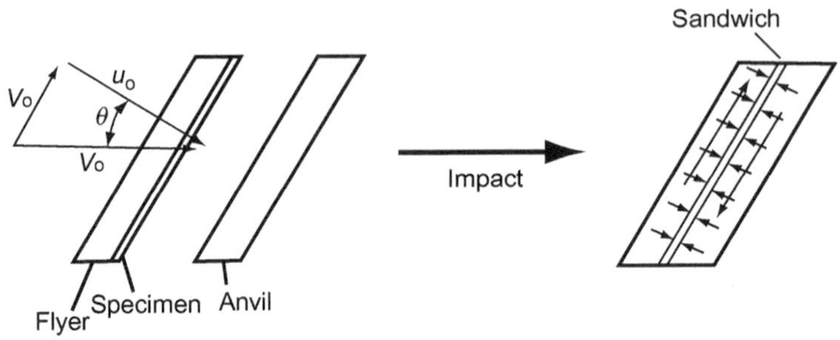

Figure 2.55 Assembly of the 'pressure-shear plate experiment', according to Clifton and Klopp [133] and Wright [9].

assembly is called a 'pressure-shear plate experiment'. Klopp et al. [132] and Clifton and Klopp [133] used it to investigate the materials concerning their bi-axial strength and adiabatic shear-failure behaviour (Figure 2.55).

At an angle θ and the corresponding shear inducing velocity v_0 under sufficiently high friction between the flyer, specimen and anvil, a shear stress will be induced in the material. The test material is fixed as a thin foil between the impactor and anvil. Through this small material thickness, very high strain rates can be reached, which are much higher than in a Hopkinson bar assembly, as stated in Klepaczko [134]. For example, a specimen thickness of 0.2 mm leads to a strain rate of $10^5\ s^{-1}$, and a thickness of 2.5 μm leads to a strain rate of up to $10^7\ s^{-1}$ [133]. Klopp et al. [132] investigated the flow shear stress behaviour of aluminium, alpha-iron and a high-strength steel under high-rate conditions. The high-strength steel shows a lower increase of flow stress with increasing strain rate in contrast to aluminium and alpha-iron. The exact perpendicular collision process is not easy to reach.

Impact-Wear Test

The impact-wear test examines the susceptibility of the material concerned with adiabatic shear failure through a combination of impact and wear loading (e.g. in the case of roller bearings), according to Zhang et al. [135] and Zhang et al. [136]. Hardened steel spheres are falling on the tested material plate, which is coated with sand to induce wear. Adiabatic shear bands can occur on the surface or under the surface region as a source of interior cracks [135]. Material wear is measured by weighing the components [137].

Cutting Tests

Another way to investigate the susceptibility of materials to adiabatic shear failure is by cutting tests in different modifications. Recht [138] has conducted machining tests of different materials such as mild steel, stainless steel, titanium, titanium alloy and aluminium. At the machining process, he measured the shear stress with a dynamometer. From the dependence of cutting speed, he determined the shear stress for each material. Up to a critical cutting speed, only uniform chip formation occurs. Above this critical cutting speed, adiabatic shear failure in the material begins, and shear bands are forming. With increased cutting speed, the shear stress is decreasing, and the shear band spacing will be larger. Titanium and titanium alloys are more prone to shearing adiabatically as is mild steel.

Molinari et al. [139] conducted static and dynamic cutting tests on Ti–6Al–4V alloy. The dynamic tests were performed with a direct Hopkinson assembly where the striker was launched by a gas gun. They measured cutting forces and determined the threshold velocity for the formation of deformed and transformed shear bands.

Halle [88] conducted high-speed cutting tests with a special specimen geometry in a rotating wheel at 20 m/s. He measured the force and displacement of the cutting process during the test. He found that a tool steel has a higher susceptibility to adiabatic shear failure than the annealed ferritic-pearlitic steel AISI 1045 (Figure 2.25). A correlation was determined between the chip formation process and the dynamic bi-axial compression/shear performance (test arrangement shown earlier in Figure 2.18) of the materials.

Akcan [140] examined three different quenched and tempered steels with a martensitic microstructure at low-tempering temperatures. He machined cylindrical specimens with a diameter of 15.7 mm and a length of 60 mm. The cutting conditions were chosen to obtain a serrated chip formation, and a white layer evolution is possible. The base material and the chips are examined concerning their existence and the width of the white layer. The width of the white layer increases with increasing cutting speed and increased flank wear. All determined chips show a white etching character. The occurrence of the white layer in the bulk material is material dependent. The hardness of the white layer in the bulk material is about 25% harder than the base material with their original treatment condition.

Blümke [87] investigated two aluminium alloys – AA 7075 and AA 7050 – as well as a tool steel 40CrMnMo7 (DIN 1.2311) for their chip formation behaviour at high-speed milling. The aluminium alloys were heat treated to underaged, peak-aged and overaged conditions. The steel was heat treated to a martensitic microstructure with low tempering temperatures, to pearlitic structure and annealed condition. The chip formation was done with a high-speed milling machine. During the milling process, the cutting force was measured by means of a dynamometer. The results show that there is a correlation between the character of chip formation and the failure strain in the dynamic bi-axial compression/shear test. Materials that fail in the compression/shear test show serrated chips, and materials that are not shearing in the compression/shear test exhibit continuous chip formation. The microstructure and the heat-treatment condition influence the character of chips. The aluminium alloys for the underaged condition show a segmented chip; the annealed and overaged conditions show continuous chips. In annealed and pearlitic conditions, the steel shows continuous chips, and in the hard martensitic condition, the steel shows serrated chips. With increased cutting speed, the rate of segmentation is dependent on the heat treatment.

Duan et al. [141] performed cutting tests for AISI 1045 steel with different heat treatments to obtain different hardness conditions. With increasing cutting speed, the cutting force decreases. The transition points of ribbon chips to serrated chip zones with deformed bands or transformed bands are visible from the cutting force diagrams.

Shot Test

A more qualitative determination of materials concerning their susceptibility can be made by using shot tests, including the projectile shot, the shape-charged test and the Taylor test.

With projectile shots, the susceptibility of target plates for adiabatic shear failure is investigated. Craig and Stock [142] investigated the adiabatic shear band propagation at shot tests with cylindrical hardened steel projectiles fired on annealed brass with a velocity range of 300–600 m/s. The brass targets were semi-infinite blocks. The projectile penetrated only a short distance. The shear bands initiated near the bottom of the holes produced and propagated through the material. At the surface of the bullet hole, white etching zones were also found.

Manganello and Abbott [143] investigated the ballistic performance of heat-treated steel plates by armour-piercing projectiles and examined the evolved white etching shear bands. With increasing tempering temperature and decreasing carbon content, the penetration resistance is decreasing. Mescall and Papirno [144] investigated the failure behaviour of a 4340 steel target and projectiles in different heat-treatment conditions at shot tests. Depending on the heat-treated condition, either the target or the projectile fractures. When both the target and the projectile had a high hardness, adiabatic shear failure in the target plate was observed.

Irwin [145] investigated the behaviour of uranium heavy alloy as armour penetrator material at ballistic tests. The penetrator shows adiabatic shear failure. The shear bands were initiated because of their low work-hardening coefficient and high stress concentration.

Dormeval [7] examined fractured flat high-hardness steel projectiles, which were shot at about 250 m/s against rigid targets. He found adiabatic shear (white etching) bands on the surface of the fracture. The projectiles were broken rapidly.

Krejci [146] has used armour steel plates of AISI 4340 and an 18/8 austenitic steel. The steels were heat treated to different conditions and then shot with shaped charges of copper. The 10 mm plates were stacked to form layered targets. The layered targets were then varied in their properties to mix the hardness. He found that in hard materials, adiabatic shear bands (white etching bands) propagated from the surface to the middle of the plate. In soft material, white etching bands were not found. Derep [147] studied the width and microstructure of the shear bands in armour steel, which was subjected to a hollow charge.

Meyers and Wittman [148] conducted tests on the influence of two steels and heat-treatment conditions on their propensity to fail by adiabatic shear failure. The materials were tested in a gas gun at velocities between 450 and 1050 m/s. The test plates were impacted with tool steel rods. The appearances of the adiabatic shear bands were examined. Similar tests at lower-impact velocities were conducted by Borvik et al. [149] on Weldox460 steel. For each target thickness, the ballistic limits were determined.

Nguyen [79] made a reversed Taylor test where a rigid disk made of Maraging steel was shot against the cylindrical metal specimen of 6 mm diameter and 50 mm length at a velocity of 220 m/s. In pure titanium, he found a transformed shear band, and in pure aluminium, he found a deformed shear band.

Couque et al. [150] investigated the strength and failure behaviour on a tungsten alloy in four different heat-treatment conditions. These conditions show a similar strength but a difference in the unnotched Charpy energy and fracture strain under tension. With a symmetric Taylor experiment (Figure 2.56), he determined the dynamic compression strength and failure behaviour. A critical loading velocity for adiabatic shear initiation was found. For these four heat-treatment conditions, ballistic tests at 1400 m/s on steel plates were conducted.

Zhu et al. [151] conducted high velocity impact tests on Al–Mg alloy and Al–Mg alloy with Ti rods reinforced. The projectiles were made of AA 2024, had a diameter of 1.2–2 mm and were shot at about 2.5 km/s. The reinforced material shows a better ballistic performance than the unreinforced material, and adiabatic shear bands were not found.

A special test to reach high deformation rates in the material were conducted by Jha et al. [152] through an explosive burst test with a miniature assembly (Figure 2.57). The microstructure concerning the formation of shear bands was examined. Forces and strains were not measured.

An overview of the used test methods for metallic materials with their advantages and disadvantages is given in Table 2.1.

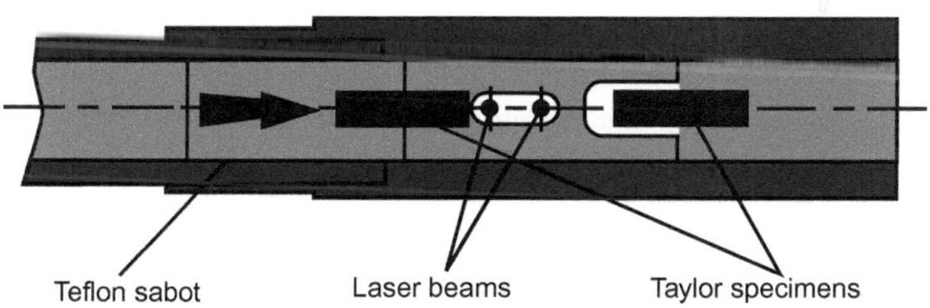

Figure 2.56 Schematic of symmetric Taylor test, according to Couque et al. [150].

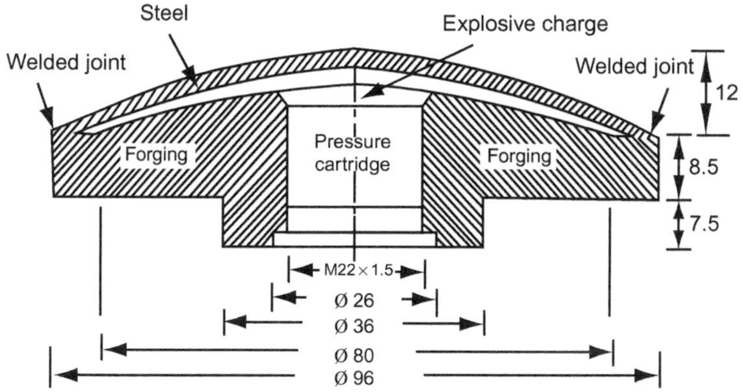

Figure 2.57 Schematic diagram of the miniature burst test, according to Jha et al. [152].

Table 2.1 Overview of Test Methods for Metallic Materials

Test Method	Advantages	Disadvantages
Main Test Techniques		
Torsion	Well-defined uni-axial stress state; homogeneous stress field for thin-walled tube specimens when ratio gauge length to diameter is minimum 1; high deformations are possible; failure initiation is material dependent; accessible surface to observe strain and temperature; shear-stress–shear-strain curves are obtainable	Buckling, bending and surface nature can influence the failure strain; determined data are not straightforward for engineers; surface quality may influence the initiation of shear failure
Hat-shaped specimen	Materials will be forced to shear; comparison of low and large deformed specimens are possible; ductile materials can be tested; hydrostatic pressure is adjustable by geometry; interrupted tests are possibly workable; very high strains and very high strain rates are achievable	No homogeneous stress and strain state; obtained shear-stress–shear-strain curves are not valuable; shear slit width and corner radii influence the measured data; failure initiation is related on discontinuity
Compression	Failure is only material dependent, not influenced by discontinuity; a macroscopically uniform stress state; specimens are easy to produce; measurement of stress–strain curves are possible; deformation and temperature pattern on the surface are observable; test results are straightforward for engineers; test assemblies are usually available; high strain rates are possible; interrupted tests are easily workable	Only quasi-brittle materials fail by adiabatic shearing; differentiation between materials is possible only for relatively hard materials; no large deformations achievable because of barrelling
Compression/shear	Failure is only material dependent, not influenced by discontinuity; certain additional shear stress amount adjustable by geometry; no influence of surface quality on failure strain; high strain rates are possible; interrupted tests are easy	Measuring of shear-stress–shear-strain curves are not intended; measuring of axial stress and axial strain is very successful; depending on specimen geometry, an upper limit of valuable deformation exists to

(*Continued*)

Table 2.1 (Continued)

Test Method	Advantages	Disadvantages
	to carry out; deformation and temperature pattern on surface are observable; wide differentiation of materials; good method to easily assess the propensity to adiabatic shear failure of materials	define a failure strain leading to adiabatic shear failure
Cylinder expansion or collapse	Failure is only material dependent, not influenced by discontinuity; homogeneous pure shear deformation until localization occurs; high strains and very high strain rates possible; large amount of material will be deformed	No defined failure locations because of complex stress distribution (smooth tube specimen without notches); test assemblies are not commonly available; only useful for a qualitative assessment of materials; no stress–strain curves obtainable
Punch	High strain and high strain rates are achievable; simple assembly to assess the materials; easy to perform; force–displacement behaviour determinable	Strong influence of geometry to determined data; obtained shear-stress–shear-strain curves are not valuable; failure initiation because of stress concentration; clearance and radii influence the results
Indenter	Initiation of adiabatic shearing is a combination of pressure and capability; good method to assess materials qualitatively; force–displacement curves achievable; interrupted tests are easily workable; easy to perform	Failure initiation because of stress concentration; no stress–strain curves obtainable; clearance and radii influence the results
Double shear	Rather used to obtain shear deformability of materials (less to assess materials concerning adiabatic shear failure); deformation and temperature evolution in situ observable	Failure initiation because of stress concentration; bending influence is large; stress–strain data depending on geometry; at large deformation, a pure shear deformation cannot be ensured
Single or double notch	High strain and high strain rates are attainable; fracture changeovers are observable; temperature deformation evolution on surface obtainable	Failure initiation because of stress concentration; special measuring techniques are necessary; stress–strain curves are not measurable

(*Continued*)

Table 2.1 (Continued)

Test Method	Advantages	Disadvantages
Special Tests		
Shear-compression specimen (SCS)	Test assemblies are usually available; high strain rates are possible; interrupted tests are easily workable; additional shear stress amount adjustable by geometry; wider differentiation of materials in contrast to pure compression; good method to easily assess the propensity to adiabatic shear failure of materials	Determined data are influenced by material (ductile or brittle) and specimen geometry; failure initiation because of stress concentration; no stress–strain curves obtainable; clearance and radii influence the results; specimen needs careful production
Inclined flyer plate	Very high strain rates attainable; additional shear stress amount adjustable by inclination of impactor/anvil assembly; failure initiation is material dependent, not influenced by discontinuity	Only low deformations achievable; special test rig necessary; thin specimens necessary; stress state and strain state are not directly comparable to classical pure compression data (conversion needed); stress–strain data attainable
Impact wear	Combination of high velocity and wear loading; investigation of surface and near-surface effects possible	No stress–strain curves achievable; failure initiation because of internal stress concentration
Cutting test	Measuring of cutting forces are possible; assessment of material behaviour through cutting velocity	Failure initiation because of stress concentration; failure strength and failure strain not directly achievable; cutting conditions influence the material behaviour
Shot test	Assessment of materials through threshold velocity; test at high loading rates and high deformations	Failure initiation because of stress concentration; no stress–strain curves achievable; only qualitative evaluation of materials
Taylor test	High velocities are achievable; measuring of strength and deformability are possible but often questioned; failure initiation is material dependent, not influenced by discontinuity	Special test rigs with high accuracy are necessary

2.2.2 Test Methods for Bulk Metallic Glasses

The test methods used for bulk metallic glasses (BMG) are in principle the same as for common metallic materials. Mainly, the compression test is used for testing BMG. A few investigations are done under compression/shear, bi-axial tension or uni-axial tensile loading.

The mechanical behaviour of BMG corresponds to that of ceramics, at least the monolithic BMG. Latest results show a more ductile behaviour because of particle reinforcement of the BMG. Because of the ceramic-like behaviour, the test methods for ceramics are also usable for BMG, especially the compression test.

Dai and Bai [153] investigated the mechanical behaviour of BMG Vitreloy-1 under static and dynamic loading velocities. They found that the material under tension and shear loading shows a brittle, linear elastic behaviour. Only several isolated shear bands can lead to a macroscopic deformation. In contrast, under compression loading, a multiple shear band formation exists, and the behaviour is elastic–plastic without strain hardening. Under compressive loading, the deformability of BMG is higher.

Martin et al. [154] performed a reverse Taylor test to investigate the failure behaviour of a BMG-matrix composite. The specimen plates were accelerated up to 244 m/s. The results show that tungsten particle reinforcement provides a positive influence on the dynamic failure behaviour. The tungsten particle acts as an obstacle for adiabatic shear band formation, and thus the fracture behaviour is delayed.

Martin et al. [155] conducted the static and dynamic behaviour of Zr-based BMG composite reinforced with tungsten particles. The behaviour under pure compression and the properties under compression/shear loading with 6° inclined cylindrical specimens were determined. Investigation of the failure evolution can be found by the interrupted test in a Hopkinson bar and in a drop-weight tower. Under this additional shear loading, the susceptibility to adiabatic shear failure is higher than under pure compression. In contrast to the monolithic BMG, the reinforced BMG shows a higher ductility.

Zhang et al. [156] conducted static compression tests on Zr-based BMG-matrix composites reinforced with different tantalum particle amounts. The largest plastic failure strain of 33% was reached with 10% reinforcement by Ta particles. The particles act as initiation sites for multiple shear band formations and block the shear band propagation. This leads to a more homogeneous distribution of shear bands. Similar investigations for Mo particles reinforced Mg-based BMG were conducted by Hsieh et al. [157]. With an increasing quantity of particles, the failure strain is increasing because of the resistance of shear bands and cracks propagation by the dispersed particles.

Zhao et al. [158] conducted static compression tests on cubic compression specimens with and without notches for a Zr-based BMG. The specimen geometry was varied to obtain the influence of notches. The ductility of the material can be improved through notches because the shear bands will be blocked by large stress gradients around the notches. This result is in contrast to that of ceramic. Further

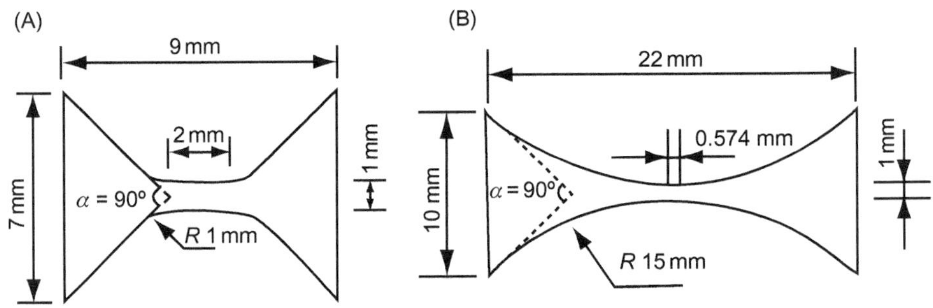

Figure 2.58 Tension specimen configurations, according to Cao et al. [160].

investigations with other materials (steel and ceramic) for comparison reasons were done by Zhao et al. [159]. It was found that high-strength steel lost its ductility with increasing stress concentration, and the ceramic material remains constant in the failure strain.

Cao et al. [160] conducted static tension tests on Zr-based BMG. Through different rolling conditions and in comparison to the as-received material, the influence of shear band formation was observed. With these pre-existing shear bands, the deformability of the material can be slightly enhanced. Two different tension specimens were used, in which at the first configuration (Figure 2.58A), the fracture occurred in the changeover to the mounting part. The second configuration (Figure 2.58B) gives a better stress distribution, and thus the failure appeared in the middle of the gauge length.

Hofmann et al. [161] conducted static tensile test on heat-treated BMG with different compositions. The titanium content was varied, which led to a different fraction of dendrites in the microstructure. Thus multiple shear band formation and propagation are possible, and the ductility under tension is increasing from zero failure strain for monolithic BMG to about 12% failure strain for heat-treated materials.

Wu et al. [162] conducted static punch tests for Zr-based BMG on plates. Through a special configuration of the punch test (Figure 2.59), the actual loading is a bi-axial tension–tension stress condition. Because of this stress state, shear band formation is possible and leads to a more ductile behaviour of BMG. This punch test seems to be a good method for studying the evolution of shear bands. It is possible to examine the shear deformability by measuring the critical shear offset. Three failure modes can occur: normal, radial or circumferential shear failure (Figure 2.60). The low-ductility condition exhibits only radial cracks (class III), the tougher condition exhibits also radial cracks and radial and circumferential shear bands (class II) and the toughest condition exhibits many circumferential cracks and many radial and circumferential shear bands (class I). Wu et al. [163] compared a Zr-based BMG with a Ti-based BMG also under bi-axial loading condition with the just described punch test. The Zr-based BMG showed a higher density of shear bands, a higher ductility and higher energy absorption than the Ti-based BMG. These bi-axial results indicate a better differentiation of BMG in contrast to

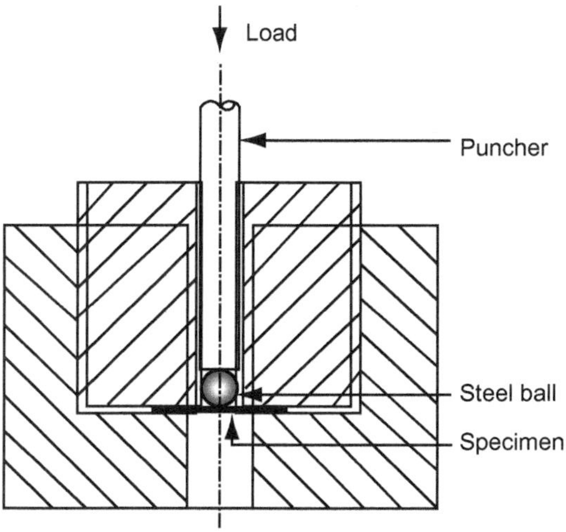

Figure 2.59 Illustration of the small punch test to investigate BMG specimen disks, according to Wu et al. [162].

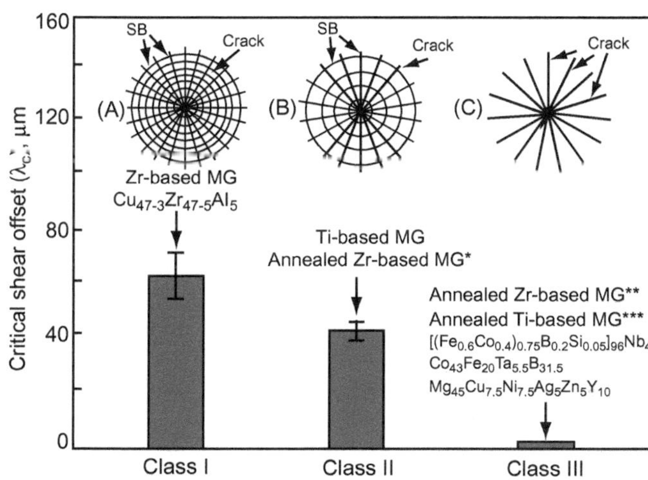

Figure 2.60 Illustration of the different failure modes of BMG materials, according to Wu et al. [163].

the classical test method such as uni-axial compression or tension. Thus this punch test with a bi-axial tensile-loading condition is an effective way to evaluate the difference in shear deformability of BMG.

2.2.3 Test Methods for Polymers

Compression Test

Amijima [164] conducted the compressive strength and fracture characteristics of fibre composites under impact loading of cylindrical specimens in an SHPB.

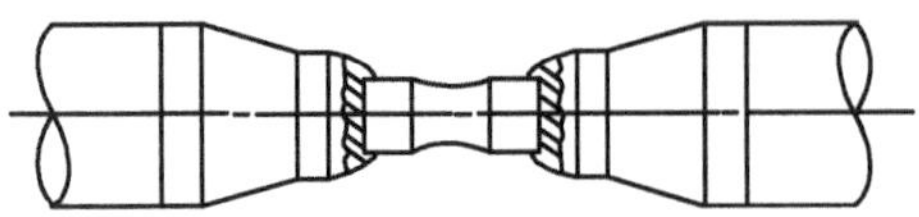

Figure 2.61 Schematic of confined pressure test set-up, according to Lankford [165].

Longitudinal and off-axis compression tests for orthogonal and unidirectional fibre composites (cloth fiber reinforced plastic (FRP) and roving FRP) were tested. Both strengths of longitudinal and off-axis composites increase with an increasing strain rate. Shear bands and fibre breakages are found for the cloth FRP.

Lankford [165] conducted compression tests in a confined and unconfined manner from a static to a dynamic loading velocity to examine the fibre-reinforced ceramic and PEEK material. The specimen geometry is shown in Figure 2.61. To prevent a premature failure, the specimens were tapered and mounted to steel caps to initiate the failure in the measuring gauge. For tests with radial confining pressure, a special pressure vessel was utilized (not indicated in the picture). The hydrostatic pressure was first raised to the desired level by pumping fluid into the pressure vessel. The axial load was then increased under displacement control at a fixed pressure. The hydrostatic pressure was measured by a pressure gauge at the fluid inlet of the pressure vessel.

Experiments are described that lead to the conclusion that there appears to exist a shear damage-controlled upper limit to the compressive strength of fibre-reinforced ceramics. At low strain rates, hydrostatic confinement increases the strength by suppressing the dilatational micro-fracture until the failure mechanism transforms to shear via micro-kink band formation. However, this mechanism saturates at a finite level. At higher strain rates, the unconfined strength is observed to reach this confined state through the dependence of the kinking process alone. On the other hand, the application of confinement in the high strain rate range cannot push the strength beyond the shear-controlled ceiling. In related experiments, the pressure strain rate dependence of the compressive flow strength of the matrices of several polymeric composites is determined. At low rates, the strength was found to increase with confinement, as has been noted by previous investigators. However, the high strain rate compressive strength was inversely reduced proportional to increasing confining pressure.

Single-Edge Specimen Test

Rittel [166] used a single-edge specimen type to investigate the crack propagation behaviour of PC and PMMA material (Figure 2.62). The tests were done in a Hopkinson bar. The test velocity was about 50 m/s at different starting temperatures. A different crack propagation behaviour (which was initiated by shear band formation) for ductile PC and brittle PMMA was determined.

Dynamic fracture (classic fracture by cracks) is commonly addressed as an isothermal phenomenon. When specific phenomena such as shear band formation and propagation are involved, the analyses include thermomechanical conversion of strain and/or fracture energy into heat (thermoplasticity). In this case, it has been

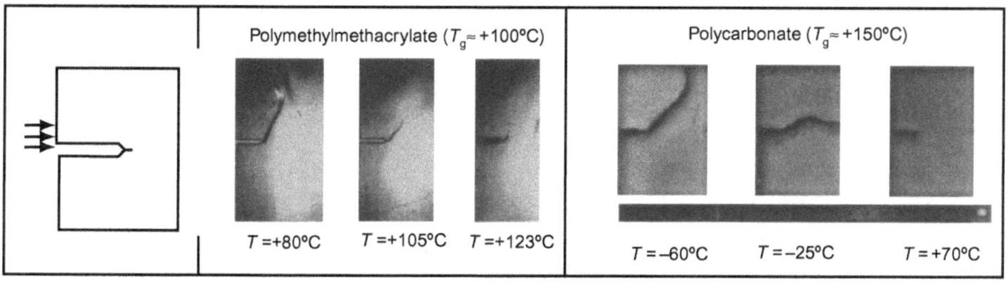

Figure 2.62 Single-edge specimens and the crack propagation behaviour for PMMA and PC, according to Rittel [166].

shown that very significant temperature increases can develop, which cause softening of the crack-tip material. In this work, thermoelastic temperature changes at the tip of the crack are implicitly neglected. In recent work, Rittel has questioned this issue and shown that for a stationary crack subjected to transient loading, adiabatic thermoelastic effects were noticeable, thus causing a large temperature drop in the elastic zone surrounding the crack-tip. In the present work, Rittel pursued this line of investigation by presenting additional experimental results about temperature changes ahead of a dynamically loaded crack in commercial polymethyl methacrylate. Mode I and mode II loading configurations were investigated. Rittel observed, as expected, that the temperature drops for mode I loading while it rises for the mode II case. In each case, the crack initiates during the phase where the temperature changes to a drop or rise. While showing that thermoelastic aspects of fracture should certainly be taken into account, the present results indicate that thermomechanical aspects in general should not be overlooked when addressing dynamic crack initiation. For these materials, only crack propagation and no adiabatic shear failure occurred.

Kalthoff [122] conducted double-edge specimen tests on epoxy resin (araldite B), comparable to the results of steels (see 'Single-Edge and Double-Edge Specimen' section). The specimen thickness of epoxy was 10 mm, and the tip radius was varied from radius = 1 mm to sharp condition as a pre-crack. The applied loading velocities were 10 m/s and higher. For all tip radii, failure occurred by tensile crack propagation. At higher velocities, above 18 m/s, the tests were not usable because the specimens were destroyed directly during contact between projectile and specimen. Because of this behaviour, further investigations at higher loading velocities could not be done to find a possible transition to a failure mode by adiabatic shearing.

Bjerke and Lambros [167] conducted dynamic fracture experiments with two specimen configurations (classical single-edge and opening-mode specimens) for PMMA and PC and loaded with projectiles by a gas gun (Figure 2.63).

The heat generated from dissipative mechanisms during shearing-dominated and crack opening-dominated dynamic fracture of polymethyl methacrylate and polycarbonate was measured with a remote-sensing technique that utilizes the detection of infrared radiation. Significant heating was detected for both materials and both

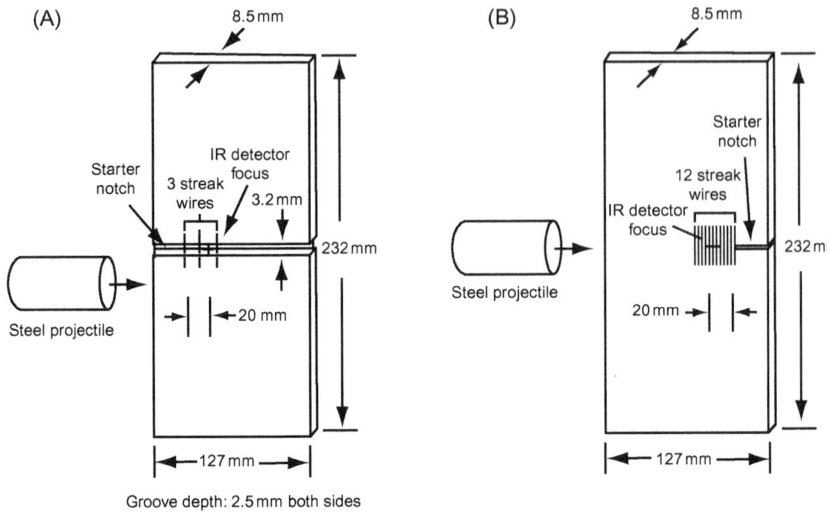

Figure 2.63 Shear-dominated (A) and opening-mode (B) specimen configurations, according to Bjerke and Lambros [167].

modes of fracture. In the shear-dominated experiments, the temperature increase at the crack-tip in polymethyl methacrylate was 85 K, the approximate increase necessary to reach the glass transition temperature. An adiabatic shear band followed by a dynamically propagating crack was observed during the shear-dominated experiments using polycarbonate. The recorded shear band temperature increase was 45 K. This was followed by an additional 100 K temperature increase from the ensuing crack, raising the temperature above the glass transition temperature. The maximum temperature increase recorded for the opening-mode experiments was 55 K for polymethyl methacrylate and 105 K for polycarbonate. The results of this study show that temperature effects are significant during the dynamic fracture of polymers. The effects are especially important in the shear-dominated case where local temperatures approach or exceed the polymer glass transition temperature.

Shear Test

Totry et al. [168] determined shear-stress–shear-strain curves of fibre-reinforced polymers for a quasi-dynamic velocity of 1 mm/min. The mechanisms of in-plane shear deformation in a cross-ply glass fibre-epoxy laminate were studied through a combination of experiments and simulations. Shear deformation parallel and perpendicular to the fibres led to very different deformation and fracture patterns (Figure 2.64).

Deformation was localized in a matrix shear band parallel to the fibres in the former, while fibre rotation and widespread plastic deformation in the matrix were

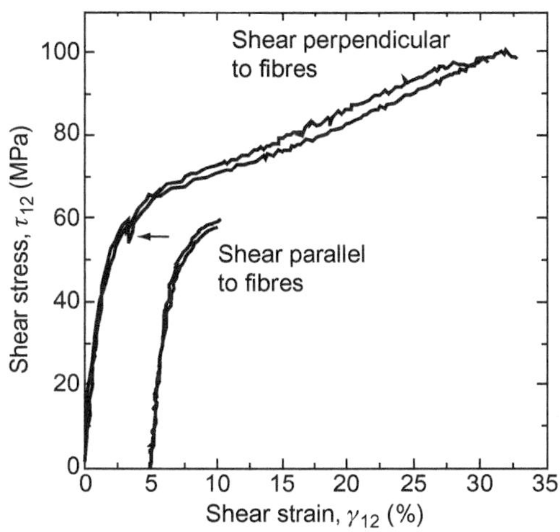

Figure 2.64 Stress–strain curves of an unidirectional laminate composite for loading direction in the fibre and perpendicular to the fibre at 1 mm/min, according to Totry et al. [168].

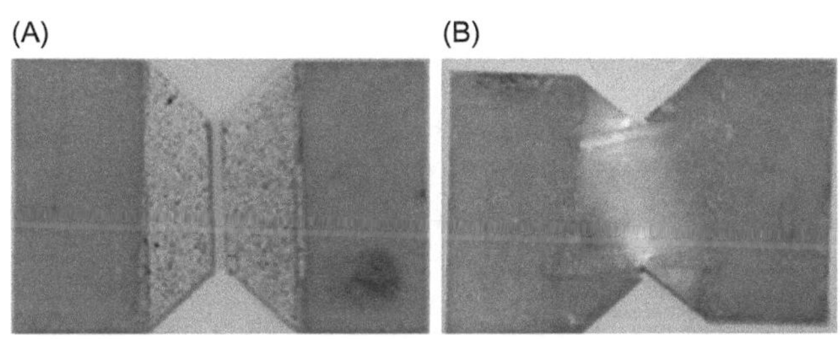

Figure 2.65 Macroscopic view of the specimens: (A) parallel to the fibre and (B) perpendicular to the fibre, according to Totry et al. [168].

dominant in the latter. The differences in the mechanical response between both orientations increased with the applied strain, and it was shown that the cross-ply laminate response was given by the averaged contribution of both. The relevance of this behaviour for the development of continuum damage mechanics models is noted. A deformed and a broken V-notch shear specimen are shown in Figure 2.65.

Flyer-Plate Test

Li et al. [169] conducted inclined planar plate impact tests for nylon 66. Inclined planar plate impact experiments are conducted by using a ∅ 57 mm keyed gas gun. Figure 2.66 shows the experimental set-up: a nylon 66 flyer of 8 mm thick hits a nylon 66 target sample of 25 mm thick with velocity u_0 at an inclined angle α in a magnetic field H. The compression and shear waves are recorded by the electromagnetic particle velocity (EMV) gauges embedded in the sample.

In this article of Li et al., nylon 66 is chosen for experimental investigation because its spherical grain structure can be observed under a polarized microscope. A rapid shear attenuation occurs when the impact velocity and inclination angle

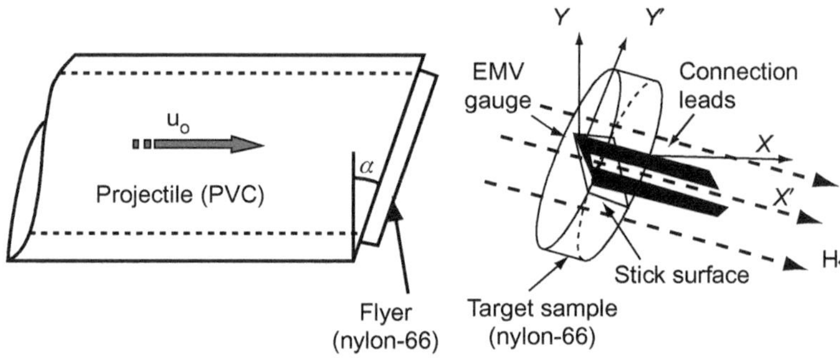

Figure 2.66 Experimental set-up for inclined flyer plate impact, according to Li et al. [169].

reach a critical value. The micro-observation of the recovered samples shows that there is a melted layer with a thickness of about 6–8 μm near the impact surface, which causes the decay of shear waves. Interestingly, there is a discontinuous layer about 2–3 μm thick above the melting layer, which indicates the melting may not be caused by the sliding and friction on the impact surface but by the heat produced inside the sample and near the surface. Further observation discloses an adiabatic shear band near the surface that causes the material failure and the shear wave decay.

The plastic deformation near the impact surface will produce a correspondingly large amount of heat. The low heat conductivity of the polymer and the impulsive loading condition lead to the formation of an adiabatic shear band. The shear bands formed beneath the surface due to the constrained condition in contact. When the loading amplitude is increasing, the shear band can transform in a melted layer.

Li and Tang [170] conducted similar investigations for polypropylene in an inclined planar plate assembly.

2.2.4 *Test Methods for Ceramics*

Ceramics and rocks are the typical prototypes for brittle materials. When they are very strongly sheared, they can also flow quasi-plastically, if the hydrostatic pressures are high enough. The cracks in the brittle case and the shear bands in the quasi-plastic case form fractal networks that reflect the detailed stress field and the history of the deformation, according to Herrmann [171].

Confined Compression Test

Lankford [172] performed static and medium velocity tests on fibre-reinforced ceramic matrix composites under confined compression conditions. The test technique is the same as Lankford [165] used (Figure 2.61). The influence of hydrostatic confinement on compressive strength and corresponding failure mechanisms

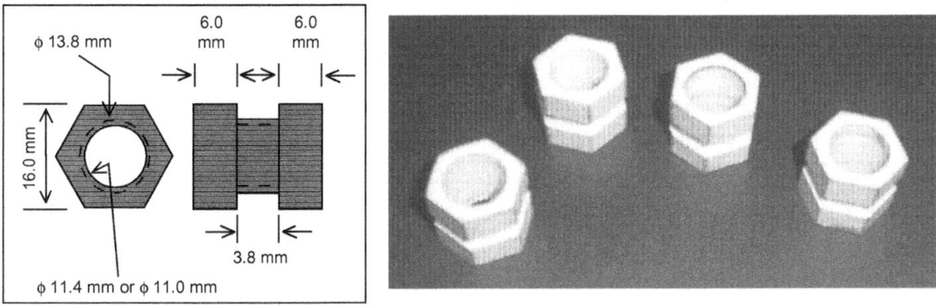

Figure 2.67 Specimen geometry for torsion, according to Bassim et al. [173].

is explored for SiC-reinforced glass-ceramics tested at different strain rates. Two composite architectures (0° and 0°/90°) are studied, and their behaviour is compared with monolithic glass-ceramic tested under similar conditions. Composite-confined pressure results are interpreted in terms of fibre buckling under quasi-static conditions and fibre kinking at high pressures, and compared with monolithic (non-composite) micro-fracture coalescence at low pressures and shear band formation under more intense confinement. In particular, dilatational fracture within the matrix dominates compressive failure at low pressures, while high pressures cause a transition to a shear-dominated mechanism based on fibre kinking.

Torsion Test

At present, there are several applications where high-strength ceramics have replaced metals that are subjected to high-speed impact from projectiles. This requires an evaluation of the behaviour of ceramics under impact at high strain rates.

Bassim et al. [173] studied the high strain-rate behaviour of alumina tested in shear using a torsional Hopkinson bar with a specimen geometry depicted in Figure 2.67. Dynamic stress–strain curves were generated to investigate deformation behaviour prior to fracture while fractography of the broken specimens was carried out to establish the mode of failure. The results of this investigation are similar to what is obtainable in metallic materials in which the mechanism of damage is controlled by strain localization and formation of adiabatic shear bands.

Depending on particle size, two mechanisms have been identified for strain localization in ceramic materials under dynamic mechanical loading. The thickness of shear bands in granular SiC and Al_2O_3 is reported to increase with particle size, while thicker shear bands are formed in pre-fractured ceramics compared to granular ceramics. In contrast to most of the previous studies on deformation and damage mechanism in ceramic materials that were conducted in compression (impact loading) or used explosive shock wave, this study is focused on deformation and damage mechanism in ceramic material in shear using torsional loading. Dynamic stress–strain curves are generated, and the fracture surface of the investigated ceramic material is evaluated using scanning electron microscopy.

The dynamic stress–strain behaviour is influenced by the strain rate and the wall thickness in a thin-walled tubular specimen. The fracture of the ceramic happens by strain localization and the occurrence of adiabatic shear bands. Particle debonding plays a major role in crack initiation and propagation along the path of shear localization.

2.2.5 Test Methods for Granular Materials

Granular media constitute an interesting field of research from the point of view of both basic science and application. The nature of the interaction between the particles leads to different deformation behaviours. One of the most apparent instabilities occurring in granular media is the formation of shear bands. Even at low strain rates, these localized shear bands can occur, according to Unger et al. [174].

Cylinder Collapse Test

Nesterenko et al. [175] studied the dynamic deformation behaviour of densified granular alumina of two different particle sizes by an explosively driven, radially symmetric collapse of a TWC. The test arrangement is shown in Figure 2.68.

The collapse test leads to a densification, deformation and localization due to shear band developments, as shown in Figure 2.69.

The densified granular alumina was used to model the flow in ballistic impact and penetration of fragmented ceramic armour. Shear localization takes place as a well-developed deformation mode at an overall radial strain of 0.2–0.4 and a strain rate of $10^4\ s^{-1}$. The following qualitative features of shear bands were established:

- Shear bands have clear boundaries, and their thickness does not depend on the initial particle size, which has a typical value of approximately 10 μm.

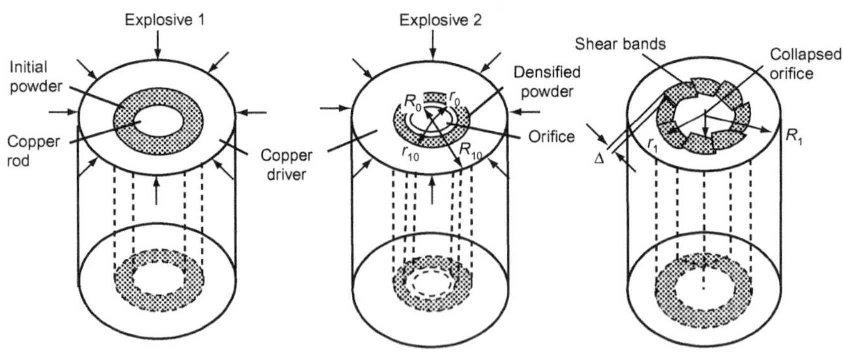

Figure 2.68 Experimental set-up of a TWC test for densification and plastic deformation of granular ceramic, according to Nesterenko et al. [175].

Figure 2.69 Test result of the ceramic after collapse, according to Nesterenko et al. [175].

- The structure of the shear bands was dependent on the initial particle size, suggesting differences in the mechanisms of flow. For the 4 μm alumina, comminution (break up) and softening of particles were observed. For the 0.4 μm particles, a peculiar structure was formed consisting of a central crack with two lateral cracks.
- Distributions of shear bands and displacement magnitudes were dependent on initial particle size.

The observed differences in powder behaviour are associated with different mechanisms of powder repacking. For large particles (approximately 4 μm), additional hardening takes place, resulting from micro-fracture and subsequent repacking of different-sized particles in the powder. The small-sized (approximately 0.4 μm) ceramic does not go through the particle-fracturing stage, and the hardening is due to 'classical' repacking.

Shih et al. [176] performed similar investigations as Nesterenko et al. [175] for granular and pre-fractured silicon carbide. The test arrangement is the same as in Figure 2.69. At large strains and high-strain-rate deformations of granular and pre-fractured silicon carbide, shear localization is an important failure mechanism. In granular SiC, the formation of shear bands occurs through particle comminution and rearrangement of comminuted particles. The shear bands in pre-fractured SiC are developed through incorporation of fragments and the erosion of the fragments.

Compression Test with Uni-Axial or Multi-Axial Loading

Desrues et al. [177] investigated the localization behaviour of dry dense sand under quasi-static conditions with cubic specimens with different geometries. They conducted multi-axial compression tests, where the shear band mechanism occurred and led to localization. The tests were performed with a true tri-axial apparatus and a bi-axial apparatus. The initiation of the localization takes place before the apex of the stress–strain curve is reached.

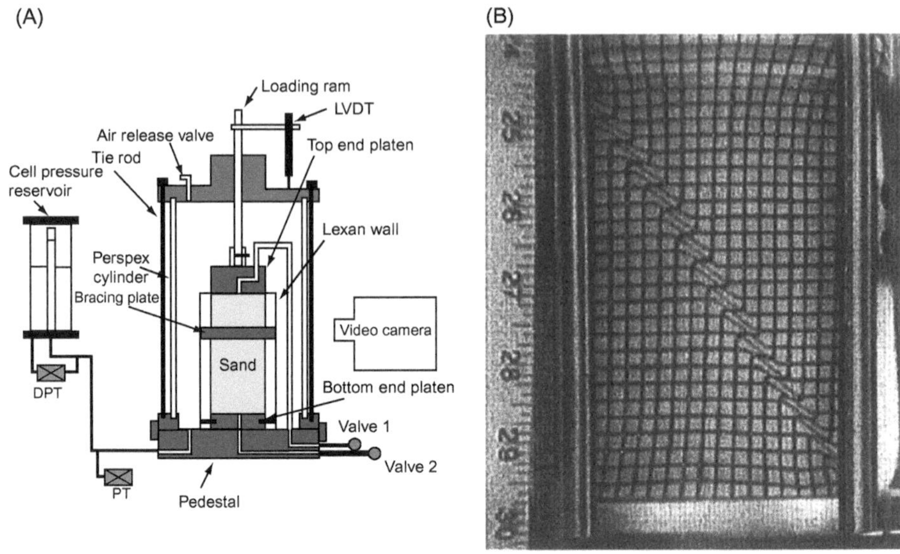

Figure 2.70 Test arrangement and tested sand specimen: (A) schematic of experiment apparatus and (B) example of digitized image of specimen, according to Alshibli and Sture [178].

Alshibli and Sture [178] performed a series of bi-axial (plane strain) experiments on three sands under low- and high-confining pressure conditions to investigate the effects of specimen density, confining pressure, sand grain size and sand grain shape on the constitutive and stability behaviour of granular materials. The used technique for loading is shown in Figure 2.70A. The specimen deformation was monitored and analysed with the help of a grid pattern (Figure 2.70B).

To investigate the effects of specimen density, confining pressure, and sand grain size and sand grain shape on the constitutive and stability behaviour of granular materials and furthermore to characterize the evaluation of the grain-scale process, which necessarily control shear band formation and growth in sands, a variety of experimental techniques have been used by Rechenmacher [179].

A new measurement technique to measure two-dimensional (2D) and three-dimensional (3D) surface displacements on plane-strain and axially symmetric sand specimens over short time steps is the non-destructive displacement-measurement technique of digital image correlation (DIC).

Insights are offered regarding the relation between strain localization and global stress–strain behaviour and the ensuing interpretations of shear banding as a hardening or softening phenomenon. Comparison of behaviours between plane-strain and the tri-axial test offer additional perspectives on the influences of 3D stresses and boundary conditions on shear banding. The experimental test set-up is outlined in Figure 2.71.

Hasan et al. [180] investigated sand in plane-strain loading with rectangular specimens. The study was focused on the formation of shear bands, which causes the failure of the specimen. A schematic view of the specimen and sub-specimens is depicted in Figure 2.72.

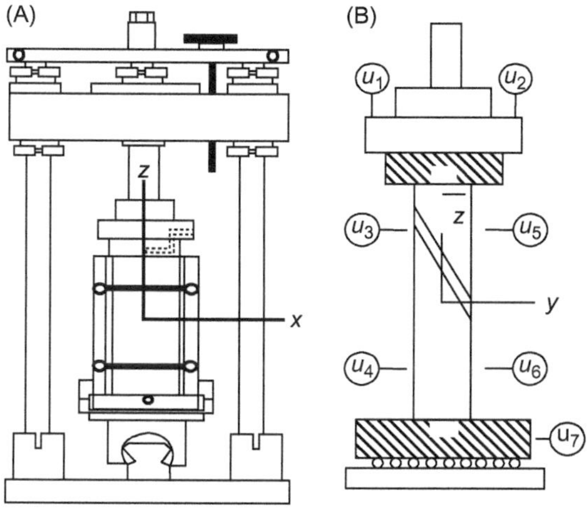

Figure 2.71 Schematic plane-strain testing apparatus: (A): out of plane view and (B) in plane view, according to Rechenmacher [179].

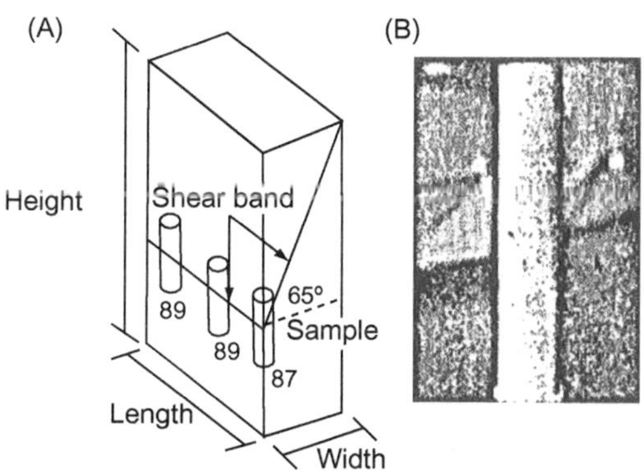

Figure 2.72 Location of the sub-sample in the main specimen (A) and the detailed view of the sub-sample (B), according to Hasan et al. [180].

A qualitative analysis of the shear band from 3D images was done and reports accurate measurements of the shear band thickness. Synchrotron micro-tomography (SMT) was used to acquire high-resolution 3D images of a sand specimen sheared under plane-strain condition.

The thicknesses of the shear bands were measured to be about 2 mm in width. Column-like structures were observed within the shear band between voids. The orientation of the particles in the shear band was found parallel to the inclination of the shear band.

Hall et al. [181] conducted tri-axial compression tests for sand. X-ray tomography monitoring during those tri-axial compression tests allows high-resolution, full-field observation of the development of the deformation. However, such images only indicate clearly the deformation when there are significant changes in material density (i.e. volume changes) that produce a change in X-ray absorption. As such, 3D volumetric DIC approaches have been developed that allow analysis of the full-strain tensor field throughout the imaged volume. These images have

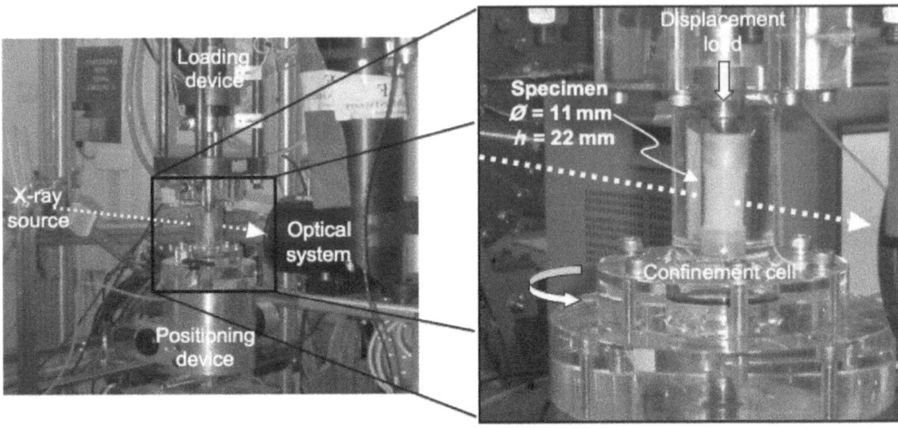

Figure 2.73 Tomography set-up for tri-axial testing at the beamline D15A at European Synchrotron Radiation Facility (ESRF). Complete set-up on the beamline (A) and zoom on the specimen inside the tri-axial cell (B), according to Hall et al. [181].

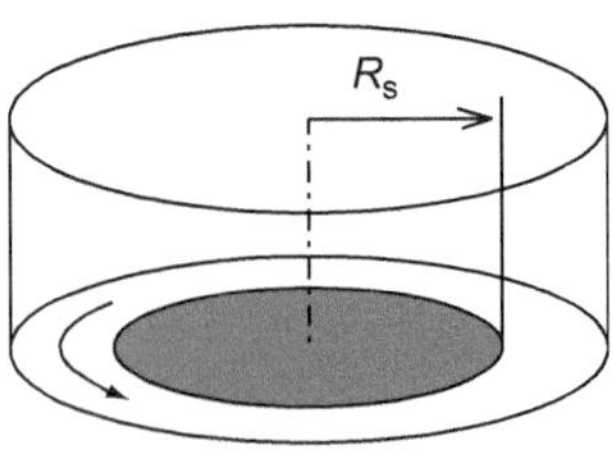

Figure 2.74 Test principle for torsion test, according to Unger et al. [174].

been analysed using 3D volumetric DIC to provide full-field displacement and strain measurements, which allowed the detection of the onset of strain localization and its timing relative to the load peak plus insight into the 3D structure of the localized zone. The experimental set-up is outlined in Figure 2.73.

Torsion

Unger et al. [174] presented a torsional test assembly for granular materials. This experiment in a modified Couette cell provided a localized shear flow in the material, away from the confining walls. The test arrangement is depicted in Figure 2.74. The rotating (white) and the stationary (grey) parts induce shear flow in the granular material held by the container. The non-trivial shape of the shear band can be measured depending on the cell geometry.

Similar investigations were performed by Utter and Behringer [182]. When dense granular materials are sheared, a shear band and an anisotropic force network will be formed. The approach to steady-state behaviour depends on the history of the packing and the agent force and the terms of the contact network. By application of a 2D Couette geometry, it is possible to investigate the history and evolution of shear bands by measuring particle trajectories and stresses during transients.

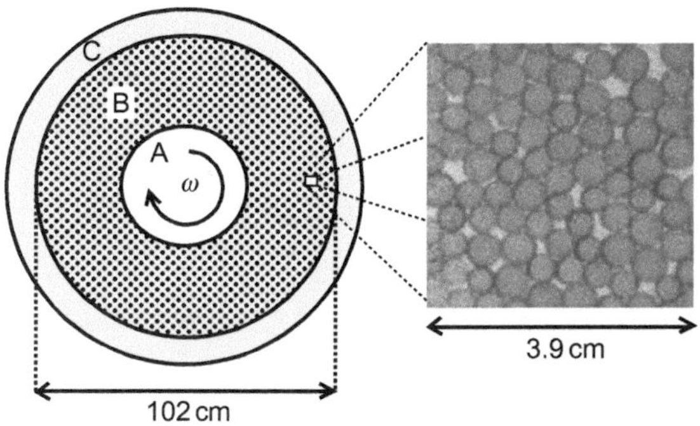

Figure 2.75 Top view of the experimental apparatus, according to Utter and Behringer [182].

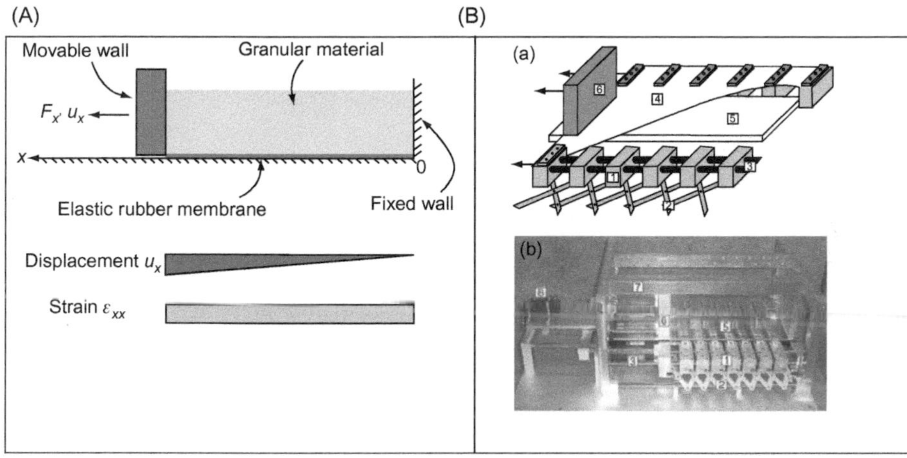

Figure 2.76 Test assembly to investigate granular materials for shear behaviour: (A) simplified model of the experimental set-up and (B) schematic view and photograph of the experimental set-up, according to Wolf et al. [183].

The test assembly is shown in Figure 2.75. The shearing wheel (A) rotates at angular frequency to shear the granular material (B). The grains are contained by a stationary outer ring (C).

Shear Test

Wolf et al. [183] determined the stress–strain behaviour of granular non-cohesive materials and the geometry of shear band systems, specified by the spacing and the inclination of the shear bands, through sandbox experiments on quartz sand in natural as well as increased gravity. The tests were performed by applying an extensional strain to specimens with different densities in the initial state. The test arrangement is depicted in Figure 2.76.

Radiography is used to determine the spacing and the inclination of the developing shear zones. DIC is applied to identify the deformation mechanism and the point of

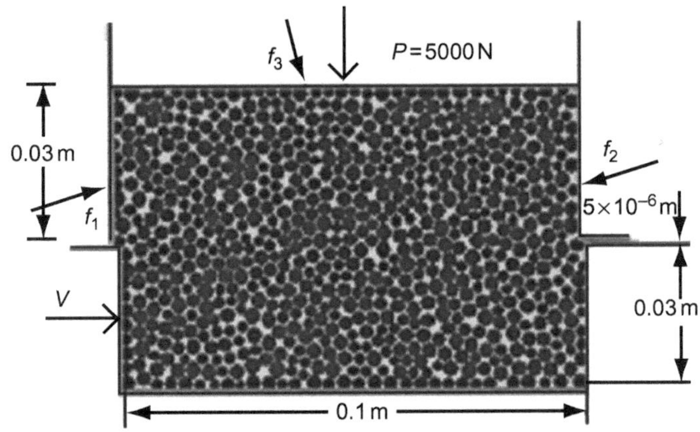

Figure 2.77 Test assembly of direct shear box, according to Zhou et al. [184].

localization. Tri-axial extension tests with respective densities are performed in order to determine the stress–strain behaviour of the granular material (sand). It is found that the shear band spacing can be correlated with the initial density of the specimen.

Zhou et al. [184] conducted direct shear tests in a shear box for granular materials and determined the shear band evolution under geometrical and different particle sizes. These investigations were done by numerical calculations. The shear assembly where the bottom part is being moved perpendicular to the upper pressure is shown in Figure 2.77.

Wu et al. [185] performed different direct shear tests for sand material. They measured the force and the displacement during the test. The effect of specimen size was evaluated by performing constant pressure tests on fine, poorly graded sand. These specimen size effects can be attributed to the thickness of the shear zone and the number of shear bands included in the shear zone. The test arrangement is depicted in Figure 2.78.

2.2.6 Test Methods for Rocks

Confined Compression Test

Ord et al. [186] conducted unconfined and confined compression tests on sandstone. They measured displacements in different directions during the test. A schematic view of the measuring technique is shown in Figure 2.79A. U1 and U2 are for axial, U3–U6 for lateral and U7 for sled displacement. With increasing confining pressure, the axial strengths are also increasing. Furthermore, the shear modulus was measured complementary. A failed specimen tested with 15 MPa side pressure is shown in Figure 2.79B.

These experimental results show unequivocally, and for the first time in a material with cohesion, that shear band formation under plane-strain conditions occurs

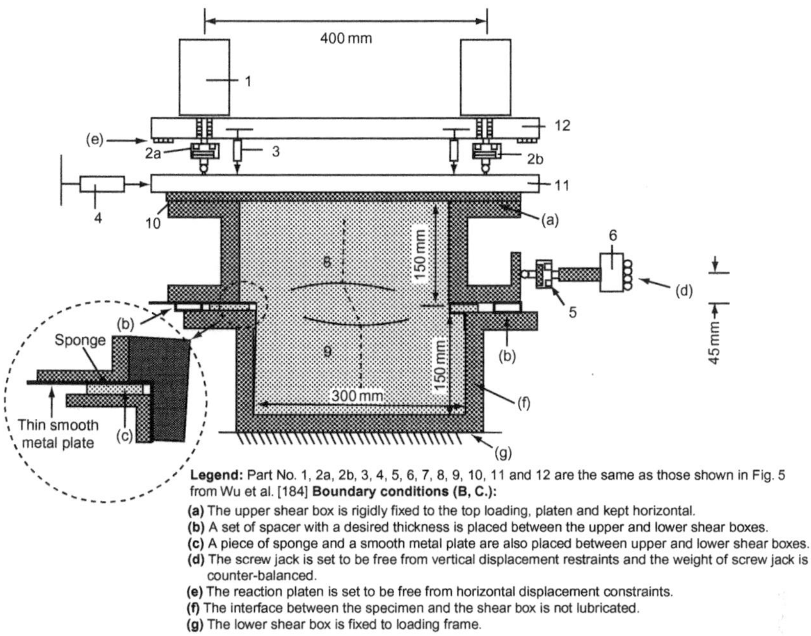

Figure 2.78 Direct shear assembly, according to Wu et al. [185].

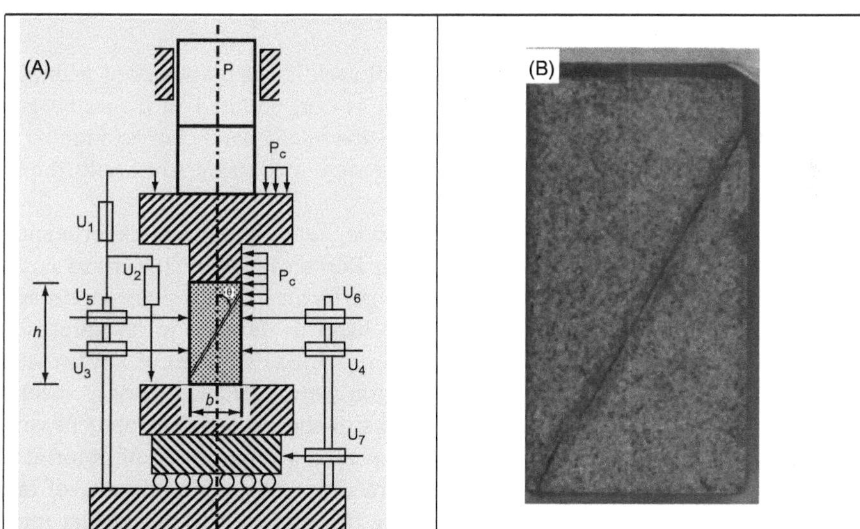

Figure 2.79 Test rig for confined compression tests with failed specimen: (A) measuring of displacements and (B) tested specimen at 15 MPa side pressure, according to Ord et al. [186].

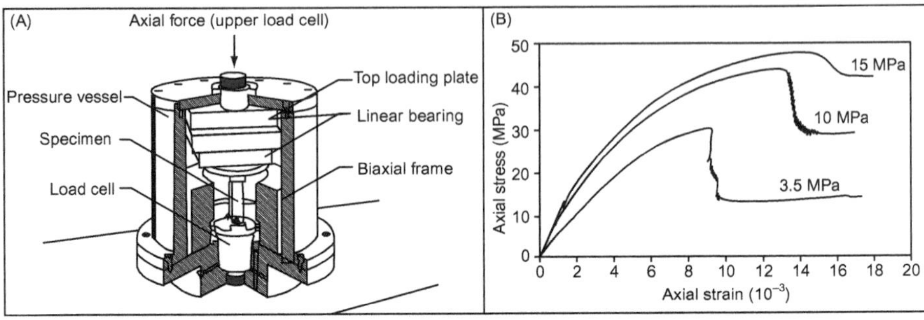

Figure 2.80 Test apparatus and measured curves of sandstone with confining pressure: (A) test apparatus for confining pressure and (B) test results, according to Labuz and Dai [187].

in the hardening regime, that is, when the hardening modulus is still positive. Initial localization is inferred to be accommodated by rigid-body grain motions through the matrix, followed by only minor grain cracking occurring before peak stress. Evaluation of the hardening modulus at shear band initiation is made on the basis of derived friction and dilatancy angles for this strongly non-associated material.

Labuz and Dai [187] conducted similar investigations as Ord et al. [186] on soft rock and sandstone. They performed tests under different side pressures, measuring force and axial as well as lateral displacements. The test apparatus and examples of measured stress–strain curves are shown in Figure 2.80. With increasing confining pressure, the maximum strength and the deformability to failure are increasing, similar to the results of Ord et al. [186].

Field observations indicate that failure in soft rock is often associated with a slip surface or shear band, where the deformation is concentrated in a narrow zone. Displacements occur with decreasing stress in the shear band. For examining the propagation of shear bands, it is useful to measure the shear stress and the slip displacement.

Riedel and Labuz [188] conducted closed-loop, servo-controlled experiments to investigate the development of a shear band in Berea sandstone at various confining pressures by using a TWC. The tests were performed with a plane-strain apparatus, which was designed to allow the shear band to develop in an unrestricted manner. The measured load and displacements provided estimates of the stress and deformation states whereby dilatancy and friction were evaluated prior to localization. Experiments were stopped at various stages of shear-band development within the strain-softening regime. The specimens displayed a progression of deformation from inception, where the shear band was characterized by a high density of intragranular micro-cracks and crushed grains to the tip, where the intragranular microcracks were significantly less dense and separated by intact grains. Decreased slip deformation towards the tip of the shear band indicated that localization developed and propagated in the plane. Thin-section microscopy showed porosity increase within the shear band was 3–4 grain diameters wide. Increased porosity did not

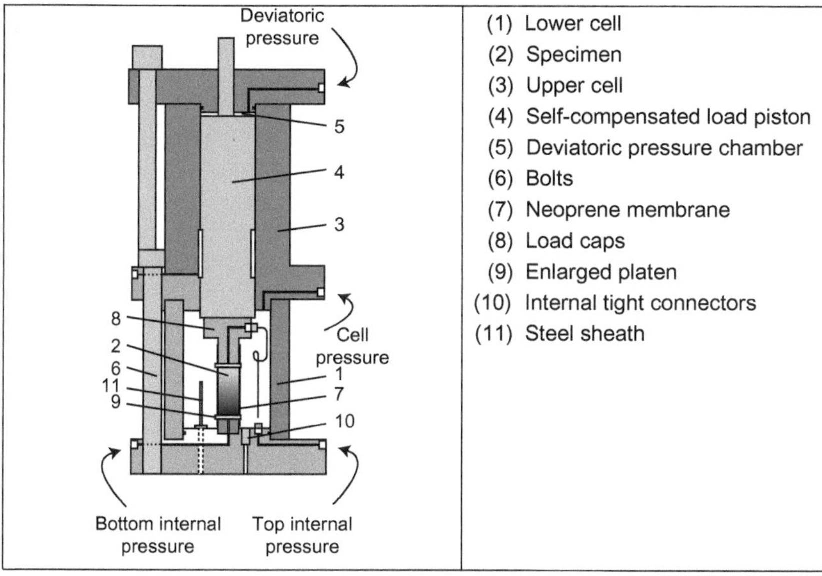

Figure 2.81 Schematic of the tri-axial test arrangement, according to Besuelle et al. [189].

extend beyond the tip of the shear band. The dilatancy of the material prior to peak stress was not related to micro-cracking of grains, which occurred in the post-peak phase and was identified with localization, i.e. the shear band.

Tri-Axial/Confined Compression Test

Besuelle et al. [189] presented results that characterize the behaviour of a dry sandstone tested in a high-pressure tri-axial apparatus. The objective of this investigation was to provide insight into the phenomenon of strain localization in soft rocks. The onset of localization was detected by means of internal measurement of axial and radial deformation. The deviatoric stress was measured in dependence on confining pressure loading. The tri-axial apparatus is outlined in Figure 2.81.

The confining pressure has a strong influence on the behaviour of the sandstone and of the localized deformation structure that emerges at failure. For confining pressures up to 40 MPa, the deviatoric strength increases with confining pressure, and the global volumetric strain dilates before failure. The angle of the shear bands with respect to the major principle stress direction increases, and the number of shear bands increases from one to several (Figure 2.82).

Torsion Test

Schmocker et al. [190] examined Dover flint to high shear strains in torsion. Tests were carried out under drained conditions at temperatures of 1250 and 1300 K,

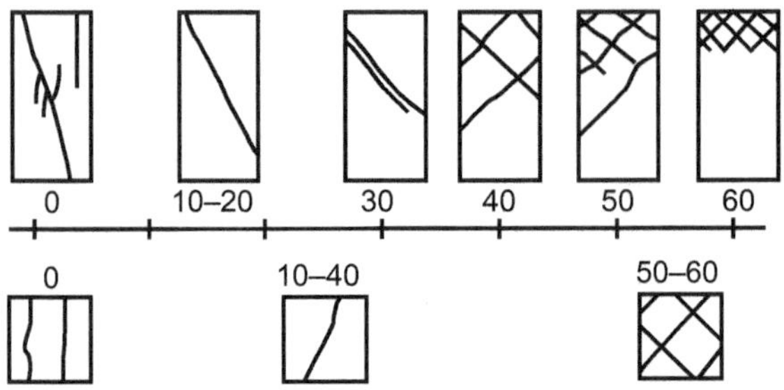

Figure 2.82 Observed shear band pattern versus confining pressure and two different specimen ratios, according to Besuelle et al. [189].

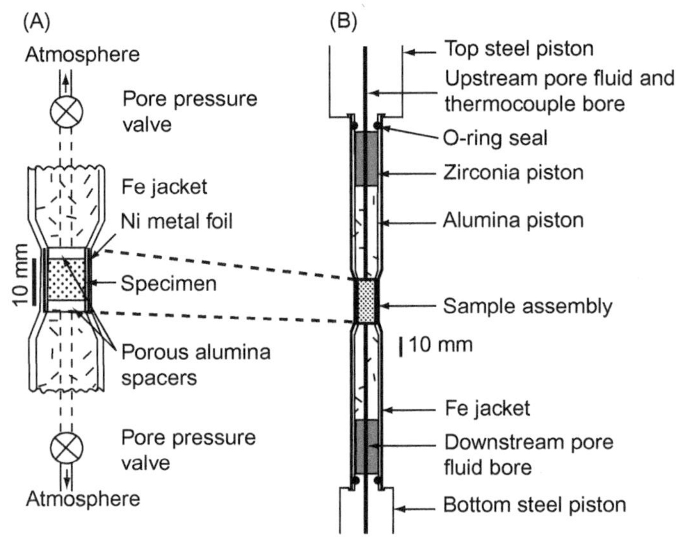

Figure 2.83 Longitudinal cross section of (A) sample and (B) piston assembly, according to Schmocker et al. [190].

350 MPa confining pressure and constant twist rate. Heat treatments prior to deformation were performed at 1300 K for different amounts of time at confining pressures of 150 or 350 MPa, resulting in an initial dilation with an isotropic distribution of water-filled pores, followed by compaction with continuous reduction in porosity and grain growth. Deformation of the heat-treated samples reached steady-state behaviour until a shear strain of $\gamma = 0.2$, followed by sudden hardening stages for shear strains up to $\gamma = 2.5$. The used test arrangement with specimen is depicted in Figure 2.83.

The shear-stress–shear-strain curves and the failure evolution were examined, as shown in Figure 2.84.

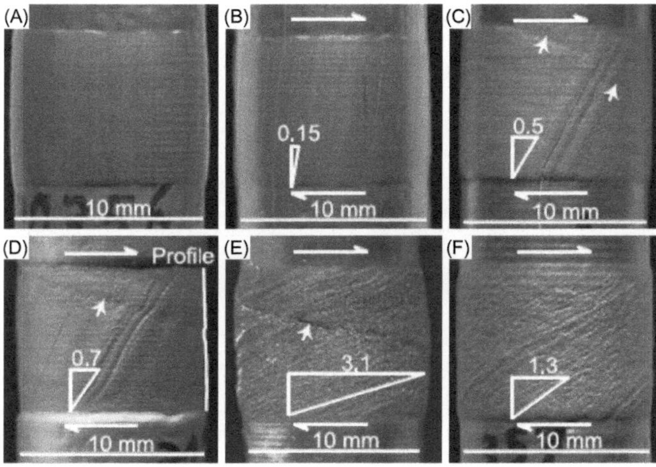

Figure 2.84 Shear band evolution of a heat-treated sample prior to the test, from uninitial state (A) to $\gamma = 3.1$ (E), (F) is other pre-heat treatment; according to Schmocker et al. [190].

2.3 Special Results for Steels

In the literature, much information can be found which supports the theory that some material properties influence the adiabatic shear-failure behaviour in a positive or negative manner. Tresca [191] and Zener and Hollomon [192] have observed that the strength and the heat capacity, as well as the thermal softening of the materials, have an important influence on the adiabatic shear behaviour.

Today, it is known that a lower strain-hardening coefficient [6,7,29], a lower strain-rate dependence [77,193], a lower heat capacity or thermal conductivity [6,7,29], a lower grain size [6,194] or a lower density of the material [193–195] promote the adiabatic shear-failure behaviour.

Additionally, a high hardness [196–199], a high strength [6,7,63,200], a high thermal softening [6,7,29], a high loading velocity [29,124,194], a high pre-deformation [6,59] and a large specimen size [201] promotes the adiabatic shear-failure behaviour too. Furthermore, additional properties exist, for example, initial temperature [127], hydrostatic stress state [8,67] and tensile loading [7]; even considering these factors, the tendency for initiation of the adiabatic shear failure is ambiguous.

The question then arises: Which material properties have the strongest influence on adiabatic shear failure for the investigated materials? Observing and measuring the adiabatic shear-failure behaviour was accomplished through the performance of materials under different experimental tests, firstly with compression and compression/shear loading. Furthermore, there is much discussion concerning high temperature compressive strength, the Culver theory [1], analytical consideration and a description of the correlation between the material properties and the adiabatic failure behaviour of the materials.

The aim of this chapter is to distinguish the material properties that have the strongest influence on the adiabatic shear-failure propensity for high-strength,

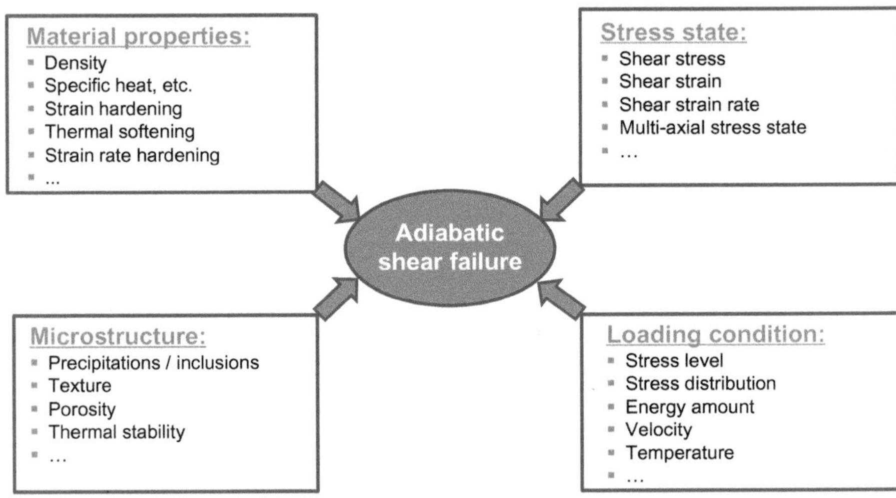

Figure 2.85 Influences on adiabatic shear failure, according to Pursche [36].

low-alloyed steels (HSLA steel). Is it possible to find correlations and associated material properties? The next issue involves the validity of the Culver theory on the observed materials and a discussion of the influence of high-temperature behaviour on the adiabatic failure process.

Published literature contains a broad range of statements concerning which properties have an influence on the initiation of the adiabatic shear failure. These results can be divided into four groups (Figure 2.85). One group deals with the stress state (shear strength, shear strain, shear rate); a second group deals with the loading conditions such as temperature, stress state, energy or velocity. The third group deals with the microstructure of the material (precipitations, inclusions, voids, etc.), and the fourth group describes the influence of the material properties such as density, heat capacity, flow stress, strain-hardening coefficient or thermal softening. This chapter concentrates on the fourth group – material properties.

There are different assumptions for the critical condition of the beginning of the adiabatic shear failure. Most authors assume it is necessary to reach a certain amount of strain [1,8,96]. Other authors contend that a certain level of strain rate is crucial for the initiation of adiabatic shear bands [134,138]. Wang et al. [78] and Xu et al. [29] describe a critical value as a consideration for strain and strain rate. Furthermore, there are assumptions that a critical energy must exist [127] or that definite stress intensity is necessary [202]. New results of Medyanik et al. [203] and Murr [204] indicate that a dynamic recrystallization (called DRX) is responsible for the initiation condition.

A well-known theory for describing adiabatic shear failure is based on the principle from the work hardening and thermal softening of the material [1,6,77,192]. These competing processes are described in (Eq. (2.1)).

$$d\sigma = \left(\frac{\partial\sigma}{\partial\varepsilon}\right)_{\dot{\varepsilon},T} d\varepsilon + \left(\frac{\partial\sigma}{\partial\dot{\varepsilon}}\right)_{\varepsilon,T} d\dot{\varepsilon} + \left(\frac{\partial\sigma}{\partial T}\right)_{\dot{\varepsilon},\varepsilon} dT \qquad (2.1)$$

Adiabatic shear failure can occur when the thermal softening of the material overcomes the strengthening due to the strain hardening and the strain-rate hardening [7,9,93,138]. The first term describes the strain hardening, the second term the strain-rate hardening and the third term the thermal softening behaviour, which acts against the first and second term. With the negligence of the second part and the differentiation with $\mathrm{d}\varepsilon$, a criterion was formulated for instability (Eq. (2.2)) [1,8,138]:

$$\left(\frac{\partial\sigma}{\partial\varepsilon}\right)_{\dot{\varepsilon},T} + \left(\frac{\partial\sigma}{\partial T}\right)_{\dot{\varepsilon},\varepsilon} \frac{\mathrm{d}T}{\mathrm{d}\varepsilon} = 0 \qquad (2.2)$$

The left term describes the influence of the strain hardening, and the right term describes the thermal softening behaviour. The failure behaviour can be imagined as a form of a convex adiabatic curve. When the equilibrium of both competing processes is reached, identical with the apex of the curve, an adiabatic shear failure can occur. From the preceding displayed softening theory, Culver [1] has evolved a plane relation (Eq. (2.3)) to predict the failure strain of the material. Culver has made three assumptions. The strain-hardening coefficient n_T from the isothermal behaviour will be substituted by the static one. The thermal softening behaviour of the materials will remain linear, and the stress relation ($\sigma_\mathrm{T}/\sigma_\mathrm{D}$) between dynamic isothermal and the dynamic adiabatic values will be neglected. Furthermore, the density ρ, the heat capacity c_p and the Taylor–Quinney factor 0.9 are included.

$$\varepsilon_\mathrm{i} = \frac{n_\mathrm{T} \cdot \rho \cdot c_\mathrm{p}}{0.9(\partial\sigma/\partial T)} \cdot \frac{\sigma_\mathrm{T}}{\sigma_\mathrm{D}} \qquad (2.3)$$

where ε_i = strain to adiabatic shear initiation.

The question is whether this theory from Culver is applicable to the investigated HSLA steels. A second objective is to determine if there are additional material properties, other than the thermal softening or hardness, which may influence the adiabatic shear-failure behaviour.

Various tests were used to determine the correlations between material properties and adiabatic shear failure. Compression tests were conducted at room temperature and at higher temperatures to measure the material's strength for the evaluation of the thermal softening of the materials versus temperature.

Dynamic tensile tests were used with a rotating wheel (1 m/s) to determine values of strength and ductility. Also hardness tests, Charpy impact and fracture toughness tests were performed.

Bi-axial compression/shear tests with inclined compression specimens were performed to verify the propensity to adiabatic shear failure of the materials. The specimens were tested at 2–3 m/s with a drop-weight device at room temperature. During the test, both the force and the displacement were measured. Figure 2.86A shows a representative selection from all of the materials with the axial engineering

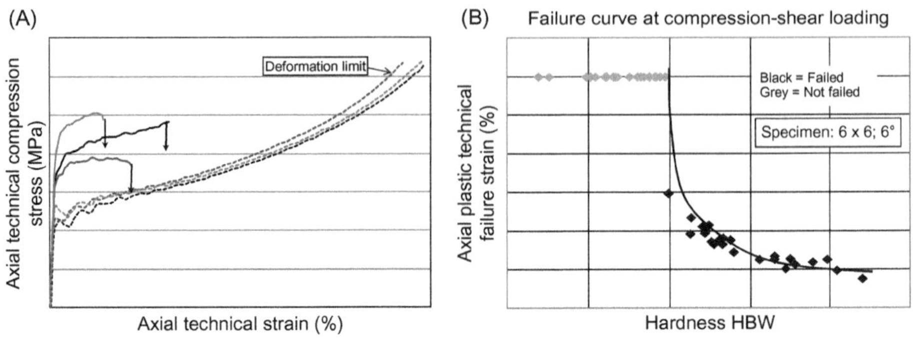

Figure 2.86 Stress–strain behaviour of the compression/shear test ($\dot{\varepsilon} = 200\ s^{-1}$) and the correlation with the hardness, according to Pursche [36] and Pursche and Meyer [37].

or technical compressive stress versus the axial technical strain. In principle, two different behaviours exist. A few steels, with lower flow stresses, exhibit a more or less homogenous deformation until the limitation of displacement is reached. The other group of materials (about half of the investigated materials) shows a sudden stress drop at lower strains. At this point, the material failed due to adiabatic shearing. The measured height of reduction for all failed materials was used to determine additional correlations. It can be seen that very different failure strengths and failure strains exist for the different steels, which cover a hardness range from HB 250 to HB 600. An initial consideration of the stress level of the material shows that high-strength steels are much more prone to adiabatic shear failure than the lower strength materials. This correlates with the hardness of the materials (Figure 2.86B). With low hardness values, no adiabatic shear failure occurred. The grey data points are indicating that no adiabatic shear failure occurred up to the deformation limit. At a certain hardness level, first-time adiabatic shear failure begins. As the material's hardness increases, the failure strain declines along with potential function. These results are valid for the HSLA steels and for the strain rate of $\dot{\varepsilon} = 200\ s^{-1}$ investigated in this study. This dependence, already cited in the literature, is hereby confirmed.

The next examinations involve (pure) compression loading at room temperature. Figure 2.87A shows a selection of true stress/true strain diagrams of the investigated materials at a strain rate of $\dot{\varepsilon} = 200\ s^{-1}$.

The material behaviour under this adiabatic condition begins at the lowest strength level with a pure strain-hardening behaviour, for example, of plain carbon steel. In the medium-strength level, for example, HSLA steel, equilibrium between strain hardening and softening exists. At the highest strength level, mainly a softening behaviour occurs. These flow stress variations, responding to the hardening or softening behaviour under dynamic compression loading, are arranged following their amounts (Figure 2.87B). Materials with a strong decline are prone to failure by adiabatic shearing. On the other hand, there is an area with only positive values, i.e. a hardening behaviour. These materials most likely do not fail, at least not in a

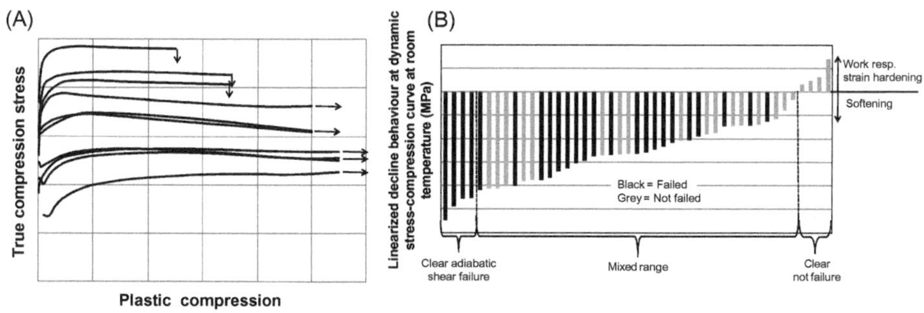

Figure 2.87 True stress/true strain curves at (pure axial) dynamic compression ($\dot{\varepsilon} = 200\ s^{-1}$) and results of softening determination, according to Pursche [36] and Pursche and Meyer [37].

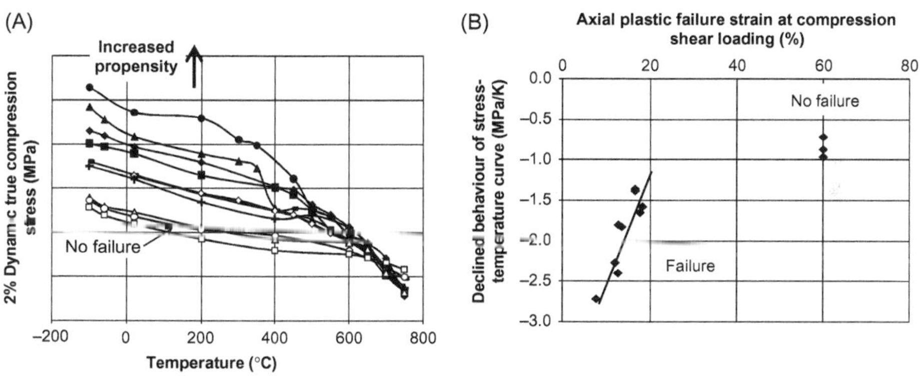

Figure 2.88 Dynamic 2% flow compression stress versus starting temperature and determined declined flow stress values at dynamic strain rate of ($\dot{\varepsilon} = 200\ s^{-1}$) between −100°C and 750°C, according to Pursche [36] and Pursche and Meyer [37].

6° inclined compression, high-rate test. In the depicted middle range, no clear dependence between the decline or softening behaviour and the occurrence of adiabatic shear failure is found. Thus, the conclusion can be made that the stress-softening behaviour only under adiabatic compression is not clearly correlated to the initiation of adiabatic shear failure, at least not for this particular quenched and tempered HSLA material group. As described in the literature, the strain-hardening coefficient, even at a strain rate of 200 s^{-1}, which has an important influence on the adiabatic failure behaviour, cannot be confirmed for this material group. There might be a tendency, but for most of the steels, a clear relation cannot be drawn.

Another investigation was to determine the flow-stress behaviour versus high temperatures under dynamic compression loading. The results exhibit the different strength levels of the different steels (Figure 2.88A). This distinct difference in flow stress lasts until about 600°C. With higher flow stress, the propensity to adiabatic shear failure is increased (arrow in Figure 2.88A). Materials with a high strength at room and at elevated temperatures are prone to adiabatic shearing.

Furthermore, these materials show a rapid loss of strength with increased temperature. These drops of strength yield to a gradient of temperature-dependent stress resistance. This gradient is considerably higher for high-strength steels. For low-strength steels, the gradient is low, and therefore these steels are not prone to adiabatic shearing, valid for the three materials depicted with no failure (not filled circles) in Figure 2.88A. The results of the determined values of the averaged decline behaviour between $-100°C < T < 750°C$ at the stress–temperature curves is shown in Figure 2.88B. There is a good correlation of the normalized strength decline value to the occurrence of adiabatic shear failure under compression/shear loading. The higher the drop in stress versus temperature (the decline behaviour), the lower the failure strain under compression/shear loading will be. Above a certain value, no failures occurred.

The determined failure strains according to Culver (Eq. (2.3)) show a good qualitative correlation (Figure 2.89A). All materials with low calculated failure strain fail under compression/shear loading, too. High values correlate to the materials with no failure. The depicted transition state is in the upper-limit range of the used compression/shear loading test configuration. For the failed states, the correlation between the experimental and the calculated failure strain (Figure 2.89B) is an absolute insufficient agreement (or no tendency). Thus, for the investigated materials, only a qualitative correlation with the softening theory of Culver exists.

The flow stress–temperature dependencies were considered to profoundly concern the adiabatic shear-failure behaviour. Two examples of dynamic flow-compression stress versus temperature relations are shown in Figure 2.90. Figure 2.90A shows a distinctive drop in strength at a certain temperature of about 400°C; Figure 2.90B shows at another certain temperature, a change in the decline behaviour. These temperature points represent a change in the softening behaviour from a lower to a high or higher decrease. These points are called ‘instability points’, for which the related temperatures and stress values are taken.

These instability temperatures were correlated to the failure strain at compression/shear loading (Figure 2.91A). In the sheared area, a linear correlation with a failure strain at a higher instability temperature exists. Above an instability

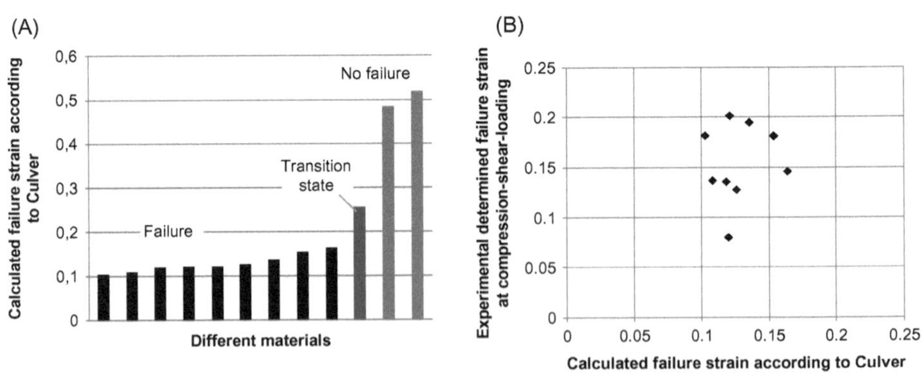

Figure 2.89 Qualitative and quantitative evaluation of the materials using Culver theory, according to Pursche [36] and Pursche and Meyer [37].

temperature of 600°C, no more failures occurred. The instability temperatures and instability stresses for all HSLA materials were then depicted (Figure 2.91B) to correlate with failure strains under compression/shear loading. With lower instability temperatures and higher instability stresses, the propensity to adiabatic shear failure increases dramatically. These results clearly show that the specific dynamic flow stress—temperature characteristics of the materials, especially the instability points, from this study are very valuable bits of information and that the use of the averaged or linearized decrease behaviour is not sufficient, respectively, in neglecting the initiation of a mechanical instability.

The previous results prove that the dynamic elevated temperature behaviour of the materials is important for the evaluation of the propensity of adiabatic shear failure. The measured compression flow-stress behaviour at elevated temperatures was used to study the stress—strain temperature field analytically. With the

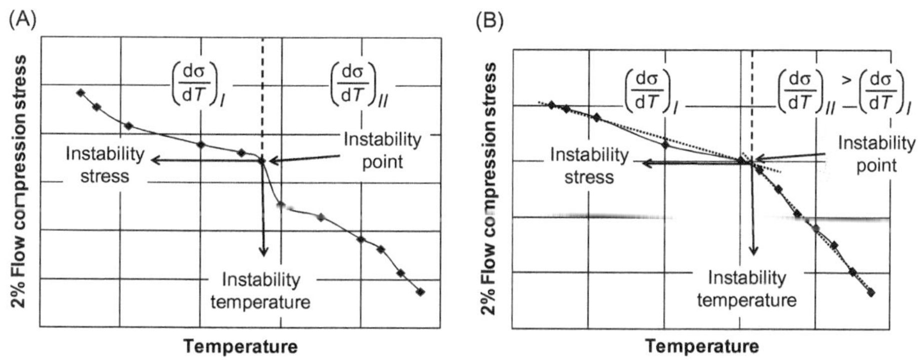

Figure 2.90 Examples for determined instability points from dynamic flow compression stress versus temperature ($\dot{\varepsilon} = 200\ s^{-1}$), according to Pursche [36] and Pursche and Meyer [37].

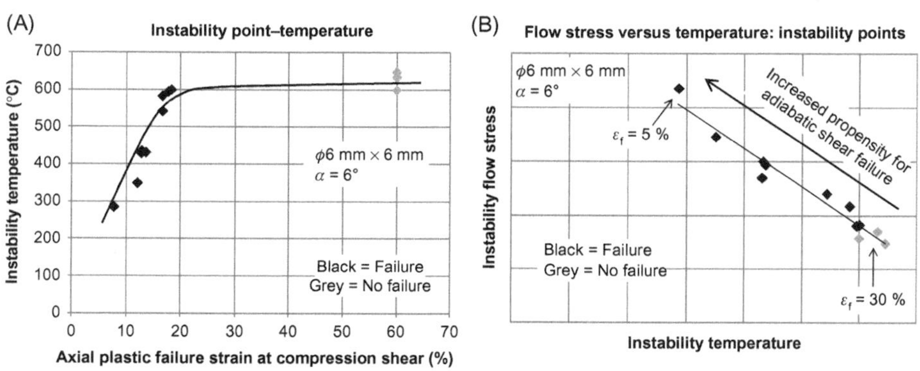

Figure 2.91 Temperature instability versus failure strain at compression/shear loading and correlation of instability temperatures and stresses for all materials, according to Pursche [36] and Pursche and Meyer [37].

(adiabatic) assumption (Figure 2.92) that the stress is a function of temperature, the temperature is a function of strain and the strain is a function of a local position. Considering slow shear deformation, there is normally a homogenous distribution of strain and thus a homogenous distribution of temperature and stress. In the case of impact loading, adiabatic deformation occurs. The related deformation field is inhomogeneous, like the strain distribution across the shear band, as shown in Figure 2.92, lower right. This certain distribution yields an inhomogeneous distribution of temperature and stress (or a consequence of). The materials used have different strength levels that evoke a different temperature rise in the specimen for the same deformation, according to Klepaczko and Rezaig [205].

This different temperature behaviour leads to a different stress-resistance behaviour across the shear band (Figure 2.93A). Feng and Bassim [206] published similar relations. The material A shows a considerable loss in stress resistance by increased temperatures and thus a development of a strong gradient (Figure 2.93A) close to

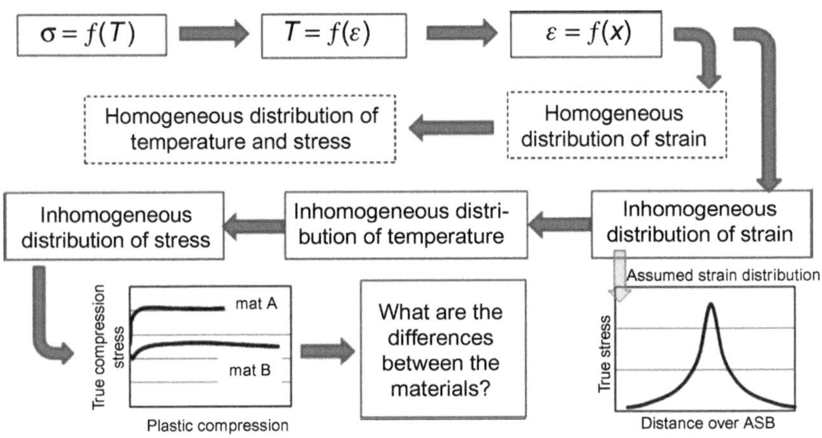

Figure 2.92 Approach for the analytical consideration, according to Pursche [36] and Pursche and Meyer [37].

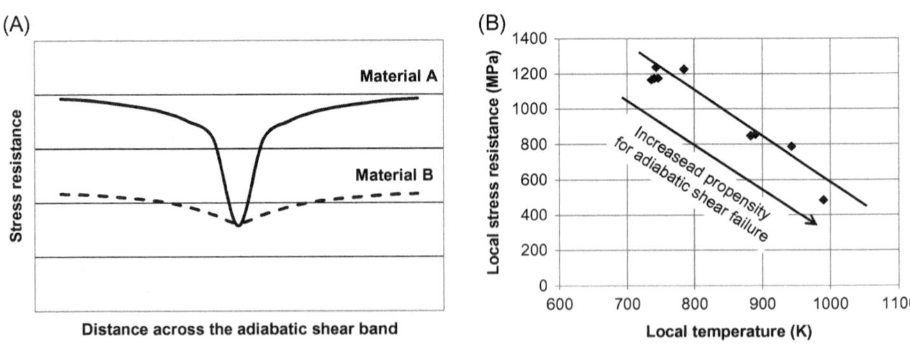

Figure 2.93 Results of the analytic consideration and correlation to adiabatic shear failure, according to Pursche [36] and Pursche and Meyer [37].

the shear band. This gradient is caused by the stress–temperature behaviour, similar to those in Figure 2.90. This gradient leads to a local increased deformation and finally to a local failure. Material B exhibits only a light loss of stress resistance. Consequently, this material is not susceptible to adiabatic shear failure. For the evaluation of the propensity to adiabatic shearing, the strength level and the intensity of the gradient are *both* important. From the analytically calculated temperature and stress gradient behaviour for all materials across the shear band, the minimum of stress resistance and the maximum of temperature (from the apex) were used to create a correlation between the calculated inner local temperature and the inner stress resistance. These determined data show a linear dependence: with an increased local temperature, the propensity to adiabatic shear failure increases, as shown in Figure 2.93B. These analytical results on an experimental basis confirm the conclusion that the stress–temperature behaviour is important for the evaluation of the propensity of adiabatic shear failures.

The compression/shear and the shear behaviour with a hat-shaped specimen, the geometry and the test procedure mentioned in Section 2.1 were compared to find any existing relations. The result is outlined in Figure 2.94. The failure strains at compression/shear loading correspond to the shear-failure behaviour for hat-shaped tests with a linear dependence. This behaviour can be explained by the influence of a similar shear-stress loading condition on the failure behaviour for both test techniques, combined with a perpendicular compression stress component in the hat configuration by a temperature-driven internal hydrostatic pressure increase.

Grady [202] defined a new criterion with a shear fracture energy per unit area (Eq. (2.4)), which is necessary for the initiation of a shear band formation. This approach is very useful for the evaluation of materials for the propensity to adiabatic shear failure, too. In contrast to Culver's theory, other main properties are incorporated, such as thermal conductivity, flow shear stress and strain rate.

$$\Gamma = \frac{\rho \cdot c}{\alpha}\left(\frac{9 \cdot \rho^3 \cdot c^2 \cdot \chi^3}{\tau_y{}^3 \cdot \alpha^2 \cdot \dot{\varepsilon}}\right)^{1/4} \tag{2.4}$$

where Γ = fracture energy per unit area, ρ = density, c = specific heat, χ = thermal conductivity, τ_y = flow shear stress, α = thermal softening and $\dot{\varepsilon}$ = strain rate.

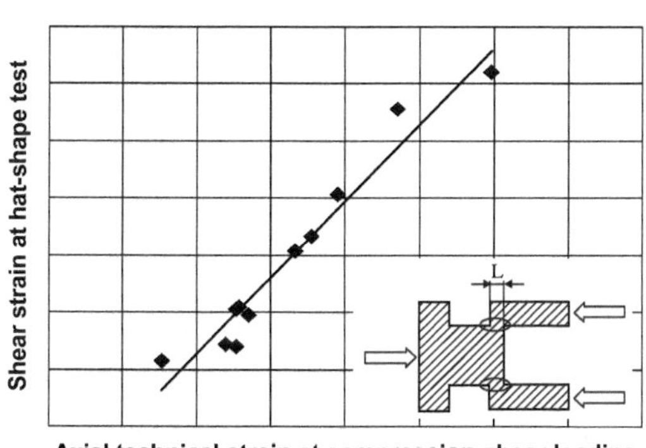

Figure 2.94 Correlation of shear failure at a hat-shaped test to the compression/shear failure, according to Pursche [36] and Pursche and Meyer [37].

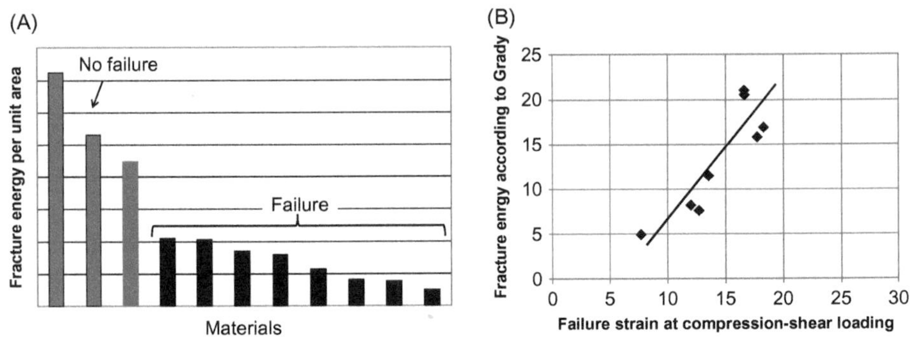

Figure 2.95 Fracture energy, according to Grady, and the correlation to adiabatic compression/shear failure, according to Pursche [36] and Pursche and Meyer [37].

Using this assumption, the determined energy values (Figure 2.95A) present a good differentiation of the tested materials. Materials with a high amount of energy consumption correspond to the 'no failure' states, and all materials with a low energy value failed under compression/shear loading. The difference between these two areas is considerable. Furthermore, in contrast to the theory of Culver [1], a quantitative correlation with the shear fractures energy of Grady [202] is possible. This fracture energy corresponds in a nearly linear agreement to the energy consumption, reduction of height with respect to the strain, until failure occurs under dynamic compression/shear loading, as shown in Figure 2.95B.

All the results of different procedures to characterize the propensity to the onset of adiabatic shear failure are summarised in Table 2.2. When a qualitative relation is fulfilled, a quantitative consideration might be reasonable, and an assessment rating can be done. With this assessment rating, a ranking of the methods to establish the propensities to the adiabatic shear-failure behaviour can be determined.

A threshold value for the transition between the areas of 'shear failure' and 'no shear failure' can be defined, already from qualitative correlations. When this correlation works well without an overlapping of data, the correlation is usable for the assessment. For the failed materials, it is possible to define a quantitative correlation. With the level of the least square coefficient R^2 of the agreement, a ranking of material properties is created according to their significance to the adiabatic shear-failure behaviour, as shown in Table 2.2.

A very good correlation to the occurrence of adiabatic shear-failure behaviour is related (No. 1 in Table 2.2) to the temperature instability, to the stress instability (No. 3 in Table 2.2) determined from the dynamic stress–temperature behaviour, the shear strain from the dynamic hat-shaped test (No. 2 in Table 2.2) and the area under the dynamic stress–temperature curve (No. 4 in Table 2.2). These are properties governed by the effect of temperature and shear deformation. Pertaining to hardness (No. 5 in Table 2.2), the failure energy according to Grady (No. 6 in Table 2.2), the dynamic compression flow stress (No. 7 in Table 2.2), the dynamic tensile strength (No. 8 in Table 2.2) and the decline behaviour of the stress–temperature curve (No. 9 in Table 2.2) are all feasible to be directly correlated to the adiabatic

Table 2.2 Assessment of Sensitive Properties to Adiabatic Shear Failure, According to Pursche [36] and Pursche and Meyer [37]

No.	Property	Qualitative Consideration	Quantitative Consideration	Assessment Rating
		Threshold Value	R^2-Value	
1	Temperature instability from dynamic stress–temperature behaviour	Yes	0.95	Very good
2	Shear failure from dynamic hat-shaped test	Yes	0.95	
3	Stress instability from dynamic stress–temperature behaviour	Yes	0.94	
4	Area under the dynamic stress–temperature curve	Yes	0.94	
5	Hardness	Yes	0.91	Good
6	Failure energy, according to Grady	Yes	0.87	
7	Dynamic compression flow stress at room temperature	Yes	0.86	
8	Dynamic tensile strength at room temperature	Yes	0.84	
9	Decline behaviour from dynamic compression-stress–temperature behaviour	Yes	0.79	
10	Decline behaviour from dynamic stress–compression behaviour of compression/shear test	Yes	0.61	Insufficient
11	Uniform elongation under dynamic loading	Yes	0.47	
12	Simplified Culver equation	Yes	0.02	–

shear behaviour. These properties are based on hardness, energy consumption and strength. It is notable and evident that material properties exist, which exhibit a better correlation to the adiabatic shear-failure behaviour than the material's hardness. The decline- or softening behaviour of the stress–strain curve under compression/shear loading (No. 10 in Table 2.2) and the uniform elongation under dynamic tensile loading (No. 11 in Table 2.2) give an insufficient correlation. The failure criterion according to Culver (No. 12 in Table 2.2) shows only a qualitative but no good quantitative correlation.

2.3.1 Summary

The aim of this chapter has been to examine the influence of the high-rate material properties on the initiation of local adiabatic shear-failure behaviour. The test results of the investigated quenched and tempered HSLA steels allowed an evaluation of the materials concerning their adiabatic shear-failure propensity. The determination of the failure strain under adiabatic conditions was performed with 6° inclined compression specimens in a compression/shear test, according to Meyer et al. [27]. These dynamic tests were carried out in a drop-weight tower with initial strain rates of 200 s^{-1}.

The most important material property for the analysis of the adiabatic failure behaviour is the temperature-softening characteristic of the material. Therefore, the strength level and the stress drop versus temperature characteristic are crucial. Mostly a so-called instability point can be defined. These stress and temperature values give a good agreement to the measured failure strain under a bi-axial adiabatic shear condition. Analytical studies with the use of the determined dynamic stress–temperature behaviour confirmed, by means of the development of the stress-resistance gradient, the strong influence of the temperature-softening behaviour on the adiabatic shear failure.

The Culver theory can only be used as a qualitative prediction for this material group. This theory is based on the linear softening behaviour.

The consideration of the shear fracture energy according to Grady gives both a qualitative as well as a good quantitative correlation to the measured shear-failure behaviour.

Additionally, hardness and strength at room temperature and the 'pure' shear capability with hat-shaped specimen are also important for the evaluation of adiabatic failure behaviour.

For several material properties, no correlation to the adiabatic shear-failure behaviour was found, such as compression-strain hardening at elevated temperatures, fracture strain and energy assumption under tensile loading, fracture toughness under mode I and Charpy impact energy. The reason is assumed to be the different loading conditions, which are not comparable. They will lead to other different failure modes.

2.4 Conclusions

The test methods to assess the adiabatic shear-failure behaviour for metallic materials are widespread. The test methods described are also valid for reinforced metals, for example, aluminium matrix composites. To properly investigate the propensity of materials to adiabatic shear failure, there are many prerequisites to keep in mind. The prerequisites are the available test materials with their associated properties, existing test arrangements, specimen geometry (failure because of discontinuity or material dependent), reachable deformation and strain rate. A concluded overview about the test methods for metallic materials is presented in Table 2.3. The test methods are listed after their frequency of use, kind of shear-failure initiation, strain rate, deformability, ascertainable shear-stress–shear-strain curves, feasibility and kind of complexity.

Table 2.3 Concluded Overview for Test Methods of Metals

Test Methods	Frequency of Use	Initiation Reason		ε	$\dot{\varepsilon}$	$\tau-\gamma$	Feasibility/ Complexity
		Discontinuity	Material				
Torsion	***	✓	✓	++	++	+++	++
Hat shape	****	✓	✓	+++	++++	+	+++
Compression	****		✓	+	+++	++	++++
Compression/ shear	**		✓	+	+++	–	++++
Cylinder explosive	***		✓	+	++++	–	++
Punch	***	✓		+++	+++	+	+++
Intender	*	✓		++	+++	+	+++
Double shear	*	✓		++	+	+	+
Edge specimen	*	✓		++	+++	–	+
SCS	*	✓		+	++	–	+++
Flyer plate	*		✓	–	++++	–	+
Cutting test	*	✓		++	+++	–	+
Shot test	**	✓		++	++++	–	++
Taylor test	*		✓	+	+++	–	+

*, seldom; **, some; ***, often; ****, very often (published); ++++, very good; +++, good; ++, acceptable; –, not possible; ✓, applied.

The overview of the test methods shows, that in the past decades a few test techniques have proved successfull the testing and assessment of materials concerning their adiabatic shear failure behaviour. There exist techniques where the failure initiation occurs either at a discontinuity or depends on material properties. The best combination of features has the torsion, compression, hat shape and compression/ shear test to gain proper results for material evaluation.

The test methods used for BMG are in principle the same as for common metallic materials. The compression test is mainly used for testing BMG. A few investigations have been done under compression/shear, bi-axial tension (correlate to punch test) or uni-axial tensile loading.

To investigate polymers (including reinforced polymers) concerning their propensity to adiabatic shear failure, there are different test possibilities such as compression, single edge, shear or flyer plate. The majority of this testing was performed under dynamic loading.

For testing the ceramics (including the reinforced ones), there are only a few test methods used, such as the compression test and the torsion test. These test methods will be conducted mainly with dynamic loading.

In contrast to the polymers and ceramics, test methods for granular materials and rocks are mainly performed under static and medium velocity loading. Except (of course) the cylinder collapse test, the tests are done under very high dynamic loading. The most used test methods for granular materials and rocks are slow-confined compression tests. For granular materials, other techniques are used, such as cylinder collapse, torsion and shear test. For rocks, a tri-axial compression and torsion test is described.

References

1. Culver, R. S. (1973). Thermal instability strain in dynamic plastic deformation. Metallurgical effects at high strain rates. Proceedings of the Conference, Albuquerque, NM; United States; 5–8 February 1973 (R. W. Rohde, Ed.), Plenum Press, New York, NY, pp. 519–529.
2. Lindholm, U. S., and Johnson, G. R. (1983). Strain-rate effects in metals at large shear strains. Materials behavior under high stress and ultra-high loading rates (J. Mescall and V. Weiss, Eds.), Plenum Press, New York, NY, pp. 61–79.
3. Lindholm, U. S., Johnson, J. R., and Hoegfeldt, J. M. (2000). High speed hydraulic torsional machines. High strain rate torsion testing. ASM handbook, mechanical testing and evaluation (vol. 8) (H. Kuhn and D. Medlin, Eds.), ASM International, Materials Park, OH, p. 448.
4. Culver, R. S. (1972). Torsional-impact apparatus. Experimental Mechanics, 12, 398–405.
5. Baker, W. E., and Yew, C. H. (1966). Strain rate effects in the propagation of torsional plastic waves. Journal of Applied Mechanics, 33(6), 917–923.
6. Rogers, H. C. (1983). Adiabatic shearing-general nature and material aspects. Material behavior under high stress and ultra high loadings rates. 29th Sagamore Army Materials conference (J. Mescall and V. Weiss, Eds.), Plenum Press, New York, Department of Materials Engineering, Drexel University, pp. 101–118.
7. Dormeval, R. (1987). The adiabatic shear phenomena. Materials at high strain rates (T. Z. Blazynski, Ed.), Elsevier, New York, NY, pp. 47–70.
8. Bai, Y., and Dodd, B. (1992). Adiabatic shear localization: occurrence, theories and applications, Pergamon Press, Oxford.
9. Wright, T. W. (2002). The physics and mathematics of ASB, Cambridge University Press, New York, NY.
10. Duffy, J., and Chi, Y. C. (1992). On the measurement of local strain and temperature during the formation of ASB. Materials Science and Engineering, A157, 195–210.
11. Giovanola, J. H. (1988). Adiabatic shear banding under pure shear loading. Part I: Direct observation of strain localization and energy dissipation measurement. Mechanics of Materials, 7, 59–71.
12. Campbell, J. D., Eleiche, A. M., and Tsao, M. C. C. (1977). Strength of metals and alloys at high strains and strain rates. Battelle colloquium. on fundamental aspects of structural alloy design (R. I. Jaffe and B. A. Wilcox, Eds.), Plenum Press, New York, NY, pp. 545–563.
13. Duffy, J., Campbell, J. D., and Hawley, R. H. (1971). On the use of a torsional split Hopkinson bar to study rate effects in 1100-0 Aluminum. Journal of Applied Mechanics, 38, 83–91.
14. Hartley, K. A., Duffy, J., and Hawley, R. H. (1985). The torsional Kolsky (split Hopkinson) bar. ASM metals handbook, mechanical testing (vol. 8, 9th ed.), ASM International, (J. Newby, Ed.), Materials Park, OH.
15. Marchand, A., and Duffy, J. (1988). An experimental study of the formation process of ASB in a structural steel. Journal of Mechanics, Physics and Solids, 36(3), 251–283.

16. Hwang, B., Lee, S., Kim, Y. C., Kim, N. J., and Shin, D. H. (2006). Microstructural development of ASB in ultra-fine-grained low-carbon steels fabricated by equal channel angular pressing. Materials Science and Engineering, A441, 308–320.
17. Mgbokwere, C. O., Nutt, S. R., and Duffy, J. (1994). Shear band formation in 4340 steel: a TEM study. Mechanics of Materials, 17, 97–110.
18. Lee, D. G., Kim, Y. G., Nam, D. H., Hur, S. M., and Lee, S. (2005). Dynamic deformation behavior and ballistic performance of Ti–6Al–4V alloy containing fine α_2 (Ti_3Al) precipitates. Materials Science and Engineering, A391, 221–234.
19. Cho, K., Chi, Y. C., and Duffy, J. (1990). Microscopic observations of ASB in three different steels. Metallurgical Transactions, 21A, 1161–1175.
20. Bai, Y., Xue, Q., Xu, Y., and Shen, L. (1994). Characteristics and microstructure in the evolution of shear localization in Ti–6Al–4V. Mechanics of Materials, 17, 155–164.
21. Bassim, M. N., and Panic, N. (1999). High strain rate effects on the strain of alloy steels. Journal of Materials Processing Technology, 92–93, 481–485.
22. Hwang, B., Kim, Y. G., Lee, S., Hwang, D. Y., and Shin, D. H. (2007). Dynamic torsional deformation behavior of ultra-fine-grained dual-phase steel fabricated by equal channel angular pressing. Metallurgical and Materials Transactions, 38A, 3007–3013.
23. Lee, W. S., Lin, C. F., and Huang, S. Z. (2006). Effect of temperature and strain rate on the shear properties of Ti–6Al–4V. Journal of Mechanical Engineering Science, Part C, 22, 127–136.
24. Lee, W. S., Lin, C. F., and Lin, Y. C. (2006). Dynamic deformation and failure behaviour of Al–Sc-alloy under shear loading. Journal de Physics IV, 135, 965–970.
25. Lartourte, F., Feinberg, Z., Mori, L. F., Olson, G. B., and Espinosa, H. D. (2010). Shear and tensile plastic behavior of austenitic steel TRIP-120 compared with martensitic steel HSLA-100. International Journal of Fracture, 162, 187–204.
26. Hartmann, K. H., Kunze, H. D., and Meyer, L. W. (1981). Metallurgical effects on impact loaded materials. Shock waves and high strain rate phenomena in metals, concepts and applications (M. A. Meyers, and L. E. Murr, Eds.), Plenum Press, New York, NY, pp. 325–337.
27. Meyer, L. W., Staskewitsch, E., and Burblies, A. (1994). Adiabatic shear failure under biaxial dynamic compression/shear loading. Mechanics of Materials, 17, 203–214.
28. Meyers, M. A., LaSalvia, J. C., Nesterenko, V. F., Chen, Y. J., and Kad, B. K. (1997). Dynamic recrystallization in high strain rate deformation. Proceedings of ReX '96: the third international conference on recrystallization and related phenomena, pp. 279–286.
29. Xu, Y., Zhag, J., Bai, Y., and Meyers, M. A. (2008). Shear localization in dynamic deformation: microstructural evolution. Metallurgical and Materials Transactions, 39A, 811–843.
30. Meyers, M. A., Subhash, G., Kad, B. K., and Prasad, L. (1994). Evolution of microstructure and shear-band formation on alpha-hcp titanium. Mechanics of Materials, 17, 175–193.
31. Meyers, M. A., Chen, Y. J., Marquis, F. D. S., and Kim, D. S. (1995). High-strain, high-strain-rate behavior of tantalum. Metallurgical and Materials Transactions, 26A, 2493–2501.
32. Beatty, J. H., Meyer, L. W., Meyers, M. A., and Nemat-Nasser, S. (1992). Formation of controlled ASB in AISI 4340 high strength steel. Shock-wave and high strain rate phenomena in materials, (M. A. Meyers, L. E. Murr and K. P. Staudhammer, Eds.), NTIS: National Technical Information Service, pp. 645–656.
33. Meyer, L. W., and Krüger, L. (2000). Shear testing with hat specimen. ASM handbook, mechanical testing and evaluation (vol. 8) (H. Kuhn and D. Medlin, Eds.), ASM International, Materials Park, OH, pp. 451–452.

34. Minnaar, K., and Zhou, M. (1998). An analysis of the dynamic shear failure resistance of structural metals. Journal of Mechanics, Physics and Solids, 40(10), 2155–2170.
35. Chung, D. T., Moon, S. K., and Yoo, Y. H. (1994). Numerical and experimental study of the formation of adiabatic shear band. Journal de Physique III, 4, 547–552.
36. Pursche, F. (2010). Specification of failure behavior of materials under compression shear loading (origin title: Spezifizierung des Versagensverhaltens von Werkstoffen bei Druck-Scher-Belastung). Doctoral thesis, Chemnitz University of Technology, Chair: 'Werkstoffe des Maschinenbaus', Director: Prof. L. W. Meyer.
37. Pursche, F., and Meyer, L. W. (2011). Correlation between dynamic material behavior and adiabatic shear phenomenon for quenched and tempered steels. Engineering Transactions, 59(2), 67–84.
38. Xue, Q. (2001). Spatial evolution of adiabatic shear localization in stainless steel, titanium and Ti–6Al–4V alloy. Doctoral thesis, University of California, San Diego, CA.
39. Krüger, L. (2001). Investigations to strength, deformation and failure behavior of titanium alloy Ti-6-22-22S in dependence of temperature, strain rate and stress concentration (origin title: Untersuchungen zum Festigkeits-, Verformungs- und Versagensverhalten der Legierung Ti-6-22-22S in Abhängigkeit von der Temperatur, der Dehngeschwindigkeit und dem Spannungszustand). Doctoral thesis, Chemnitz University of Technology, Chair: 'Werkstoffe des Maschinenbaus', Director: Prof. L. W. Meyer.
40. Meyer, L. W., Krüger, L., and Halle, T. (2001). Strength and failure characterisation and behaviour under impact loading. Impact engineering and application, Proceedings of the 4th international symposium on impact engineering, vol. 1, 16–18 July, 2001, Kumamoto, Japan, (A. Chiba, S. Tanimura, and K. Hokamoto, Eds.), Elsevier, pp. 91–98.
41. Couque, H. (2003). A hydrodynamic hat specimen to investigate pressure and strain rate dependence on adiabatic shear band formation. Journal de Physique IV, 110, 423–428.
42. Teng, X., Wierzbicki, T., and Couque, H. (2007). On the transition from adiabatic shear banding to fracture. Mechanics of Materials, 39, 107–125.
43. Meyers, M. A., Xu, Y. B., Xue, Q., Pérez-Prado, M. T., and McNelley, T. R. (2003). Microstructural evolution in adiabatic shear localization in stainless steel. Acta Materialia, 51, 1307–1325.
44. Clos, R., Schreppel, U., and Veit, P. (2003). Temperature, microstructure and mechanical response during shear-band formation in different metallic materials. Journal de Physique, 110(4), 111–116.
45. Xue, Q., and Gray III, G. T. (2006). Development of adiabatic shear bands in annealed 316L stainless steel. Part I: Correlation between evolving microstructure and mechanical behavior. Metallurgical and Materials Transactions, 37A, 2435–2446.
46. Xue, Q., Cerreta, E. K., and Gray III, G. T. (2007). Microstructural characteristics of post shear localization in cold-rolled 316L stainless steel. Acta Materialia, 55, 691–704.
47. Xue, Q., Cerreta, E. K., and Gray III, G. T. (2006). Influence of explosive-driven shock pre-straining on the microstructural evolution and shear localization of 304 and 316L stainless steel. Shock compression of condensed matter 2005, Proceedings of the Conference of the American Physical Society Topical Group on Shock Compression of Condensed Matter (M. D. Furnish, M. Elert, T. P. Russell, and C. T. White, Eds.), American Institute of Physics, Baltimore, MD (USA), pp. 783–786.
48. Lins, J. F. C., Sandim, H. R. Z., Kestenbach, H. J., Raabe, D., and Vecchio, K. S. (2007). A microstructural investigation of adiabatic shear bands in an interstitial free steel. Materials Science and Engineering, A457, 205–218.

49. Lee, W. S., Liu, C. Y., and Chen, T. H. (2008). Adiabatic shearing behavior of different steels under extreme high shear loading. Journal of Nuclear Materials, 374, 313–319.
50. Peirs, J., Verleysen, P., Degrieck, J., and Coghe, F. (2010). The use of hat-shaped specimens to study the high strain rate shear behavior of Ti–6Al–4V. International Journal of Impact Engineering, 37, 703–714.
51. Peirs, J., Verleysen, P., and Degrieck, J. (2010). High-strain-rate shear testing applied to Ti–6Al–4V. Advanced Materials Research, 89–91, 437–442.
52. Lee, W. S., Chen, T. H., Lin, C. F., and Lu, G. T. (2010). Adiabatic shearing localisation in high strain rate deformation of Al–Sc alloy. Materials Transactions, 51(7), 1216–1221.
53. Chen, Y. J., Meyers, M. A., and Nesterenko, V. F. (1999). Spontaneous and forced shear localization in high-strain-rate deformation of tantalum. Material Science and Engineering, A268, 70–81.
54. Kad, B. K., Gebert, M., Kassner, M. E., and Meyers, M. A. (2006). Microstructural evolution and grain refinement in HCP-Zr shear bands. Journal de Physique IV, 134, 1137–1144.
55. Andrade, U., Meyers, M. A., Vecchio, K. S., and Chokshi, A. H. (1994). Dynamic recrystallization in high-strain, high-strain-rate plastic deformation of copper. Acta Metallurgica et Materialia, 42(9), 3183–3195.
56. Meyer, L. W., and Krüger, L. (2000). Drop-weight compression shear testing. ASM handbook, mechanical testing and evaluation (vol. 8) (H. Kuhn and D. Medlin, Eds.), ASM International, Materials Park, OH, pp. 452–454.
57. Kolsky, H. (1949). An investigation of the mechanical properties of materials at very high rates of loading, Imperial Chemical Industries Limited, Butterwick Research Laboratories, Welwyn, Herts, pp. 676–700.
58. Belk, J. A., and Watson, C. H. (1994). The dynamic compression test applied to tungsten alloy. Materials and Manufacturing Processes, 9(6), 1155–1161.
59. Sabih, A., Elwazri, A. M., Nemes, J. A., and Yue, S. (2006). A workability criterion for the transformed ASB phenomena during cold heading of 1038 steel. Journal of Failure and Prevention, 6, 97–105.
60. Sabih, A., and Nemes, J. A. (2009). Internal ductile failure mechanism in steel cold heading process. Journal of Materials Processing Technology, 209, 4292–4311.
61. Ansart, J. P., and Dormeval, R. (1988). Adiabatic shearing in martensitic steels. Impact loading and dynamic behavior of materials (vol. 2) (C. Y. Chiem, H. D. Kunze, and L. W. Meyer, Eds.), DGM Informationsgesellschaft, Oberursel, pp. 775–782.
62. Meyer, L. W. (1984). Dynamic behaviour of thermomechanically treated ultra high strength steel under tensile and compressive loading, ASME, HERF'84, New York, NY, pp. 245–252.
63. Odeshi, A. G., Al-Ameeri, S., and Bassim, M. N. (2005). Effect of high strain rate on plastic deformation of a low alloy steel subjected to ballistic impact. Journal of Materials Processing Technology, 162–163, 385–391.
64. Xu, Y. B., Zhong, W. L., Chen, Y. J., Shen, L. T., Liu, Q., Bai, Y. L., and Meyers, M. A. (2001). Shear localization and recrystallization in dynamic deformation of 8090 Al–Li alloy. Materials Science and Engineering, A 299, 287–295.
65. Nakkalil, R., Hornaday, J. R., and Bassim, M. N. (1991). Characterization of the compression properties of rail steels at high temperature and strain rates. Materials Science and Engineering, 141(2), 247–260.
66. Meunier, Y., Sangoy, L., and Pont, G. (1988). Metallurgical aspects of adiabatic shear phenomena in armor steels with perforation. Impact loading and dynamic behavior of

materials (vol. 2) (C. Y. Chiem, H. D. Kunze, and L. W. Meyer, Eds.), DGM Informationsgesellschaft, Oberursel, pp. 711–717.
67. Hanina, E., Rittel, D., and Rosenberg, Z. (2007). Pressure sensitivity of adiabatic shear banding in metals. Applied Physics Letters, 90, 021915-1–021915-4.
68. Lee, W. S., and Lin, C. F. (1998). Plastic deformation and fracture behaviour of Ti–6Al–4V alloy loaded with high strain rate under various temperatures. Materials Science and Engineering, A241, 48–59.
69. Wei, Z., Yu, J., Li, J., Li, Y., and Hu, S. (2001). Influence of stress condition on adiabatic shear localization of tungsten heavy alloys. International Journal of Impact Engineering, 26, 843–852.
70. Jinxu, L., Shukui, L., Xiaoqing, Z., Zhaohui, Z., Haiyun, Z., and Yingchun, W. (2008). Adiabatic shear banding in a tungsten heavy alloy processed by hot-hydrostatic extrusion and hot torsion. Scripta Materialia, 59, 1271–1274.
71. Meyer, L. W., Hockauf, M., Hohenwarter, A., and Schneider, S. (2008). Ultimate strength of a tungsten heavy alloy after severe plastic deformation at quasi-static and dynamic loading. Materials Science Forum, 584–586, 405–410.
72. Rittel, D., Wang, Z. G., and Merzer, M. (2006). Adiabatic shear failure and dynamic stored energy of cold work. Physical Review Letters, 96, 075502-1–075502-4.
73. Dai, L. H., Liu, L. F., and Bai, Y. L. (2004). Effect of particle size on the formation of adiabatic shear band in particle reinforced metal matrix composites. Materials Letters, 58, 1773–1776.
74. Odeshi, A. G., Owolabi, G. M., and Bassim, M. N. (2007). Effects of particulate reinforcement and strain-rates on deformation and fractures behavior of Aluminum 6061 T6 under high velocity impact. Materialwissenschaft und Werkstofftechnik, 38(2), 66–69.
75. Wu, G. H., Zhu, D. Z., Chen, G. Q., Jiang, L. T., and Zhang, Q. (2008). Adiabatic shear failure of high reinforcement content aluminum matrix composites. Journal of Materials Science, 43, 4483–4486.
76. Lee, W. S., Liu, C. Y., and Sun, T. N. (2005). Dynamic impact response and microstructural evolution of inconel 690 superalloy at elevated temperatures. International Journal of Impact Engineering, 32, 210–223.
77. Dormeval, R. (1988). The adiabatic shear phenomena. Impact loading and dynamic behavior of materials (vol. 1) (C. Y. Chiem, H. D. Kunze, and L. W. Meyer, Eds.), DGM Informationsgesellschaft, Oberursel, pp. 43–56.
78. Wang, L. L., Bao, H. S., and Lu, W. X. (1988). The dependence of adiabatic shear banding in strain-rate, strain and temperature. Journal de Physique, C3(3), 207–214, tome 49.
79. Nguyen, C. H. (1997). Analysis of some thermal instability criteria in the adiabatic shear banding process. Lataillade, J. L. (Ed.) Journal de Physique III Volume 07, Number C3, August 1997 EURODYMAT 1997 – 5th International Conference on Mechanical and Physical Behaviour of Materials under Dynamic Loading, Toledo, Spain, 22–26. September 1997, pp. 849–854.
80. Odeshi, A. G., Bassim, M. N., and Al-Ameeri, S. (2006). Effect of heat treatment on adiabatic shear bands in a high-strength low alloy steel. Materials Science and Engineering, A419, 69–75.
81. Odeshi, A. G., Al-Ameeri, S., Mirfakhraei, S., Yazdani, F., and Bassim, M. N. (2006). Deformation and failure mechanism in AISI 4340 steel under ballistic impact. Theoretical and Applied Fracture Mechanics, 45, 18–24.
82. Odeshi, A. G., and Bassim, M. N. (2008). Evolution of adiabatic shear bands in a dual-phase steel at very high strain rates. Materials Science and Engineering, A488, 235–240.

83. Meyer, L. W. (1991). Adiabatic shear failure at biaxial dynamic compression/shear loading. Euromech 282.
84. Meyer, L. W., and Staskewitsch, E. (1992). Adiabatic shear failure of the titanium alloy Ti−6Al−4V under biaxial dynamic compression/shear loading. Titanium 1992, science and technology (vol. 1993) (F. H. Froes and I. Caplan, Eds.), The Minerals, Metals & Materials Society, Warrendale, PA (USA), pp. 1939−1946.
85. Meyer, L. W., and Halle, T. (2004). Comparison of different methods to measure shear and combined shear under high speed loading. Fifth international symposium of impact engineering (ISIE), University of Cambridge, Cambridge, UK, 11−15 July, 2004.
86. Meyer, L. W., Krüger, L., and Abdel-Malek, S. (1999). Adiabatic shear processes (origin title: Adiabatische Schervorgänge). Materialprüfung, 41, 31−35.
87. Blümke, R. (2001). Influence of microstructure on the chip formation at high speed cutting (origin title: Gefügeeinfluss auf die Spanbildung beim Hochgeschwindigkeitsfräsen). Doctoral thesis, Department of Material and Geological Science, Technical University of Darmstadt.
88. Halle, T. (2005). Correlation between high speed cutting and mechanical material behaviour at high strain of rates (origin title: Zusammenhänge zwischen Spanvorgängen und dem mechanischen Werkstoffverhalten bei hohen Dehnungsgeschwindigkeiten). Doctoral thesis, Book series material behaviour vol. 1, Chemnitz University of Technology, Chair: 'Werkstoffe des Maschinenbaus', Director: Prof. L. W. Meyer.
89. Meyer, L. W., Krüger, L., Gooch, W., and Burkins, M. (1997). Analysis of shear bands effects in titanium relative to high strain rate laboratory/ballistic impact test. Journal de Physique IV, 7, 1997, Colloque C3, Supplement au Journal de Physique III, DYMAT.
90. Wei, Z., Yu, J., Hu, S., and Li, Y. (2000). Influence of microstructure on adiabatic shear localization of pre-twisted tungsten heavy alloy. International Journal of Impact Engineering, 24, 747−758.
91. Shockey, D. A., and Erlich, D. C. (1981). Metallurgical influences on shear band activity. Shock waves and high strain rate phenomena in metals, concepts and applications (M. A. Meyers and L. E. Murr, Eds.), Plenum Press, New York, NY, pp. 249−261.
92. Wittman, C. L., Meyers, M. A., and Pak, H. R. (1990). Observations of an ASB in AISI 4340 steel by high-voltage TEM. Metallurgical Transactions, 21A, 707−716.
93. Meyers, M. A. (1994). Dynamic behavior of materials, Wiley-Interscience, John Wiley & Sons, New York, NY.
94. Grady, D.E., and Hightower, M.M. (1992). Natural fragmentation of exploding cylinders. Shock-wave and high-strain-rate phenomena in materials, Published/Created New York: Marcel Dekker, 1992. (M. A. Meyers, L. E. Murr and K. P. Staudhammer, Eds.) 713−721.
95. Staker, M. R. (1980). On adiabatic shear band determination by surface observations. Scripta Metallurgica, 14, 677−680.
96. Staker, M. R. (1981). The relation between adiabatic shear instability strain and material properties. Acta Metallurgica, 29, 683−689.
97. Pak, H. R., Wittman, C. L., and Meyers, M. A. (1986). High voltage TEM of shear bands in titanium and 4340 steel. Mechanical Engineering, 52, 749−760.
98. Bayi, H., Qingdong, D., Changsheng, H., Desheng, W., and Haibo, H. (1994). Microanalysis of adiabatic shear fracture in explosive-filled cylinders. American Institute of Physics Conference Proceedings, 309, 1229−1232.
99. Sunwoo, A. J., Becker, R., Goto, D. M., Orzechowski, T. J., Springer, H. K., Syn, C. K., and Zhou, J. (2006). Adiabatic shear band formation in explosively driven Fe−Ni−Co alloy cylinders. Scripta Materialia, 55, 247−250.

100. Goto, D. M., Becker, R., Orzechowski, T. J., Springer, H. K., Sunwoo, A. J., and Syn, C. K. (2008). Investigation of the fracture and fragmentation of explosively driven rings and cylinders. International Journal of Impact Engineering, 35, 1547–1556.
101. Nesterenko, V. F., Bondar, M. P., and Ershov, I. V. (1994). Instability of plastic flow at dynamic pore collapse. High-pressure science and technology 1993 (S. C. Schmidt, Ed.), AIP, New York, NY, p. 1172.
102. Nesterenko, V. F. (2000). Thick-walled cylinder testing. High strain rate shear testing. ASM handbook, mechanical testing and evaluation (vol. 8) (H. Kuhn and D. Medlin, Eds.), ASM International, Materials Park, OH, pp. 455–457.
103. Nesterenko, V. F., Meyers, M. A., and Wright, T. W. (1998). Self-organization in the initiation of adiabatic shear bands. Acta Materialia, 47(1), 327–340.
104. Gu, Y. B., and Nesterenko, V. F. (2007). Dynamic behavior of HIPed Ti–6Al–4V. International Journal of Impact Engineering, 34, 771–783.
105. Bondar, M. P., Pervukhina, O. L., Nesterenko, V. F., and Luk'yanov, Y. L. (1998). Development of the titanium structure in explosive collapse of thick-walled cylinders. Combustion, Explosion, and Shock Waves, 34(5), 590–597.
106. Xue, Q., Meyers, M. A., and Nesterenko, V. F. (2002). Self-organization of shear bands in titanium and Ti–6Al–4V. Acta Materialia, 50, 575–596.
107. Xue, Q., Nesterenko, V. F., and Meyers, M. A. (2000). Self-organization of adiabatic shear bands in Ti, Ti–6Al–4V and stainless steel. Shock compression of condensed matter. AIP conference proceedings, pp. 431–434.
108. Dowling, A. R., Harding, J., and Campbell, J. D. (1970). The dynamic punching of metals. Journal of Institute of Metals, 98, 215–224.
109. Moss, G. (1981). Shear strains, strain rates and temperature changes in adiabatic shear bands. Shock waves and high strain rate phenomena in metals, concepts and applications (M. A. Meyers and L. E. Murr, Eds.), Plenum Press, New York, NY, pp. 299–312.
110. Zurek, A. K. (1994). The study of adiabatic shear band instability in a pearlitic 4340 steel using a dynamic punch test. Metallurgical and Materials Transactions, 25A, 2483–2489.
111. Roessig, K. M., and Mason, J. J. (1999). Adiabatic shear localization in the dynamic punch test. Part I: Experimental investigation. International Journal of Plasticity, 15, 241–262.
112. Ryttberg, K., Knutson-Wedel, M., Dahlmann, P., and Nyborg, L. (2009). Microstructural evolution during fracture induced by high strain rate deformation of 100Cr6 steel. Journal of Materials Processing Technology, 209(7), 3325–3334.
113. Chou, P. C., Hashemi, J., Chou, A., and Rogers, H. C. (1991). Experimentation and FE simulation of ASB in controlled penetration impact. International Journal of Impact Engineering, 11(3), 305–321.
114. Meyer, L. W., and Manwaring, S. (1986). Critical adiabatic shear strength of low alloyed steel under compressive load. Metallurgical applications of shock-wave and high-strain-rate phenomena, Marcel Dekker, INC., New York and Basel, pp. 657–674.
115. Irick, E., Heinol, C., Clayton, T., Hashemi, J. F., Cardenas-Garcia, J. F., and Sadhneni, R. (1995). Numerical and experimental investigation of adiabatic shear bands in metals under low-velocity impact conditions. Journal of Materials Engineering and Performance, 4(6), 709–716.
116. Mason, C., and Worswick, M. J. (2001). Adiabatic shear in annealed and shock-hardened iron and in quenched and tempered 4340 steel. International Journal of Fracture, 111, 29–51.

117. Campbell, J. D., and Ferguson, W. G. (1970). The temperature and strain rate dependence of shear strength of mild steel. Philosophical Magazine, 21, 63–82.
118. Harding, J., and Huddart, J. (1979). The use of the double-notch shear test in determining the mechanical properties of uranium at very high rates of strain. Mechanical properties of materials at high rates strains (J. Harding, Ed.), Institute of physics conference series No. 47, pp. 49–61.
119. Klepaczko, J. R. (1994). An experimental technique for shear testing at high and very high strain rates. The case of a mild steel. International Journal of Impact Engineering, 15(1), 25–39.
120. Ferguson, W. G., and Campbell, J. D. (2000). Double-notch shear testing, double-notch shear testing and punch loading. ASM handbook, mechanical testing and evaluation (vol. 8) (H. Kuhn and D. Medlin, Eds.), ASM International, Materials Park, OH, pp. 450–451.
121. Olson, G. B., Mescall, J. F., and Azrin, M. (1981). Adiabatic deformation and strain localization. Shock waves and high strain rate phenomena in metals, concepts and applications (M. A. Meyers and L. E. Murr, Eds.), Plenum Press, New York, NY, pp. 221–247.
122. Kalthoff, J. F. (2000). Modes of dynamic shear failure in solids. International Journal of Fracture, 101, 1–31.
123. Rosakis, A. J., Ravichandran, G., and Zhou, M. (1995). Real-time experimental observations of two-dimensional dynamic shear band growth. The American Society of Mechanical Engineering, Plastic and fracture instabilities in materials, AMD-vol. 200, MD-vol. 57, pp. 95–100.
124. Guduru, P. R., Rosakis, A. J., and Ravichandran, G. (2001). Dynamic shear bands: an investigation using high speed optical and infrared diagnostics. Mechanics of Materials, 33, 371–402.
125. Rittel, D., Wang, Z. G., and Dorogoy, A. (2008). Geometrical imperfection and adiabatic shear banding. International Journal of Impact Engineering, 35, 1280–1292.
126. Rittel, D., Wang, Z. G., and Merzer, M. (2006). Some experiments on adiabatic shear failure. Journal de Physique IV, 134, 835–838.
127. Wang, Z. G., and Rittel, G. (2008). Thermomechanical aspects of adiabatic shear failure of AM50 and Ti–6Al–4V alloys. Mechanics of Materials, 40(8), 629–635.
128. Asay, J. R., and Kerley, G. I. (1987). The response of materials to dynamic loading. International Journal of Impact Engineering, 5(1–4), 69–99.
129. Barker, L. M., Chhabildas, L. C., Trucano, T. G., and Asay, J. R. (1990). High gas pressure acceleration of flier plates – experimental techniques. International Journal of Impact Engineering, 10(1–4), 67–80.
130. Mescheryakov, Y. I., and Atroshenko, S. A. (1995). Dynamic recrystallization in shear bands. Metallurgical and materials applications of shock-wave and high strain-rate-phenomena (L. E. Murr, K. P. Staudhammer, and M. A. Meyers, Eds.), Elsevier, Amsterdam, pp. 443–450.
131. Kanel, G. I., Razorenov, S. V., and Fortov, V. E. (2004). Shock-wave phenomena and the properties of condensed matter (L. Davison and Y. Horie, Eds.), Springer-Verlag, New York, NY. ISBN: 0-387-20572-1.
132. Klopp, R. W., Clifton, R. J., and Shawki, T. G. (1985). Pressure-shear impact and the dynamic viscoplastic response of metals. Mechanics of Materials, 4, 375–385.
133. Clifton, R. J., and Klopp, R. W. (2000). Pressure-shear plate impact testing. ASM handbook, mechanical testing and evaluation (vol. 8), ASM International, Materials Park, OH, pp. 230–239.

134. Klepaczko, J. R. (1998). Remarks on impact shearing. Journal of Mechanics, Physics and Solids, 46(10), 2139–2153.
135. Zhang, B., Shen, W., Liu, Y., Tang, X., and Wang, Y. (1997). Microstructures of surface white layer and internal white adiabatic shear band. Wear, 211, 164–168.
136. Zhang, B., Shen, W., Liu, Y., and Zhang, R. (1998). Adiabatic shear bands in impact wear. Journal of Materials Science Letters, 17, 765–767.
137. Zhang, B., Liu, Y., Shen, W., Wang, Y., Tuang, X., and Wang, X. (1996). A study on the behavior of adiabatic shear bands in impact wear. Wear, 198, 287–292.
138. Recht, R. F. (1964). Catastrophic thermoplastic shear. Journal of Applied Mechanics, 31(2), 189–193.
139. Molinari, A., Musquar, C., and Sutter, G. (2002). Adiabatic shear banding in high speed machining of Ti–6Al–4V: experiments and modeling. International Journal of Plasticity, 18, 443–459.
140. Akcan, S., Shah, S., Moylan, S. P., Chhabra, P. N., Chandrasekar, S., and Yang, H. T. Y. (2002). Formation of white layers in steel by machining and their characteristics. Metallurgical and Materials Transactions, 33A, 1245–1254.
141. Duan, C., Li, H., Yu, H., and Wang, M. (2010). Study on cutting force under serrated chip formation and adiabatic shear instability during high speed turning of AISI 1045 hardened steel. International conference on mechanic automation and control engineering, MACE 2010, pp. 3290–3293.
142. Craig, J. V., and Stock, T. A. C. (1970). Microstructural damage adjacent to bullet holes in 70-30 brass. Journal of Australian Institute of Metals, 15(1), 1–5.
143. Manganello, S. J., and Abbott, K. H. (1972). Metallurgical factors affecting the ballistic behavior of steel targets. Journal of Materials, 7, 231–239.
144. Mescall, J., and Papirno, R. (1974). Spallation in cylinder-plate impact. Experimental Mechanics, 14(7), 257–266.
145. Irwin, G. J. (1972). Metallographic interpretation of impacted ogive penetrators. DREV R-652/72, Defense Research Establishment, Valcartier, Quebec, Canada.
146. Krejci, J., Brezina, J., and Buchar, J. (1992). On the substructure of adiabatic shear bands. Scripta Metallurgica et Materialia, 27, 611–616.
147. Derep, J. L. (1987). Microstructure transformation induced by adiabatic shearing in armour steel. Acta Metallurgica, 35(6), 1245–1249.
148. Meyers, M. A., and Wittman, C. L. (1990). Effect of metallurgical parameters on shear band formation in low-carbon steels. Metallurgical Transactions, 21, 3153–3164.
149. Borvik, T., Leinum, J. R., Solber, J. K., Hopperstad, O. S., and Langseth, M. (2001). Observations on shear plug formation in Weldox460E steel plates impacted by blunt-nosed projectiles. International Journal of Impact Engineering, 25, 553–572.
150. Couque, H., Nicolas, G., and Altmayer, C. (2007). Relation between shear banding and penetration characteristics of conventional tungsten alloy. International Journal of Impact Engineering, 34, 412–423.
151. Zhu, D., Chen, G., Wu, G., Kang, P., and Ding, W. (2009). Hypervelocity impact damage to Ti–6Al–4V meshes reinforced Al-6Mg alloy matrix composites. Materials Science and Engineering A, 500, 43–46.
152. Jha, A. K., Diwakar, V., and Sreekumar, K. (2010). Metallurgical analysis of burst tested aluminium alloy AA 2219 miniature assembly. Indian Journal of Engineering and Materials Sciences, 17, 67–71.
153. Dai, L. H., and Bai, Y. L. (2008). Basic mechanical behaviors and mechanics of shear banding in BMG's. International Journal of Impact Engineering, 36, 704–716.

154. Martin, M., Kecskes, L., and Thadhani, N. N. (2008). High-strain-rate dynamic mechanical behavior of a bulk metallic glass composite. Journal of Materials Research, 23(4), 998–1008.
155. Martin, M., Meyer, L. W., Kecskes, L., and Thadhani, N. N. (2009). Uniaxial and biaxial compressive response of a bulk metallic glass composite over a range of strain rates and temperatures. Journal of Materials Research, 24(1), 66–78.
156. Zhang, Q., Zhang, W., Xie, G., and Inoue, A. (2007). Unusual plasticity of the particulate-reinforced Cu–Zr-based bulk metallic glass composites. Materials Transactions, 48(9), 2542–2544.
157. Hsieh, P. J., Yang, L. C., Su, H. C., Lu, C. C., and Jang, J. S. C. (2010). Improvement of mechanical properties in MgCuYNdAg bulk metallic glasses with adding Mo particles. Journal of Alloys and Compounds, 504S, 98–101.
158. Zhao, J. X., Qu, R. T., Wu, F. F., Li, S. X., and Zhang, Z. F. (2010). Enhanced plastic deformation in a metallic glass induced by notches. Philosophical Magazine Letters, 90(12), 875–882.
159. Zhao, J. X., Wu, F. F., Qu, R. T., Li, S. X., and Zhang, Z. F. (2010). Plastic deformability of metallic glass by artificial macroscopic notches. Acta Materialia, 58, 5420–5432.
160. Cao, Q. P., Liu, J. W., Yang, K. J., Xu, F., Yao, Z. Q., Minkow, A., Fecht, H. J., Ivanisenko, J., Chen, L. Y., Wang, X. D., Qu, S. X., and Jiang, J. Z. (2010). Effect of pre-existing shear bands on the tensile mechanical properties of a bulk metallic glass. Acta Materialia, 58, 1276–1292.
161. Hofmann, D. C., Suh, J. Y., Wiest, A., Duan, G., Lind, M. L., Demetriou, M. D., and Johnson, W. L. (2008). Designing metallic glass matrix composites with high toughness and tensile ductility. Nature, 451(28), 1085–1090.
162. Wu, F. F., Zhang, Z. F., Shen, J., and Mao, S. X. (2008). Shear deformation and plasticity of metallic glass under multiaxial loading. Acta Materialia, 56, 894–904.
163. Wu, F. F., Zhang, Z. F., Shen, J., and Mao, S. X. (2008). Shear deformability capability of different metallic glasses. Journal of Materials Research, 23(10), 2662–2667.
164. Amijima, S., and Fujii, T. (1980). Compressive strength and fracture characteristics of fiber composites under impact loading. Economic Computation and Economic Cybernetics Studies and Research, 1, 399–413.
165. Lankford, J. (1997). Shear versus dilatational damage mechanisms in the compressive failure of fibre-reinforced composites. Composites A, 28, 215–222.
166. Rittel, D. (2000). Experimental investigation of transient thermoplastic effects in dynamic fracture. International Journal of Solids and Structures, 37, 2901–2913.
167. Bjerke, T., and Lambros, J. (2002). Heating during shearing and opening dominated dynamic fracture of polymers. Experimental Mechanics, 42(1), 107–114.
168. Totry, E., Gonzales, C., LLorca, J., and Molina-Aldareguia, J. M. (2009). Mechanisms of shear deformation in fiber-reinforced polymers: experiments and simulations. International Journal of Fracture, 158, 197–209.
169. Li, T., Tang, Z., and Cai, J. (2007). Micro-observation of shear wave attenuation mechanism in nylon-66. Materials Letters, 61, 1436–1438.
170. Li, T., and Tang, Z. (2008). Experimental observations and mechanism analysis of the shear failure near the impact interface for polypropylene under combined compression and shear loading, Advances in heterogeneous material mechanics 2008, 2nd International conference on heterogeneous material mechanics (ICHM 2008), pp. 511–514.
171. Herrmann, H. J. (1995). Some new results on fracture. Physica A, 221, 125–133.

172. Lankford, J. (1994). The effect of hydrostatic pressure and loading rate on compressive failure of fiber-reinforced ceramic-matrix composites. Composite Science and Technology, 51, 537–543.
173. Bassim, M. N., Odeshi, A. G., and Bolduc, M. (2008). High strain rate study of ceramics using Hopkinson bar system. International Journal of Modern Physics B, 22 (9–11), 1147–1152.
174. Unger, T., Török, J., Kertész, J., and Wolf, D. E. (2004). Shear band formation in granular media as a variational problem. Physical Review Letters, 92, 1–4.
175. Nesterenko, V. F., Meyers, M. A., and Chen, H. C. (1996). Shear localization in high-strain-rate deformation of granular alumina. Acta Materialia, 44(5), 2017–2026.
176. Shih, C. J., Nesterenko, V. F., and Meyers, M. A. (1997). Shear localization and comminution of granular and fragmented silicon carbide. Journal de Physique IV, 7, 577–582, Colloque C3, Supplément au Journal de Physique III d'août.
177. Desrues, J., Lanier, J., and Stutz, P. (1985). Localization of the deformation in tests on sand sample. Engineering Fracture Mechanics, 21(4), 909–921.
178. Alshibli, K. A., and Sture, S. (2000). Shear band formation in plane strain experiments of sand. Journal of Geotechnical and Geoenvironmental Engineering, 495–503.
179. Rechenmacher, A. L. (2006). Grain-scale processes governing shear band initiation and evolution in sands. Journal of the Mechanics and Physics of Solids, 54, 22–45.
180. Hasan, A., Alshibli, K., Asce, M. P. E., Jaeger, H., Rivers, M., and Eng, P. (2008). Visualization of shear band in sand using synchrotron micro-tomography. Geotechnical Special Publication, No. 179, March 9–12, 2008, American Society of Civil Engineers, New Orleans, LA, pp. 1028–1035.
181. Hall, S. A., Lenoir, N., Viggiani, G., Desrues, J., and Bésuelle, P. (2009). Strain localisation in sand under triaxial loading: characterization by X-ray micro tomography and 3D digital image correlation. International symposium on computational geomechanics ComGeo I, April 29–May 1, 2009, Juan-les-Pins, France.
182. Utter, B., and Behringer, R. P. (2004). Transients in sheared granular matter. European Physical Journal E, 14, 373–380.
183. Wolf, H., König, D., and Triantafyllidis, Th. (2006). The influence of the stress–strain behavior of non-cohesive soils on the geometry of shear band systems under extensional strain. Engineering Structures, 28, 1760–1773.
184. Zhou, Q., Shen, H., Helenbrook, B. T., and Zhang, H. W. (2009). Scaled dependence of direct shear tests. Chinese Science Bulletin, 54, 4337–4348.
185. Wu, P. K., Matsushima, K., and Tatsuoka, F. (2008). Effects of specimen size and some other factors on the strength and deformation of granular soil in direct shear tests. Geotechnical Testing Journal, 31(1), 2008, ASTM, Paper ID GTJ100773, pp. 45–64.
186. Ord, A., Vardoulakis, I., and Kajewski, R. (1991). Shear band formation in Gosford sandstone. International Journal of Rock Mechanics and Mining Sciences & Geomechanics Abstracts, 28(5), 397–409.
187. Labuz, J. F., and Dai, S. T. (2000). Residual strength and fracture energy from plane strain testing. Journal of Geotechnical and Geoenvironmental Engineering, 126(10), 882–889.
188. Riedel, J. J., and Labuz, J. F. (2007). Propagation of a shear band in sandstone. International Journal for Numerical and Analytical Methods in Geomechanics, 31, 1281–1299.
189. Besuelle, P., Desrues, J., and Raynaud, S. (2000). Experimental characterisation of the localization phenomenon inside a Vosges sandstone in a triaxial cell. International Journal of Rock Mechanics and Mining Sciences, 37, 1223–1237.

190. Schmocker, M., Bystricky, M., Kunze, K., Burlini, L., Stünitz, H., and Burg, J. -P. (2003). Granular flow and Riedel band formation in water-rich quartz aggregates experimentally deformed in torsion. Journal of Geophysical Research, 108(B5), 2242, pp. 2-1_2-16.
191. Tresca, H. (1879). Sur la fluideite et lécoulement des corps solides. Annales du Conservatoire des arts et metiers, 41, 153–160 Tome XI 1er fasc.
192. Zener, C., and Hollomon, J. H. (1944). Effect of strain rate upon plastic flow of steel. Journal of Applied Physics, 15, 22–32.
193. Wang, X. B. (2007). Adiabatic shear localization for steels based on Johnson-Cook-Model and second- and fourth-order gradient plasticity models. Journal of Iron and Steel Research International, 14(5), 56–61.
194. Pintat, T., Scholz, B., Kunze, H. D., and Vöhringer, O. (1988). The influence of carbon content and grain size on energy consumption during adiabatic shearing. Journal de Physique, C3(9), 237–244, tome 49.
195. Edwards, M. (2006). Properties of metals at high rates of strain. Materials Science and Technology, 22(4), 453–462.
196. Mescall, J. F. (1986). On the relative roles of strain-hardening and thermal softening in ASB. Metallurgical application of shock-wave and high-strain-rate phenomena, mechanical engineering (vol. 52) (L. E. Murr and K. P. Staudhammer, Eds.), Dekker, New York, pp. 689–704.
197. Barry, J., and Byrne, G. (2002). Chip formation, acoustic emission and surface white layers in hard machining. Annals of the CIRP, 51/1, 65–70.
198. Bedford, A. J., Wingrove, A. L., and Thompson, K. R. L. (1974). The phenomenon of adiabatic shear deformation. The Journal of the Australian Institute of Metals, 19(1), 61–73.
199. Meunier, Y., Rouy, R., and Moureaud, J. (1992). Survey of adiabatic shear phenomena in armor steels with perforation. Shock-wave and high strain rate phenomena in metals (M. A. Meyers, L. E. Murr, and K. P. Staudhammer, Eds.), Marcel Dekker, Inc., New York, pp. 637–644.
200. Yellup, J. M., and Woodward, R. L. (1980). Investigation into the prevention of adiabatic shear failure in high strength armour materials. Res Mechanica, 1, 41–57.
201. Herzig, N. (2008). Determination and description of the scaled flow, strain hardening and failure behaviour of special materials (origin title: Erfassung und Beschreibung des skalierten Fließ-, Verfestigungs- und Versagensverhalten ausgewählter metallischer Werkstoffe). Doctoral thesis, book series vol. 4, Material behaviour, Chemnitz University of Technology, Chair: 'Werkstoffe des Maschinenbaus', Director: Prof. L. W. Meyer.
202. Grady, D. E. (1994). Dissipation in adiabatic shear bands. Mechanics of Materials, 17, 289–293.
203. Medyanik, S. N., Liu, W. K., and Li, S. (2007). On criteria for dynamic adiabatic shear band propagation. Journal of the Mechanics and Physics of Solids, 55, 1439–1461.
204. Murr, L. E. (2007). Applications of extreme deformation. Materials Technology, 22(4), 193–199.
205. Klepaczko, J. R., and Rezaig, B. (1996). A numerical study of adiabatic shear banding in mild steel by dislocation mechanics based constitutive relations. Mechanics of Materials, 24, 125–139.
206. Feng, H., and Bassim, M. N. (1999). Finite element modeling of the formation of ASB in AISI 4340 steel. Material Science and Engineering, A266, 255–260.

3 Nanostructural and Microstructural Aspects of Shear Localization at High-Strain Rates for Materials

Yongbo Xu[1] *and Marc A. Meyers*[2,3]

[1]Shenyang National Laboratory for Materials Science, Institute of Metal Research, Chinese Academy of Sciences, Shenyang, People's Republic of China, [2]Department of Mechanical and Aerospace Engineering, [3]Department of Nanoengineering, University of California, San Diego, CA

Nomenclature[1]

α' martensite phase
σ_f fracture strength
n work-hardening exponent
γ strain
$\dot{\gamma}$ strain rate
γ_c average critical strain
δ half-width of the shear band
θ_* temperature rise in the band
$\vec{F}$ force acting on a particle (a mechanical stress induced diffusion)
∇V gradient of the potential energy field
$\vec{v}$ mean diffusion velocity
J flux along a grain boundary with thickness δ and depth L_2 (cross-sectional area $L_2\delta$)
D diffusion coefficient
C concentration of the mobile species (expressed in terms of mass per unit volume)
θ grain-boundary rotation angle
dV volume transferred
L instantaneous length of the segment of the grain boundary in the rotational recrystallization model
L_1 initial length of the grain boundary in the rotational recrystallization model
ρ material density
δD_{GB} grain-boundary diffusion coefficient
M grain-boundary mobility (the proportionality constant)
d grain size
D_2 average diameter of the recrystallized grains

[1] Other symbols used are defined in the text.

Adiabatic Shear Localization. DOI: http://dx.doi.org/10.1016/B978-0-08-097781-2.00003-4

γ_c critical strain stopping the subgrain rotation
η boundary energy density
$\boldsymbol{D}$ equivalent diameter of the cell structure (or subgrain)
$\boldsymbol{D}_B^V$ vacancy diffusion coefficient
$\boldsymbol{D}_{B_0}$ grain-boundary diffusion coefficient
Q_B activation energy of the grain-boundary diffusion
λ coefficient of the heat conduction
c heat capacity
$\dot{\varepsilon}$ minimum strain rate occurring in serrated flow
Q activation energy for diffusion
C frequency factor

3.1 Introduction

The occurrence of localized shear bands, as an inevitable outcome of thermoplastic instability induced under high strain and high-strain rate, has been a topic of great interest to scientists and engineers, and thus a great deal of investigation has been made experimentally and theoretically since its first recognition made by Zener and Hollomon [1]. Indeed, there was an earlier observation by Tresca in the nineteenth century, during deformation of platinum. Because of its scientific and technological importance, it poses an interesting challenge to researchers, and much attention has been received during the past several decades. The number of contributions on this topic is of the order of 1200, and there has been a monotonic increase from the 1980s to 2007, with a plateau being reached of ~200 papers per year. This is a substantial and global effort. This chapter, with less than 200 references, is not comprehensive and emphasizes our own work, with mention of other contributions. This should not be understood as an effort to ignore the important literature but a necessity of space limitation.

Mechanicians have focused their efforts on the macro-description of the constitutive models, developing the criteria required for the onset and propagation of thermal plastic instability [2–12]. Most of these approaches consist of a combination of mechanical and thermal instability analysis. On the other hand, however, materials scientists have focused on the microstructural and nanostructural aspects of localized shear deformation, concerning the effects of the materials and structures on the formation of shear localization. There are a number of extensive reviews [13–19] and special articles [20–34] published on the microstructural aspects of shear banding.

In this chapter, we will focus on the observation and analysis of the shear localization evolution and microstructures in metals and alloys, although in various conditions, other materials such as some polymers, granular media and metallic glasses are sensitive to adiabatic shearing. We focus here on imposed strain-rate regimes of 10^{-3} to $10^6\ s^{-1}$ at ambient temperature and on results of research carried out over the past 20 years. The emphasis is placed on the mechanical conditions required for the shear localization and microstructural evolution induced by dynamic torsional, impact compression, as well as the explosive collapse of thick-walled cylinders (TWCs).

3.2 Modern Techniques for Microstructural Characterization

The physical, chemical and, in particular, mechanical behaviours of metallic and non-metallic materials are sensitive to the microstructure and nanostructure. In other words, these properties of the materials are closely connected to their structures, including morphology, crystal structure and chemical composition at the micro scale. Scientists and engineers have been sparing no effort to manufacture and improve research tools for the characterization of microstructures of the materials accompanying the advance of science and technology. In the long history of the development of characterization instruments, perhaps the greatest impact is the emergence of the transmission electron microscope (TEM), which has become one of the primary tools for characterizing the nano-, micro- and meso-structures of materials because of its high-resolution capabilities. For example, the commercial TEM is now routinely available with point resolution better than 2 Å (0.2 nm), and the high-resolution electron microscope (HREM) is now capable of producing interpretable information approaching 1 Å provided the sample examined is sufficiently thin [35]. The development of TEM as a magnifying tool has taken place since the end of 1940s. Since the 1950s, bright-field, dark-field and diffraction contrast have been extensively applied to structural analysis, leading to an in-depth understanding of the structure of materials, especially direct observation of the crystal defects such as dislocations and stacking faults in thin crystals [36]. In addition to these, small-angle diffraction, high-resolution diffraction, as well as convergent beam electron diffraction and other techniques can also be chosen for various purposes. However, the narrow region of the shear bands (of the order of one to several micrometres) renders the microscopic examination difficult. In spite of its high resolution, TEM still appears to have limitations for investigating the microstructures of the bands because the perforation produced using the common electro-polishing method does not usually coincide with the band area. On the other hand, the choice of the field observed on the specimen is significantly restricted. Therefore, improper and even mistaken information introduced from the thin foil is inevitable sometimes, leading to confusion in interpreting images. Electron probe microanalysis (EPMA), developed based on the TEM in 1960s, has been applied rapidly to the micro-area of chemical composition determination approaching 1 μm. Scanning electron microscopy (SEM), based on the EPMA, appeared in 1965. It has revolutionized material-related fractographic studies where comparisons of fracture surfaces may be made with archival results to identify sources of failure and consequent failure mechanisms. A recent great headway is the coupling of SEM with electron backscatter diffraction (EBSD), particularly the use of EBSD in conjunction with a field-emission SEM (FESEM). EBSD is a relatively new technique and has been widely employed to characterize subgrain structures recently. The advantages of EBSD related to optical microscopy as the quantitative

Table 3.1 Electron-Based Techniques and Their Application in Microstructure/Nanostructure Characterization

Information	Imaging	Diffraction, Contrast Effects	Micro-Chemical	Analyzers
Trans.-E	TEM	HEED, Kikuchi	–	TEM
	STEM	HEED, Kikuchi	–	TEM, SEM
CELE	ECTEM	–	CELS	TEM
BSE	SEM	ECP, EBSD	BSC-Z	EPMA, SEM
SE	SEM	ECP	–	EPMA, SEM
AE	SEM	ECP	AEC-Z	EPMA, SEM
Auger-E	SEM	–	AES	AES, SEM
LEE	–	LED	–	LEDS
C-Ray	XRMS	Kossel	CXES	EPMA, SEM, TEM

Trans.-E, transmission electron; CELE, characteristic energy lost electron; BSE, backscatter electron; SE, secondary electron; AE, absorbed electron; Auger-E, Auger electron; LEE, low energy electron; C-Ray, characteristic X-ray.

metallographic method are higher spatial resolution, more accurate data and more complete structure characterization. In particular, EBSD in an FESEM can reduce the angular resolution to $\sim 0.01°$ and allows quantitative analysis of grain and sub-grain size as small as $\sim 0.2\ \mu m$. Humphreys [37] has compared and summarized the advantages and disadvantages of EBSD relative to other metallographic techniques. The advantages of EBSD over the TEM are as follows:

- The microstructural analysis can be made of the bulk materials rather than thin foils, avoiding the production of improper and even mistaken information arising from the observation.
- Information is accurately and easily obtained from a specific area of interest, which is important and powerful for the narrow size of shear bands.
- Rapid and automated acquisition and analysis of diffraction patterns, which is not routinely available in the TEM, enables many thousands of grain/subgrains to be characterized from a single map.

The disadvantages of EBSD relative to TEM might include the following:

- Poorer spatial resolution cannot image individual dislocations and defects.
- Longer time is required for data acquisition.

Techniques arising from the modern instruments for microstructure/nanostructure characterization are summarized in Table 3.1.

3.3 The Process of Shear Localization Evolution

The evolution of shear localization has stimulated a considerable number of theoretical and experimental investigations. Among the analytical approaches, the following are noteworthy:

- Criteria proposed by Zener and Hollomon [1], Recht [2] and Culver [3] are exclusively based on the effects of work hardening, thermal softening and strain-rate sensitivity.

- Perturbation analyses of Molinari and Clifton [38], Bai [4] and Grady and Kipp [39] represent a significant advance over the simpler first-generation analyses.
- Finite element formulation in various forms (e.g. Kuriyama and Meyers [40]).
- Spacing of shear bands, which was theoretically treated by Grady and Kipp [39], Wright and Ockendon [41] and Molinari [42].

Additional noteworthy studies are by Clifton et al. [43], Bai et al. [44], Shawki and Clifton [45] and Liao and Duffy [46]. A number of scaling laws have been obtained by Backman et al. [47], Bai [48] and Dodd et al. [49]. However, the analysis of localized plastic shear deformation is still limited by a lack of critical comparisons of theory and experiment. Some experimental attempts to measure the process of shear localization have been made. Costin et al. [50] are known to have made the first measurement of the temperature history of shear localization, and later Hartley et al. [51] have improved their infrared technique and measured the temperature distribution. They assumed that the localized shear process was divided into three consecutive stages. In the first stage, the grid lines incline but remain straight, implying that the deformation is homogeneous. In the second stage, the grid lines become slightly curved, indicating inhomogeneous deformation. In the third stage, the grid lines appear discontinuous, implying shear-band formation. Marchand and Duffy [21] and Giovanola [52] have independently observed the transient deformation field of the shear process by high-speed photography and a grid pattern. More recently, Guduru et al. [53] obtained detailed in situ temperature profiles at the shear-band tips in a C-300 Maraging steel. They focused on the tips of the band using a 2D infrared radiation (IR) array with an acquisition rate of 10^6 frames per second. They observed that the front contained 'hot spots' with a spacing of $\sim$0.25 to 1 mm, which they attributed to velocity effects. They report local temperature rises of up to 600 K.

All of these works certainly have blazed a trail in the experimental study of the localized process, but they have not revealed the evolution process in microstructure occurring during shearing and have not clarified the relationship of the mechanical parameters with the microstructures in the band. The main reason for this is that the original split Hopkinson torsional bar (SHTB) could not be used to study the microstructure evolution occurring during localization. Bai, Xue and co-workers [25,54,55] modified the original SHTB, which enabled them to combine ingeniously the measurements of the mechanical parameters (stress and strain) at successive points on the $\tau-\gamma$ curve with correlated microstructural observation and to follow the formation and evolution processes of the bands. Figure 3.1 shows four interrupted tests on the four stress–strain curves with four prescribed loading durations labelled a, b, c and d for Ti–6Al–4V alloys subjected to torsional loading at an average strain rate of $3.1 \times 10^3\ s^{-1}$ [25]. The curves overlap, the difference being the maximum strain that each specimen undergoes. There is a node in the curve at which the average critical strain is 0.160. The specimen with terminal point a is loaded for 450 μs, showing a maximum strain of 0.160, and it represents the state near the critical point. There is no kink on the curve, implying the specimen is still in homogeneous deformation, and no microstructural change is observed during this regime of homogeneity. Beyond this point, the stress drops slowly. The specimen with terminal point b, loaded for 550 μs, attained a nominal

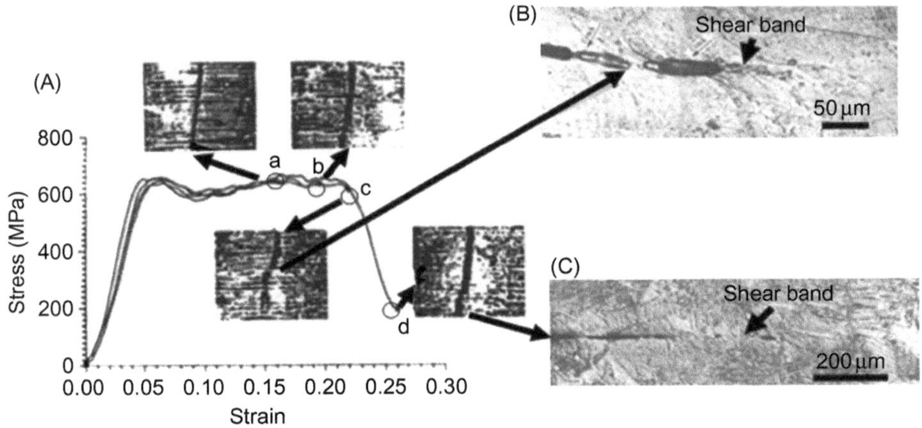

Figure 3.1 Stress−strain response of Ti−6Al−4 V alloys subjected to dynamic torsional loading and the corresponding patterns of the inside scribe lines [25].

strain of 0.196 before unloading. Local maxima and minima in the stress−strain curve for this sample are suggestive of inhomogeneous deformation. Observation displays the distinct occurrence of shear localization; the width of the shear zone is about 44 μm, and maximum shear strain is 1.07, which is five times the average value. A tiny elliptical void can be seen within the shear zone (Figure 3.1B). More interesting is that the width of the local shear zone is much narrower than the size of the grain, but its extension covers several grains. Loaded for 650 μs, a specimen with terminal point c gained a nominal strain of 0.219. Severe localized shear bands appeared. Obviously, the micro-voids have grown, and some of them have coalesced into a large crack (Figure 3.1C). They even extend from one grain to another. There are still some uncracked segments in the band. Near the crack tip, the band is 20 μm wide, attaining a strain of 0.214. The occurrence of the band does not significantly affect the capacity of the material for continued deformation (point c). In fact, the steep drop in the shear stress appears only after point c. A specimen with terminal point d is subjected to a larger average strain of 0.26 but much lower stress, only one-third of that at the critical point. A long crack has formed along the band (Figure 3.1C). The half-width of the band is about 10 μm. Typically, the sudden drop in the $\tau-\gamma$ curve has been described as a catastrophic failure due to the localized shearing. These observations of both stress−strain response behaviour and related microstructure suggest that the critical mechanism governing the loss in load-carrying capacity is the coalescence of the micro-cracks (voids) within the band, rather than shear-band formation. Similarly, Figure 3.2 shows the localized process of shear deformation in Al−Li alloy subjected to dynamic torsion with an average strain rate of about $2.3 \times 10^3\ s^{-1}$ [56]. It was found that when the average strain is ~0.10, corresponding to a loading time of 40 μs, the deformation appears to be homogeneous (Figure 3.2A), and there is no shear localization to be recognized

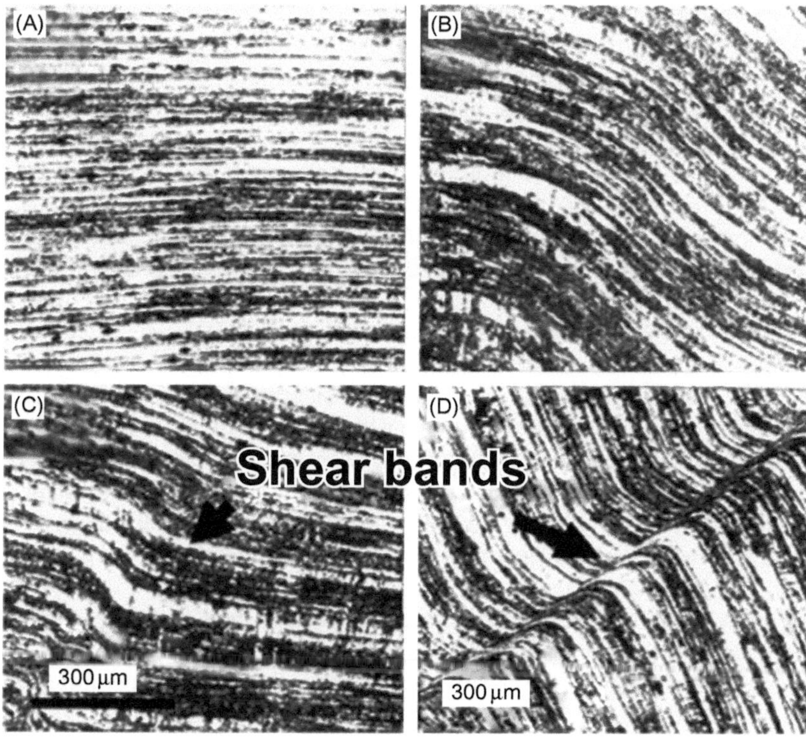

Figure 3.2 Evolution of localized deformation in Al–Li alloy subjected to dynamic impact compression [56].

on the cross section of the specimen, implying that the specimen is still work hardening. However, as the average strain increases from 0.10 to 0.17, corresponding to a loading time of 80 μs, localized deformation starts to occur (Figure 3.2B). Beyond this point, the specimen appears to work soften as the loading time increases, and localized deformation becomes more apparent as shown in Figure 3.2C. When the loading time is ~120 μs, the shear band with a width of 25 μm is shown in Figure 3.2D. It is clear that the evolution of localized deformation is a rapid progressive process during which the localization becomes more apparent, and the width of the band becomes gradually narrower.

3.4 Conditions Required for Shear-Band Formation

Determining the criteria for shear-band formation has been one of the most interesting research challenges. For more than three decades, approaches have been focused on theoretical treatments, which consist of a combination of mechanical instability analysis with a thermal model. These analyses generally concur that after the critical strain corresponding to the macroscopic maximum in load on a stress–strain curve of the material has been achieved, the shear bands form. Based on the thermoplastic instability in simple shear, Bai [4] and Timothy et al. [57] carried out the linear perturbation analysis and suggested a maximum load required for the formation of the shear bands. Molinari [58] provided a more advanced non-linearized perturbation analysis. Wright [59] extended this treatment and proposed a double instability/localization criterion. In this criterion, the maximum in the load corresponded to the instability. Localization was a separate event that followed instability. The strain between the two could be significant. Staker [60] examined the effect of the steel-tempering temperature on the susceptibility to transformed band formation. He proposed that there exists a critical strain for the formation of the transformed bands that depends on the heat-treatment condition of the steel. Taking thermal softening and conduction into account, Recht [2] proposed a critical strain rate as a criterion for the occurrence of the shear band. Rogers [14] pointed out in his review that 'the above results bear on the problem of whether or not a "critical strain rate" exists for adiabatic shearing in a given material'. He pointed out again that 'large strain can be achieved quasi-statically in steel without transformed band formation; hence, provided that a minimum strain is achieved, there must exist a strain rate above which the removal of heat from the region of deformation is sufficiently limited that the temperature can rise above that needed for transformation to occur – a critical strain rate'. Zurek [61] found that the average strain of 0.5 and the associated average strain rate of $1.8 \times 10^4\ s^{-1}$ were sufficient to induce an adiabatic shear instability in 4340 steel, implying that both strain and strain rate should be required for white etching band formation. Bai [62] proposed a criterion describing possible instability in simple shear deformation combining thermal softening, current stress, density, specific heat, work hardening, thermal conductivity and current strain rate implying that the condition for band formation should include both the stress and strain rate. This is verified by the experimental results of the aluminium and titanium alloys [56,63].

Xu et al. [56] showed that two types of bands can be formed in an aluminium–lithium alloy subjected to dynamic impact compression. It reveals that as the strain rate increases, localized deformation develops gradually, and when the strain rate reaches $1.75 \times 10^3\ s^{-1}$, the deformed shear bands appear first as shown in Figure 3.3A, and as deformation proceeds, the width of the band becomes gradually narrower. When the strain rate is $\sim 2.0 \times 10^3\ s^{-1}$, the white bands occur (Figure 3.3B). The results measured are listed in Table 3.2 [56]. The critical strain is 0.14 for deformed band formation, at a given strain rate ($1.6 \times 10^3\ s^{-1}$). As the strain increases from 0.14 to 0.17, a white etching band appears. This result is verified by Li et al. [66]. They found that the shear bands began to form when both strain and strain rate reached the critical values of 23% and $2.8 \times 10^3\ s^{-1}$, respectively.

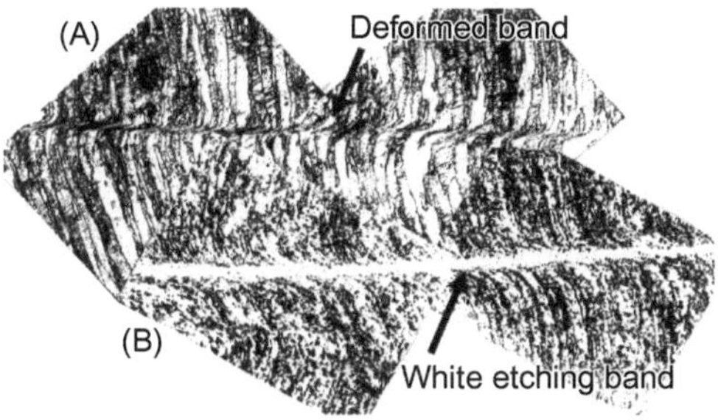

Figure 3.3 The deformed and white etching bands generated during dynamic impact compression testing in Al–Li alloy [56].

Table 3.2 Critical Strain and Strain Rate for Shear-Band Formation in Al–Li Alloys [56]

	Peak-Aged Alloy	Underaged Alloy
Strain rate (s^{-1})	1,600	2,000
Critical strain for deformed band	0.14	0.17
Critical strain for white etching band	0.17	0.21

3.5 Deformed and Transformed Bands

The shear bands have been classified into deformed bands and transformed bands, on the basis of their appearance in the metallographic section. The transformed band is also often referred to as a white etching band or white shear band in steels and has received much attention because it was suggested that the phase transformation temperature is reached in the band, supporting the thermoplastic instability theory of shear localization. The white etching bands have been reported mainly in steels [1,61,67–70], titanium and titanium alloys [22,23,30,71], aluminium alloys [56] and magnesium alloys [72].

The evidence for occurrence of phase transformation in the bands in steels seems to rise from the white etch in a nital, a well-defined width, distinct boundaries between the band and the matrix and very high hardness. Trent [73] made the earliest observation of the white etching bands in steel and found that these bands appeared to be white or slightly yellow on the sheared surface. He attributed to the occurrence of deformation-induced transformation. Some investigators refer to it as a dislocation cell structure, while others refer to it as consisting of extremely fine grains [69,74]. Zener and Hollomon [1] assumed that the white etching band was caused by a rapid quenching from the high temperature. Scott et al. [74] pointed out that the white etching bands could not show the tempering characteristics of conventional martensite, even when heated to temperatures above normal tempering temperatures. Heating to above the austenite temperature, however, could cause a disappearance of the white etching bands and the formation of a structure indistinguishable from the remainder of the specimen. Based on the selected area electron diffraction (SAED) analysis, they suggested that the white band was supersaturated ferrite. Further investigations by TEM and X-ray show that the band area was a typical looking martensite [69], a fine-grained equi-axed delta ferrite and a martensite with carbides [27] and untempered martensite [1]. TEM examinations by Beatty et al. [75] demonstrated that the shear-band region was ferrite with the nanograins. We now know that the white etching effect and the high hardness can be caused by the formation of an ultrafine-grained structure, which can, by virtue of the Hall–Petch effect, reach higher levels than the surrounding matrix that is deformed homogeneously. However, phase transformations can and do occur, and the resulting phases in the shear bands are the result of the following:

- Thermodynamics of phase stability
- Kinetics of phase transformations
- Shear deformation effect, introducing high densities of defects
- Cooling rate to ambient temperature.

For instance, one of the first observations of deformed and transformed bands, by TEM, is by Me-Bar and Shechtman [64]. (The latter is the 2011 recipient of the Chemistry Nobel prize for his discovery of quasi-crystals.) Figure 3.4A shows the diffraction pattern in a deformed band, which clearly shows the ringed and arced appearance that was later attributed to an ultrafine-grained structure. The spots are clear and sometimes smeared, evidence of deformed grains and a small grain size. Similar results were obtained by Grebe et al. [65] in commercially pure titanium. They identified the grain structure as equi-axed grains with approximate diameter of 300 nm. It is interesting that, upon observation in optical microscopy, the bands in Ti give the appearance of being 'transformed'. Grebe et al. [65] indexed the rings in Figure 3.4B to the hexagonal structure. Thus, no transformation was obtained.

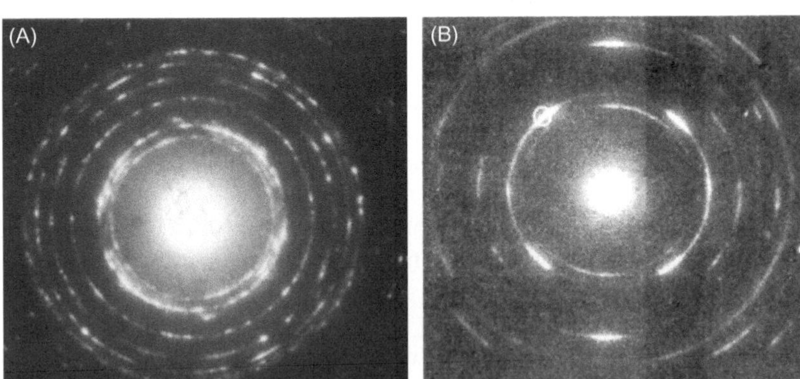

Figure 3.4 (A) First diffraction pattern from the Ti–4Al–4V shear band; spelcimen shows appearance of transformed band, but the diffraction pattern indicates breakup of microstructure [64] and (B) first diffraction pattern from Ti showing arced circles characteristic of ultrafine-grained size [65].

When discussing the phase transformation in the band, the critical role of the temperature rise in transformation should be mentioned. A number of authors proposed that the maximum temperature during localized shearing may reach several hundred degrees above that of the surrounding matrix. This is usually inferred indirectly from metallurgical evidence, and the shear band is then proposed to be rapidly cooled by the surrounding bulk material when plastic deformation ceases, and cooling rates as high as 10^7 K/s have been calculated [22]. However, some measurements have been made directly [21,24,30]. These authors used the IR detection system to determine that the temperature distribution of the band region was in the range of 440–550°C for the Ti–6Al–4V [30] and 450–590°C for steel [23,30]. They concluded, therefore, that there was no clear evidence to suggest that the material within the band had undergone a phase transformation. Giovanola [52] pointed out that there is no conclusive direct evidence for occurrence of phase transformation in the white bands. The only important point of common agreement is that the structure in the white bands is composed of very fine grains. Rogers [14] mentioned that 'whether there is a single crystallographic structure for the transformed bands in steel or whether there are a variety of structures dependent on the particular material and the deformation condition is still a very open question'. In addition to etching behaviour, there are two almost universally recognized characteristics of the transformed bands in steels: their very high hardness and their extremely fine structure'. The investigation made in Al–Li alloys has shown that deformed bands form first (Figure 3.3A), corresponding to a critical strain of 0.14, and as deformation proceeds, localization gradually becomes apparent. Finally, a white etching band occurs on the base of the deformed band corresponding to a critical strain of 0.17 (Figure 3.3B). Timothy [15] pointed out that the formation of a 'transformed' band appears to correspond to an advanced stage of adiabatic strain localization in a given metal, with a deformed band representing an earlier stage in the process. The results mentioned here are supported by the recent result in high-speed machining of the high-strength steel [76].

A number of investigations have shown that whether the bands are deformed or transformed, their presence is generally indicated by the different etching responses in the band of material in metallographic cross sections. Figure 3.5, imaged by

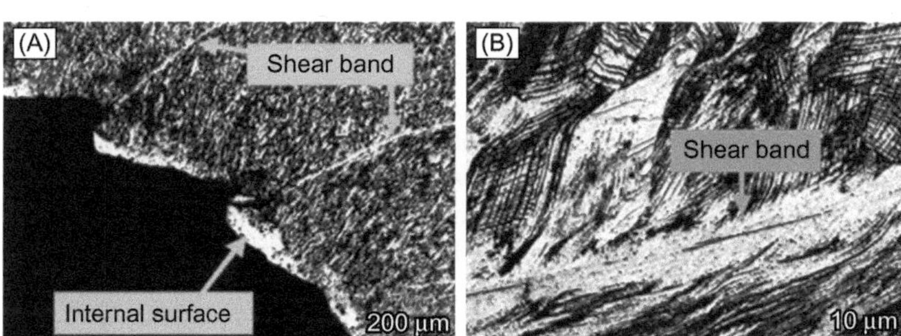

Figure 3.5 White etching bands in 304 stainless steel imaged by CLSM at low (A) and high (B) magnifications [20].

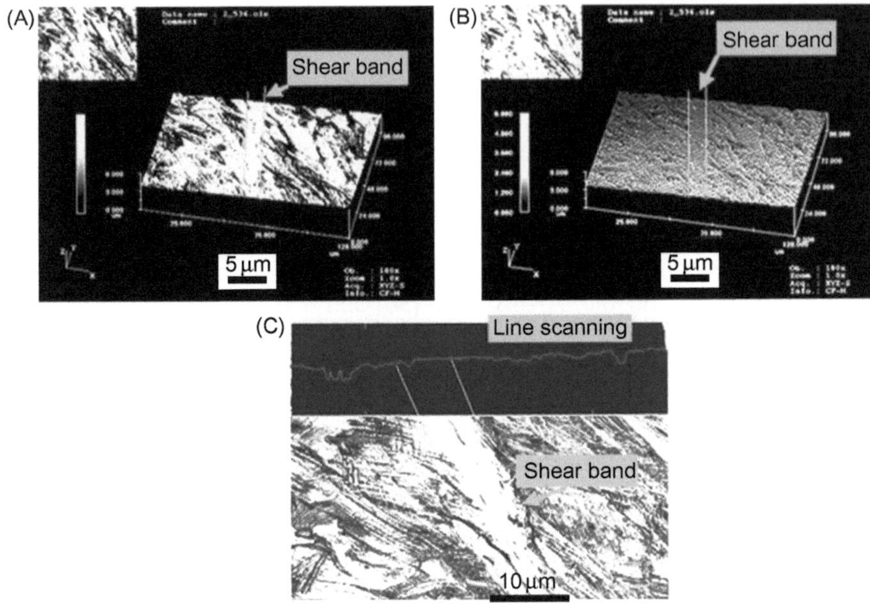

Figure 3.6 Shear bands imaged by CLSM: (A) scanning mode, (B) profile scanning and (C) line scanning across the band [19].

confocal laser scanning microscopy (CLSM), shows the white shear bands in 304 stainless steel subjected to explosive collapse testing in the thick-walled cylinder (TWC) geometry. One of the possible reasons for the 'white' colour of the band is the occurrence of a martensite that is too hard to be etched, causing the white colour (the α'-martensite phase transformation does occur in the band; this will be shown in Section 3.6). To verify this, profile line scanning across the band was performed by CLSM. If the band is lightly etched, then the profile-scanning line should appear to be a protrusion when the scanning is across the band (Figure 3.6). The profile scanning is selected along the horizontal line (Figure 3.6C). The line does not appear to be a protrusion, implying that the structure in the band is not difficult to etch. EBSD analysis shows [77–79] that the structures in the band undergo intensive localized deformation with large strains and ultra-high-strain rates, forming substructures highly elongated along the band direction; these substructures form low-angle boundaries of about 2–15° in misorientation. This is proposed to be the possible reason for the white colour of the band. In other words, the white etching band is an unlikely indication of the phase transformation; in fact, the observed substructures in the white etching bands differ by a degree not observable by ordinary optical microscopy or by SEM. Both the ordinary light and the secondary electron wave could not 'see' the substructure difference in orientation in the bands. Therefore, the band appears to be white when viewed by ordinary light under optical microscope or secondary electron wave in SEM. So, it is reasonable to propose that the white etching is unlikely to be due to the occurrence of

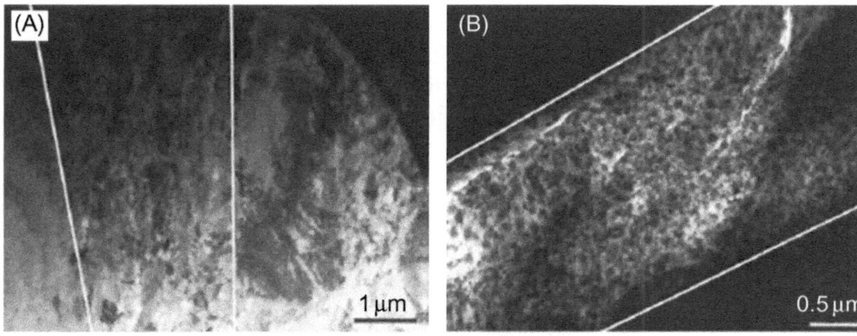

Figure 3.7 Adiabatic shear bands in magnesium alloy AM60B target impacted by steel projectile at a velocity of 0.5 km/s: (A) 'deformed' bands and (B) 'transformed' bands. Boundaries of bands are marked by white lines.

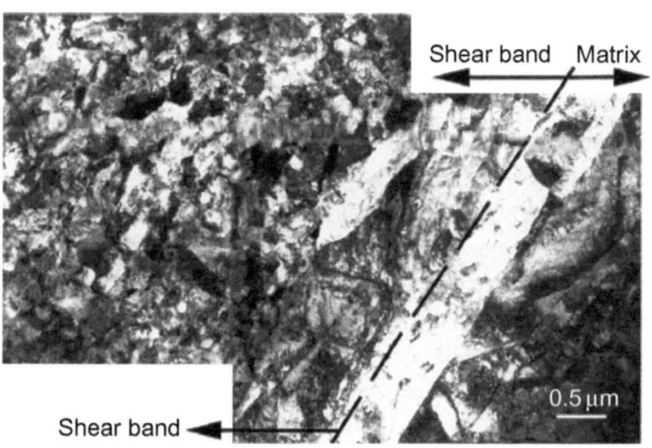

Figure 3.8 Shear band (left), matrix (right) and the matrix/shear band (dashed region) interface for Zircaloy deformed in a hat shaped geometry. The interface between adjacent regions is sharp.
Source: From Ref. [76].

phase transformation in the bands. This analysis is verified further by EBSD studies in Fe–Cr–Ni monocrystal and 304 stainless steel [77–79].

The recent results obtained by Zou et al. [72] on magnesium alloys confirm the generality of the structures in shear bands. There is a critical strain above which the structure reorganizes itself into nanometre or ultrafine-grain sizes. In the case of the AM60B magnesium alloy, the grain size in the 'transformed' band is in the range of 50–100 nm. This is shown in Figure 3.7. Figure 3.7A shows the band that is called 'deformed' and exhibits the effect of deformation. On the other hand, in Figure 3.7B, for a larger strain, the deformed structure is replaced by one consisting of nanosized grains. This structure is called 'transformed', but the only transformation it underwent is the size and configuration of the grains. Zou et al. [72] observed the characteristic increase in hardness, but it is due to the reduction in grain size. The structure of the nanocrystalline shear band remains hexagonal. Figure 3.8 shows the same phenomenon in Zircaloy [80]. The structure inside the shear band, on the left side, consists of ultrafine grains with ~200 nm diameters. The interface between band and surrounding material is clear.

Recent investigations have shown, however, that phase transformation can occur in both the deformed and transformed bands; this will be described in the following section.

3.6 α'-Martensite Transformation in Shear Bands

Recent studies have shown that phase transformations do occur, certainly in the shear bands in 304 stainless steel [32], Fe–Cr–Ni monocrystal [20] and Ti–6Al–4V alloys [81]. Figure 3.9 is a shear band, and its microstructure observed in 304 stainless steel is subjected to the cylindrical collapse (TWC) test with a strain rate of $10^4\ s^{-1}$. Figure 3.10 is the TEM image taken from a field in the band shown in Figure 3.9, indicating that a α'-martensite transformation occurs. This kind of phase transformation particularly generates at the intersection between the shear band and twins. Further analysis shows that the $(\bar{1}10)$ planes of the α'-martensite are coherent with the $(\bar{1}11)$ planes of the parent austenite and parallel to each other. The direction, [110] of the α'-martensite, is parallel to the [211] of the austenite [32], for example $(1\ 1\ 0)_{\alpha'}$ // $(1\ 1\ 1)_{\gamma}$; $[0\,0\,\bar{2}]_{\alpha'}$ // $[0\,2\,\bar{2}]_{\gamma}$; $[\bar{1}12]_{\alpha'}$ // $[1\ 3\ 1]_{\gamma}$; $[1\ 1\ 0]_{\alpha'}$ // $[211]_{\gamma}$, which is in accordance with the Nishiyama orientation. From this, it is concluded that the (110) of the α'-martensite nucleates along the {111} of the austenite. Essentially, these results confirm earlier investigations made by Staudhammer et al. [82] in dynamic tensile testing of the same materials. These α'-martensite laths nucleate preferentially at twin-band intersections in the area of localized strain. They have been identified by Murr et al. [83] and Kestenbach et al. [84] in connection with shock compression. These are the twinning and slip planes, and, thus, their intersections provide the nucleus [85]. A similar phase transformation was also observed in the shear bands in Fe–15%Cr–15%Ni monocrystal subjected to the dynamic explosive loading [20,86].

Figure 3.11A is a bright-field image taken from a primary α-grain in the band of the Ti–6Al–4V alloy, and Figure 3.11B is its combined electron diffraction pattern. Figure 3.11C and D shows the bright-field and dark-field images obtained

Figure 3.9 Shear band generated in cylindrical collapsed specimen of 304 stainless steel subjected to global $\varepsilon_{eff} = 0.92$ [32].

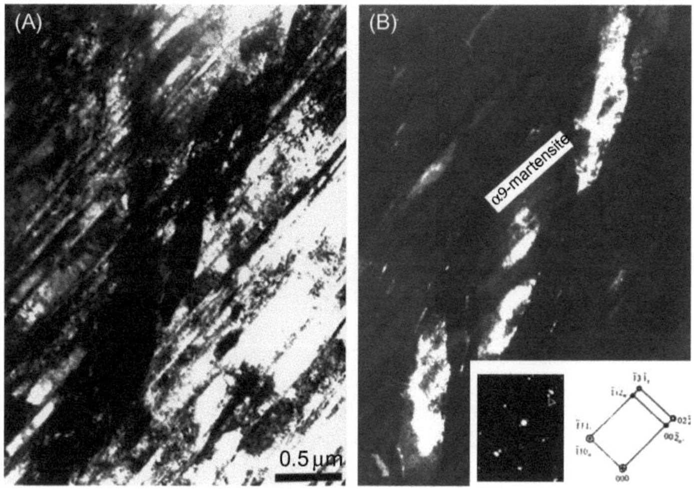

Figure 3.10 TEM bright-field (A) and dark-field (B) images, showing α′-martensite phase transformation formed at the intersection between the twins and the shear bands in 304 stainless steel subjected to explosive collapsed testing [32].

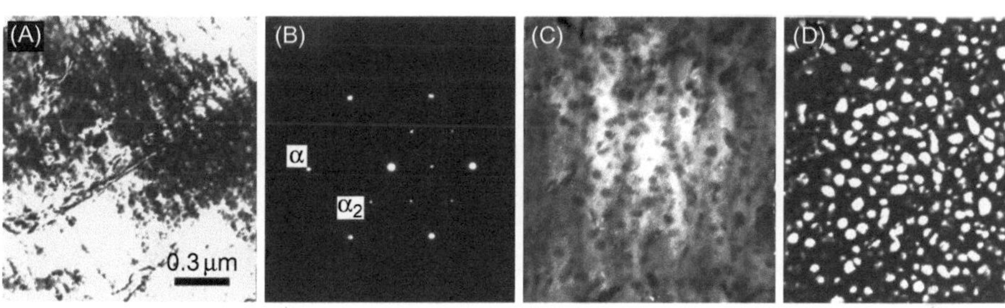

Figure 3.11 TEM, showing bright-field images (A) of the high-strain-rate deformation structure in Ti−6Al−4V alloy, and its [0001] zone axis diffraction pattern (B), and dark-field images obtained by using strong diffraction spots $\{2\bar{2}00\}_{\alpha}$ (C) and weak super-lattice reflection $\{1\bar{1}00\}\alpha_2$ (D), respectively [71].

by the strong diffraction (α) and the weak super-lattice reflection $\{1\bar{1}00\}_{\alpha 2}$, respectively. Analysis shows that the α_2 phases (Ti_3Al) are proposed to be transformed from the α-matrix, and both the α_2 phases and the α-matrix are completely coherent [71].

An interesting observation by Wang et al. [87] in a Ti−3Al−5Mo−4.5V alloy in the two-phase α-β microstructure is that the α phase recrystallized into ultrafine grains, whereas the β phase, passing the transformation temperature, recrystallized into an α′ structure. The β phase disappeared inside the alloy. This is additional

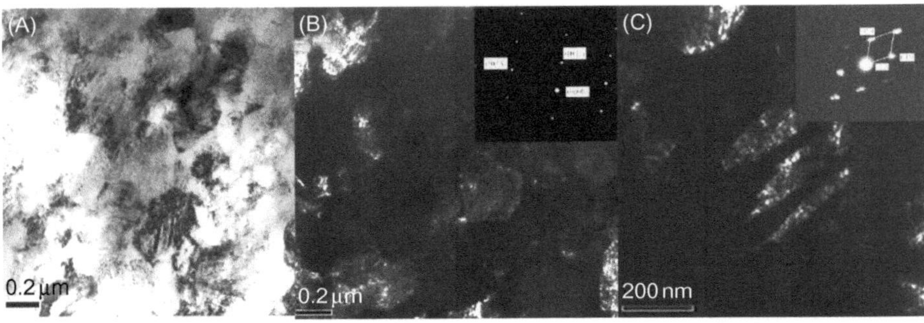

Figure 3.12 TEM microstructure and SAD patterns showing the microstructure in the centre of a shear bands in TiAlMoV alloy initially in the two-phase α-β structure: (A) is the bright-field image of the microstructure, and (B) and (C) are the dark-field images of α and α′ and their corresponding SAD patterns (inserts).
Source: From Ref. [87].

evidence that phase transformation can occur in conjunction with recrystallization. The bright-field and dark-field TEM micrographs together with diffraction patterns are shown in Figure 3.12.

3.7 Effect of the Microstructures on the Tendency for Localized Shear Bands

It is generally accepted that a material will be sensitive to localized shearing if it is characterized by a low strain hardening, a low specific heat and a high thermal softening. This was implicit in the first analysis carried out by Zener and Hollomon [1]. Rogers [14] pointed out that the deformed bands are likely to form in iron and carbon steels when the structure is coarse. Also, as the rate of deformation decreases, there is a greater tendency for the shear bands to be of the deformed type and a transformed portion to extend from the region of the most intense shear to a point at which the local temperature has not exceeded the transformation temperature. Costin et al. [88] found that shear bands developed in all high strain-rate tests with cold rolled steel and that none formed in hot rolled steel when deformation was limited to about 30% strain. These steels showed very similar strain-rate sensitivities and thermal softening rates. Shawki [89] predicted that the shear bands would form in hot rolled steel deformed dynamically, if a nominal strain of 80–100% was imposed; they considered that the reason for the absence of localized shear in the hot rolled steel could be attributed to the low strain used. This proposal was justified by the results of Hartley et al. [51]. However, they found that the shear bands in hot rolled steel are narrower than that in cold rolled steel.

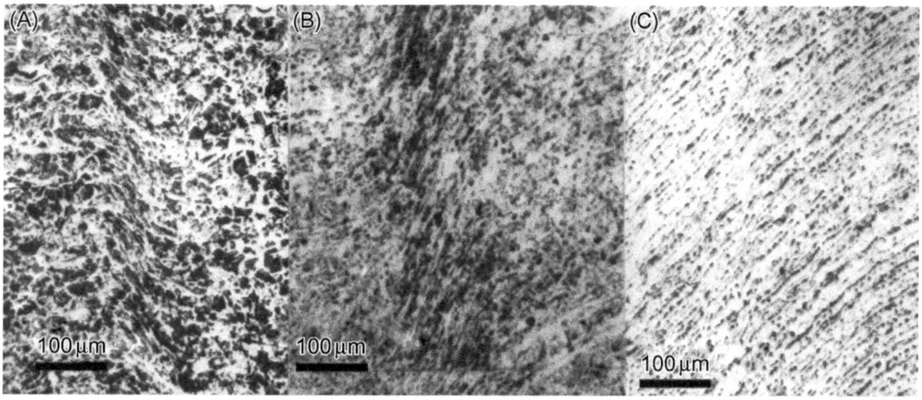

Figure 3.13 Shear bands in the low-carbon steels: (A) quenched condition, (B) quenched and tempered and (C) normalized [26].

Table 3.3 Mechanical Properties and Structural Parameters of the Steels Tested [26]

Microstructure	σ_f (MPa)	n (s^{-1})	γ	γ_c	W_m (μm)	ΔT (°C)
$F + M_q$	699	0.179	650	0.30	75–100	210
$F + M_p$	649	0.180	610	0.36	130–150	265
F + P	498	0.22	1,500	0.83	350–400	275

σ_f, fracture strength; n, work hardening; γ, strain rate; γ_c, average critical strain; W_m, width of the band; ΔT, the rise in temperature in the band.

Xu et al. [26] studied the tendency of the low-carbon steels with different structures to shear localization and found that the critical strain required for the occurrence of shear localization is strongly dependent on the strength of the steels at a given strain rate. Figure 3.13 shows that the critical strain values required for the shear-band formation for the quenched, quenched and tempered and normalized steels are 0.30, 0.36 and 0.83, respectively, implying that the steel with quenched martensite is most susceptible to shear deformation. The higher the strength of the steels, the easier it is for the band to occur (Table 3.3). The quenched steel with high strength and low work-hardening capability does develop a pronounced band with a width of 70–100 μm (Figure 3.13A). However, the width of the bands in the normalized steel with higher capability of the work hardening is much larger than those of the other two steels, ~350–400 μm (Figure 3.13C). The width in the quenched and tempered steel is in between them (Figure 3.13B). It should be pointed out that the boundary between the bands and the matrix for these steels is not well defined.

One more example is shown in Figure 3.14, displaying the effect of the microstructure on the tendency for shear deformation in an aluminium matrix composite reinforced with different sized SiC particles [90]. It is found that the composite

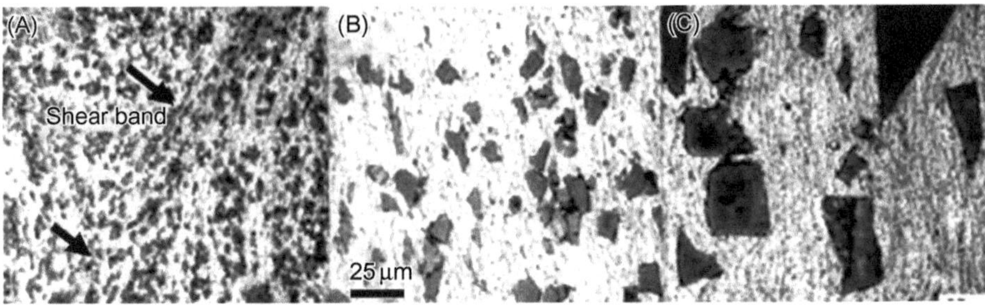

Figure 3.14 Deformation pattern in Al-17% SiC_p composite with an average particle size of 3 μm (A), 13 μm (B) and 17 μm (C) produced during high-strain-rate loading. Note that material with smaller particles (3 μm) develops a well-defined localized shear band (A), and shear localization in the materials with larger particles (13 and 17 μm) does not develop well (Figure 3.10B and C).

reinforced with smaller particles (3 μm) develops well-defined shear bands as shown in Figure 3.14A, and those with larger SiC_p of 13 (Figure 3.14B) and 17 μm (Figure 3.14C) in size tend to deform inhomogeneously, so localized shear was not well developed. This implies that the particles may play a significant role in formation and development of shear localization, and the smaller the particle size, the easier shear-band formation, which coincides with the results obtained by Lee et al. [91]. The tendency of the composite reinforced with small particles to form shear localization is greater than that of the other two materials with large particles, when the volume fraction of the particles in the composite remains the same. This size-dependent deformation localization behaviour of particle-reinforced metal-matrix composites is confirmed by recent research results [92]. In Chapter 7, Walley discusses shear localization in materials [19]. Shih et al. [93] studied shear localization in granular SiC and observed the fragmentation of the material within the shear bands. The existence and propagation of strain in shear bands benefits from small particles, smaller than the width of the shear band.

3.8 Crystalline to Amorphous Transition Within Shear Bands

Figure 3.15B shows a bright-field image taken from an area of the shear band (Figure 3.15A) observed in Fe–Cr–Ni single crystal subjected to the cylindrical collapse (TWC) test with a strain rate of $10^4\ s^{-1}$. It is seen that both amorphous and nanosized grain regions can be recognized. The amorphous area is different from the crystalline lattice, and the corresponding diffusion halo-ring (insert) and its radius coincides with that of a Fe–Cr–Ni single crystal. On the other hand, the combined diffraction pattern consists of the spots and rings (insert), indicating

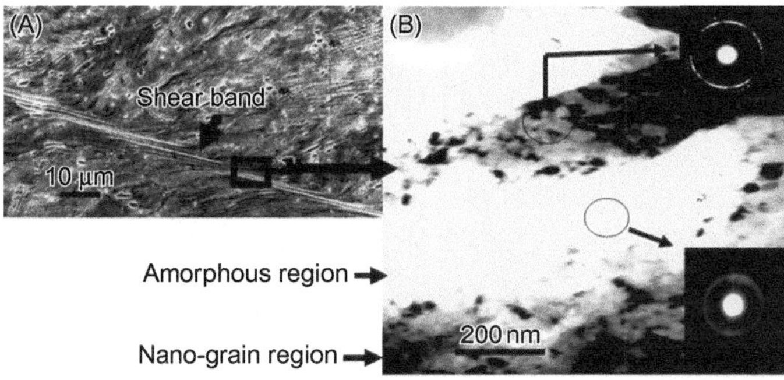

Figure 3.15 The shear-band pattern (A) and a bright-field image taken from a square in the band (B).

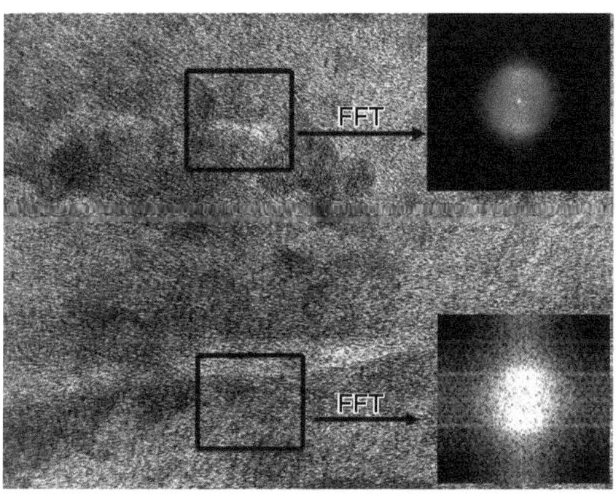

Figure 3.16 HREM image from the amorphous region, and the FFTs performed on different areas marked by squares.

occurrence of the nanosized grains. Figure 3.16, an HREM image from the amorphous area, shows a disordered lattice, which is verified further by fast Fourier transformation (FFT) (inserts) from different areas marked by the squares, confirming the absence of crystalline lattice. These provide strong evidence for the lack of crystalline symmetry in the shear band induced during dynamic deformation of this material.

The TEM (Figure 3.17A) and HREM (Figure 3.17B) images from the ultrafine-grained region of the band in Figure 3.15 show the nanograins clearly. EBSD was used to characterize the microstructure inside the band. Figure 3.18A and B show Euler maps, and Figure 3.18C is a Kikuchi contrast taken from the band shown in Figure 3.15A. It is clear that in addition to equi-axed recrystallized grains, the area in grey is not resolved, which implies the absence of the crystalline lattice symmetry [94]. These results are supported by the first observation by Meyers et al. [32], identifying the crystalline to amorphous transition in bands in 304 stainless steel

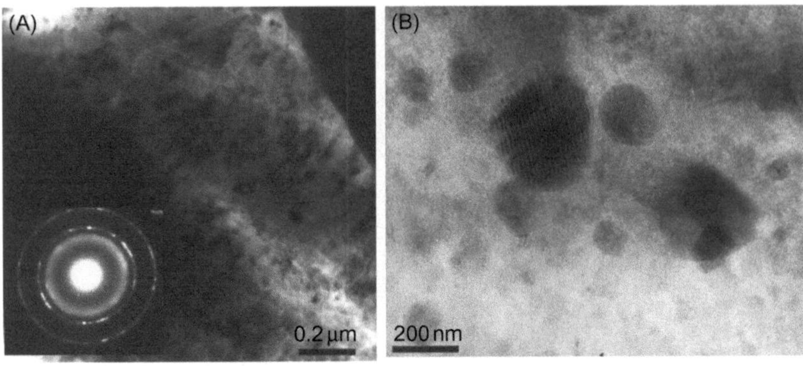

Figure 3.17 Images from the nanosized grain region by TEM (A) and HREM (B).

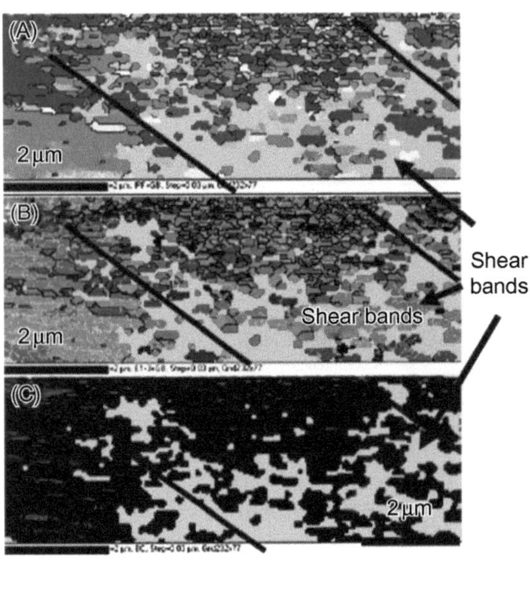

Figure 3.18 Euler maps (A and B), and Kikuchi contrast (C) obtained in the shear band.

under dynamic deformation, and further verified by the more recent work of Li et al. [95], who found the coexistence of both amorphous and nanocrystalline phases in a shear band in twinning induced plasticity (TWIP) steel subjected to ballistic impact with a high strain rate of $\sim 10^5\ s^{-1}$ as shown in Figure 3.19.

One of possible reasons for this could arise from either the surface contamination or the uneveness of the sample tested. In fact, the occurrence of the equi-axed grains in the same area as that in grey (Figure 3.18) implies that surface contamination or uneveness of the sample is ruled out. From these analyses, it is reasonable to propose that the transition of crystalline to amorphous structures within the bands can occur during dynamic deformation.

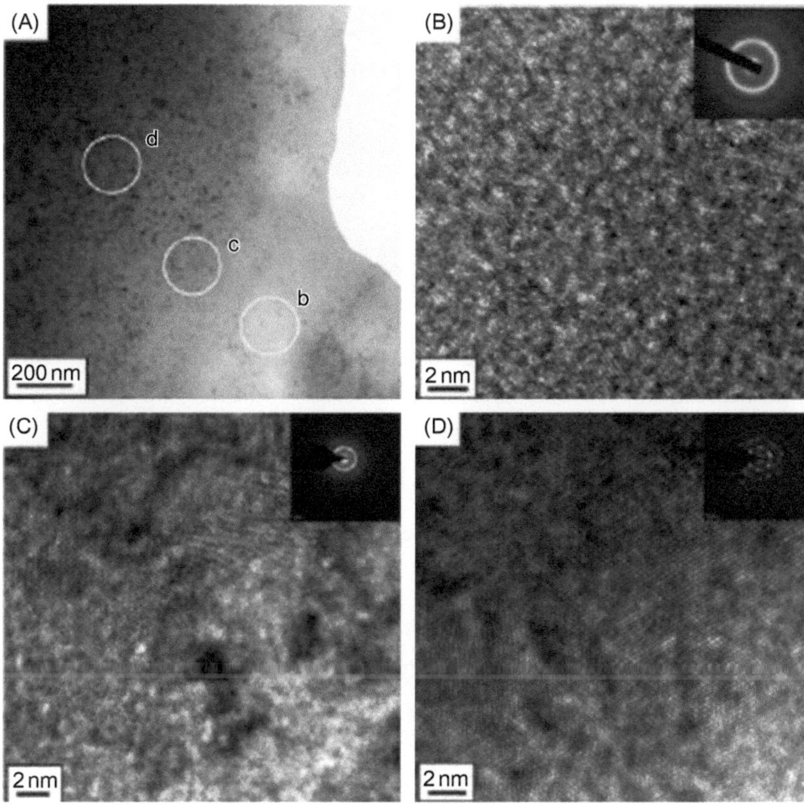

Figure 3.19 Bright-field image (A) from a shear band in TWIP steel, and HREM images showing the amorphous phase (B), the transition region of crystalline to amorphous (C) and nanocrystalline (D).
Source: From Ref. [95].

3.9 Static/Dynamic Recrystallization

3.9.1 Observations by EBSD

Recrystallization in the shear bands arising from high-strain-rate loading has been documented well and reported in steels [32,96,97], titanium [98], copper [99,100] and tantalum [101,102]. It is generally accepted that the deformation and associated temperature rise in the band are therefore sufficient to induce new recrystallized grains. Figure 3.20A shows the nanograins with an average size of 100 nm in diameter in the band of 304 stainless steel. The large grains outside the band with about 50–100 μm in diameter are also shown (Figure 3.20B). Recently, EBSD has been used to characterize structures in shear bands in steel [77–79,97] and Ti alloy [103]. Figure 3.21 shows the Euler (A) and Kikuchi (B) contrasts and the pole

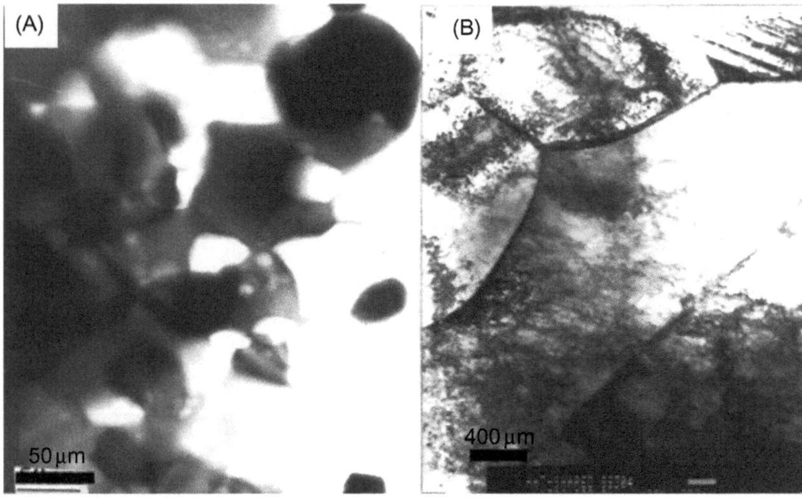

Figure 3.20 Nanograins (50–100 nm in diameter) inside (A) and large grains (50–100 μm) outside (B) bands [32].

figure (C) contrasts of Fe–15%Cr–15%Ni monocrystal, respectively [77]. A band that is 2 μm wide and a great number of distortion-free and equi-axed grains are apparent. This type of equi-axed structure is proposed to be the result of the occurrence of static recrystallization in the band [79]. The pole figures (Figure 3.21C) obtained in a region crossing the band show that the orientation (<100>) of the red area outside the band is in good agreement with that of the starting single crystal orientation (<100>). The substructures (in orange) below the band have an orientation (<111>) that is parallel to normal direction (ND) in the {111} pole figure, and the substructures (in red) have an orientation (<100>) that is parallel to ND, therefore, $<111>//<100>$. However, misorientation between them (in orange and blue) is typical of low-angle boundaries (1–3°) [81]. The average size of the grains in the band determined by linear intercept is about 200 nm. A similar structure has also been observed in the band of 304 stainless steel as shown in Figure 3.22 [81]. From these observations, one can see that the grains in the band are basically equi-axed, implying that these new recrystallized grains grow isotropically, after nucleation, and they are no longer oriented along the shear deformation direction. It is reasonable to propose, therefore, that the equi-axed grains in the band should be the result of static recrystallization. Recent observations show that another mechanism of recrystallization in the band operates in Fe–Cr–Ni monocrystal as shown in Figure 3.23. By comparing Figure 3.23 with Figures 3.21 and 3.22, it is seen immediately that the grains shown in Figure 3.23 are not equi-axed but elongated along the shear direction [78]. This indicates strongly that after nucleation, the nuclei of the new recrystallized grains grow preferentially under the stress along the shear direction. In other words, these elongated grains

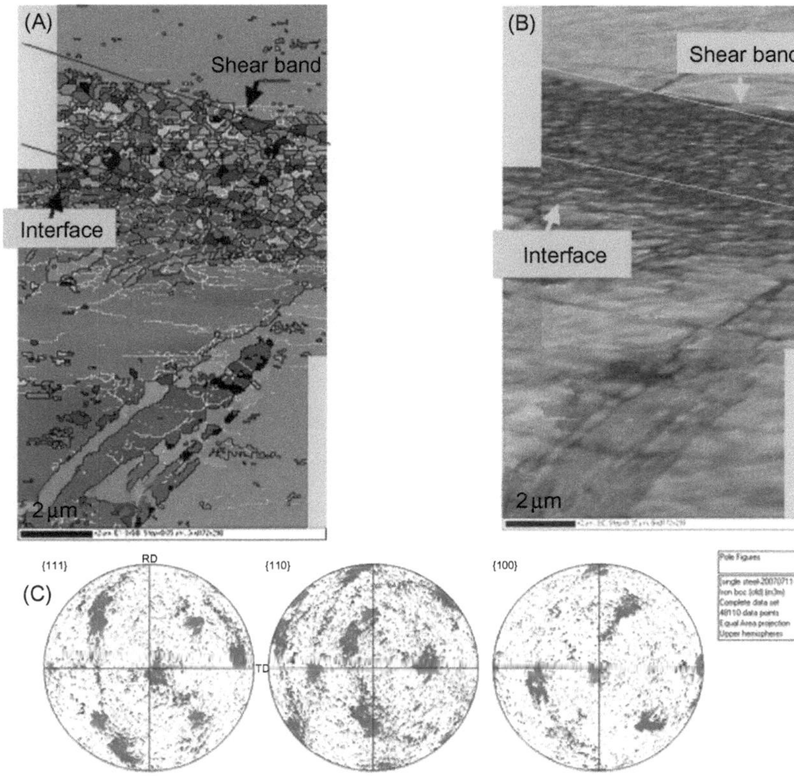

Figure 3.21 Euler (A) and Kikuchi (B) contrast and pole figure (C) of Fe−15%Cr−15%Ni monocrystal subjected to explosive collapsed loading [77].

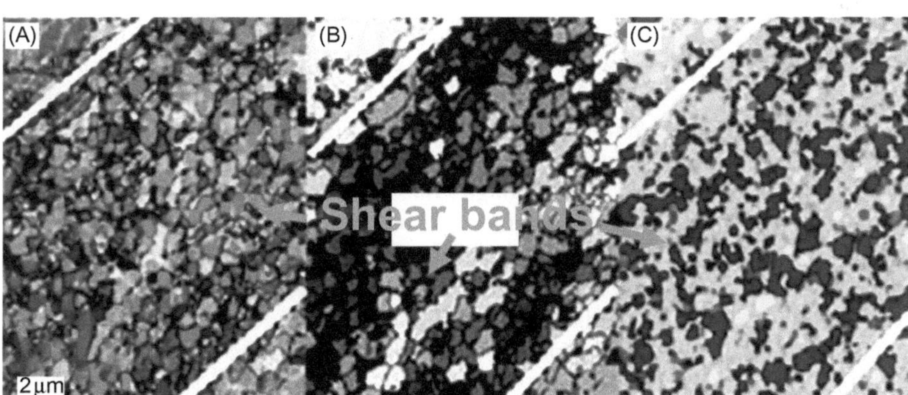

Figure 3.22 EBSD maps inside a band in 304 stainless steel: (A) Euler contrast map, (B) Kikuchi contrast map and (C) deformation map showing the equi-axed recrystallized grains in the band [81].

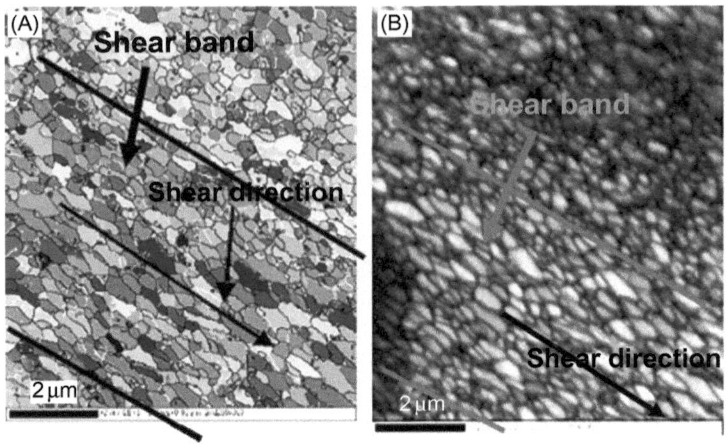

Figure 3.23 EBSD maps of recrystallized grains in the band in Fe−Cr−Ni monocrystal subjected to explosive collapsed loading: (A) Euler and (B) Kicuchi contrasts [78].

along the shear direction were developed simultaneously with localized deformation, and, therefore, these elongated grains are proposed to be dynamically recrystallized [78].

3.9.2 Analysis of Microstructure Evolution

Grain-Boundary Rotation

Meyers et al. [104] proposed a grain-boundary rotation mechanism to describe the process of recrystallization occurring during deformation. For this mechanism to be viable, it has to be shown that grain-boundary rotation can be accomplished within the time/temperature window created by plastic deformation. The relaxation of the rotated grains into a more equi-axed micro-crystalline structure can occur by minor rotations of the grain boundaries. This is shown in Figure 3.24A, where it is shown that a rotation of 30° of the boundaries transforms elongated segmented subgrains into an equi-axed structure. If each longitudinal grain-boundary segment rotates by an angle $\theta = 30°$, the equi-axed structure will be created. This can be accomplished by the flux of atoms along the grain boundary, which can occur at rates that are orders of magnitude higher than in the bulk. A general form of Fick's law expressed in terms of a potential energy gradient must be used [105]. This is a critical element of the model: a mechanical stress induces diffusion. We start with force $\vec{F}$, acting on a particle,

$$\vec{F} = \nabla V \tag{3.1}$$

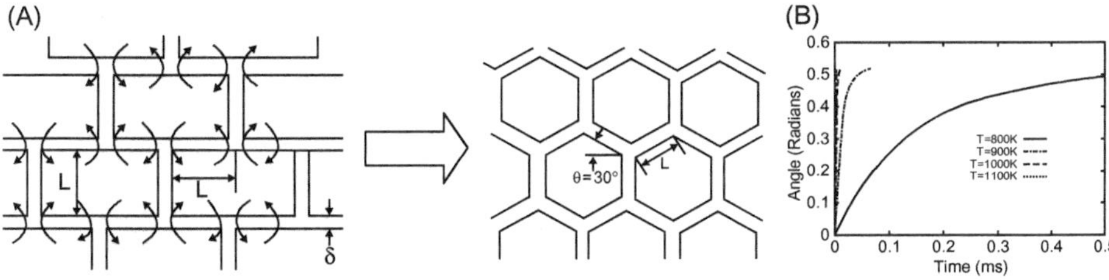

Figure 3.24 (A) Representation of the rotational recrystallization mechanism leading from elongated subgrains to equi-axed ultrafine grains through rotation of grain boundaries. (B) Calculated rotation of boundaries (from a grain size of 100 nm as a function of time for three temperatures) [32,78].

where ∇V is the gradient of the potential energy field. The mean diffusion velocity $\vec{v}$ is the product of the mobility M by this force: $\vec{v} = M\vec{F}$. The flux along a grain boundary with thickness δ and depth L_2 (cross-sectional area $L_2\delta$) is

$$J = L_2\delta CMF = \left(\frac{L_2\delta DC}{kT}\right)F \tag{3.2}$$

where D is the diffusion coefficient and C is the concentration of the mobile species (expressed in terms of mass per unit volume).

The rotation of the boundaries is driven by the minimization of the interfacial energy [106]. The force exerted by the grain boundaries is equal to

$$F = \gamma\left(1 - 2\cos\frac{\theta_3}{2}\right)L_2 \tag{3.3}$$

where γ is the interfacial energy per unit area.

The relationship between the grain-boundary rotation and the volume flow, dV, is (this is computed per thickness, L_2)

$$\frac{L^2\,\mathrm{d}\theta}{4L_2} = \mathrm{d}V \tag{3.4}$$

where dV is the volume transferred and L is the instantaneous length of the segment. However,

$$L = \frac{L_1}{\cos\theta} \tag{3.5}$$

where L_1 is the initial length. Thus,

$$\frac{d\theta}{dt} = \frac{4\cos^2\theta}{L_2{L_1}^2}\frac{dV}{dt} = \frac{4\cos^2\theta}{L_2{L_1}^2\rho}\frac{dm}{dt} \tag{3.6}$$

where dm/dt is the mass change, which is the rate of volume change multiplied by ρ, the density.

The rate of mass change is the flux, and we have, substituting Eq. (3.2),

$$\frac{d\theta}{dt} = \frac{4\cos^2\theta}{L_2{L_1}^2\rho}J = \frac{4\cos^2\theta}{L_2{L_1}^2\rho}\left(\frac{L_2\delta DC}{kT}\right)F \tag{3.7}$$

Substituting F from Eqn. 3.3:

$$\frac{d\theta}{dt} = \frac{4\cos^2\theta}{{L_1}^2\rho}\frac{\delta DC}{kT}\gamma(1-2\sin\theta)L_2 \tag{3.8}$$

We consider equi-axed grains and $L_2 \approx L_1$:

$$\frac{4\delta D\gamma}{L_1 kT}t = \int_0^\theta \frac{d\theta}{\cos^2\theta(1-2\sin\theta)} \tag{3.9}$$

Integrating, we arrive at an expression,

$$t = \frac{L_1 kT}{4\delta D\gamma}f(\theta) \tag{3.10}$$

where

$$f(\theta) = \frac{\tan\theta - \frac{2}{3}\cos\theta}{(1-2\sin\theta)} + \frac{4}{3\sqrt{3}}\ln\frac{\tan\frac{\theta}{2} - 2 - \sqrt{3}}{\tan\frac{\theta}{2} - 2 + \sqrt{3}} + \frac{2}{3} - \frac{4}{3\sqrt{3}}\ln\frac{2+\sqrt{3}}{2-\sqrt{3}} \tag{3.11}$$

A step-by-step derivation is provided by Meyers et al. [107]. The most important parameter in Eq. (3.10) is the grain-boundary diffusion coefficient presented in the following equation [107] for stainless steel:

$$\delta D_{GB} = 2.0\times10^{-16}\exp\left[\frac{-167\ \text{kJ/mole}}{RT}\right](\text{m}^3/\text{s}) \tag{3.12}$$

For the grain rotation calculation, the grain-size diameter is taken as 200 nm. For $T = 800$, 900, 1000 and 1100 K, the plots are given in Figure 3.24B. The grain-boundary width, variously taken as 0.5–1 nm, is not needed because it is embedded into Eq. (3.12). The rate of rotation decreases with increasing θ and asymptotically approaches 30°. The temperature rise inside the band was not calculated in this study because of the uncertainty in the shear strain. Nevertheless, it has been done before [32], and the preceding values are within the range of values to be expected. Meyers et al. [32] estimated the temperature to be equal to ~800 K for a shear strain of 10 and equal to ~1000 K for a shear strain of 35 [32]. Figure 3.24B shows that the rotation of a 200 nm grain from a rectangular to equi-axed takes place in 0.5 ms. This is on the order of the deformation time. Thus, it is feasible to assume that the equi-axed structure develops during deformation.

Dynamic Deformation/Adiabatic Shear-Band Formation

Figure 3.24B shows that for 1100 K, grain rotation of 30° takes place in 1.4 μs, whereas with decreasing temperature, it progressively takes a longer time. For example, for a rotation of 30°, it takes 0.9 ms at 800 K compared with <1.4 μs at 1100 K. The calculations predict significant rotations of the boundary within the deformation time (~1–64 μs) at temperatures between 900 and 1100 K for micro-grain sizes of 0.1–0.3 μm. Thus, the reorientation of the grain boundaries can take place during plastic deformation. This does not exclude the possibility of reorientation/accommodation of the grain boundaries during cooling. The existence of both elongated and/or rectangular subgrains scattered in the centre of shear bands in 316 stainless steel [108], with a composition not too different from the alloy used in this study (12% Ni, 17% Cr), is strong evidence for the mechanism of equi-axed grain formation by the rotation of boundaries described previously.

Grain-Boundary Mobility

The grain-boundary mobility can be calculated using the analysis developed by Rath and Hu [109]. The driving force on a grain boundary can be simply estimated by equating the force acting on a curved segment of dS due to the grain-boundary energy, γ. The total force acting on the grain boundary in the y direction (Figure 3.25A) is

$$F = 2\gamma \sin\frac{\mathrm{d}\theta}{2} \approx \gamma \mathrm{d}\theta \text{ (because } \theta \text{ is considered small)} \tag{3.13}$$

Therefore,

$$\text{force/area} = \frac{\gamma\, \mathrm{d}\theta}{R\, \mathrm{d}\theta} = \frac{\gamma}{R} \tag{3.14}$$

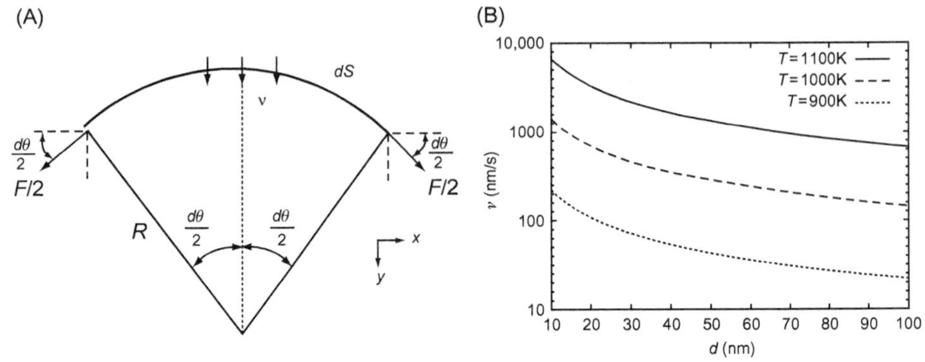

Figure 3.25 (A) Schematic representation of a grain boundary with curvature $1/R$. (B) Calculated velocity of a grain-boundary migration from the Rath-Hu equation [109] at three temperatures: 900, 1000 and 1100 K.

The velocity of the grain boundary is proportional to the force acting on the segment dS,

$$v \propto \frac{\gamma}{R} = M\frac{\gamma}{R} \tag{3.15}$$

where M, grain-boundary mobility, is the proportionality constant. The mobility is exponentially dependent on temperature through an Arrhenius-type relationship [109]:

$$M = M_0 \exp\left(\frac{-Q}{RT}\right) \tag{3.16}$$

For steel [110,111]: $Q = 140$ kJ/mol, $M_0 = 3.5 \times 10^{-7}$ m^4/J/s and $\gamma = 0.835$ J/m^2.

Assuming $d/2 = R$, where d is the grain size, grain-boundary velocity is plotted against grain size in Figure 3.25B for three different temperatures, 900, 1000 and 1100 K. For $d = 100$ nm (the lower limit of grain sizes observed in shear bands), $v = 22$ nm/s at 900 K. The plots at 1000 and 1100 K exhibit higher velocities: $v = 141$ and 652 nm/s, respectively. Thus, a grain-boundary movement of 22 nm at 900 K can be expected in 1 s. This cannot be sufficient to reorganize the deformation grain configuration. In contrast, for a shear band, the cooling takes place in time scales of fractions of milliseconds. For 1 ms and a grain size of 100 nm ($v = 0.01$ m/s), a displacement of 10 nm is obtained. Thus, this simple mobility calculation shows that no significant grain growth occurs on cooling and that recrystallization occurs by rotation of the boundary. Rath and Hu [109] use a similar expression, except for the exponent, m, which depends on the grain-boundary misorientation and grain-boundary purity:

$$v = M\Delta F^m = M\left(\frac{\gamma}{R}\right)^m \tag{3.17}$$

In their experiments, m varies from 1 to 4. This can be attributed to grain-boundary solutes and misorientation.

Kinetics of Grain Growth

Dynamic recrystallization is currently considered to be the nucleation and growth of new grains. They nucleate either homogeneously or heterogeneously along the elongated substructural boundaries of the deformed material. Li et al. [111,112] proposed that the growth of these new nuclei of recrystallized grains should be related to both strain rate and atomic thermal migration. For the former case, the growth rate is associated with strain, and it can be expressed by the following equation:

$$\dot{D}_2 = \frac{\alpha\mu b}{2\eta}D_2^2\sqrt{\frac{2\rho_m}{3}}\dot{\gamma} \quad \text{or} \quad \frac{dD_2}{d\gamma} = \frac{\alpha\mu b}{2\eta}D_2^2\sqrt{\frac{2\rho_m}{3}} \tag{3.18}$$

In Eq. (3.18), $D_2 = D_1\left(\frac{1}{1 - B(\gamma - \gamma_c)D_1}\right)$ is an average diameter of the recrystallized grains, where $B = \frac{\alpha\mu b}{2\eta}\sqrt{\frac{2\rho_m}{3}}$, γ_c is the critical strain stopping the subgrain rotation, η is the boundary energy density, D is the equivalent diameter of the cell structure (or subgrain), $\alpha\mu b^2$ is the energy per unit length of dislocation (α: a constant of 0.5−1.0 [113]), δ is the cell wall width, ρ_m is the mobile dislocation density, γ is the strain and $\dot{\gamma}$ is the strain rate. On the other hand, formation of a shear band results in a temperature rise in the band, which also leads to the growth of recrystallized grains. The growth rate of the recrystallized grains has to be related to the atomic thermal migration, and it is expressed by:

$$\frac{dD_3}{dt} = (D_{B_0} + k_1\gamma D_B^V t)\left(\frac{2\eta}{D_3} + \frac{2\tau_i^2}{E}\right)\frac{\Omega}{\delta kT} \tag{3.19}$$

where $D_{B_0} = D_0' \exp(-Q_B/RT)$ and $D_B^V = D_0'' \exp(-Q_B/RT)$, D_B^V is the vacancy diffusion coefficient in the grain boundary, D_{B_0} is a grain-boundary diffusion coefficient, $\Omega = b^3$ is the atomic volume, Q_B is the activation energy of the grain-boundary diffusion, D_0' and D_0'' are the pre-exponential factors, R is the gas constant and T is the absolute temperature.

The effects of both the high-strain rate and the atomic thermal migration should be taken into account simultaneously while we discuss the nucleation and growth of dynamically recrystallized grains. The former is an athermal process, and the latter is related to the thermal diffusion. Li et al. [111] superimposed these two effects and obtained the grain growth rate during localization in the bands, which is expressed as follows:

$$\frac{dD}{dt} = (D_{B_0} + k_1\dot{\gamma}D_B^V t)\left(\frac{\eta}{D} + \frac{\tau_i^2}{E}\right)\frac{2\Omega}{\delta kT} + \frac{\alpha\mu b}{2\eta}D^2\sqrt{\frac{2\rho_m}{3}}\dot{\gamma} \tag{3.20}$$

This equation can be solved by using a numerical method [112]. The calculation shows that the new grain diameter (D) increases almost linearly with time (t), and after several microseconds, it shifts to an exponential growth. However, according to the Gurney equation, the current test is over in 2.4 ms, so grain growth would be extremely low at this point, suggesting that any real growth is not really happening during the loading of the test and is occurring purely thermally post-mortem. This calculation essentially arrives at the same result as that of Li et al. who calculated [111,112] the recrystallized grain size in a Ni–Cu–Al alloy, and the result (the grain size ~13–200 nm in diameter) is in reasonable agreement with TEM observations. This implies that the theoretical approach based on the kinetics of the recrystallized grains during dynamic deformation may be also valid in conditions of our test.

3.10 Deformation and Microstructure Within the Bands

3.10.1 Ultra-High-Strain-Rate Deformation

It is known that once localized shear deformation has commenced, steep strain and strain rate as well as temperature rise will appear in the shear bands. Li [111] calculated the deformation behaviour in the white band in α-Ti shown in Figure 3.26, where one can see that the grains were sheared for a relative movement; for example, the displacements from a to a′, b to b′ and c to c′ occur during localization. From this, it is found that the width of the band is 12 μm, and the average shear strain measured is 5.2. According to the equation of Dodd and Bai [114], the half-width of the shear band is

$$\delta = \left(\frac{\lambda\theta_*}{\sigma_*\dot{\gamma}_*}\right)^{\frac{1}{2}} \tag{3.21}$$

and the temperature rise in the band obtained by assuming that 90% of deformation work is converted into heat is

$$\theta_* = \frac{0.9\tau_*\gamma_*}{\rho c} \tag{3.22}$$

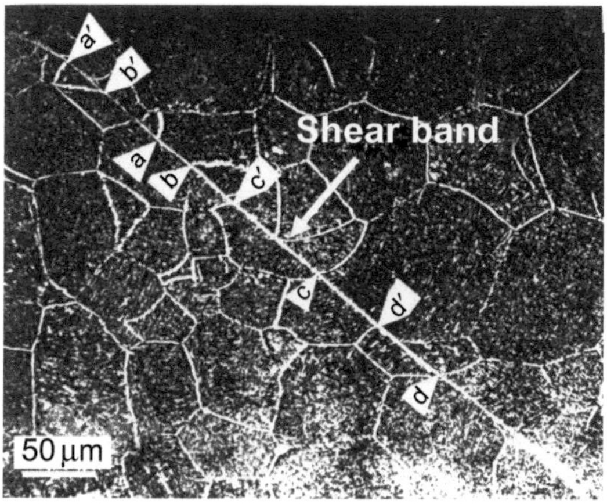

Figure 3.26 White shear band in α-Ti developed during dynamic compression [111].

The time required for shear-band formation can be obtained by the following equation, from Eqs (3.21) and (3.22):

$$t = \gamma_* / \dot{\gamma}_* = \frac{\rho c}{0.9\lambda} \delta^2 \quad (3.23)$$

where $\tau_*, \dot{\gamma}_*$ are the shear stress and strain rate, θ_* is the temperature rise in the band, λ is the coefficient of heat conduction, ρ is the material density and c is the heat capacity. The parameters used in Eq. (3.23) are $\rho = 4.5\ g/\mathrm{cm}^3$, $c = 0.473$ J/g K and $\lambda = 14.63$ W/m K for α-Ti. Therefore, $t = 0.1616\delta^2$, where the units of the t are microseconds, and the unit of the δ is micrometre. From Eq. (3.23), it is found that the time required for formation of the band is 5.8 μs, and, therefore, the strain rate that the shear band underwent is $9 \times 10^5\ \mathrm{s}^{-1}$, which is higher by two orders of magnitude than the average strain rate ($2 \times 10^3\ \mathrm{s}^{-1}$) required for shear-band formation, implying that formation of the shear band may accompany an abrupt increase in strain rate. In other words, the material within the band underwent an ultra-high-strain-rate deformation. This is confirmed further in Al–Li alloy [56]. Xing et al. [115] developed a numerical model that simulated the evolution of shear banding from several finite amplitude disturbances (FADs) in both temperature and strain rate and found that the shear banding process beyond plastic instability consists of two stages: inhomogeneous shearing and true shear banding. In the first stage, localized deformation occurs, and then localization is gradually restricted into a narrow band companying a burst in strain rate; both temperature and strain are raised as shown in Figure 3.27.

Giovanola [52] found that shear localization occurs in two sequential stages in VAR 4340 steel: during the first stage, the strain rate ($10^4\ \mathrm{s}^{-1}$) jumps by more than an order of magnitude to values larger than $10^5\ \mathrm{s}^{-1}$, then approaching

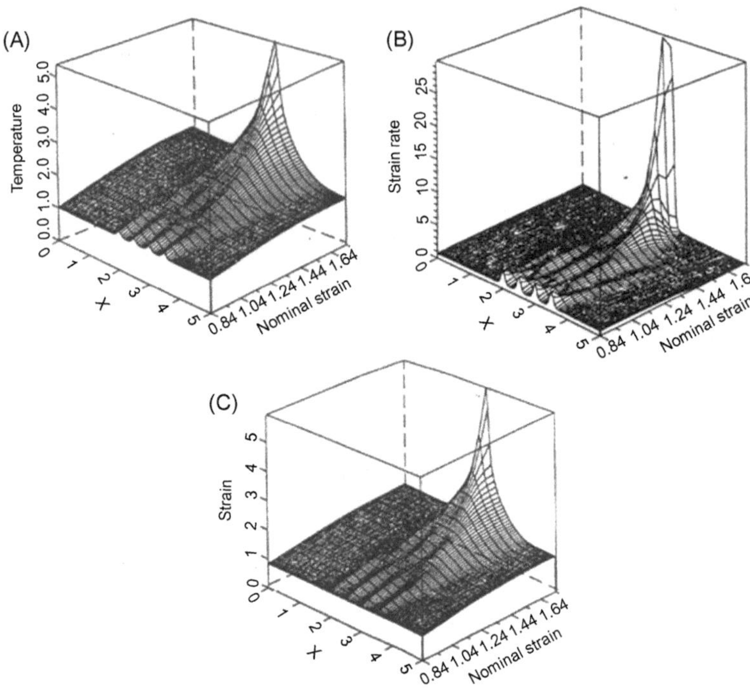

Figure 3.27 Simulation of evolution of localized deformation: (A) temperature, (B) strain rate and (C) strain [115].

$1.4 \times 10^6\ s^{-1}$ during second localization. The deformation behaviour with ultra high-strain rate in the band mentioned earlier was in good agreement with the numerical simulation [52,115].

A number of investigations show that a polycrystalline material with high-strain-rate sensitivity will appear to deform by a mode of super-plastic flow. First, the structures in the bands are very fine. Second, the temperature in the bands may reach and even exceed the melting temperature. These provide the conditions for super-plastic deformation of materials. Ashby et al. [116] proposed a constitutive equation to describe super-plastic flow of a material under high-strain rate. They suggested that when polycrystalline material is deformed at temperatures above 0.4 T_m (the melting temperature), one possible mode of super-plastic flow is a 'diffusion-accommodated flow'. Dodd and Bai [114] pointed out that, in the process of machining, although the average strain rate may be quite low, the strain rate in the bands of shear may be markedly higher. Murr et al. [117] have also suggested that the shearing deformation actually achieved inside an adiabatic band is extremely large, and the mechanism involves dynamic recrystallization (DRX) and super-plastic flow. Indeed, it has been proposed that the extensive plastic deformation undergone by shaped charges is connected to the nanocrystalline grain structure generated by the deformation of the liner [118].

3.10.2 Formation of the Shear-Band Network

In some cases, one can find large and complex networks of shear bands. To understand the formation of the band networks, Li [111,112] designed a special specimen with the shape of conical frustum of Ti alloy deformed dynamically. He found that localized deformation occurred on the cone, and the region deformed was limited to a thin conical shell, so that different types of shear-band networks were observed on the longitudinal (Figure 3.28A) and transverse (Figure 3.28B) sections through the central axis. From these observations, he pointed out that the adiabatic shear bands were formed initially at the location of the maximum shear stress, and the actual deformation localization zone was confined in a conical shell in this kind of specimen with axial symmetry [112,163].

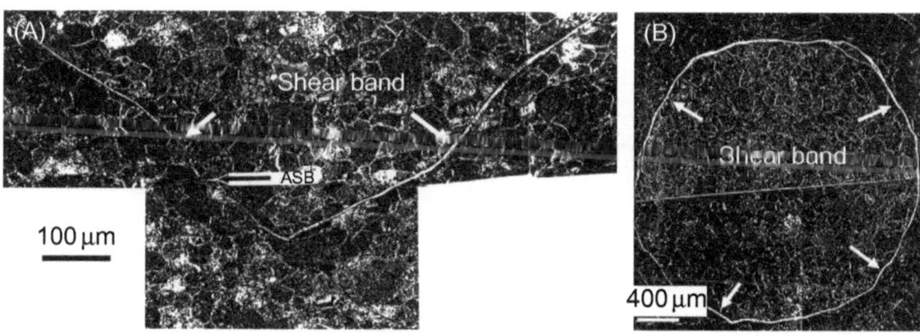

Figure 3.28 Complex shear-band networks formed in Ti alloy subjected to impact compression by a split-Hopkinson bar: the shear-band networks observed on the longitudinal (A) and transverse (B) sections through the central axis [111,112].

3.10.3 Deformation Structures

It is generally accepted that the tendency for dislocation cross-slip and dislocation pile-up is weakened when the stacking fault energy decreases in face-centered cubic (FCC) metals. Thus, the dislocations may extend into two partial dislocations, and there is a strip of stacking fault between them. Figure 3.29 shows a shear band (Figure 3.29A) and its magnifications (Figure 3.29B and C) taken from the circle area in Figure 3.29A. It is characterized by twins and stacking faults as well as partial dislocations. Analysis shows that these dislocations are the partials with a Burgers vector of $\frac{1}{6}\left[\bar{1}\bar{1}2\right]$. In addition, it is found that the stacking faults may overlap each other during movement on the parallel slip planes, and the intrinsic and extrinsic stacking faults are separated by the partial dislocations [81].

The effect of shock on a microstructure is known and has been documented by Murr [119]. Lee et al. [120] have investigated the microstructure of the band in

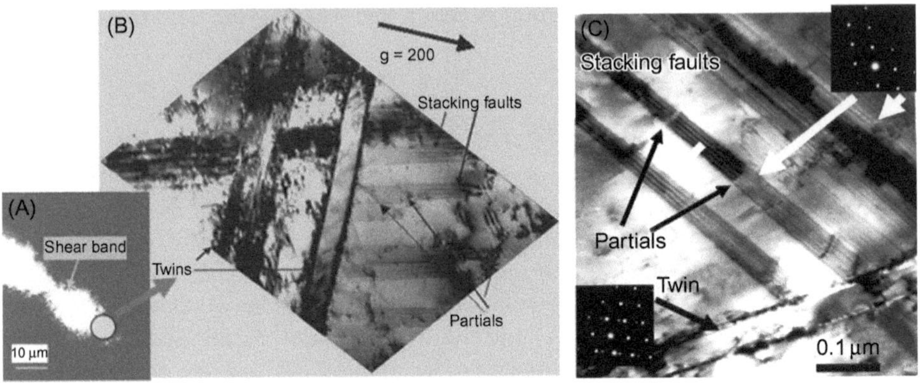

Figure 3.29 TEM images of a shear band at (A) low magnification, and dislocation structures of the band taken from the circle area in the band are shown in (B and C) [77].

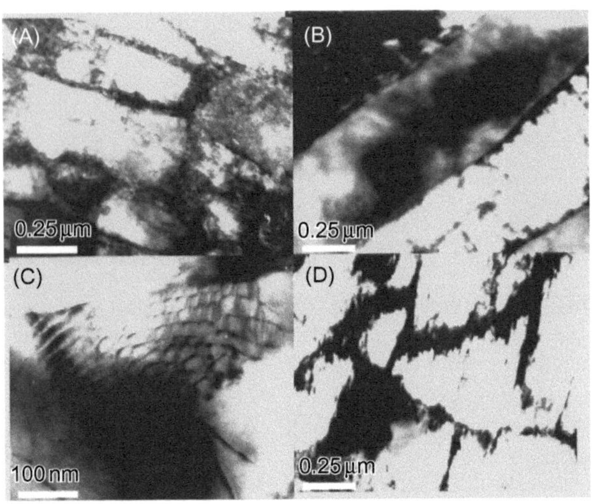

Figure 3.30 A set of the TEM images, showing the dislocation structure in the low-carbon steel subjected to dynamic torsional loading: elongated dislocation cells along the band and high dislocation density at the cell walls (A and B); dislocation network (C) and avalanche of cells (D) [26].

HY-100 steel and found that the centre of the bands contains a mixture of the highly elongated subgrains and fine equi-axed cells with high dislocation density, and that the misorientations of 80% of random cells have angles with less than about 5°. They proposed that the primary metallurgical process occurring in the bands is dynamic recovery, resulting in highly elongated subgrains as well as more equi-axed cellular structures. Meyers et al. [121] found that the equi-axed cells are replaced by the elongated cells and that these elongated cells inside the band break down and are replaced by small grains with a relatively low dislocation density as the strain is increased. Therefore, they attributed this kind of microstructure to dynamic recrystallization. Figure 3.30 is an image taken from a shear band in a

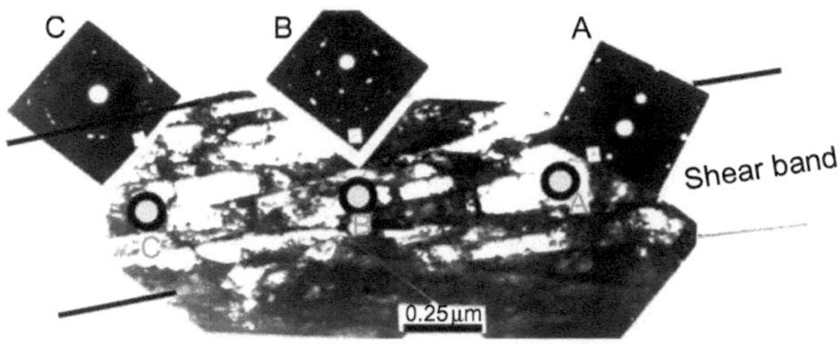

Figure 3.31 Microstructure within the shear band generated during dynamic loading in low-carbon steel [26].

low-carbon steel with the structure of the ferrite–cementite [26]. Several features are noteworthy. First, the α-ferrite in the band formed tangled structure and dislocation cells. The elongation and arrangement of the cells tend to be aligned along the shear direction, and the cell walls are full of dense dislocations as shown in Figure 3.30A. Second, the interfaces between the ferrite and cementite, as the sources, could release a great deal of dislocations (Figure 3.30B). Analysis indicates that the majority are dislocations with Burgers vector 1/2[111] lying on a plane $(1\bar{1}0)$ and $1/2[1\bar{1}\bar{1}]$ lying on a plane (110). These dislocations may interact with each other to form a stable network (Figure 3.30C) expressed by $1/2[111] + 1/2[1\bar{1}\bar{1}] \rightarrow [100]$. It might be expected that the formation of the tangles, cells and networks of the dislocations could be responsible for the work hardening of the material during dynamic deformation. Third, the avalanche of the dislocation cells observed in the bands corresponds to the sharp drop of the load in the $\tau - \gamma$ response curve shown in Figure 3.30D. Fourth, the strain distribution is heterogeneous from place to place along the band. This can be seen clearly from the comparison of the selected area diffraction pattern (SADPs) from the different regions in the bands as shown in Figure 3.31. A relatively simple pattern obtained from region A (marked by a circle) implies that the crystallographic nature of this area can be explained by slip deformation. On the other hand, however, in region C, it shows a spotty ring pattern, indicating the operation of multi-slip systems, arising from different subgrains with different suborientations. This examination reveals that the shear bands are highly localized deformation regions that may cross through many grains, implying that the shear bands may start on a governing slip system within one grain arising from the crystallographic slip, and then penetrate into adjacent grains by cooperative slip events that may involve a number of slip systems. This kind of cooperative slip may be assisted by both the stress concentration at the grain boundaries and the temperature rise in the band during localization.

Twinning is one of the basic plastic deformation modes in metals and alloys, in particular, under high-strain-rate loading. Figure 3.32 shows a structure composed of intersecting twins in a lightly deformed area outside the band in 304 stainless steel subjected to the cylindrical collapse (TWC) test with a strain rate of $10^4\ s^{-1}$. In Figure 3.32A, the thin foil orientation is close to [100], providing a perpendicular pattern of twins, which are on {111}. In Figure 3.32B, an incipient shear band is shown. The twins were distorted during localization. These twins have formed prior to the large deformation within the bands. They can be formed during the

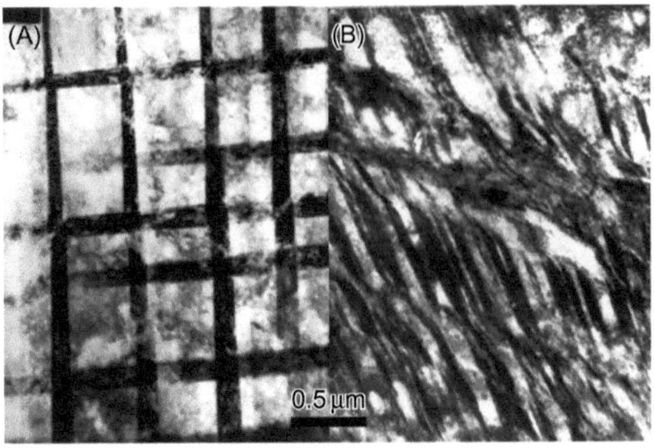

Figure 3.32 Primary twins inside (A) and high-order twins outside the band (B) [32].

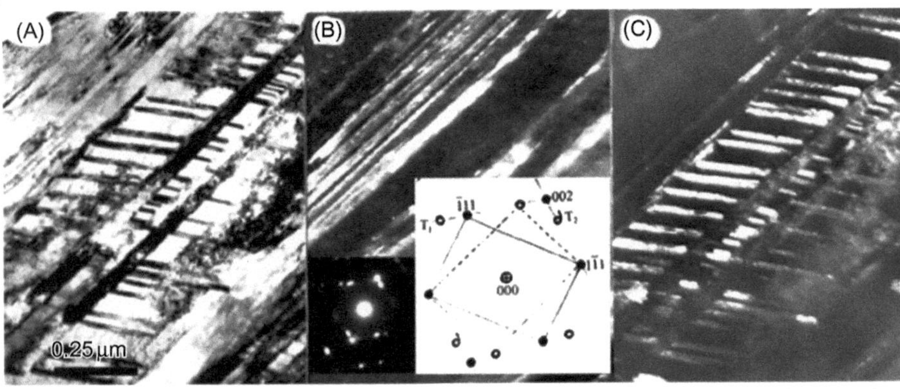

Figure 3.33 (A) BF image of the twins; (B) and (C) DF images of the first-order twins and second-order twins and corresponding diffraction patterns and their index (insert) [32].

passage of the first shock pulse, which precedes the implosion of the cylinder. Figure 3.33A shows the twins in greater detail. In addition to the first-order twins with a spacing of 0.1–0.2 μm, there are second-order twins with a much lower spacing (Figure 3.33C). The second-order twins imaged by HREM are shown in Figure 3.34. The atomic planes (111) and their change in orientation inside the twins are clear. These two-set twins are shown with a thickness of 20 nm each. A more detailed example is shown in Figure 3.35, displaying twinning and multi-order twinning structures within the bands produced in Ti–6Al–4V alloy subjected to explosive collapsed loading [81]. Figure 3.35A is a bright-field image, and Figures 3.35B and C show dark-field images by spots T_1 and T_2. All of these results indicate that twinning is a favoured dynamic and continuously proliferated process for a material under high-strain-rate loading.

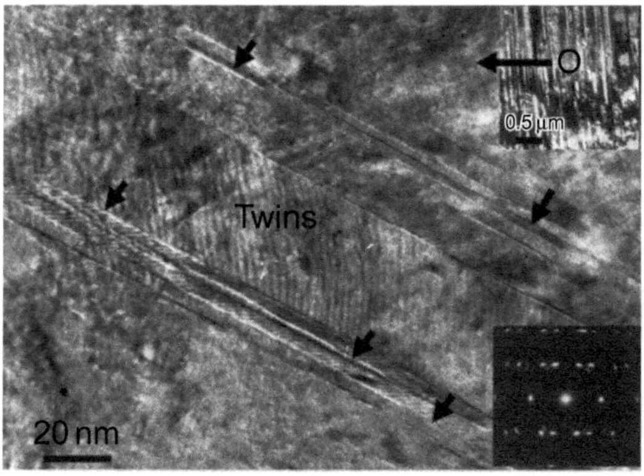

Figure 3.34 Multi-order deformation twins imaged by HREM [32].

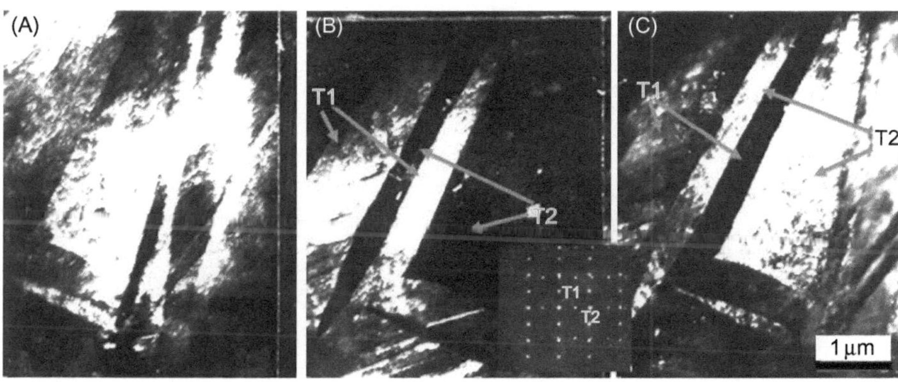

Figure 3.35 Multi-order deformation twins produced during explosive collapse of a TWC in Ti–6Al–4V alloy [81].

3.10.4 Damage and Fracture

The interrupted tests show that the sudden drop appearing in the $\tau-\gamma$ curve, leading to the fracture, is caused by initiation and coalescence of the micro-cracks along the bands, rather than the formation of the shear bands [25,26]. The localized shear deformation may promote further the nucleation, growth and coalescence of the micro-cracks, and therefore accelerate the fracture of material along the bands. The degree of the structural damage in the shear bands, leading to final failure, can be assessed by SEM as shown in Figure 3.36. The number of cracks per square millimetre, which is an indication of the degree of the structural damage, was determined as a function of the distance from the boundary between the shear bands and matrix to the centre of the band of the dual-phase (DP) steel subjected to dynamic torsional loading. It can be seen that the nearer the centre of the band, the bigger the structural damage extent [34]. This is verified by the result of Al–Li alloy [56]

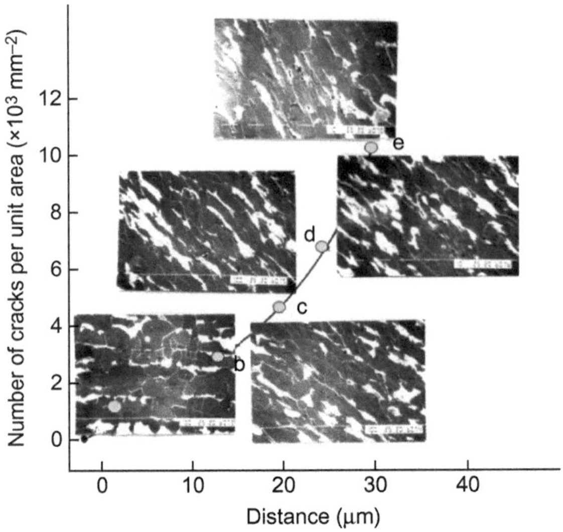

Figure 3.36 The relationship between the number of the cracks per square millimetre and the distance from the boundary of the band to the matrix in the shear band in low-carbon steel [34].

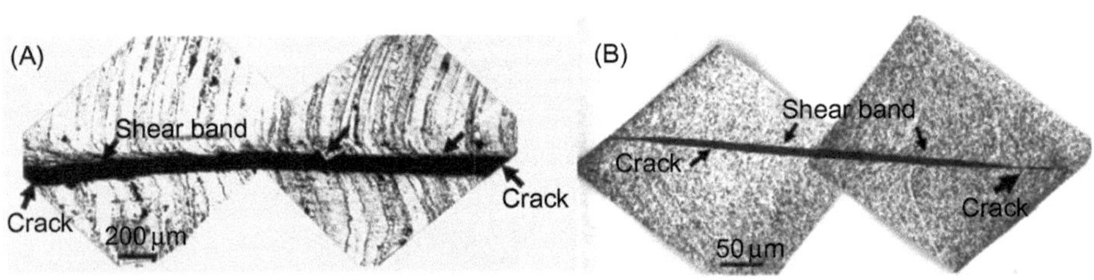

Figure 3.37 Fracturing along the shear bands in Al–Li alloy [56] (A) and SiC/Al composite (B) [90].

(Figure 3.37A) and Al/SiC composite [90] (Figure 3.37B), where the final break is obvious along the shear band. The fracture of a material subjected to dynamic loading was classified into both ductile and brittle modes [14,18,122]. Rogers reported [14] that if the fracture is ductile, it almost certainly occurred during adiabatic deformation when the band was hot and weak. On the other hand, brittle fracture occurs subsequently to the termination of deformation and the quenching of the hot band by the matrix material to form the hard, brittle, transformed structure. However, a number of observations of the fracture surface reveal that whether the fracture is along deformed-type bands in quenched martensite steels [26] or the white shear bands in Ti-55 alloys [63], their fracture topography appeared to be ductile characterized by shear dimples on the fracture surface. The material in the bands subjected to a large accumulated strain under high-strain rate, resulting in an abrupt temperature rise, seems to suffer an annealed or temper treatment. Therefore, the fracture morphology in the quenched steel deformed at high-strain rate appears to be similar to those in the annealed or tempered steels, and thus it is reasonable to propose that in addition to the stress-state condition, the temperature rise within the band plays an important role in failure of materials along the bands [14,26]. The temperature rise in the band is so high that it can cause recrystallization or even melting, leading to softening of material in comparison with that in the adjacent matrix. Timothy et al. [23] proposed that void formation in adiabatic shear bands in titanium alloys was intimately associated with thermal softening and local melting of the metal in the bands. This is the reason that, in all cases, the fracture surface topography observed is entirely of a ductile nature, irrespective of whether it is occurring in the deformed or white etching bands in quenched and tempered steels. Figure 3.38A shows voids that nucleated and grew inside the shear band in a Ti–6Al–4V alloy [31,81]. If, subsequent to void nucleation, there is a combined action under tension and shear stresses, these voids grow until their edges reach the boundary of the shear band. Then they do not grow into the surrounding material because of its higher yield strength. The voids became gradually ellipsoidal, as they elongate along the shear bands. This sequence is shown in Figure 3.38B. As they grow, they eventually coalesce, forming a crack. Zhou et al. [70] measured the velocity of propagation of the shear bands in a Ti–6Al–4V alloy impacted at a velocity of 50 m/s. Their propagation velocities were 5–50 m/s. Xue et al. [123] used the thick-walled cylinder (TWC) method, which had an initial wall velocity of 200 m/s and obtained a propagation velocity of 556 m/s, inferred from their experimental results. This great difference can be explained through Mercier and Molinari's analysis [124]: the imposed velocity externally determines the shear-band velocity V through the equation

$$V = \frac{\sigma_f}{\rho c} g\left(\frac{\lambda}{h}\frac{t}{h}\right) \tag{3.24}$$

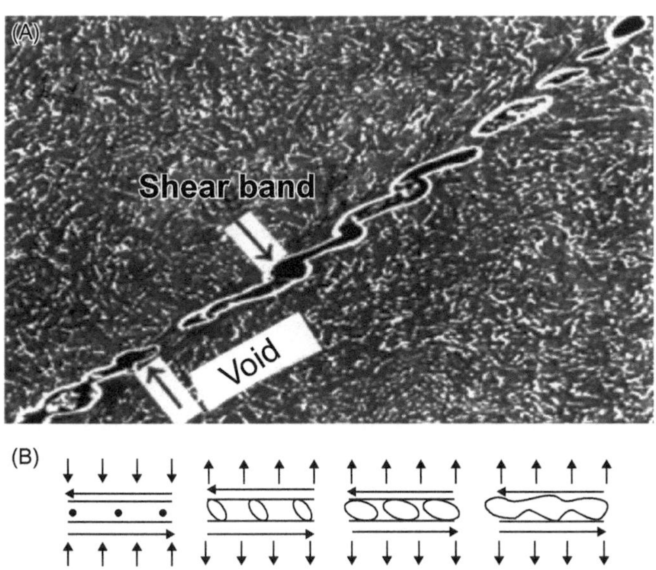

Figure 3.38 Initiation and coalescence inside bands (A) and schematic representation of sequence of events leading to ductile failure (B) [31,81].

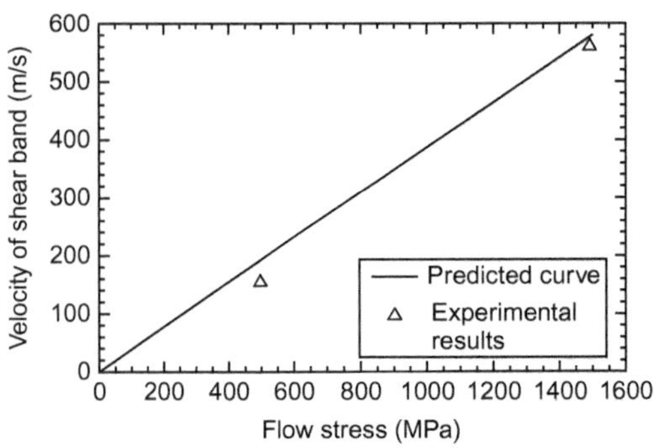

Figure 3.39 Shear-band velocity as a function of flow stress for Ti alloy; comparison of Mercier–Molinari's theory [124] with results by Xue et al. [31].

where σ_f is the material flow stress, c is the heat capacity, ρ is the density and g is a function of λ/h, the normalized process zone length, and t/h, the normalized shear-band thickness. The effect of flow stress on the velocity of propagation of the shear band is shown in Figure 3.39. Two experimental points are given: one for Ti and one for Ti–6Al–4V, with a flow stress approximately triple the one for pure Ti. There is a linear relationship between flow stress and V, for the same external conditions, and g. The experimental results by Xue et al. [123] compare well with the Mercier–Molinari predictions.

3.11 Spacing and Self-Organization of the Shear Bands [31,125]

Most of the current theories on prediction of spacing of shear bands are based on the analysis of a single shear band along its propagating direction.

We briefly describe the evolution of multiple adiabatic shear bands in commercially pure titanium and Ti–6Al–4V alloy through the radial collapse technique of a thick-walled cylinder (TWC) under high-strain-rate deformation. Shear-band initiation, propagation and spatial distribution were examined under increasing global strains. As an illustration, Figure 3.40A shows the pattern of helicoidally shear bands in a Ti–6Al–4V specimen, whereas Figure 3.40B shows the size and spatial distribution of shear bands at a global effective strain of 0.264. The shear bands nucleate at the internal boundary of the specimens and construct a periodical distribution at an early stage. The shear bands undergo bifurcation as they progress in their spiral trajectory and as their spacing increases. The shear bands are favoured initiation sites for failure, which occurs by void nucleation and growth, and then coalescence inside the thermally softening regions. The evolution of the morphology of the voids is determined by the restrictions imposed by the bands. Figure 3.41 shows that the shear-band spacing is quite dependent on the material, being much lower in 304 stainless steel than in Ti–6Al–4V. The differences of mechanical response between the two alloys are responsible for significant differences in the evolution of the shear-band patterns. The same differences are observed between titanium and Ti–6Al–4V. Figure 3.42 shows how the shear-band spacing and its evolution determine the

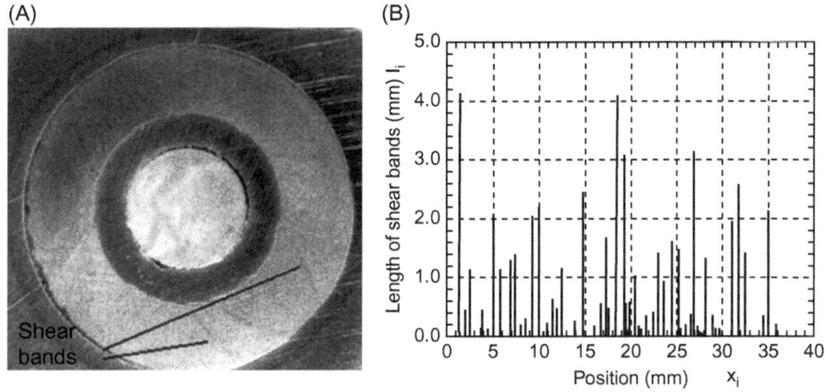

Figure 3.40 Shear-band pattern of Ti–6Al–4V alloy at well-developed stage $\varepsilon_{\text{eff}} = 0.254$: (A) shear-band pattern and (B) spatial distribution [31].

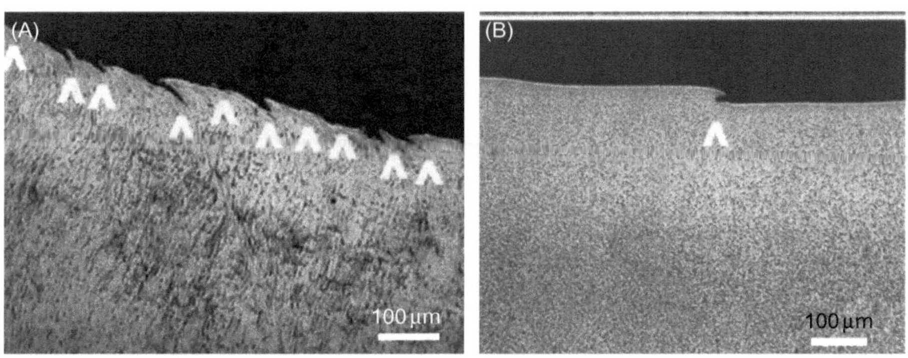

Figure 3.41 Comparison of spacing for stainless steel (A) and Ti–6Al–4V alloy (B) at initial stage [31].

fragmentation of a cylindrical specimen. The number of shear bands initiated in Ti (spacing of 0.18 mm) is considerably larger than in Ti–6Al–4V (spacing of 0.53 mm). The experimentally obtained shear-band spacings are compared with theoretical predictions [39,41,42] in Table 3.4. The shear-band interactions are more complex than previously thought, and their spacing cannot be predicted by the one-dimensional perturbation theories [41,42]. The Grady–Kipp theory cannot accommodate the increased spacing as the shear-band size increases because it is also one-dimensional. This was done by Xue et al. [31,125]. Thus, the treatment needs to incorporate the following elements:

- *Rate of nucleation of shear bands*: The probability of nucleation, $p(v_o, s_o)$, in a reference volume v_o, or surface s_o, was successfully described by a Weibull distribution in which the stress was replaced by strain as the independent variable. Parameters defining the distribution are a critical strain for nucleation, mean nucleation strain and Weibull

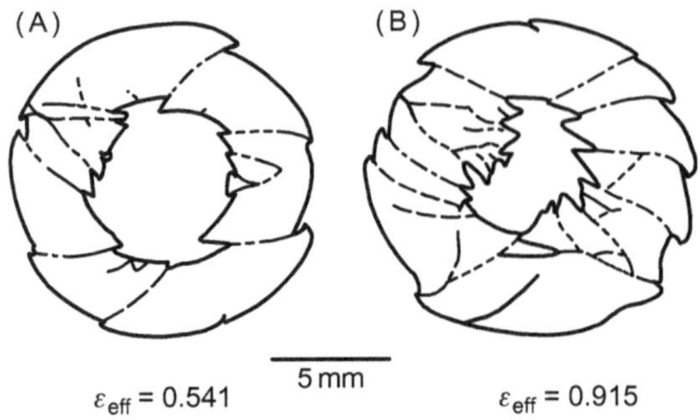

Figure 3.42 Evolution of the shear-band pattern in Ti–6Al–4V alloy [31].

Table 3.4 Prediction and Experimental Results for Stainless Steel and Titanium (WO: Wright Ockendon; GK: Grady Kipp; MO: Molinari)

Spacing (mm)	Experimental Data Initial Level	Experimental Data Developed	L_{WO} (mm)	L_{GK} (mm)	L_{MO} (mm)
SS 304L	0.12	3.2	0.17	2.62	0.16
CP Titanium	0.69	2.57	0.52	3.3	0.36

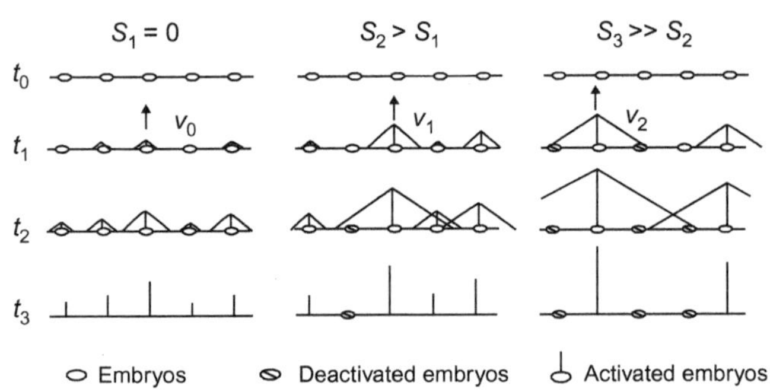

Figure 3.43 Two-dimensional representation of concurrent nucleation and shielding. S is a shielding parameter; as S increases, the release of stress produced by growing shear bands deactivates embryos. For $S = 0$, all embryos are activated, and shear-band spacing is small. For a large value of shielding S_3 (right side), only a fraction of embryos are activated.

modulus. There can also be shielding at the nucleation stage, depending on the relative values of the rate of nucleation and rate of growth.

- *Rate of growth or velocity of propagation*: This is an important factor in their self-organization. Shear bands compete among themselves and gradually change their patterns. A 'Darwinian' natural selection takes place, and a large number of small bands evolve gradually into a smaller number of large bands, due to the shielding of stresses produced during growth. Such evolution of shear-band pattern occurs under a homogeneously distributed pressure acting on the external boundary of the cylindrical specimen. This is a typical manifestation of self-organization. This is schematically rendered in Figures 3.43 and 3.44.

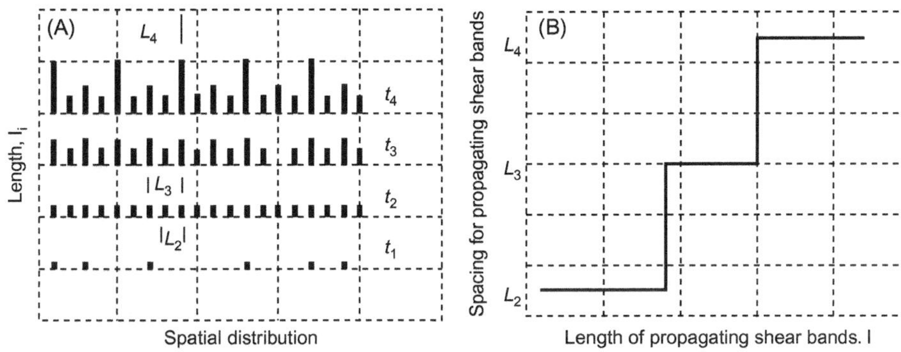

Figure 3.44 (A) Schematic diagram of the evolution of shear-band spacing at different levels. t_1, random initiation; t_2, self-organization into 'periodic' pattern among nuclei; t_3, some shear bands grow faster suppressing others; t_4, self-organization of developed shear bands. (B) Spacing of propagating shear bands as a function of length.

This theory was developed by Xue et al. [31,125] and applied to Ti, Ti–6Al–4V and 304SS. It clearly needs additional work.

3.12 Effect of Dynamic Strain Aging

Many alloys (steels, titanium and aluminium alloys) that undergo dynamic strain aging are also prone to shear localization. What is the effect of the changes in thermal softening and strain-rate sensitivity on the development of shear bands?

Dynamic strain aging is a phenomenon that has its origin in the interaction of mobile dislocations with solute atoms. Cottrell [126,127] proved that solute atoms could form atmospheres around moving dislocations and retard their motion. He proposed that the minimum strain rate at which serrated flow initiates is

$$\dot{\varepsilon} = C \exp\left(\frac{-Q}{kT}\right) \approx C_1 = 10^9 \text{cm}^{-2}, D \tag{3.25}$$

where Q is the activation energy for diffusion, C is a frequency factor, and D is the diffusion coefficient. The unit of D is cm^2/s.

The trapping of mobile dislocations by solute atoms produces a range of mechanical effects: serrations in the stress–strain curve (Portevin–Le Chatelier effect [128,129]), increased work-hardening rate, decreased ductility (the blue brittleness of steels) and a change in thermal softening, which leads, in extreme cases, to thermal hardening.

Dynamic strain aging has been observed in a number of alloys: steels [130,131], aluminium alloys [132], Ni–C alloys [133,134] and titanium alloys [135].

The literature abounds with dynamic strain aging measurements. The calculations are carried out herein for a low-carbon steel. The experimental results by

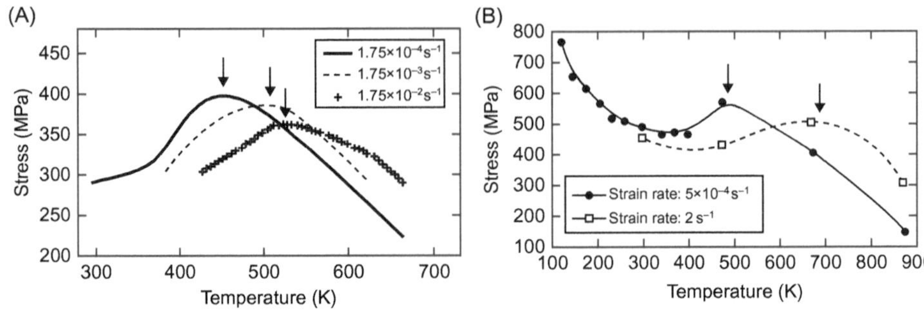

Figure 3.45 (A) Effect of temperature on flow stress (5% plastic strain) of AISI 1008 steel [137] and (B) effect of temperature on flow stress of AISI 1020 steel [138].

Bergstrom and Roberts [136] (0.035 wt%C), Li and Leslie [137] (AISI 1008, 1020, 1035) and Gilat and Wu [138] (AISI 1020) serve as a basis for establishing the experimental parameters for DSA. Figure 3.45A shows the flow stress (at 5% plastic strain) for AISI 1008 steel [137] as a function of temperature at strain rates of 1.75×10^{-4}, 1.75×10^{-3} and $1.75 \times 10^{-2}\ s^{-1}$. Two effects are seen: (a) the flow stress increases with temperature in the region of 350–500 K and (b) the peak temperature increases with the strain rate. Similar results are reported by Gilat and Wu (Figure 3.45B). The same effect is observed: a shift in the hump to higher temperatures as the strain rate is increased. The experiments carried out by Gilat and Wu [138] extend to a much higher strain rate, and the hump is at 720 K for a strain rate of $2\ s^{-1}$. It is also known that the temperature for the hump is dependent on the strain rate but independent of the carbon content. The carbon content was varied from 0.035% [136] to 0.2% [138].

There are two effects that will modify the susceptibility to shear-band formation under these conditions:

1. The thermal softening decreases, with actually thermal hardening. This will oppose the formation of shear bands. However, this is followed by an accelerated softening on the right side of the hump.
2. The strain-rate sensitivity becomes negative in a special region. This will produce and enhance susceptibility to shear localization.

The exact nature of this interaction needs to be better understood, but it can be stated that if thermal softening is increased and the strain-rate sensitivity is decreased (or even become negative), the tendency for shear localization is enhanced. This area needs additional research.

3.13 Shear Bands in Nanocrystalline Metals and Metallic Glasses

The formation of shear bands in nanocrystalline, ultrafine-grained and glassy metals (especially, BMGs) is discussed here. These materials are characterized by

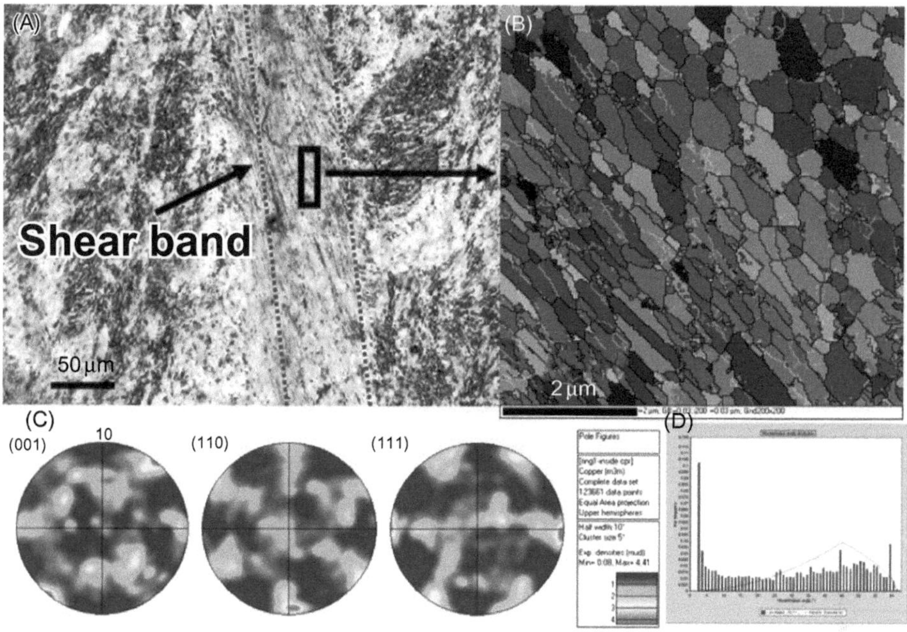

Figure 3.46 (A) SEM showing shear band with width ~50 μm, (B) EBSD image showing equi-axed and elongated recrystallized grains, (C) the pole figure and (D) the grain-size distribution.

low work hardening and are prone to shear localization, even when the external rate of load application is low.

Ultrafine-grained copper was subjected to the hat-shaped experiments, and forced bands were produced by Mishra et al. [139]. On close examination, the structure of these bands was quite different from that of the starting material, which has a grain size of ~200–400 nm. Figure 3.46 shows the width of the band, ~50 μm, and its structure as revealed by SEM–EBSD. The material in the band clearly underwent recrystallization, and the grains grew to a diameter much larger than the original ones. In contrast, forced bands obtained earlier in copper were much thicker. Shock-hardened copper had bands of ~200 μm.

The grains in the band are both elongated with an average long/short axis ratio of about 2 (Figure 3.46B) and equi-axed with an average size of ~500 nm in diameter. Most grains have high-angle boundaries of 15–60° (Figure 3.46D), and there is no evident texture with a maximum of pole density of about ~5.2 (Figure 3.46C) [140]. These observations confirm further the previous proposal mentioned in Section 3.9 concerning the static/dynamic recrystallization mechanism operated in Fe–Cr–Ni monocrystals produced under high strain and high-strain rate.

The Bai–Dodd equation (Eq. (3.21)) was applied to predict the thicknesses of the bands. The results, shown in Figure 3.47, are quite different. There is a large

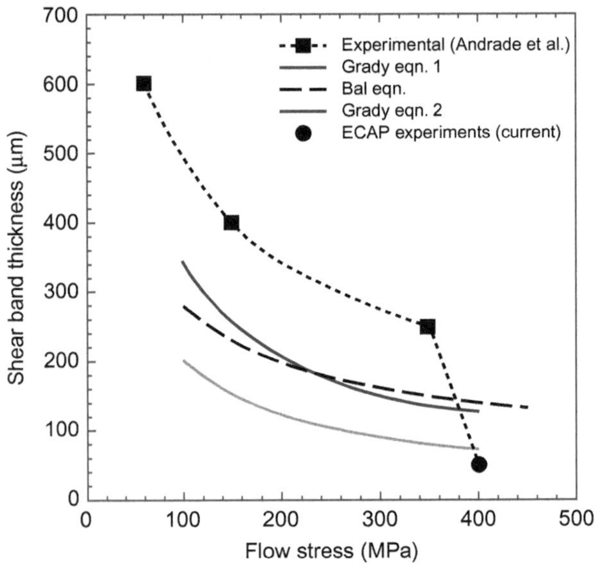

Figure 3.47 Shear-band thickness obtained from experiments as compared to prediction by Bai's and Grady's [141,142] equations. *Source*: From Ref. [139].

drop for the ultra-fine grained (UFG) copper. One of the reasons for this difference is thought to be the highly unstable grain structure that undergoes softening at a higher rate than the others. The Bai' equation does not contain a thermal softening term. Grady [141] proposed an equation that does. It is the term α in

$$\delta = \left(\frac{9\rho^3 C_p^2 \lambda^3}{\tau^3 \alpha^2 \dot{\gamma}}\right)^{1/4} \tag{3.26}$$

where λ is the thermal conductivity, and T, τ and $\dot{\gamma}$ are the temperature, stress and strain rate, respectively, inside the band. The thermal softening rate, α, assumes linear behaviour:

$$\tau = \tau_0(1 - \alpha T) \tag{3.27}$$

Grady [142] later proposed another equation, where the term 9 is replaced by 16. Figure 3.45 shows the prediction from these equations and the as-measured shear-band thickness from Cu experiments. Wright [59] states that the agreement of Bai's equation with experimental results is usually within a factor of 2. For the experiments carried out by Andrade et al. [143], there is a reasonable correspondence with both Bai's [44] and Grady's [141] equations. The trend of the experimental data parallels the two equations. Although both Bai's and Grady's equations predict a decrease in band thickness with increasing flow stress, the precipitous drop for the UFG copper has to be attributed to an additional mechanism for the loss of mechanical stability. This can be due to the increased propensity of thermal softening in equal channel angular pressing (ECAP) samples. In the original formulation of Grady [141], this parameter assumes a linear thermal softening with the melting point marking the zero strength value. To simulate a more drastic thermal softening of the UFG copper, consistent with the microstructural instability, it is assumed that the full softening is achieved at 400 K. The results are plotted as Eq. (2) in Figure 3.47. Thus, the agreement with the Grady equation for the UFG specimen can be significantly improved by modifying the linear softening parameter, α.

The process of dynamic plastic deformation (DPD) pioneered by Lu and coworkers produces grain sizes significantly below those reached in either ECAP or high-pressure torsion (HPT). The deformation process is applied at 77 K and at a high-strain rate, on the order of $10^2\ \mathrm{s}^{-1}$. This increases the Zener–Hollomon (Z–H) parameter of the process and consequently decreases the resulting grain size because the recrystallized grain size is found to correlate well with the Z–H parameter as it incorporates both temperature and strain rate:

$$\ln Z = \frac{1}{T}\ln\dot{\gamma} \tag{3.28}$$

These experiments were first conducted in pure copper by Li et al. [144] and then in copper–aluminium alloys by Hong et al. [145]. Plastic deformation in this regime produced a profusion of twins in addition to the ultrafine-grained structure. A third component of microstructural refinement set in at higher strains because of the lower temperature and higher strain rate: adiabatic shear-band formation. These are shown for Cu (Figure 3.48) and Cu–Al (Figure 3.49). The grain sizes are significantly lower than those obtained by Andrade (~200–400 nm), by virtue of the higher Z–H parameter. An increasing tendency of deformation twinning and shear

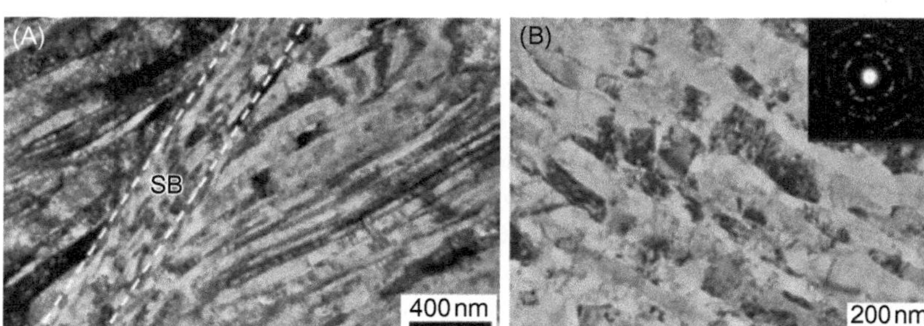

Figure 3.48 Shear band (A) produced in structure consisting of nanosized twins (B) generated by successively deforming copper at LN temperature dynamically ($10^2\ \mathrm{s}^{-1}$); the process is called DPD.
Source: From Ref. [144].

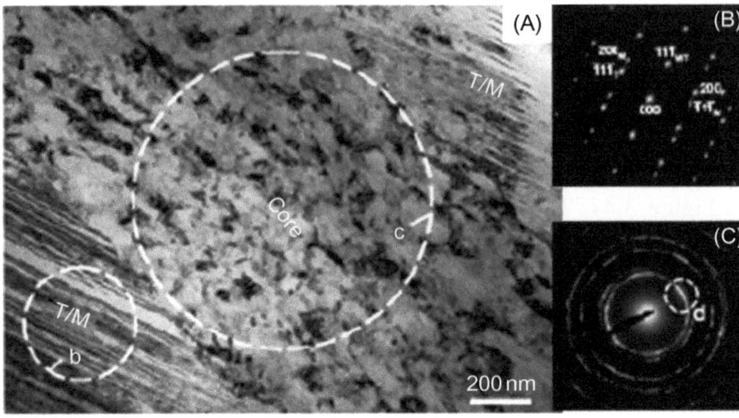

Figure 3.49 Shear band in a Cu–Al alloy subjected to DPD at 77 K; the shear band traverses a nanoscale structure characterized by twins with a spacing of 20 nm; grain size in shear band is ~60 nm. Diffraction patterns of (B) twinned region and (C) shear band. *Source*: From Ref. [141].

banding was observed at higher strains. For strain $e = 2.1$, a mixed nanostructure is formed in the DPD Cu bulk sample with nanoscale T/M lamellae making up about 33% of the volume, and nanosized grains making up about 67%. The nanograins can be classified into three types in terms of their origin: (a) nanograins derived from fragmentation of nanoscale T/M lamellae with an average transverse size of about 47 nm; (b) nanograins in shear bands with an average transverse size of about 75 nm; and (c) nanograins derived from dislocation cells with an average transverse size of about 121 nm. For Cu–Al the grain sizes are even lower, and these are indeed truly nanocrystalline structures. The mechanism by which a shear band forms according to Hong et al. [145], dissolving away the twin lamellar structure, is shown in the schematic sequence of Figure 3.50.

Metallic glasses deform primarily by the initiation and propagation of shear bands. The evolution of metallic glasses into BMGs enhanced the interest in these materials in a most significant manner. This was a development that owes a great deal to Johnson and co-workers [146,147] and more recently to Inoue [148]. Shear localization is observed frequently under both quasi-static and dynamic loading. The total uniform strain in BMGs is rather small. The release of the elastic energy stored in the system is the driving energy. There are two views on the softening mechanism leading to localization: thermal softening, observed by Lewandowski and Greer [149], and free-volume coalescence softening, proposed by Spaepen [150]. Dai et al. [151–153] performed experiments and calculations that indicate that both mechanisms are operative in the $Zr_{41.2}Ti_{13.8}Cu_{12.5}Be_{22.5}$ BMG.

Wright et al. [154,155] observed that the shear bands in metallic glasses loaded quasi-statically propagate a certain distance and then stop. This was accompanied by significant load drops. Some of these are shown in Figure 3.51.

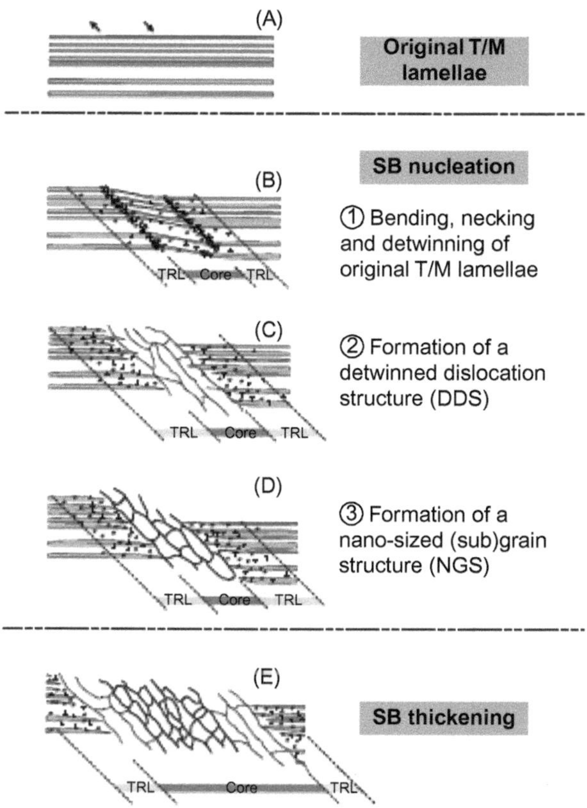

Figure 3.50 Sequence by Hong et al. [145] for the formation of shear bands in nanotwinned Cu–Al alloy.

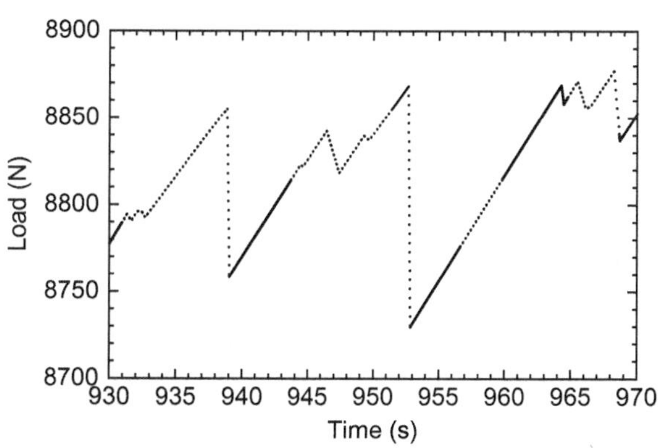

Figure 3.51 Load drops associated with shear-band formation in Pd40Ni40P20. *Source*: From Ref. [154].

They observed the failure surface, and there was a clear evidence for melting, from spheroidal droplets. They calculated the temperature rise but concluded that it was low and could only reach significant levels during fracture. However, there are parameters in the calculations that are not well known, and assuming a rapid

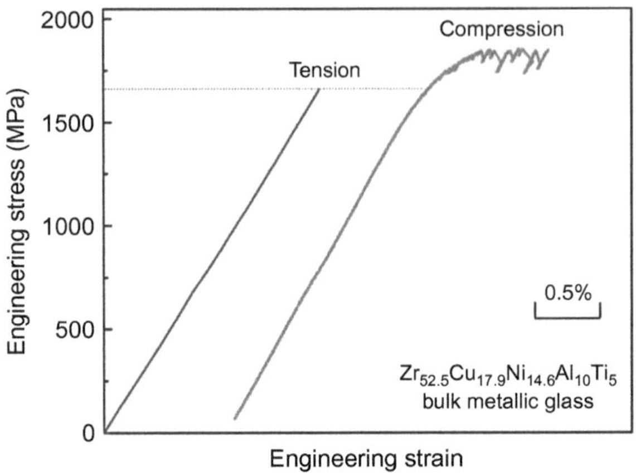

Figure 3.52 Engineering stress–strain curves of $Zr_{52.5}Cu_{17.9}Ni_{14.6}Al_{10}Ti_5$ (at.%) BMG in tension and compression; difference in strength ~15%. Notice load drops in the compression curve. *Source*: From Ref. [156].

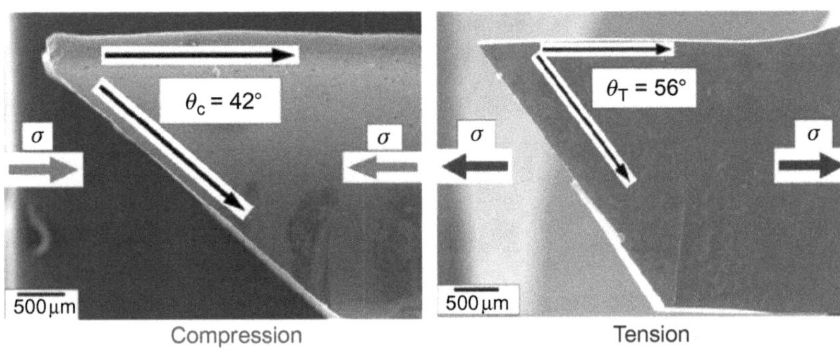

Figure 3.53 Typical shear fracture modes of Zr-based BMG [$Zr_{52.5}Cu_{17.9}Ni_{14.6}Al_{10}Ti_5$ (at.%)].

propagation of the shear band under adiabatic conditions, significantly larger values are possible. Unknowns are the thickness of the shear band and the velocity of propagation.

Experiments by Zhang and co-workers [156,157] show that there is little difference between the compressive and tensile strength of BMGs. Two typical curves are shown in Figure 3.52 for $Zr_{52.5}Cu_{17.9}Ni_{14.6}Al_{10}Ti_5$. The compressive strength is slightly (~15%) higher than the tensile strength. Load drops can be seen in compression testing, of the same nature as the ones observed by Wright et al. [154,155]. The failure in both tension and compression occurs by shearing, as demonstrated by the fractured specimens shown in Figure 3.53. The deviation in the angles from 45° was explained by Zhang et al. [158] in terms of a new failure criterion, modified from Mohr–Coulomb.

We suggest here that a possible explanation for the differences between the compressive and tensile responses can be due to thermal effects in the shear band. Figure 3.54 shows a sequence of events for a possible scenario in compression

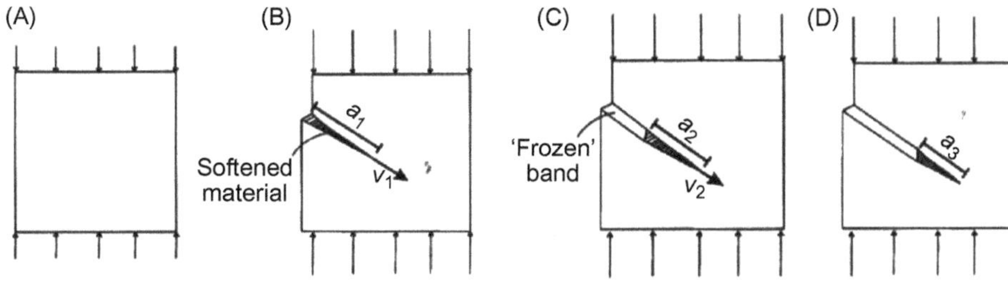

Figure 3.54 Schematic sequence showing how a shear band initiates, propagates and arrests in a metallic glass subjected to quasi-static compression.

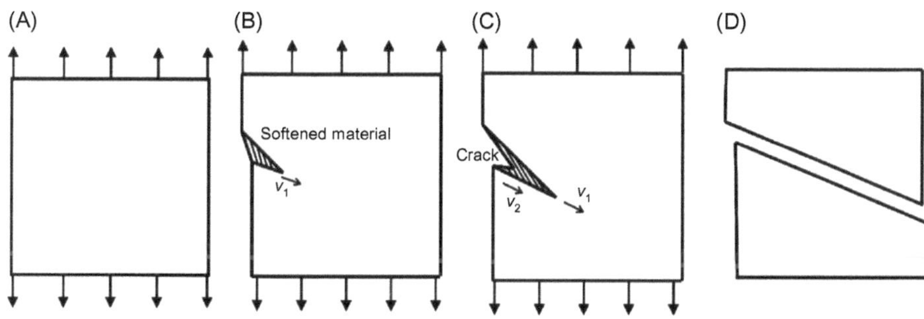

Figure 3.55 Schematic sequence showing how a shear band initiates in tension and how a crack propagates along the softened path eventually overcoming band and fracturing BMG.

testing. A shear band initiates at the surface of the specimen (Figure 3.54B) and propagates at a velocity V_1. The thermally softened region has a length a_1. As the band propagates down (Figure 3.54C), the region away from the band cools down and can 'freeze' back. This reduces the driving energy for the band and reduces the length of the softened region to a_2. Thus, one would expect an associated reduction in propagation velocity to V_2. This velocity is reduced to zero in Figure 3.54D, when the softened region length reaches a critical size a_3. Each formation and arrest event leads to a load drop. Thus, the BMG exhibits a limited ductility in compression, propitiated by the successive formation and 'freezing' of shear bands.

On tensile deformation, the situation is different. After a shear band forms, softening the material, a crack can easily initiate at the surface and follow the advance of the shear band. In this scenario, shown by the sequence of Figure 3.55, the specimen fractures after the first band is initiated and propagated.

The scenarios presented in Figures 3.54 and 3.55 require thermal softening inside the band and support the hypothesis of thermal softening as an important drive for shear localization. The TEM results by Jiang and Atzmon [159] on Al90Fe5Gd5 are intriguing. They observed nanocrystallization in both the shear

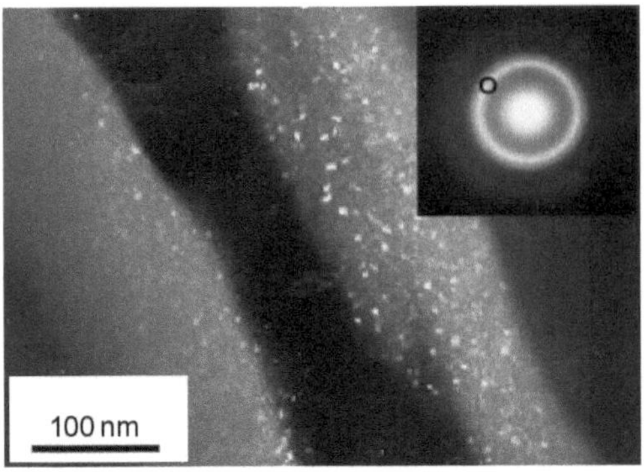

Figure 3.56 Nanocrystallization in shear band formed in compression deformation of Al90Fe5Gd5 BMG [153].

band and fracture surface. They attributed the precipitation of nanocrystallites in the compressive region in the shear bands to a kinetic effect due to the uniformly distributed free volume and the formation of the nanocrystallites at the fracture surfaces to adiabatic heating induced by fracture. The nanocrystals, having a few nanometres size and evenly distributed inside the shear band, formed in compression are shown in Figure 3.56.

It should be emphasized that different mechanisms operate under different conditions in different BMGs. Hence, the occurrence of thermal softening in no manner diminishes the importance of free-volume softening, which might dominate under specific conditions.

It is interesting that the shear-band spacing also plays a role in deformation of metallic glasses. Conner et al. [160] measured the spacing in quasi-static bending of a Zr-based BMG and found it to be ~0.1 mm initially. Eventually, a few shear bands grow and dominate the process as in the case of dynamic deformation, discussed in Section 3.11.

3.14 Summary

In this chapter, we presented the results of individual and collaborative investigations on the microstructural aspects of shear localization in materials under high strain and high-strain rates with emphasis on our contributions. We also reviewed and discussed some important phenomena such as the evolution process of shear localization; the experimental criterion for shear-band formation; possible mechanisms for the generation of the deformed and transformed bands; dynamic/static recrystallization occurrence; phase transformation and transition of the crystalline lattice to amorphous phase; super high-strain-rate deformation and microstructures within the band; and shear localization in ultrafine-grained and nanostructured metals as well as metallic glasses.

Extreme deformation under high strain and high-strain rate is unique, enormously complex and beyond the realm of technology that utilizes common stress–strain diagrams. Deformation at rates involving explosions, penetrating shatter, fragmentation and so on are also impossible to monitor and require innovative approaches to interpret the phenomena and behaviour of matter in periods of several millionths of a second. A number of scientific questions still remain open, and further investigations and measurements are needed.

- Real-time experimental examination of the 2D deformation in a propagating shear band has two components:
  - Propagation velocity as a function of applied stress. The experiments developed by Zhou et al. [70] reveal an extremely important characteristic of shear bands: their propagation velocity. More systematic experimentation is needed to understand how the velocity is related to the constitutive response in metals and alloys. Is it possible to develop a generalized theory? Is the Mercier–Molinari [124] theory applicable over a broad spectrum?
  - The temperature rise plays a key role in the formation of the shear bands but has not been determined precisely at present.
- Thermal softening at high rates. The microstructural changes that are well documented undoubtedly affect the thermal softening. However, the exact nature of this softening is virtually unknown. The thickness and spacing of bands are directly dependent on this softening. The constitutive approach by Molinari and Ravichandran [161] might be a good effort in that direction.
- A better understanding of the rate sensitivity of work hardening is needed for improved constitutive description. The Los Alamos MTS model pioneered by Kocks and co-workers [162] incorporates this effect through a dislocation evolution term, but the physical basis needs experimental verification.
- Concerning the ultrafine-grain size within the shear bands, what are the contributions of rotational recrystallization, post-deformation recrystallization and grain growth processes on the recovered microstructure?
- The self-organization of the bands is a new topic of study requiring both experimental and theoretical investigations. Further experimental investigation is needed to provide the modelling of such complicated behaviour.
- Both shear-band bifurcation and the development of a quantitative treatment of a shear-band toughness (initially proposed by Grady [141,142]) are desirable and necessary areas of research.

Advanced sample preparation and characterization tools as well as diagnostics are significantly accelerating our understanding of the nature of the phenomenon. The focused ion beam (FIB), TEM, HREM and FESEM–EBSD techniques will play an important role in this endeavour.

Finally, although there is rather good relation between theoretical analysis and experimental results, the experimental data that can be directly compared with these analyses to verify and guide model development of shear localization are still scarce. Materials researchers need to provide clear descriptions of the nano/micro/mesostructural evolution, thus enabling mechanicians performing theoretical analyses to arrive at more complete constitutive descriptions. Mechanics and materials researchers should come together more often to accelerate the understanding of the phenomena.

Acknowledgements

This research was supported by the National Natural Science Foundation of China Grant No. 50071064, 19891180-2 and 19392300; by The Chinese Academy of Sciences under Special Grant 87-52; by the US Army Sciences Office MURI programs at UCSD in the 1988–1998 period; by the US National Science Foundation Division of Engineering (Institute for Mechanics and Materials). The Chinese Academy of Sciences is acknowledged by MAM for having granted him a Visiting Professorship for Senior International Scientists at the Institute of Metal Research, in Schenyang.

Collaborations and discussions with former students and colleagues, Drs Z.C. Li, T.W. Wright, W.L. Zhong, J.Q. Yu, Y.J. Chen, Q. Xue, J.C. La Salvia, X.L. Huang, G.T. Gray, D. Kong, H. Li, E. Cerreta, B. Kad, M. Ehlers, U.R. Andrade, J.H. Beatty, T. Perez-Prado, C.L. Wittman, Profs K. Lu, L. Lu, V.F. Nesterenko, L.W. Meyer, K.S. Vecchio, Q. Li, T. Shen, Z. Ling, J.H. Zhang, G. Subhash, Z.F. Zhang and Z.G. Wang are greatly appreciated. Professor G. Ravichandran generously contributed to the last section on suggested areas for future research and shared his wisdom with the authors. Especially, Dr W.L. Zhong, Dr J.Q. Yu, Dr Z.C. Li, Dr H.J. Yang, Dr B. Kad and Prof. Q. Li have joined this joint program and provided a number of micrographs and much experimental data for this paper. Their contributions to this project are gratefully acknowledged. All photographs were taken by TEM, HREM and EBSD in the State Key Laboratory of Fatigue and Fracture for Materials, Atomic Image Laboratory of Chinese Academy of Sciences, Oak Ridge National Laboratory and UCSD Electron Microscopy Center.

References

1. Zener, C. and Hollomon, J.H. (1944). Effect of strain rate upon plastic flow of steel. J. Appl. Phys. 15, 22–32.
2. Recht, R.F. (1964). Catastrophic thermoplastic shear. J. Appl. Mech. 31, 189–193.
3. Culver, R.F. (1973). Thermal instability strain in dynamic plastic deformation. In Metallurgical Effect at High Strain Rates (R.W. Rohde, B.M. Butchern, J.R. Holland and C.H. Karnes eds.). Plenum Press, New York, NY, p. 519.
4. Bai, Y.L. (1981). A criterion for thermo-plastic shear instability. In Shock Waves and High-Strain-Rate Phenomena (L.E. Murr and M.A. Meyers eds.). Plenum Press, New York, NY, p. 277.
5. Burns, T.J. and Trucano, T.G. (1982). Instability in simple shear deformation of structural softening materials. Mech. Mater. 1, 313–326.
6. Pan, J. (1983). Perturbation analysis of shear strain shear in rate sensitive materials. Int. J. Solids Struct. 19, 153–164.
7. Semiatin, S.L., Staker, M.S. and Jonas, J.J. (1984). Plastic instability and flow localization in shear at high rates of deformation. Acta Mater. 32, 1347.
8. Wu, F.H. and Freund, L.B. (1984). Deformation trapping due to thermoplastic stability in one dimensional wave propagation. J. Mech. Phys. Solids 32, 119–132.
9. Freessengeas, C. and Molinari, A. (1987). Instability and localization of plastic flow in shear at high strain rates. J. Mech. Phys. Solids 35, 185–211.
10. Drew, D.A. and Flaherty, J.E. (1984). In Phase Transformation and Material Instabilities in Solids, M.E. Gurtin (ed.). Academic Press, Orlando, FL, p. 37.

11. Tvergaard, V.J. (1987). Effect of yield surface curvature and void nucleation on plastic flow localization. J. Mech. Phys. Solids 35, 43.
12. Anderson, M.A., Fleck, N. and Johnson, K.L.J. (1990). Localization of plastic deformation in shear due to microcracks. J. Mech. Phys. Solids 38, 681.
13. Lemonds, J. and Needleman, A. (1986). Finite element analysis of shear localization in rate and temperature dependent solids. Mech. Mater. 5, 339–361.
14. Rogers, H.C. (1979). Adiabatic plastic deformation. Annu. Rev. Mater. Sci. 9, 283–311.
15. Timothy, S.P. (1987). The structure of adiabatic shear bands in metals: a critical review. Acta Mater. 35, 301–306.
16. Murr, L.E. and Pizana, C. (2007). Dynamic recrystallization: the dynamic deformation regime. Metall. Mater. Trans. 38A, 2611–2628.
17. Dormeval, R. (1987). The adiabatic shear phenomenon, In "Materials at High Strain Rates", ed. by T.Z. Blazynski, London and New York. Elsevier Applied Science Publishers LTD, pp. 47–67.
18. Meyers, M.A. (2001). Encyclopedia of Materials: Science and Technology. Elsevier, pp. 7093–7103.
19. Malley, S.M. (2007). Shear localization: a historical overview. Metall. Mater. Trans. 38A, 2629–2654.
20. Xu, Y.B., Zhang, J.H., Bai, Y.L. and Meyers, M.A. (2008). Shear localization in dynamic deformation: microstructural evolution. Metall. Mater. Trans. 39A, 811–843.
21. Marchand, A. and Duffy, J. (1988). An experimental study of the formation process of adiabatic shear bands in a structural steel. J. Mech. Phys. Solids 36, 251–283.
22. Me-Bar, Y. and Shechtman, D. (1983). On the adiabatic flow of Ti–6Al–4V ballistic targets. Mater. Sci. Eng. 58, 181–188.
23. Timothy, S.P. and Hutchings, I.M. (1985). Initiation and growth of microfractures along adiabatic shear bands in Ti–6Al–4V. Mater. Sci. Technol. 1, 526–530.
24. Hartley, K.A., Duffy, J. and Hawley, A.H. (1987). Measurement of the temperature profile during shear band formation in steels deforming at high strain rates. J. Mech. Phys. Solids. 35, 283–301.
25. Bai, Y.L., Xue, Q., Xu, Y.B. and Shen, L.T. (1994). Characteristics and microstructure in the evolution of shear localization in Ti–6Al–4V alloy. Mech. Mater. 17, 155–164.
26. Xu, Y.B., Bai, Y.L. and Shen, L.T. (1996). Formation, microstructure and development of the localized shear deformation in low-carbon steels. Acta Mater. 44, 1917–1926.
27. Derep, J.L. (1987). Microstructure transformation induced by adiabatic shearing in armor steel. Acta Mater. 35, 1245–1249.
28. Giovanola, J.H. (1988). Adiabatic shear banding under pure shear loading, Part II: Fractographic and metallographic observations. Mech. Mater. 7, 73–87.
29. Cho, K., Chi, Y.C. and Duffy, J. (1990). Microscopic observation of adiabatic shear bands in three different steels. Metall. Mater. Trans. 21A, 1161–1175.
30. Liao, S.C. and Duffy, J. (1998). Adiabatic shear bands in a Ti–6Al–4V titanium alloy. J. Mech. Phys. Solids 46, 2201–2231.
31. Xue, Q., Meyers, M.A. and Nesterenko, N.F. (2002). Self-organization of shear bands in titanium and Ti–6Al–4V alloy. Acta Mater. 50, 575–596.
32. Meyers, M.A., Xu, M.A., Xue, Q., Perez-Prado, M.T. and McNelley, T.R. (2003). Microstructural evolution in adiabatic shear localization in stainless steel. Acta Mater. 51, 1307–1325.
33. Batra, R.C. and Zhang, G.M. (2008). Modified smoothed particle hydrodynamics (MSPH) basis functions for meshless methods, and their application to axisymmetric Taylor impact test. J. Comput. Phys. 227, 1962–1981.

34. Xu, Y.B., Wang, Z.G., Huang, X.L., Shen, L.T. and Bai, Y.L. (1989). Microstructure of shear localization in low carbon ferrite-pearlite steel. Mater. Sci. Eng. A114, 81–87.
35. Meyers, M.A., Armstrong, R.W. and Kirchner, H. (1999). The Mechanics-Materials Fundamentals and Linkages (eds. Meyers, M.A., Armstrong, R.W., and Kirchner H. John Wiley & Sons, INC. New York, Chichester, Wanheim, Brisbane, Toronta, Singapore. pp. 1–15(Chapter).
36. Guo, K.H. (1978). Electron Optical Microanalysis. Scientific Press, China, pp. 1–89.
37. Humphreys, F.J. (2001). Characterization of fine-scale microstructures by electron backscatter diffraction (EBSD). J. Mater. Sci. 36, 3833–3854.
38. Molinari, A. and Clifton, R.J. (1987). Analysis characterization of shear localization in thermoviscoplastic materials. Transaction of the ASME, Journal of Applied Mechanics 54, 806–812.
39. Grady, D.E. and Kipp, M.E. (1987). The growth of unstable thermoplastic shear with application to steady-wave shock compression in solids. J. Mech. Phys. Solids 35, 95.
40. Kuriyama, S. and Meyers, M.A. (1986). Numerical modeling of the propagation of an adiabatic shear band. Metall. Mater. Trans. 17A, 443.
41. Wright, T.W. and Ockendon, H. (1996). A scaling law for the effect of inertia on the formation of adiabatic shear bands. Int. J. Plast. 12, 927.
42. Molinari, A. (1997). Collective behavior and spacing of adiabatic shear bands. J. Mech. Phys. Solids 45, 1551.
43. Clifton, R.J., Duffy, J., Hartley, K.A. and Ghawki, T.G. (1984). On critical conditions for shear band formation at high-strain rates. Scr. Mater. 18, 443–448.
44. Bai, Y.L., Cheng, C. and Yu, S.B. (1986). On evolution of thermo-plastic shear band. Acta Mech. Sinica. 2, 1–7.
45. Shawki, T.G. and Clifton, R.J. (1989). Shear band formation in thermal viscoplastic materials. Mech. Mater. 8, 13–43.
46. Liao, S.C. and Duffy, J. (1998). Adiabatic shear bands in a Ti–6Al–4V titanium alloy. J. Mech. Phys. Solids 46, 2201–2231.
47. Backman, M.E., Finnegan, S.A., Shulz, J.C., and Pringle, J.K. (1986). In Metallurgical Applications of Shock-Wave and High Strain-rate phenomena, eds. L.E. Murr, K.P. Staudhammer and M.A. Meyers, Marcel Dekker, New York and Basle, pp. 675.
48. Bai, Y.L. (1989). Evolution of thermo-visco-plastic shearing. In Proceedings of Materials at High Rates of Strain (J. Harding, ed.), JOP Publishing LTD., Bristol, pp. 99–100.
49. Dodd, B. and Bai, Y.L. (1987). Ductile Fracture and Ductility with Application to Metalworking. Academic Press, London.
50. Costin, L.F., Crisman, E.E., Hartley, R.H., and Duffy, J. (1979). In Proceedings of the 2nd Conference on Mechanical Properties at High Rates of Strain (J. Harding ed.). Institute of Physics, Briston and London, p. 90.
51. Hartley, K.A., Duffy, J. and Hawley, R.H. (1987). Measurement of the temperature profile during shear band formation in steels deformation at high strain rates. J. Mech. Phys. Solids 35, 283–301.
52. Giovanola, J.H. (1988). Adiabatic shear banding under pure shear loading. Part I: Direct observation of shear localization and energy dissipation measurements. Mech. Mater. 7, 59–71.
53. Guduru, P.R., Rosakis, A.J. and Ravichandran, G. (2001). Dynamic shear band: an investigation using high speed optical and inferred diagnostics. Mech. Mater. 33, 371–402.
54. Xue, Q., Shen, L.T. and Bai, Y.L. (1995). A modified split Hopkinson torsional bar in studying shear localization. Meas. Sci. Technol. 6, 1557–1566.

55. Xue, Q., Shen, L.T. and Bai, Y.L. (1955). Elimination of loading reverberation in the split Hopkinson torsional bar. Rev. Sci. Instrum. 66, 5298–5304.
56. Xu, Y.B., Zhong, W.L., Chen, Y.J., Shen, L.T., Liu, Q., Bai, Y.L. and Meyers, M.A. (2001). Shear localization and recrystallization in dynamic deformation of 8090 Al–Li alloy. Mater. Sci. Eng. 299A, 287–295.
57. Timothy, S.P. and Hutchings, J.M. (1984). In Proceedings of 3rd International Conference on Mechanical Properties at High Rates of Strain (J. Harding ed.). Institute of Physics, Bristol, p. 397.
58. Molinari, A. (1988). Shear band analysis. On Nonlinear Phenomena in Material Science. Aussois France, 10–18 Sept. (G. Martin and L.P. Kubin eds.). Trans. Tech. Publications, Aedermannsdor, Switzerland, Vol. 3–4, pp. 447–468.
59. Wright, T.W. (2002). The Physics and Mathematics of Shear Bands, Cambridge Monographs on Mechanics, Cambridge University Press.
60. Staker, M.R. (1981). The relation between adiabatic shear instability strain and material properties. Acta Mater. 29, 683–689.
61. Zurek, A.K. (1994). The study of adiabatic shear band instability in a pearlite 4340 steel using a dynamic punch test. Metall. Mater. Trans. 25A, 2483–2489.
62. Bai, Y.L. (1982). Thermo-plastic instability in simple shear. J. Mech. Phys. Solids 30, 195–207.
63. Xu, Y.B., Yu, J.Q., Shen, L.T. and Bai, Y.L. (2000). The thermoplastic shear localization in titanium alloys during dynamic deformation. Mater. Sci. Technol. 16, 609–611.
64. Me-Bar, Y. and Shechtman, D. (1983). On the adiabatic flow of Ti-6Al-4V ballistic targets. Mater. Sci. Eng. 58, 181–188.
65. Grebe, A., Pak, H.-R. and Meyers, M.A. (1985). Adiabatic shear localization in titanium Ti6Al4V alloy. Metall. Mater. Trans. 16, 761–775.
66. Li, Q., Xu, Y.B. and Bassim, M.N. (2004). Dynamic mechanical behavior of pure titanium. J. Mater. Process. Technol. 155, 1889.
67. Welsh, N.C. (1957). Frictional heating and its influence on the wear of steel. J. Appl. Phys. 28, 960–968.
68. Cho, K., Chi, Y.C. and Duffy, J. (1990). Microscopic observation of adiabatic shear bands in three different steels. Metall. Mater. Trans. 21A, 1161–1174.
69. Meyers, M.A. and Wittman, C.L. (1990). Effect of metallurgical parameters on shear band formation in low carbon (approximately 0.20 wet. pct) steel. Metall. Mater. Trans. 21A, 3153–3164.
70. Zhou, M., Rosakis, A.J. and Ravichandran, G.J. (1996). Dynamically propagating shear bands in impact-loaded prenotched plates. 2. Numerical simulations. J. Mech. Phys. Solids 44, 981–1006.
71. Xu, Y.B. and Meyers, M.A. (2003). Microstructural evolution of localized shear bands induced during explosion in Ti–6Al–4V alloy. J. Mater. Sci. Technol. 19, 385–387.
72. Zou, L., Zhen, L., Xu, C.Y. and Shao, W.Z. (2011). Characterization of adiabatic shear bands in AM60B magnesium alloy under ballistic impact. Mater. Charact. 62, 496–502.
73. Trent, E.M. (1941). The formation and properties of materials on the surface of rope wire. J. Iron Steel Inst. 143, 401–419.
74. Scott, D., Loy, B. and Mills, G.H. (1967–1968). Inst. Mech. Eng. Proc. 181, 30.
75. Beatty, J.H., Meyer, L.W., Meyers, M.A. and Nemat-Nasser, S. (1992). Shock-Wave and High-Strain-Rate Phenomena in Materials. Marcel Dekker, New York, NY, pp. 645–656.

76. Duan, C.Z. (2004). Doctoral Thesis. Study on Microscopic Mechanism of Adiabatic Shear Behavior in Orthogonal Cutting of High Strength Steel. Dalian University of Technology.
77. Xu, Y.B., Yang, H.J. and Meyers, M.A. (2007). Dynamic recrystallization in the shear bands of Fe−Cr−Ni monocrystal: electron backscatter diffraction characterization. Scr. Mater. 58, 691−694.
78. Yang, H.J., Xu, Y.B., Seki, Y., Nesterenko, V.F. and Meyers, M.A. (2009). Analysis and characterization by electron backscatter diffraction of microstructural evolution in the adiabatic shear bands in FE-Cr-Ni alloys. J. Mater. Res. 24, 2617−2627.
79. Yang, H.J., Zhang, J.H., Xu, Y.B. and Meyers, M.A. (2008). Microstructural characterization of the shear bands in Fe−Cr−Ni single crystal by EBSD. J. Mater. Sci. Technol. 24, 819−828.
80. Kad, B.K., Gebert, J.-M., Perez-Prado, M.T., Kassner, M.E. and Meyers, M.A. (2006). Ultra-grain-sized zirconium by dynamic deformation. Acta Mater. 54, 4111−4127.
81. Xu, Y.B., Bai, Y.L. and Meyers, M.A. (2006). Deformation, phase transformation and recrystallization in the shear bands induced by high-strain rate loading in titanium and its alloys. J. Mater. Sci. Technol. 22, 737−746.
82. Staudhammer, K.P., Frantz C.E., Hecker, S.S., and Murr L.E. (1981). Effect of strain rate on deformation-induced martensite in 304 stainless steel. In Shock Waves and High-Strain-Rate Phenomena in Metals and Alloys. (L.E. Murr, K.P. Staudhammer, M.A. Meyers, eds.) Amsterdan-Lausanne-NewYork-Oxford-Shannon-Tokyo. Addison Dekker, pp. 91−111.
83. Murr, L.E. and Ross, M.F. (1968). Thermal recovery of explosive shock-loading stainless steel. Philos. Mag. 18, 281−295.
84. Kestenbach, H.J. and Meyers, M.A. (1976). The effect of grain size on the shock-loading response of 304-type stainless steel. Metall. Trans. 7A, 1943−1950.
85. Olson, G.B. and Cohen, M.A. (1972). Mechanism for the strain-induced nucleation of martensitic transformation. J. Less Common Met. 28, 107−118.
86. Meyers, M., Cao, B.Y., Nesterenko, V.F., Benson, D. and Xu, Y.B. (2004). Shear localization-martensitic transformation in Fe−Cr−Ni monocrystal. Metall. Trans. 35A, 2575−2586.
87. Wang, B.F. and Yang, Y. (2008). Microstructure evolution in adiabatic shear band in fine-grain-sized Ti−3Al−5Mo−4.5V alloy. Mater. Sci. Eng. A 473, 306−311.
88. Costin, L.S. and Duffy, J. (1979). The effect of loading rate and temperature on the initiation of fracture in a mild rate-sensitive steel. J. Eng. Mater. Technol. 101, 258−264.
89. Shawki, T.G. (1994). An energy criterion for the onset of shear localization in thermal viscoplastic materials. 1. Necessary and sufficient conditions. J. Appl. Mech. Trans. ASME 61, 530−537.
90. Xu, Y.B., Ling, Z., Wu, X. and Bai, Y.L. (2002). Evolution of thermoplastic shear localization and related microstructures in Al/SiCp composites under dynamic compression. J. Mater. Sci. Technol. 18, 504−508.
91. Lee, S., Cho, K.M., Kim, K.C. and Choi, W.B. (1993). Adiabatic shear band formation in Al-SiC composite. Metall. Mater. Trans. 24A, 895.
92. Zhou, M. (1998). The growth of shear bands in composite microstructures. Int. J. Plast. 14, 733−754.
93. Shih, C.J., Nesterenko, V.N. and Meyers, M.A. (1998). High-strain rate deformation and comminution of SiC. J. Appl. Phys. 9, 4660−4671.
94. Xu, Y.B., and Meyers, M.A. (2011). Transition of crystalline lattice to the amorphous phase in the shear bands induced under high-strain rate of Fe−Cr−Ni single crystal. Unpublished work.

95. Li, N., Wang, Y.D., Peng, R.L., Sun, X., Liaw, P.K., Wu, G.L., Wang, L. and Cai, H. N. (2011). Localized amorphism after high-strain rate deformation in TWIP steel. Acta Mater. 59, 6369–6377.
96. Meyers, M.A. and Park, H.-K. (1986). Observation of an adiabatic shear band in titanium by high-voltage transmission electron microscoy. Acta Mater. 34, 2493.
97. Lins, J.F.C., Sandim, H.R.Z., Kestenbach, H.-J., Raabe, D. and Vecchio, K.S. (2007). A microstructural investigation of adiabatic shear bands in an interstitial free steel. Mater. Sci. Eng. A457, 205–218.
98. Chichili, D.R., Ramesh, K.T. and Hempker, K.J. (1998). The high-strain rate response of alpha-titanuim: experiments, deformation, mechanics and modeling. Acta Mater. 46, 1025–1043.
99. Murr, L.E., Niou, C.S., Pappu, S., Rivas, J.M. and Quinones, S.A. (1995). Leds in ultra-high strain- rate deformation. Phys. Status. Solidi. 149, 253–274.
100. Hines, J.A., Vecchio, K.S. and Ahzi, S. (1998). A model for microstructure evolution in adiabatic shear bands. Metall. Trans. 29A, 191–203.
101. Nesterenko, V.F., Meyers, M.A., LaSalvia, J.C., Bondar, M.P., Chen, Y.J. and Lukyanov, Y.L. (1997). Shear localization and recrystallization in high-strain, high-strain rate deformation of tantalum. Mater. Sci. Eng. A229, 23–41.
102. Meyers, M.A., Chen, Y.J., Marquis, F.D.S. and Kim, D.S. (1995). High-strain, high-strain rate behavior of tantalum. Metall. Mater. Trans. 26A, 2493–2501.
103. Martinez, F., Murr, L.E., Ramirez, A., Lopea, M.I. and Gaytan, S.M. (2007). Dynamic deformation and adiabatic shear microstructures associated with ballistic plug formation and fracture in Ti-6Al-4V targets. Mater. Sci. Eng. A454, 581–589.
104. Meyers, M.A., LaSalvia, J.C., Nesterenko, V.F., Chen, Y.J., and Kad, B.K. (1997). Recrystallization and related phenomena. In Proceedings of Rex'96 (J.I. McNelley, ed.), Monterey, CA, p. 279.
105. Li, J.C.M. (1995). TMS Annual Meeting. Unpublished results.
106. Murr, L.E. (2001). Interfacial Phenomena in Metals and Alloys. Addison-Wesley, London and Amsterdam, p. 1.
107. Meyers, M.A., Nesterenko, V.F., LaSalvia, J.C. and Xue, Q. (2001). Shear localization in dynamic deformation of materials: microstructural evolution and self-organization. Mater. Sci. Eng. A317, 204–225.
108. Xue, Q., Bingert, J.F., Henrie, B.L. and Gray, III, G.T. (2007). EBSD characterization of dynamic shear band regions in pre-shocked and as-received 304 stainless steels. Mater. Sci. Eng. A473, 279–289.
109. Rath, B.B. and Hu, B.B. (1972). In Nature and Behavior of Grain Boundaries (H. Hu ed.). Plenum Press, New York, NY, p. 405.
110. Humphreys, F.J. and Hatherly, M. (1995). Recrystallization and Related Annealing Phenomena. Pergamon Press, Oxford.
111. Li, Q., Xu, Y.B., Lai, Z.H., Shen, L.T. and Bai, Y.L. (2000). Dynamic recrystallization induced by plastic deformation at high strain rate in a Monel alloy. Mater. Sci. Eng A276, 250–256.
112. Li, Q. (1999). Dynamic Mechanical Behavior, Plastic Deformation Localization and Formation of Ultrafine Grain Structures in Metallic Materials. Post-Doctoral Report. Institute of Metal Research, Chinese Academy of Sciences, Shenyang, China.
113. Hirth, J.P. and Lothe, J. (1968). Theory of Dislocations. McGraw-Hill, New York, NY.
114. Dodd, B. and Bai, Y.L. (1989). Width of adiabatic shear band formed under combined stresses. Mater. Sci. Eng. 5, 557–559.

115. Xing, D., Bai, Y.L., Chen, C.M. and Huang, X.L. (1991). On post instability processes in adiabatic shear in rot rolled steel. J. Mech. Phys. Solids 39, 1017–1042.
116. Ashby, M.F and Verrall, R.A. (1973). Diffusion-accommodated flow and super plasticity. Acta Mater. 21, 149–163.
117. Murr, L.E., Trillo, E.A., Pappu, S. and Kennedy, C. (2002). Adiabatic shear bands and examples of their role in severe plastic deformation. J. Mater. Sci. 37, 3337–3360.
118. Chokshi, A.H. and Meyers, M.A. (1990). The prospects for superplasticity at high-strain rates: preliminary considerations and an example. Scr. Mater. 24, 605–610.
119. Murr, L.E. (1987). Metallurgical effect of shock and high-strain-rate loading. In Materials at High Strain Rates (T.E. Blazynski ed.). Elsevier, London and New York, pp. 1–46.
120. Lee, S., Cho, K.-M., Lee, C.S. and Choo, W.Y. (1993). Microstructural study of adiabatic shear formed by ballistic impact in an HY-100 steel. Metall. Mater. Trans. 24A, 2217–2224.
121. Meyers, M.A., Andrade, U.R. and Chokshi, A.H. (1995). The effect of grain-size on the high-strain, high-strain-rate behavior of copper. Mech. Mater. Trans. A26, 2881–2893.
122. Brandon, D.G. (1980). A comment on fracture mechanism maps. Scr. Mater. 14, 555–557.
123. Xue, Q. and Gray, G.T. (2006). Development of adiabatic shear bands in annealed 316L stainless steel. Part I: Correlation between evolving microstructure and mechanical behavior. Metall. Mater. Trans. 37A, 2435–2446.
124. Mercier, S. and Molinari, A. (1998). Steady-state shear band propagation under dynamic conditions. J. Mech. Phys. Solids 46, 1463–1495.
125. Xue, Q., Meyers, M.A. and Nesterenko, N.F. (2004). Self organization of shear bands in stainless steel. Mater. Sci. Eng. A384, 35–46.
126. Cottrell, A.H. (1953). Theory of dislocations. Progress in Metal Physics 4, 205–264.
127. Cottrell, A.H., Hunter, S.C. and Nabarro, F.R.N. (1953). Electrical interaction of a dislocation and a solute atom. Phil. Mag. 44, 1064–1067.
128. Portevin, A. and Le Chatelier, F.C.R. (1923). Acad. Sci. 176, 507.
129. Portevin, A. and Le Chatelier, F. (1924). Trans. Am. Soc. Steel Treat. 530, 457.
130. McGregor, C.W. and Fisher, J.C.J. (1945). J. Appl. Mech. 12, A-217.
131. Baird, J.D. (1963). Iron Steel Inst Lond. 36, 186, 326, 368, 400, 450.
132. Morris, J.G. (1974). Mater. Sci. Eng. 13, 101.
133. Mulford, R.A. and Kocks, U.F. (1979). New observations on the mechanisms of dynamic strain aging and of jerky flow. Acta Metall. 27, 1125–1134.
134. Kocks, U.F., Cook, R.E. and Mulford, R.A. (1985). Strain aging and strain hardening in Ni-C alloys. Acta Metall. 33, 623–638.
135. Nemat-Nasser, S., Guo, S. and Cheng, J.Y. (1999). Mechanical properties and deformation mechanisms of a commercially pure titanium. Acta Mater. 47, 3705–3720.
136. Bergstrom, Y. and Roberts, W. (1973). Dynamic strain aging of Alpha-iron effects of strain rate and nitrogen content in Jerk-flow region. Acta Metall. 21, 741–745.
137. Li, C.C. and Leslie, W.C. (1978). Effects of dynamic strain aging on the subsequent mechanical properties of carbon steels. Metall. Trans. 9, 1765–1775.
138. Gilat, A. and Wu, X. (1997). Plastic deformation of 1020 steel over a wide range of strain rates and temperatures. Int. J. Plast. 13, 611–632.
139. Mishra, A., Martin, M., Thadhani, N.N., Kad, B.K., Kenik, E.A. and Meyers, M.A. (2008). High-strain-rate response of ultra-fine-grained copper. Acta Mater. 56, 2770–2783.
140. Xu, Y.B., Yang, H.J., and Meyers, M.A. (2011). Dynamic/Static Recrystallization in Forced Shear Bands of Ultra-Fine Grained Copper Under High-Strain Rate Loading. Unpublished work.

141. Grady, D.E. (1994). Dissipation in adiabatic shear bands. Mech. Mater. 17, 289–293.
142. Grady, D.E. (2007). In TMS symposium on dynamic behavior of materials. TMS, Warrendale, PA.
143. Andrade, U.R., Meyers, M.A., Vecchio, K.S. and Chokshi, A.H. (1994). Dynamic recrystallization in high-strain, high-strain-rate plastic deformation of copper. Acta Metall. Mater. 42, 3183–3195.
144. Li, Y.S., Tao, N.R. and Lu, K. (2008). Microstructural evolution and nanostructure formation in copper during dynamic plastic deformation at cryogenic temperatures. Acta Mater. 56, 230–241.
145. Hong, C.S., Tao, N.R., Huang, X. and Lu, K. (2010). Nucleation and thickening at shear band in nano-scale twin/matrix lamellae of a Cu–Al alloy processed by dynamic plastic deformation. Acta Materialia 58, 3103–3116.
146. Peker, A. and Johnson, W.L. (1993). A highly processable metallic glass $Zr_{41.2}Ti_{13.8}Cu_{12.5}Ni_{10.0}Be_{22.5}$. Appl. Phys. Lett. 63, 2342–2344.
147. Johnson, W.L. (1999). Bulk glass-farming metallic alloys: Science and Technology. MRS Bull. 24(10), 42.
148. Inoue, A. (2000). Stabilization of metallic supercooled liquid and bulk amorphous alloys. Acta Mater. 48, 279–306.
149. Lewandowski, J.J. and Greer, A.L. (2006). Temperature rise at shear bands in metallic glasses. Nat. Mater. 5, 15–18.
150. Spaepen, F. (2006). Metallic glasses: Must shear bands be hot? Acta Mater 5, 7–8.
151. Dai, L.H., Liu, L.F. and Bai, Y.L. (2002). Formation of adiabatic shear band in metal matrix composites. Inter. Sol. Struct. 41, 5979–5993.
152. Liu, L.F., Dai, H., Bai, Y.L. and Wei, B.C. (2005). Initiation and propagation of shear bands in Zr-based bulk metallic glass under quasi-static and dynamic shear loadings. J. Non-Cryst. Solids 351, 3251–3270.
153. Dai, L.H., Yan, M., Liu, L.F. and Bai, Y.L. (2005). Adiabatic shear banding instability in bulk metallic glasses. Appl. Phys. Lett. 87, 14196.
154. Wright, W.J., Saha, R. and Nix, W.D. (2001). Localized heating during serrated plastic flow in bulk metallic glasses. Mat. Trans. 42, 642–649.
155. Wright, W.J., Schwarz, R.B. and Nix, W.D. (2001). Deformation mechanisms of the Zr40Ti14Ni10Cu12Be24 bulk metallic glass. Mater. Sci. Eng. A 319, 229–232.
156. Qu, R.T., Eckert, J. and Zhang, Z.F. (2011). Tensile fracture criterion of metallic glass. J. Appl. Phys. 109, 083544.
157. Zhang, Z.F., Eckert, J. and Schultz, L. (2003). Difference in compressive and tensile fracture mechanisms of $Zr_{59}Cu_{20}Al_{10}Ni_{8}Ti_{3}$ bulk metallic glass. Acta Mater. 51, 1157–1166.
158. Zhang, Z.F., He, G. and Eckert, J. (2005). Unified tensile fracture criterion. Phys. Rev. Lett. 94, 094301.
159. Jiang, W.H. and Atzmon, M. (2003). The effect of compression and tension on shear-band structure and nanocrystallization in amorphous Al90Fe5Gd5 : a high-resolution transmission electron microscopy study. Acta Mater. 51, 4095–4105.
160. Conner, R.D., Li, Y., Nix, W.D. and Johnson, W.L. (2004). Shear band spacing under bending of Zr-based metallic glass plates. Acta Mater. 52, 2429–2434.
161. Molinari, A. and Ravichandran, G. (2005). Mech. Mater. 37, 737–752.
162. Kocks, U.F. and Mecking, H. (2003). Prog. Mater. Sci. 48, 171–273.
163. Li, Q., Xu, Y.B. and Bassim, M.N. (2003). Dynamic mechanical properties in relation to adiabatic shear band formation in titanium alloy-Ti17. Mater. Sci. Eng. 358, 128–133.

4 Analysis of Adiabatic Shear Bands by Numerical Methods

Romesh C. Batra

Department of Engineering Science and Mechanics, Virginia Polytechnic Institute and State University, Blacksburg, VA

Nomenclature

ρ mass density in the present configuration
ρ_0 mass density in the reference configuration
J determinant of the deformation gradient
$F_{i\alpha}$ components of deformation gradient
x_i coordinates of a point in the present configuration
X_α coordinates of a point in the reference configuration
$T_{i\alpha}$ components of the first Piola–Kirchhoff stress tensor
v_i components of velocity
b_i components of body force per unit mass
e specific internal energy
Q_α components of heat flux per unit area in the reference configuration
r supply of internal energy per unit mass
Ω region occupied by the body in the reference configuration
$\partial\Omega$ boundary of Ω
θ temperature of a material point
N_α components of the outward unit normal to the boundary
f_i components of the present force per unit undeformed area
t time
ϕ empty set
h prescribed heat flux per unit undeformed area
$\dot{\theta}$ time rate of change of θ
c specific heat
ϕ_i components of test function
η scalar test function
H^1 collection of real valued functions whose first-order derivative is square integrable
$S_m \otimes S_\theta$ tensor product between sets S_m and S_θ
ψ_A finite element basis function for node A for displacement degree of freedom
ζ_A finite element basis function for node A for temperature degree of freedom
d_{iA} displacement of node A along x_i-axis
θ_A temperature of node A

Adiabatic Shear Localization. DOI: http://dx.doi.org/10.1016/B978-0-08-097781-2.00004-6

$\boldsymbol{M}_{\mathbf{AB}}$ mass matrix
$F^{m}_{\mathrm{A}i}$ load vector component of node A along x_i-axis
$\boldsymbol{H}_{\mathbf{AB}}$ heat capacity matrix
F^{θ}_{A} load vector of node A for the thermal problem
Ω_e region occupied by element e in the reference configuration
Ω_M master element of sides $2 \times 2 \times 2$
ξ_f coordinates of a point of the master element with respect to local coordinate axes
N_{A} shape function for node A
Δt time-step size
s shear stress
$s_{\mathbf{max}}$ maximum shear stress at a point
w specimen width
k thermal conductivity
γ_{p} plastic shear strain in the simple shearing problem
μ shear modulus
σ_0 yield stress in quasi-static simple shear deformations
$\delta_{\mathbf{ASB}}$ adiabatic shear bandwidth
$\gamma_{\mathbf{avg}}$ average shear strain in the specimen deformed in simple shear
D_{ij} components of the strain-rate tensor
σ_{ij} components of the Cauchy stress tensor
I magnitude of the deviatoric strain-rate tensor
s_{e} effective (von Mises) stress
α coefficient of thermal expansion
D^{e} elastic part of the strain-rate tensor
D^{p} plastic part of the strain-rate tensor
$\mathbf{s}$ deviatoric stress tensor derived from the Cauchy stress tensor
σ_m von Mises stress
ε^p_m effective plastic strain
f porosity at a point
$\mathbf{q}$ heat flux per unit area in the current configuration
$\mathbf{W}$ spin tensor
$\boldsymbol{\sigma}^{\nabla}$ Jaumann rate of the Cauchy stress tensor $\boldsymbol{\sigma}$
E Young's modulus
ν Poisson's ratio
$\mathbf{1}$ identity tensor

4.1 Introduction

Other chapters in this volume describe experimental and analytical works on adiabatic shear banding, and constitutive relations for a microporous thermo-elasto-viscoplastic material. No attempt is made here to describe all numerical methods that have been used to delineate adiabatic shear bands (ASBs). Rather, the focus is on illuminating a numerical technique, namely the finite element method (FEM), commonly used to analyse initial-boundary-value problems (IVBPs) for non-linear thermo-elasto-viscoplastic problems, and illustrating some results obtained using this technique. The interested reader is referred to books [1–4] and review articles

[5–7] for studying earlier works on ASBs, and to books [8–10] for studying the FEM. Suffice it to say that the phenomenon was discovered experimentally by Tresca [11] in 1878 and by Massey [12] in 1921 during high strain-rate compressive deformations of a steel block. Kalthoff [13–15] observed ASBs in a rectangular plate with two identical prenotches parallel to the short side and impacted on the notched side by a circular cylindrical projectile made of the same material as the plate with the projectile diameter equal to the distance between the two notches. For low-impact speeds, a crack initiated from the notch tip indicative of brittle failure, but at high-impact speeds an ASB initiated; the transition in the failure mode also depended on the notch-tip radius and strain hardening, strain-rate hardening and thermal-softening characteristics of the plate material. This problem has been numerically studied by Zhou et al. [16,17], Batra and co-workers [18–21] as well as Wang and Liu [22] who have cited several other works. However, work related to the Kalthoff problem is not discussed here because of space limitations.

An ASB usually precedes shear fracture in ductile materials. Whereas it is desirable to delay the initiation and growth of ASBs in most structures, their occurrences in high kinetic energy penetrators are promoted because it is believed that they enhance the penetration of the rod into a target by continuously making the penetrator conically nose shaped.

4.2 Brief Overview of the FEM

Equations, in the referential description of motion, governing thermo-mechanical deformations of a body are the following:

$$\text{Balance of mass: } \rho J = \rho_0, \quad \text{in } \Omega \tag{4.1}$$

$$\text{Balance of linear momentum: } \rho_0 \dot{v}_i = T_{i\alpha,\alpha} + \rho_0 b_i \quad \text{in } \Omega \tag{4.2a}$$

$$\text{Balance of moment of momentum: } T_{i\alpha} F_{j\alpha} = T_{j\alpha} F_{i\alpha} \quad \text{in } \Omega \tag{4.2b}$$

$$\text{Balance of internal energy: } \rho_0 \dot{e} = -Q_{\alpha,\alpha} + T_{i\alpha} \dot{F}_{i\alpha} + \rho_0 r \quad \text{in } \Omega \tag{4.3}$$

Here, ρ is the present mass density of a material particle whose mass density in the reference configuration is ρ_0, $J = \det[F_{i\alpha}]$ is the Jacobian, $F_{i\alpha} = \partial x_i / \partial X_\alpha$, x_i is the present position of the material particle that occupied place X_α in the reference configuration, $T_{i\alpha}$ is the first Piola–Kirchhoff stress tensor, v_i is the velocity, a superimposed dot indicates the material time derivative, b_i is the body force per unit mass, a comma followed by α denotes the partial derivative with respect to X_α, e is the specific internal energy, Q_α is the heat flux per unit reference area, r is the supply of internal energy per unit mass and a repeated index implies summation over the ranges 1, 2 and 3 of the index. Equations (4.1)–(4.3) are written in rectangular

Cartesian coordinates, and Ω is the region occupied by the body in the reference configuration. These equations are supplemented by constitutive relations that characterize the material of the body, and initial and boundary conditions such as those listed here:

$$x_i(X_\alpha, t) = \bar{x}_i, \quad X_\alpha \in \partial_1\Omega \tag{4.4}$$

$$\theta(X_\alpha, t) = \bar{\theta}, \quad X_\alpha \in \partial_3\Omega \tag{4.5}$$

$$T_{i\alpha}(X_\beta, t)N_\alpha = f_i, \quad X_\beta \in \partial_2\Omega \tag{4.6}$$

$$-Q_\alpha(X_\beta, t)N_\alpha = h, \quad X_\beta \in \partial_4\Omega \tag{4.7}$$

$$\partial_1\Omega \cap \partial_2\Omega = \phi, \quad \partial_1\bar{\Omega} \cup \partial_2\bar{\Omega} = \partial\Omega$$

$$\partial_3\Omega \cap \partial_4\Omega = \phi, \quad \partial_3\bar{\Omega} \cup \partial_4\bar{\Omega} = \partial\Omega$$

$$x_i(X_\alpha, 0) = x_i^0(X_\alpha), \quad \dot{x}_i(X_\alpha, 0) = \dot{x}_i^0(X_\alpha), \quad \theta(X_\alpha, 0) = \theta^0(X_\alpha) \tag{4.8}$$

Boundary conditions (4.4) and (4.5) are usually called essential, and Eqs (4.6) and (4.7) natural. In them, N_α is an outward unit normal to the boundary $\partial\Omega$ of Ω in the reference configuration, and functions x_i^0, θ^0, $\dot{x}_i^0$, $\bar{x}_i$, $\bar{\theta}$, f_i and h are presumed to be given. We will postulate constitutive relations later, and merely state for the time being that $T_{i\alpha}$, Q_α and e may be functions of time histories of $F_{i\alpha}$, $\dot{F}_{i\alpha}$, θ and $\theta_{,\alpha}$ with θ being the absolute temperature of a material particle. The constitutive relation for $T_{i\alpha}$ is such that the balance of the moment of momentum (4.2b) is identically satisfied; it is thus not considered hereafter. Equation (4.3) can alternatively be written as:

$$\rho_0 c\dot{\theta} = -Q_{\alpha,\alpha} + T_{i\alpha}^{\text{ne}}\dot{F}_{i\alpha} + \rho_o r \tag{4.9}$$

where $T_{i\alpha}^{\text{ne}}$ is the non-equilibrium part of the first Piola–Kirchhoff stress tensor,

$$T_{i\alpha}^{\text{ne}} = T_{i\alpha} - T_{i\alpha}(\mathbf{F}, \mathbf{0}, \theta, \mathbf{0}), \quad \dot{e} = T_{i\alpha}(\mathbf{F}, \mathbf{0}, \theta, \mathbf{0})\dot{F}_{i\alpha} + c\dot{\theta} \tag{4.10a, 4.10b}$$

and c is the specific heat, which may depend upon $\mathbf{F}$ and θ. Thus, Eqs (4.1)–(4.3) involve second-order derivatives with respect to X_α of unknown functions $x_i(X_\alpha, t)$ and $\theta(X_\alpha, t)$.

For $i = 1, 2, 3$, let $\phi_i : \bar{\Omega} \to \mathbb{R}$ be a smooth function such that $\phi_i = 0$ on $\partial_1\Omega$. Taking the inner product of both sides of Eq. (4.2a) with ϕ_i, integrating the resulting equation over Ω, using the divergence theorem on the first term on the right-hand side and boundary condition (4.6), we arrive at the following integral equation:

$$\int_\Omega \rho_0 \dot{v}_i \phi_i \, \mathrm{d}\Omega = \int_{\partial_2\Omega} f_i \phi_i \, \mathrm{d}\Gamma - \int_\Omega T_{i\alpha}\phi_{i,\alpha} \, \mathrm{d}\Omega + \int_\Omega \rho_0 b_i \phi_i \, \mathrm{d}\Omega \tag{4.11}$$

Let $\eta : \overline{\Omega} \to \mathbb{R}$ be a smooth function such that $\eta = 0$ on $\partial_3\Omega$. Multiplying both sides of Eq. (4.9) with η, integrating the resulting equation over Ω, using the divergence theorem on the first term on the right-hand side and boundary condition (4.7), we get the following:

$$\int_\Omega \rho_0 c\dot{\theta}\eta \, d\Omega = \int_{\partial_4\Omega} h\eta \, d\Gamma + \int_\Omega Q_\alpha \eta_{,\alpha} \, d\Omega + \int_\Omega (T_{i\alpha}^{\text{ne}}\dot{F}_{i\alpha} + \rho_0 r)\eta \, d\Omega \tag{4.12}$$

We note that different quantities in Eqs (4.11) and (4.12) depend on only time histories of first-order derivatives with respect to X_α of unknown functions $x_i(X_\alpha, t)$ and $\theta(X_\alpha, t)$.

Let

$$H^1 = \left\{ \phi | \phi : \overline{\Omega} \to \mathbb{R} H^1 \int_\Omega \phi_{,\alpha}\phi_{,\alpha} \, d\Omega < \infty \right\} \tag{4.13}$$

$$\begin{aligned}
T_m &= \{\phi_i | \phi_i \in H^1, \ \phi_i = 0 \text{ on } \partial_1\Omega\} \\
S_m &= \{\phi_i | \phi_i : \overline{\Omega} \times (0, \bar{t}) \to \mathbb{R}, \ \phi_i(\cdot, t) \in H^1, \ \phi_i(X_\alpha, t) = \bar{x}_i(X_\alpha, t) \text{ on } \partial_1\Omega\} \\
T_\theta &= \{\phi | \phi \in H^1, \ \phi = 0 \text{ on } \partial_3\Omega\} \\
S_\theta &= \{\phi | \phi : \overline{\Omega} \times (0, \bar{t}) \to \mathbb{R}, \ \phi(\cdot, t) \in H^1, \ \phi(X_\alpha, t) = \bar{\theta}(X_\alpha, t) \text{ on } \partial_3\Omega\} \\
S &= S_m \otimes S_\theta, \ T = T_m \otimes T_\theta
\end{aligned} \tag{4.14}$$

Then, a weak formulation of the given problem can be stated as follows. Find $(x_i, \theta) \in S$ such that Eqs (4.11) and (4.12) hold for every $(\phi_i, \eta) \in S$. One can think of $(\phi_i, \eta) \in S$ as representing virtual displacements and virtual temperature fields. Functions in S and S have the same smoothness (differentiability) requirements and differ only in boundary conditions on $\partial_1\Omega$ and $\partial_3\Omega$. When $S \neq S$, the weak formulation is called Petrov–Galerkin. Let $g \in S$ be a fixed function. Then, every function $u \in S$ can be written as $u = g + v$ for some $v \in S$. Equivalently, $S = S \oplus \{g\}$. Regarding g as a known function, the problem reduces to finding a $v \in S$ such that Eqs (4.11) and (4.12) hold for every $(\phi_i, \eta) \in S$, and the weak formulation is called Bubnov–Galerkin (or simply Galerkin in many books).

Let $S^n \subset S$ and $S^n \subset S$ be finite-dimensional sets. Then an approximate solution of the given problem is functions $(x_i^n, \theta^n) \in S^n$ such that Eqs (4.11) and (4.12) hold for every $(\phi_i^n, \eta^n) \in S^n$ with $T_{i\alpha}$, Q_α and $T_{i\alpha}^{\text{ne}}$ now evaluated from (x_i^n, θ^n) rather than from (x_i, θ).

Let $\psi_1, \psi_2, \ldots, \psi_n$ be basis functions in T_θ^n and $\zeta_1, \zeta_2, \ldots, \zeta_n$ be basis functions in T_m^n. Then,

$$\eta^n(X_\alpha) = \psi_A(X_\alpha)\eta_A, \quad A = 1, 2, \ldots, n \tag{4.15a}$$

$$\theta^n(X_\alpha, t) = \psi_A(X_\alpha)\theta_A(t) \tag{4.15b}$$

$$\phi_i^n(X_\alpha) = \zeta_A(X_\alpha)\phi_{Ai}, \quad i = 1, 2, 3 \tag{4.15c}$$

$$x_i^n(X_\alpha, t) = \zeta_A(X_\alpha)d_{Ai}(t) \tag{4.15d}$$

Recall that a repeated index implies summation over the range of the index. Substitution from Eq. (4.15) into Eqs (4.11) and (4.12) and using the fact that these equations must hold for all choices of η_A and ϕ_{Ai}, we arrive at the following set of coupled ordinary differential equations (ODEs) in time:

$$M_{AB}\ddot{d}_{Bi} = F^m_{Ai} \tag{4.16}$$

$$H_{AB}\dot{\theta}_B = F^\theta_A \tag{4.17}$$

$$F^m_{Ai} = \int_{\partial_2\Omega} f_i\zeta_A \, d\Gamma - \int_\Omega T_{i\alpha}\zeta_{A,\alpha} \, d\Omega + \int_\Omega \rho_0 b_i \zeta_A \, d\Omega \tag{4.18a}$$

$$F^\theta_A = \int_{\partial_4\Omega} h\psi_A \, d\Gamma + \int_\Omega Q_\alpha \psi_{A,\alpha} \, d\Omega + \int_\Omega G\psi_A \, d\Omega \tag{4.18b}$$

$$G = T^{ne}_{i\alpha}\dot{F}_{i\alpha} + \rho_0 r \tag{4.18c}$$

$$M_{AB} = \int_\Omega \rho_0 \zeta_A \zeta_B \, d\Omega \tag{4.18d}$$

$$H_{AB} = \int_\Omega \rho_0 c \psi_A \psi_B \, d\Omega \tag{4.18e}$$

In these equations, $T_{i\alpha}$, Q_α and $T^{ne}_{i\alpha}$ are functions of unknown fields $x^n_i(X_\alpha, t)$ and $\theta^n(X_\alpha, t)$ or equivalently of $d_{Ai}(t)$ and $\theta_B(t)$.

From initial condition (4.8), we can similarly derive

$$\begin{aligned} M_{AB}d_{Bi}(0) &= \overset{\circ}{F}_{Ai} \\ M_{AB}\dot{d}_{Bi}(0) &= \dot{\overset{\circ}{F}}_{Ai} \\ H_{AB}\theta_B(0) &= \overset{\circ}{F}{}^\theta_A \end{aligned} \tag{4.19}$$

and thus evaluate initial conditions for integrating, with respect to time t, ODEs (4.16) and (4.17).

Recall that the natural boundary conditions have been embedded in the weak formulation of the problem, and the essential boundary conditions are satisfied by either selecting a function g or another equivalent method.

Because of using the referential description of motion, a weak formulation of Eq. (4.1) is not needed, and the mass matrix can be evaluated only once. The present mass density can be determined from the computed displacement field and Eq. (4.1).

Let Ω be divided into the union of disjoint subdomains Ω_e satisfying

$$\Omega = \bigcup_{e=1}^{m} \Omega_e; \quad \Omega_e \cap \Omega_f = \phi, \quad e \neq f; \quad \overline{\Omega}_e \cap \overline{\Omega}_f \neq \phi$$

where Ω_e and Ω_f are adjacent subdomains, and $\overline{\Omega}_e$ includes boundaries of Ω_e. That is, adjacent subdomains share at most a common boundary. These subdomains are called finite elements (FEs), and they generally have straight boundaries (e.g., Ω_e is a straight line in one-dimensional (1D) space, is a triangle or a quadrilateral in a plane and is a cube or a tetrahedron in a 3D space). Points selected on the boundaries of FEs are called nodes. The collection of FEs and nodes is called an FE mesh or simply a mesh.

In the FE work, the number of basis functions $\{\psi_{\mathrm{A}}\}$ (or $\{\zeta_{\mathrm{A}}\}$) equals the number of nodes in the mesh. The basis functions are simple polynomials, and the basis function ψ_{A} corresponding to node A equals 1 at node A, and 0 on all elements that do not meet at node A. Thus, ψ_{A} equals zero at all nodes except node A and has a compact support. Equations (4.15c) and (4.15d) imply that $d_{\mathrm{A}i}(t)$ and $\theta_{\mathrm{A}}(t)$ equal, respectively, $x_i(t)$ and $\theta(t)$ at node A. The restrictions of basis functions to an FE are called shape functions. For the problem being studied here, the number of shape functions for an element equals the number of nodes on the element, and the shape function corresponding to node A of element Ω_e equals 1 at node A, 0 at all other nodes and vanishes on all sides of the element Ω_e that do not pass through node A. Generally, shape functions and basis functions are simple polynomials; the former are defined on an FE, and the latter are defined on the entire domain Ω.

The integrals appearing in Eqs (4.18a) and (4.18b) are evaluated numerically. To illustrate the procedure, we consider one such integral:

$$I = \int_{\Omega} f \, \mathrm{d}\Omega = \sum_e \int_{\Omega_e} f \, \mathrm{d}\Omega \equiv \sum_e I^e \tag{4.20}$$

$$I^e = \int_{\Omega_e} f \, \mathrm{d}\Omega \tag{4.21}$$

Let

$$T^e : \Omega_M \to \Omega_e; \quad \Omega_M = [-1, 1] \times [-1, 1] \times [-1, 1] \tag{4.22}$$

be a one-to-one, continuously differentiable and invertible map with continuously differentiable inverse. Here, we have tacitly assumed that Ω_e is a 'brick' or cubic element. We express T^e as

$$X_\alpha = X_\alpha(\xi_f) \tag{4.23}$$

and require that

$$\det \left[\frac{\partial X_\alpha}{\partial \xi_f} \right] > 0 \quad \text{for every} \quad \xi_f \in \Omega_M \tag{4.24}$$

With

$$J = \det\left[\frac{\partial X_\alpha}{\partial \xi_f}\right] \tag{4.25}$$

the integral (4.21) can be written as:

$$I^e = \int_{\Omega_e} f \, dX_1 \, dX_2 \, dX_3 = \int_{\Omega_M} fJ \, d\xi_1 \, d\xi_2 \, d\xi_3 = \int_{-1}^{1} d\xi_3 \int_{-1}^{1} d\xi_2 \int_{-1}^{1} fJ \, d\xi_1 \tag{4.26}$$

Recall that for a polynomial of degree $2n+1$,

$$\int_{-1}^{1} f \, ds = \sum_{i=1}^{n+1} W_i f(s_i) \tag{4.27}$$

Here, $s_1, s_2, \ldots, s_{n+1}$ are sampling points or quadrature points and $W_1, W_2, \ldots, W_{n+1}$ are the corresponding weights. The sampling points are roots of

$$P_{n+1}(s) = 0$$

and

$$W_i = \int_{-1}^{1} L_i(s) \, ds \tag{4.28}$$

where P_{n+1} is a Legender polynomial of degree $n+1$ and L_i is a Lagrange polynomial of degree n associated with point s_i and is defined by

$$L_i(s) = \prod_{\substack{j=1 \\ j \neq i}}^{n+1} (s - s_j) \Bigg/ \prod_{\substack{j=1 \\ j \neq i}}^{n+1} (s_i - s_j) \tag{4.29}$$

Thus, when the integrand fJ in Eq. (4.26) is a polynomial, I^e can be evaluated exactly by using an appropriate quadrature rule in each coordinate direction. However, in general, the integrand is not a polynomial, and all integrals in Eqs (4.18a), (4.18b), (4.18d) and (4.18e) are evaluated approximately by employing a quadrature rule.

We note that the mapping T^e in Eq. (4.22) is generally taken to be

$$X_\alpha = \sum_{\mathrm{A}=1}^{8} N_{\mathrm{A}}(\xi_1, \xi_2, \xi_3) X_\alpha^{\mathrm{A}} \tag{4.30a}$$

$$N_{\mathrm{A}}(\xi_1, \xi_2, \xi_3) = \frac{1}{8}(1 + \xi_1^{\mathrm{A}}\xi_1)(1 + \xi_2^{\mathrm{A}}\xi_2)(1 + \xi_3^{\mathrm{A}}\xi_3) \tag{4.30b}$$

where $(\xi_1^{\mathrm{A}}, \xi_2^{\mathrm{A}}, \xi_3^{\mathrm{A}})$ are coordinates $(\pm 1, \pm 1, \pm 1)$ of node A of element Ω_M (a cube of side 2) in the local coordinate system, and X_α^{A} coordinates in the global coordinate system of the node of Ω_e that corresponds to node A of Ω_M.

A field variable, for example, temperature θ defined on Ω_e, can be expressed as a function of ξ_1, ξ_2 and ξ_3 through the transformation in Eq. (4.23). To evaluate constitutive quantities, we need gradients of $x_i(X_\alpha, t)$ and $\theta(X_\alpha, t)$ with respect to X_α. To solve Eq. (4.30a) for ξ_1, ξ_2 and ξ_3, we note that

$$\begin{aligned} \mathrm{d}X_\alpha &= \frac{\partial N_{\mathrm{A}}}{\partial \xi_i}\mathrm{d}\xi_i X_\alpha^{\mathrm{A}} \qquad \text{(summed on A and } i\text{)} \\ &= \frac{\partial N_{\mathrm{A}}}{\partial \xi_i} X_\alpha^{\mathrm{A}} \frac{\partial \xi_i}{\partial X_\beta}\mathrm{d}X_\beta \end{aligned}$$

Therefore,

$$\left[\frac{\partial \xi_i}{\partial X_\beta}\right] = \left[\frac{\partial N_{\mathrm{A}}}{\partial \xi_i} X_\beta^{\mathrm{A}}\right]^{-1}$$

Thus,

$$\frac{\partial \theta}{\partial X_\alpha} = \frac{\partial \theta}{\partial \xi_i}\frac{\partial \xi_i}{\partial X_\alpha} \tag{4.31}$$

can be computed.

4.2.1 Lumping of the Mass Matrix

The bandwidth of the consistent mass matrix computed from Eq. (4.18d) depends on the node-numbering scheme. To reduce computational effort, we generally approximate it by a lumped (or diagonal) matrix; the same is also done for the consistent heat capacity matrix given by Eq. (4.18e). The use of node points as quadrature points to numerically evaluate Eq. (4.18d) will give a diagonal mass matrix but will result in some zero diagonal entries for axisymmetric problems. Zero or negative diagonal entries in mass matrix can have disastrous consequences. The following two techniques are commonly used to obtain a lumped mass matrix:

- *Row-sum technique*: Elements in each row of the consistent mass matrix are summed, and the result is placed on the diagonal. It can sometimes produce negative masses.
- *Special-lumping technique*: Entries of the lumped mass matrix are set proportional to corresponding diagonal elements of the consistent mass matrix with the constant of proportionality selected to conserve the total mass. The positive definiteness of the consistent mass matrix requires that its diagonal entries be positive.

4.2.2 Numerical Integration of ODEs

Equations (4.16) and (4.17) are non-linear, coupled ODEs in $d_{i\mathrm{B}}$ and θ_B because their right-hand sides are also functions of $d_{i\mathrm{B}}$ and θ_B. In ASB problems, Jaumann's rate of Cauchy's stress tensor is usually expressed as a linear function of the elastic part of the strain-rate tensor. Invariably, one postulates evolution laws for internal variables describing microstructural changes such as porosity, dislocations and damage in the body. Timescales for these processes may differ by several orders of magnitude, which make the system of equations very stiff. For stiff ODEs, the ratio of the maximum eigenvalue to the minimum eigenvalue is very large.

For example [23],

$$\begin{aligned} \dot{u} &= 998u + 1998v \\ \dot{v} &= -999u - 1999v \end{aligned} \tag{4.32}$$

with initial conditions

$$u(0) = 1, \quad v(0) = 0 \tag{4.33}$$

have the solution

$$\begin{aligned} u(t) &= 2\mathrm{e}^{-t} - \mathrm{e}^{-1000t} \\ v(t) &= -\mathrm{e}^{-t} + \mathrm{e}^{-1000t} \end{aligned} \tag{4.34}$$

The term e^{-1000t} requires a time step of $\leq 1/1000$ for the method to be stable. However, this term makes negligible contributions to values of u and v for $t > 0$. The generic problem with stiff equations is that for the forward-difference integration scheme to be stable, we need to follow the variation in the solution on the shortest timescale. We can overcome this by evaluating the time derivative with a backward-difference scheme. For example, consider

$$\dot{y} = -cy \tag{4.35}$$

Then,

$$y_{n+1} = \begin{cases} y_n/(1 + c\Delta t), & \text{backward-difference} \\ y_n(1 - c\Delta t), & \text{forward-difference} \end{cases} \tag{4.36}$$

where $y_n \simeq y(t_n)$. Clearly the forward-difference method is unstable if $\Delta t > 2/c$, but the backward-difference method is unconditionally stable (i.e., stable for all values of Δt). We sacrifice accuracy by using large time steps, but we maintain stability.

For a system of linear equations with constant coefficients,

$$\dot{\mathbf{y}} = -\mathbf{C}\mathbf{y} \tag{4.37}$$

where $\mathbf{C}$ is a positive definite matrix; the explicit method is stable only if

$$\Delta t < \frac{2}{\lambda_{\max}} \tag{4.38}$$

where $\lambda_{\max}$ is the largest eigenvalue of $\mathbf{C}$. The backward-difference scheme gives

$$\mathbf{y}_{n+1} = (\mathbf{1} + \mathbf{C}\Delta t)^{-1}\mathbf{y}_n \tag{4.39}$$

and is stable for all values of Δt. For a system of non-linear equations,

$$\dot{\mathbf{y}} = \mathbf{f}(t, \mathbf{y}) \tag{4.40}$$

The backward differencing gives

$$\begin{aligned} \mathbf{y}_{n+1} &= \mathbf{y}_n + \Delta t\mathbf{f}(t_{n+1}, \mathbf{y}_{n+1}) \\ &\simeq \mathbf{y}_n + \Delta t\left[\mathbf{f}(t_{n+1}, \mathbf{y}_n) + \left.\frac{\partial \mathbf{f}}{\partial \mathbf{y}}\right|_{\mathbf{y}_n} \cdot (\mathbf{y}_{n+1} - \mathbf{y}_n)\right] \end{aligned} \tag{4.41}$$

and requires the inversion of $\left(\mathbf{1} - \Delta t \frac{\partial \mathbf{f}}{\partial \mathbf{y}}\right)$, which is computationally expensive. By averaging the explicit and the implicit first-order methods, we get a second-order accurate method:

$$\mathbf{y}_{n+1} = \mathbf{y}_n + \frac{\Delta t}{2}\left[\mathbf{f}(t_{n+1}, \mathbf{y}_n) + \left.\frac{\partial \mathbf{f}}{\partial \mathbf{y}}\right|_{\mathbf{y}_n} \cdot (\mathbf{y}_{n+1} - \mathbf{y}_n) + \mathbf{f}(t_n, \mathbf{y}_n)\right] \tag{4.42}$$

One can use either LSODE (Livermore Solver for ODEs) or VODEPK (Variable Coefficient ODE Solver package) to numerically integrate a stiff set of ODEs. These solvers adaptively adjust the time-step size to compute a solution of the ODEs within prescribed absolute and relative tolerances. In strain localization problems, the time-step size drops dramatically after the deformation begins to localize, indicating an increase in the stiffness of the ODEs.

4.2.3 Satisfaction of Essential Boundary Conditions

Instead of explicitly using the function g, Eqs (4.16) and (4.17) are assembled for all nodes and are then modified as follows to satisfy the essential boundary conditions prescribed at node A. From the given time history of the displacement at node A, one computes its acceleration at any time t. For a lumped mass matrix, the displacement boundary condition at node A is satisfied when $F_{\mathrm{A}i}$ is replaced by $M_{\mathrm{AA}}(\ddot{d}_{\mathrm{A}i})$ (no sum on A). One follows this procedure for every node where an essential boundary condition is given. For a consistent mass matrix, one can use the penalty method in which M_{AA} is replaced by $M_{\mathrm{AA}} + \lambda$ and $F_{\mathrm{A}i}$ by $\lambda \ddot{d}_{\mathrm{A}i}$, where λ is 10^5 times the largest entry in the mass matrix. The accuracy in the satisfaction of the essential boundary condition can be improved by increasing the value of λ.

4.2.4 Interpretation of the FE Solution

The computed nodal values of displacements and temperatures are reasonably accurate. However, stresses, heat flux and other quantities involving derivatives of displacements and temperatures are evaluated at quadrature points, which are generally in the interior of Ω_e. Then, they can be extrapolated or interpolated to other points of interest, for example, see Refs [24–26].

4.2.5 Factors Affecting Quality of the Approximate Solution

The following list of factors that affect the quality of the solution is not exhaustive:

1. *Dimensionality of the finite-dimensional space*: The quality of the approximate solution generally improves with an increase in the dimensionality (number of nodes) of the finite-dimensional space (recall the convergence to the function of its Fourier series with an increase in the number of terms in the series).

2. *Choice of basis functions*: For localization problems, lowest order basis functions are recommended.
3. *FE mesh design*: Adaptively refined meshes with element size inversely proportional to a scalar measure of deformation within an element are recommended. Care should be taken to properly grade the mesh to avoid a very large element abutting an extremely small one.
4. *Number of quadrature points used to evaluate integrals*: The CPU time increases with an increase in the number of quadrature points. However, too few quadrature points may result in hourglass or spurious modes.
5. *Consistent/lumped mass matrix*: Consistent mass matrices generally require a smaller time-step size for stability than lumped mass matrices.
6. *Explicit/implicit time integration scheme*: Numerical integration schemes generally introduce time-period errors and dissipation. For the 1D linear elastic problem, a uniform FE mesh, a lumped mass matrix and the explicit (central-difference) integration scheme with $\Delta t = h/c$ give exact values of nodal displacements. Here, h is the element size and c is the speed of the wave in the bar.
7. *Tolerances*: Tolerances in solving non-linear algebraic equations if an implicit time integration scheme is used.
8. *Time-step size*: As $\Delta t \rightarrow 0$, an approximate solution of Eqs (4.16) and (4.17) converges to the analytical solution of the discretized problem and not of the continuous problem defined by Eqs (4.1)–(4.3).
9. *Points where stresses/fluxes are computed for post-processing of results*: It is recommended that fluxes be computed at quadrature points, at the element centroid or at the Barlow points [24–26].
10. *Time integration*: The time integration scheme should maintain objectivity of the rate form of constitutive relations.

4.3 Analysis of 1D Shear Band Problems

Marchand and Duffy [27] analysed the initiation and growth of ASBs by twisting thin-walled tubes that had varying thickness because of machining errors. As a first approximation, torsional deformations of the tube can be regarded as equivalent to simple shearing deformations of a representative volume element situated near the tube's midsection. The material presented in this section has been taken from Refs [28–30]; other investigations are cited in these articles. The goals here are to define ASB initiation criterion and bandwidth, and characterize effects of different viscoplastic constitutive equations, thermal conductivity, inertia forces and defect size.

We choose a fixed set of rectangular Cartesian coordinate axes with origin at the lower surface of the block and the direction of shearing along the x-axis, for example, see Figure 4.1. We assume that particles move only in the x-direction, and a particle's x-velocity, v, and temperature rise, θ, are functions of y only. These are found by solving

$$\rho w(y)\dot{v} = (w(y)s)_{,y}, \quad 0 < y < H \tag{4.43}$$

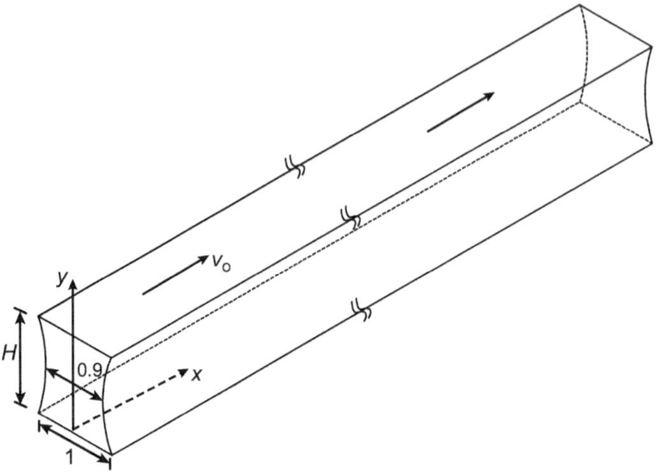

Figure 4.1 Schematic sketch of the simple shearing problem studied.

$$\rho c(\theta)w(y)\dot{\theta} = [w(y)k(\theta)\theta_{,y}]_{,y} + w(y)s\dot{\gamma}_{\mathrm{p}}, \quad 0<y<H \tag{4.44}$$

$$\dot{s} = \mu(\theta)(v_{,y} - \dot{\gamma}_{\mathrm{p}}) \tag{4.45}$$

$$\dot{\gamma}_{\mathrm{p}} = g(s, \gamma_{\mathrm{p}}, \theta) \tag{4.46}$$

Here, s, γ_{p}, w, k and μ represent, respectively, the shear stress, plastic strain, the block thickness, the thermal conductivity and the shear modulus. We have used Fourier's law of heat conduction and assumed an additive decomposition of the shear strain rate into elastic and plastic parts. The evolution of the shear stress is governed by the rate form of Hooke's law; all of the plastic working is converted into heating and have neglected effects, if any, of thermal stresses. The last assumption is reasonable because deformations are driven by the shear stress. We note that simple shear deformations are isochoric, i.e., volume preserving. Thus, the mass density does not change during the deformation process. We solve the problem for four different viscoplastic relations in Eq. (4.46) (i.e., function g).

We non-dimensionalize variables as follows:

$$\begin{aligned}
&\bar{y} = y/H, \quad \bar{t} = v_0 t/H, \quad \bar{w} = w/H, \quad \bar{\theta} = \theta/\theta_0, \quad \theta_0 = \sigma_0/\rho c_R \\
&\bar{s} = s/\sigma_0, \quad \bar{\rho} = \rho v_0^2/\sigma_0, \quad \bar{k} = k/(\rho v_0 c_R H) \\
&\bar{c} = c(\theta)/c_R, \quad \bar{\mu} = \mu/\sigma_0, \quad c_R = c(\theta_R), \quad \theta_R = 300\ \mathrm{K}
\end{aligned} \tag{4.47}$$

Here, v_0 is the steady value of the shearing velocity applied to the top surface of the block when its lower surface is kept fixed. Furthermore, $\bar{t}$ equals the average strain induced in the specimen at time t, θ_0 is the temperature rise at a point due to plastic strain of 1 when deformations are locally adiabatic and the flow stress is kept constant at σ_0, and $\overline{\rho}$ is the ratio of the kinetic energy density to the plastic work done per unit volume at constant flow stress σ_0 and plastic strain of 2. In terms of non-dimensional variables, Eqs (4.43)–(4.46) remain unchanged except that dimensional quantities are replaced by non-dimensional ones, and the problem domain becomes $0<\overline{y}<1$. Henceforth, in this section, only non-dimensional variables are used, and superimposed bars have been dropped.

For initial and boundary conditions, we take the following:

$$\begin{aligned} &\theta(y,0)=0, \quad v(y,0)=0, \quad s(y,0)=0, \quad \gamma_p(y,0)=0 \\ &\theta_{,y}(0,t)=0, \quad \theta_{,y}(1,t)=0, \quad v(0,t)=0 \\ &v(1,t)=t/0.01, \quad 0\le t\le 0.01 \\ &\quad =1, \quad t\ge 0.01 \end{aligned} \tag{4.48}$$

That is, the block is initially stress free, undeformed, at rest and has a uniform temperature, normalized to be zero. The overall deformations of the block are taken to be adiabatic, and the lower surface is at rest, whereas the upper surface is assigned a velocity that increases from 0 to 1 in a non-dimensional time of 0.01 and then stays at 1.0. We note that the prescribed x-velocity at surfaces $y=0$ and $y=1$ is neither symmetric nor antisymmetric about the midsurface, $y=1/2$. The block is taken to be thinnest at $y=1/2$ and thickest at $y=0$, 1, with the thickness variation given by

$$w(y)=w_0\left[1+\frac{\delta}{2}\sin\left(\frac{1}{2}+2y\right)\pi\right] \tag{4.49}$$

Marchand and Duffy [27] reported nearly 10% variation in the thickness of steel tubes they tested in torsion. Our choice of locating the thinnest section at $y=1/2$ is for convenience only and should not affect results.

4.3.1 Viscoplastic Flow Rules

Wright–Batra Relation

Wright and Batra [31] modified the Litonski relation [32] to account for elastic unloading of a material point and postulated the following:

$$\dot{\gamma}_p=\Lambda s \tag{4.50}$$

$$\Lambda=\max\left[0,\left(\left(\frac{s}{(1-\upsilon\theta)\left(1+\frac{\psi}{\psi_0}\right)^n}\right)^{1/m}-1\right)\Big/bs\right] \tag{4.51}$$

$$\dot{\psi}=s\dot{\gamma}_p/(1+\psi/\psi_0)^n \tag{4.52}$$

Here, ψ may be thought of as an internal variable that describes work hardening of the material. Equation (4.52) implies that the rate of growth of ψ is proportional to plastic working. In Eq. (4.51), $(1 - \upsilon\theta)$ describes the material softening due to its being heated up, b and m characterize its strain-rate sensitivity and ψ_0 and n describe its work hardening. Equations (4.50) and (4.51) imply the following:

$$\dot{\gamma}_p = 0 \quad \text{if } s \leq (1 - \upsilon\theta)(1 + \psi/\psi_0)^n \tag{4.53}$$

Thus, if the local state given by (s, ψ, θ) lies inside or on the loading surface, $s = (1 - \upsilon\theta)(1 + \psi/\psi_0)^n$, then the plastic strain rate vanishes, and the material deforms elastically. However, if the point (s, ψ, θ) is outside the loading surface, then the plastic strain rate is adjusted so that the local state lies on the yield surface, $s = (1 - \upsilon\theta)(1 + \psi/\psi_0)^n(1 + b\dot{\gamma}_p)$. Besides σ_0, which has been used to non-dimensionalize quantities, five parameters, υ, b, m, ψ_0 and n, are needed to specify the viscoplastic response of the material.

Bodner–Partom Relation

Bodner and Partom [33] assumed that there is no loading surface and plastic strain rate, $\dot{\gamma}_p$, albeit very small at low values of s, is always positive. Their constitutive relation can be written as:

$$\dot{\gamma}_p = D_0 \exp\left[-\frac{1}{2}\left(\frac{z^2}{3s^2}\right)^n\right], \quad n = \frac{a}{T} + b \tag{4.54}$$

$$z = z_1 - (z_1 - z_0)\exp(-mW_p) \tag{4.55}$$

$$\dot{W}_p = s\dot{\gamma}_p \tag{4.56}$$

Here, T is the absolute temperature of a material particle, W_p is the plastic work done, z is an internal variable and D_0 is the limiting value of $\dot{\gamma}_p$, usually taken as $10^8\ \text{s}^{-1}$. Besides D_0, we need values of a, z_1, z_0, m and b to characterize the material.

Johnson–Cook Relation

Johnson and Cook [34] tested 12 materials in simple shear and compression at different strain rates and found that

$$\begin{aligned} r &= \left[\left(\frac{s}{(A + B\gamma_p^n)(1 - \overline{T}^m)} - 1.0\right)\right] \\ \dot{\gamma}_p &= \dot{\gamma}_0 \exp\left(\frac{r}{C}\right), \quad r \geq 0 \\ &= 0, \quad r < 0 \end{aligned} \tag{4.57}$$

$$\overline{T} = (\theta - \theta_0)/(\theta_m - \theta_0) \tag{4.58}$$

describe well the test data. For θ_0 equal to the ambient temperature and $\dot{\gamma}_0 = 1/sec$, they tabulated values of A, B, n, m, θ_m and C for 12 materials. One can regard $r = 0$ as the loading surface.

Power Law

Klopp et al. [35] and Marchand and Duffy [27] have described the stress–strain curve for simple shear loading by:

$$s = s_0 \left(\frac{\gamma}{\gamma_y}\right)^n \left(\frac{\dot{\gamma}_p}{\dot{\gamma}_0}\right)^m \left(\frac{\theta}{\theta_0}\right)^\nu \tag{4.59}$$

where γ_y is the strain at yield in a quasi-static simple shear test at $\dot{\gamma}_0 = 10^{-4}\ s^{-1}$, parameters n and m characterize the strain and strain-rate hardening of the material and $\nu < 0$ is its thermal softening.

4.3.2 Determination of Values of Material Parameters

Batra and Kim [28] assumed values of material parameters to be temperature independent, ignored effects of phase transformations and determined their values by solving an initial-boundary-value problem (IBVP) that closely simulated test conditions of Marchand and Duff [27] by ensuring that the computed shear stress/shear strain curve matched well the test data. Values of material parameters so determined are not unique because the underlying IBVP may not have a unique solution. Values of the quasi-static yield stress and the strain hardening exponent were kept close to those reported by Marchand and Duffy [27]. For HY100 steel, $\rho = 7860\ kg/m^3$, $\sigma_0 = 405$ MPa, $k = 49.73\ W/(m^2\ °C)$ and $c = 473$ J/(kg °C); values of other material parameters are listed below:

- *Wright–Batra relation*: $v = 0.00185\ °C^{-1}$, $\psi_0 = 4.86$ MPa, $m = 0.0117$, $n = 0.107$, $b = 10^4$ s.
- *Bodner–Partom relation*: $D_0 = 10^8\ s^{-1}$, $z_1 = 3.95\sigma_0$, $z_2 = 3.21\sigma_0$, $m = 5/\sigma_0$, $a = 1200$ K and $b = 0$.
- *Johnson–Cook relation*: $\dot{\gamma}_0 = 3300\ s^{-1}$, $A = 0.45$, $B = 1.433$, $C = 0.0227$, $n = 0.107$, $m = 0.7$, $\theta_m = 1500$ K and $\theta_0 = 300$ K.
- *Power law*: $\dot{\gamma}_0 = 10^{-4}\ s^{-1}$, $m = 0.0117$, $n = 0.107$, $\nu = -0.75$, $\theta_0 = 300$ K, $\gamma_y = 0.012$.

For an average strain rate of $3300\ s^{-1}$, Figure 4.2 shows the computed shear stress versus shear strain curves with the above given values of material parameters. Because of the magnified vertical scale, differences among these curves are exaggerated. Values of the maximum shear stress and the average strain at which they occur equal (1.60, 0.25), (1.55, 0.27), (1.53, 0.27) and (1.50, 0.26),

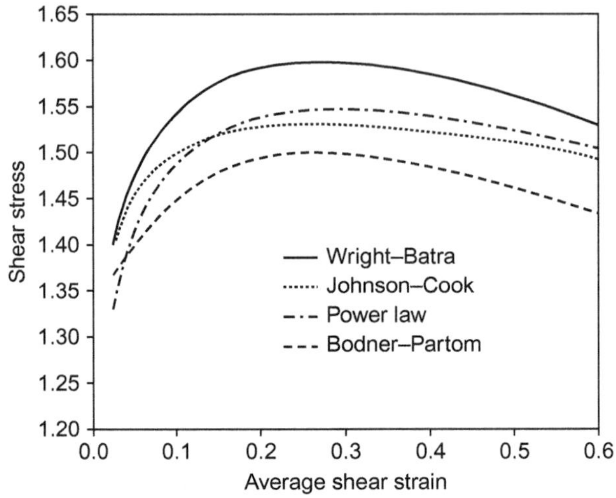

Figure 4.2 Shear stress versus shear strain curves for homogeneous simple shearing deformations of HY100 steel deformed at a strain rate of 3300 s^{-1}. *Source*: From Ref. [30].

respectively, for the Wright–Batra, the power law, the Bodner–Partom and the Johnson–Cook relations.

4.3.3 Results

Computational Considerations

Governing Eqs (4.43) and (4.44) with function g given by one of the flow rules are highly non-linear and difficult to solve analytically under side conditions given as Eq. (4.48); their approximate solution has been computed numerically by using the FEM as described in Section 4.2. The software LSODE is used to integrate with respect to time t the resulting non-linear coupled ODEs. The block thickness at the centre is taken to be 90% of that at the edges.

ASB Initiation Criterion

Batra and Kim [36] proposed the following ASB initiation criterion. An ASB initiates at a material point when the shear stress there has dropped to 90% of its maximum value at that point; the material point is deforming plastically and deformations in its neighbourhood are inhomogeneous. We recall that in Marchand and Duffy's [27] experimental work, the torque dropped when an ASB initiated. Figure 4.3A and B shows the dependence on the nominal strain rate, of the average shear strain, γ_{I}, corresponding to the maximum shear stress, and to the average shear strain, $\gamma_{0.9}$, when an ASB initiates. For the Bodner–Partom relation, γ_{I} and $\gamma_{0.9}$ monotonically increase with an increase in the nominal strain rate, $\dot{\gamma}_{\mathrm{avg}}$. For the other three viscoplastic relations, the dependence of γ_{I} and $\gamma_{0.9}$ on $\dot{\gamma}_{\mathrm{avg}}$ is not monotonic.

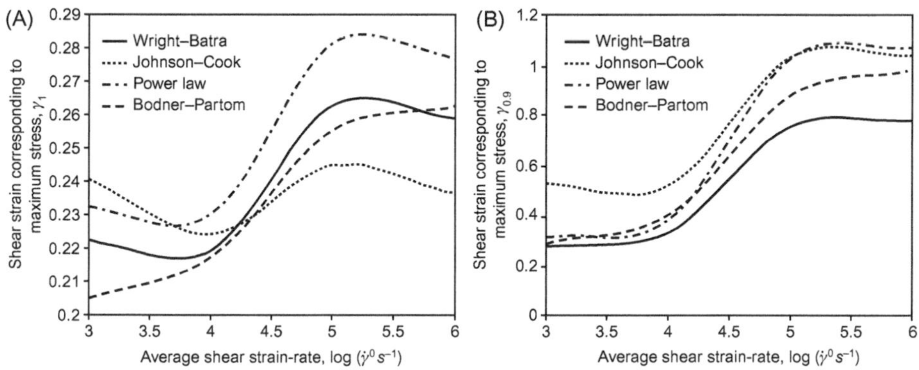

Figure 4.3 For the four viscoplastic relations, dependence on the average shear strain rate of the shear strain corresponding to the maximum shear stress and to that when the shear stress has dropped to 90% of its maximum value.
Source: From Ref. [30].

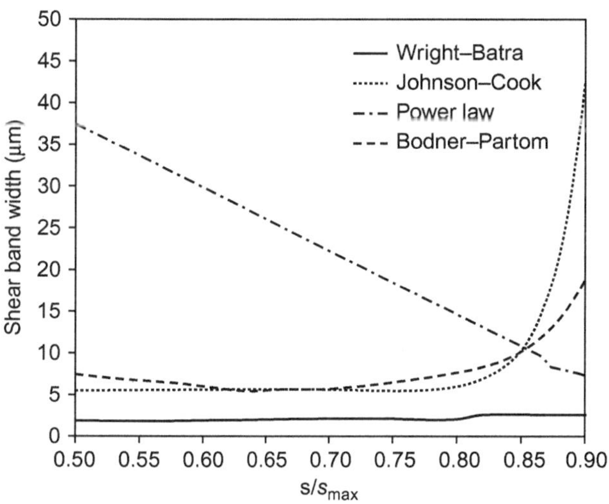

Figure 4.4 For the four viscoplastic relations, evolution of the shear band width while the shear stress is dropping.
Source: From Ref. [30].

ASB Width

In their torsional tests, Marchand and Duffy [27] called the width, δ_{ASB}, of the intensely deformed region of uniform plastic strain as the ASB width. Batra and Chen [30] equated δ_{ASB} to the width of the region over which the plastic strain varied by no more than 10% of its maximum value in that region. In Figure 4.4, we have plotted δ_{ASB} as a function of s/s_{max} for the four viscoplastic relations and $\dot{\gamma}_{avg} = 1000\ s^{-1}$. Whereas for the Wright–Batra and the Johnson–Cook relations, δ_{ASB} slowly decreases with the drop of the shear stress at the band centre; for the power law, it increases. For the Bodner–Partom relation, δ_{ASB} increases slightly when the shear stress drops from $0.7s_{max}$ to $0.5s_{max}$. Thus, even though the ASB

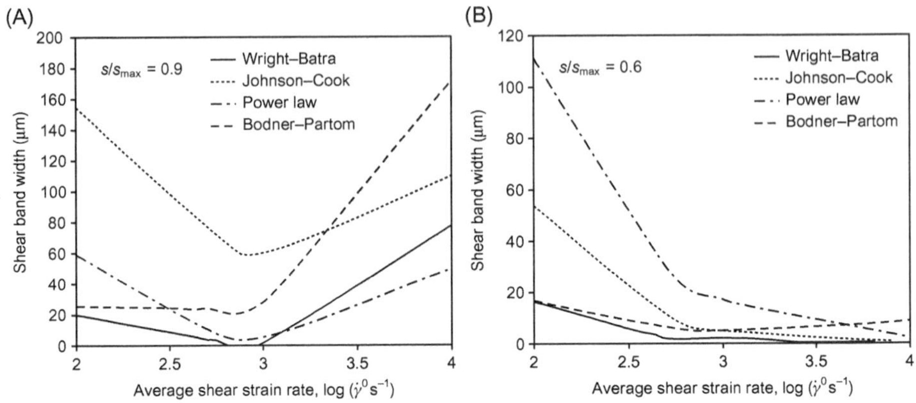

Figure 4.5 For the four viscoplastic relations, dependence of the shear band width on the nominal strain rate for two values of s/s_{max} (or of time): (A) $s/s_{max} = 0.9$ and (B) $s/s_{max} = 0.6$. *Source*: From Ref. [30].

initiates when $s/s_{max} = 0.9$, its full development depends on the viscoplastic relation employed. Dilellio and Olmstead [37] used asymptotic methods to analyse boundary layers near edges of a specimen deformed in simple shear and modelled the material response by a power law type strain hardening and an exponential thermal-softening relation; their computed time history of evolution of δ_{ASB} qualitatively agrees with that in Figure 4.4 for the power law.

For the four constitutive relations, Figure 4.5A and B exhibits the dependence of δ_{ASB} on $\dot{\gamma}_{avg}$ for $s/s_{max} = 0.9$ and 0.6. In each case, δ_{ASB} first decreases with an increase in $\dot{\gamma}_{avg}$ from $10^2\ s^{-1}$ to about $10^{2.8}\ s^{-1}$ and then increases. For $s/s_{max} = 0.9$ and $\dot{\gamma}_{avg} = 10^2\ s^{-1}$, δ_{ASB} computed with the Wright–Batra relation is the minimum, δ_{min}, and that with the Johnson–Cook relation, the maximum, δ_{max}, and $\delta_{max}/\delta_{min} = 7$. For $s/s_{max} = 0.66$ and $\dot{\gamma}_{avg} = 1600\ s^{-1}$, Marchand and Duffy [27] found that δ_{ASB} varied between 20 and 55 μm around the circumference of the tube; Batra and Chen's [30] computations give $\delta_{ASB} = 2.3$, 5.5, 14 and 6 μm for the Wright–Batra, the Johnson–Cook, the power law and the Bodner–Partom relations.

Effect of Thermal Conductivity

For $\dot{\gamma}_{avg} = 3300\ s^{-1}$, the Johnson–Cook relation and thermal conductivity, $k = 0$, 5, 50, 500 and 5000 W/(m °C), Figure 4.6 displays the average shear stress, s_{avg}, and the homologous temperature, $\overline{T}$, versus the average shear strain, γ_{avg}, in the specimen; results for the other two constitutive relations are qualitatively similar to those shown in Figure 4.6. The value of γ_{avg} at which the s_{avg} begins to drop increases a little with an increase in the value of k. However, the rate of stress drop decreases dramatically as the value of k is increased from 50 to 500 W/(m °C) as compared to that when k is increased from 5 to 50 W/(m °C). The rate of temperature rise is steepest for $k = 0$ and decreases with an increase in the value of k. As the value of

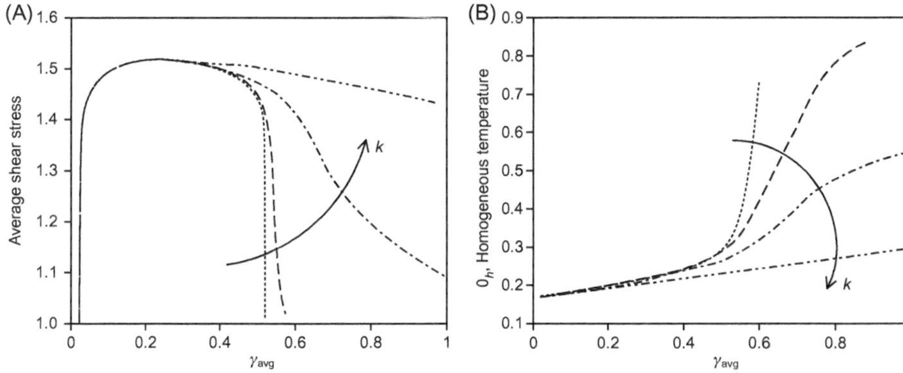

Figure 4.6 For five values of thermal conductivity and the Johnson–Cook relation, evolution with the average strain of the average shear stress (A) and the homologous temperature (B).
Source: From Ref. [38].

k increases, the heat conducted away from the central, hotter region to the colder, outer parts of the specimen increases, and the rate of temperature rise at the specimen centre decreases. Because of global adiabatic conditions assumed, the temperature everywhere in the specimen increases. Thus, the pre-localization response of a material is essentially the same whether or not heat conduction is considered, but the post-localization response is noticeably influenced by the consideration of thermal conductivity.

Remarks

Results presented in subsection 4.3.3 suggest that, for most materials, except possibly for copper whose thermal conductivity is high, it is reasonable to assume that the deformations are locally adiabatic prior to the initiation of a shear band. However, the post-localization deformations are strongly influenced by thermal conductivity. Batra [39,40] has shown through numerical experiments that the consideration of inertia forces delays the ASB initiation. Batra and Kim [29] found that for large values of the thermal-softening coefficient, υ, the sudden collapse of the shear stress within the band results in an unloading elastic wave propagating outwards from the shear-banded region. Batra and Kim [32] have analysed the effect of the defect size on the initiation and growth of shear bands in 12 materials. They found that the bandwidth, computed when the shear stress has dropped to 95% of its peak value, does not correlate well with the thermal conductivity of the material. The band begins to grow rapidly when the shear stress has dropped to 90% of its peak value. Larger defects result in more severe localization of deformation for the same percentage drop in the shear stress. However, the defect size influences very little the homologous temperature when the shear stress either has attained its maximum value or has dropped to 85% of its maximum value.

Kim and Batra [41] have accounted for the dependence of material properties upon the temperature, and they found that the bandwidth and the average strain at which an ASB forms decrease with a decrease in the initial temperature of the specimen.

4.4 Adaptive Mesh Refinement for 2D Problems

Because of intense deformations of the material in an ASB, the FE mesh is severely distorted, and an interior angle of an FE can become either too small or too large. Thus, the deforming region should be frequently remeshed to capture late stages of ASB development. One way to accomplish this is to adaptively refine the mesh. Here we discuss two such techniques for 2D problems; the *h*-method in which the element size is varied [42] and the *r*-method wherein the mesh topology (number of elements and nodes connecting various elements) is kept fixed, but locations of nodes are varied [43].

4.4.1 The h-Method

A coarse mesh is used to numerically solve the problem. This mesh is refined so that

$$a_e = \int_{\Omega_e} I \, \mathrm{d}\Omega, \quad e = 1, 2, \ldots, n_{\mathrm{el}} \tag{4.60}$$

is nearly the same for each FE Ω_e. In Eq. (4.60), I is the second invariant of the deviatoric strain-rate tensor and n_{el} equals the number of elements in the coarse mesh. Because points where the solution exhibits sharp gradients are unknown *a priori*, the starting coarse mesh may be chosen as uniform. The motivation for making a_e the same over each element Ω_e is that within the region of localization of deformation, values of I are very high as compared to those in the remaining region. Other variables such as the temperature rise, the maximum principal strain and the equivalent strain, which are also quite large within the band, will be suitable replacements for I in Eq. (4.60). The topology of the refined mesh will depend on the variable used in Eq. (4.60). To refine the mesh, we find the following:

$$\overline{a} = \frac{1}{n_{el}} \sum_{e=1}^{n_{el}} a_e, \quad \varepsilon_e = \frac{a_e}{\overline{a}}, \quad h_e = \frac{\overline{h}_e}{\varepsilon_e} \quad \text{and} \quad H_n = \frac{1}{N_e} \sum_{e=1}^{N_e} h_e, \quad n = 1, 2, \ldots, n_{ed} \tag{4.61}$$

Here, $\overline{h}_e$ is the size of element Ω_e in the coarse mesh, N_e equals the number of elements meeting at node n and n_{ed} equals the number of nodes in the coarse mesh. We refer to H_n as the nodal element size at node n.

To generate a new mesh, we first discretize the boundary by following the procedure proposed by Cescotto and Zhou [44]. Let AB be a segment of the contour to be discretized, s the arc length measured from point A, and H_A and H_B the nodal element sizes for nodes located at points A and B, respectively. Knowing the values of H at nodes in the coarse mesh, we define on AB a piecewise linear continuous function $H(s)$ that takes values at nodes given in Eq. (4.61). To discretize AB for the new mesh, we start from point A if $H_A < H_B$; otherwise, we start from point B. Let A be the starting point. We first find temporary positions of nodes on segment AB by using the following recursive procedure. Assume that points 1, 2,..., k have been found. Then the temporary location of point $(k+1)$ is given by:

$$s_{k+1} = s_k + \frac{1}{2}\left[H(s_k) + H(s^*_{k+1})\right] \tag{4.62}$$

where

$$s^*_{k+1} = s_k + H(s_k) \tag{4.63}$$

Referring to Figure 4.7, the preceding procedure will result in the following four alternatives: $a = b = 0$, $a > b$, $b \leq a$, $a = b \neq 0$. If $a = b = 0$, then the temporary locations of nodes are their final positions. Depending on whether $a < b$ or $b \leq a$, nodes 2 to p or 2 to $p+1$ are moved, the displacement of a node being proportional to the value of H there, so that either node p or node $(p+1)$ coincides with point B. This determines final positions of nodes on segment AB.

Having discretized the boundary, we use the concept of an advancing front [45–47] to generate elements. An advancing front consists of straight line segments that are available to form a side of an element. Thus, to start with, it consists of the discretized boundary. We choose the smallest line segment (say side AB in

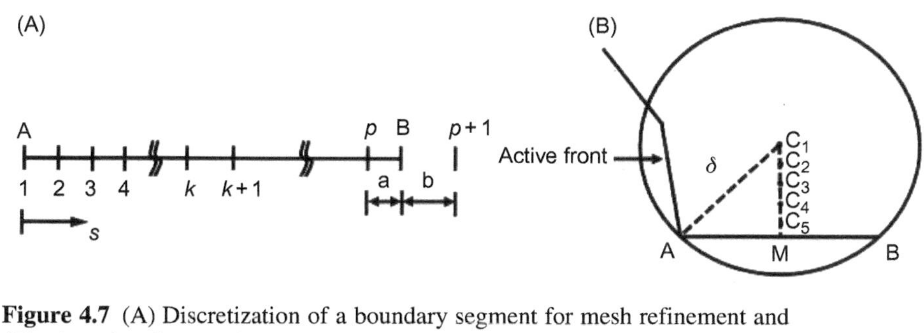

Figure 4.7 (A) Discretization of a boundary segment for mesh refinement and (B) advancing front and new element generation.

Figure 4.7) connecting two adjoining nodes and determine the nodal element size $H_M \equiv H(s_M) = (H_A + H_B)/2$ at the midpoint M of AB. We set

$$\delta = \begin{cases} 0.8\overline{AB} & \text{if} \quad H_M < 0.8\overline{AB} \\ H_M & \text{if} \quad 0.8\overline{AB} \leq H_M \leq 1.4\overline{AB} \\ 1.4\overline{AB} & \text{if} \quad 1.4\overline{AB} < H_M \end{cases} \tag{4.64}$$

and find point C_1 at a distance δ from points A and B (cf. Figure 4.7). Here $\overline{AB}$ equals the length of segment AB, and numbers 0.8 and 1.4 can be changed to generate elements of different sizes. We search for all nodes on the active front that lie inside the circle with centre at C_1 and radius δ. At the end of this list are added points C_1, C_2, C_3, C_4 and C_5, which lie on C_1M and divide it into five equal parts. We next determine the first point C in the list that satisfies the following three conditions:

1. Area of triangle ABC is > 0.
2. Sides AC and BC do not cut any of the existing sides in the front.
3. If any of the points $C_1, C_2, \ldots, C_5$ are chosen, that point is not too close to the front.

The triangle ABC is an element in the new mesh. If C is one of the points C_1, $C_2, \ldots, C_5$, then a new node is also created. The advancing front is updated by removing the line segment AB from it and adding line segments AC and CB to it. The element-generation process ceases when there is no side left in the active front.

We determine values of solution variables at a newly created node by first finding out to which element in the coarse mesh this node belongs and then interpolating values of solution variables at this node. This process and that of searching for line segments and points in the previous element-generation technique require considerable CPU time. These operations are optimized to some extent by using the heap list algorithm (e.g., see Ref. [48]) for deleting and inserting new line segments, quad tree structures, linked lists for searching line segments and points and also for interpolation of solution variables at newly created nodes.

4.4.2 The r-Method

The goal here is to reposition nodes so that a_e defined by Eq. (4.60) is nearly the same for each element Ω_e. Having solved the problem with an initial mesh it is refined as follows. We begin with either the horizontal or the vertical boundary and relocate nodes on it; we describe the process starting with the left vertical edge. After having repositioned nodes on it, we do the same on the almost vertical curve that passes through nodes next to the left vertical side, and continue the process till we reach the right vertical edge. The procedure is then repeated beginning with the top or the bottom horizontal edge and going to the other end.

Referring to Figure 4.8, let AB be the curve on which nodes are to be relocated. We plot the distribution of I on AB with abscissa as the distance of a point from

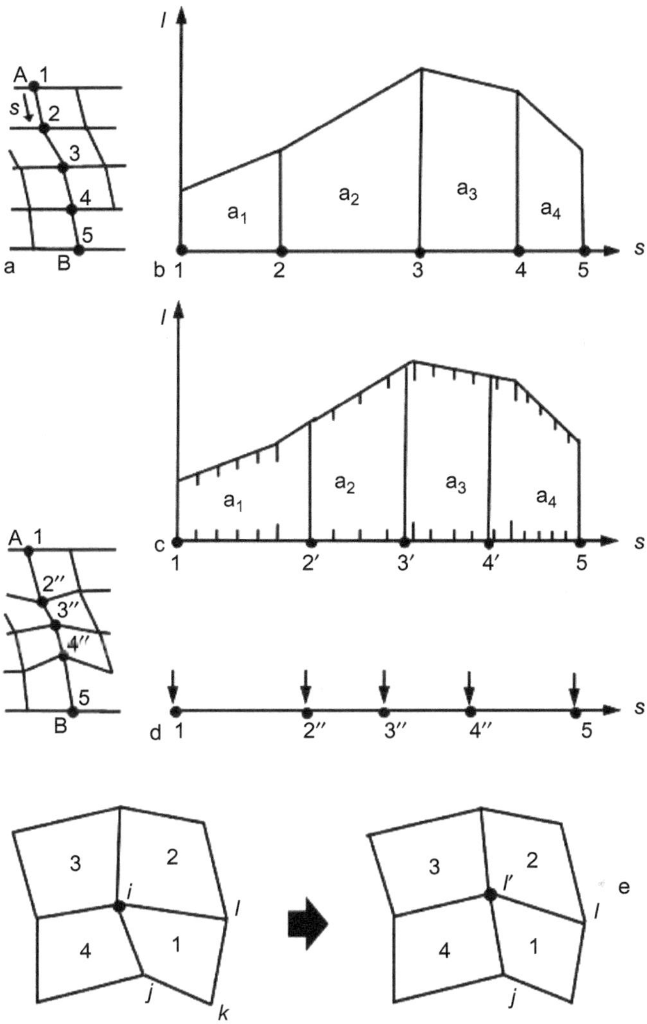

Figure 4.8 (a) Curve AB on which nodes are to be relocated; (b) distribution of I on curve AB; (c) temporary position on curve AB of relocated nodes; (d) reposition nodes on curve AB; (e) relocation of an interior node to smoothen out the generated mesh.

A measured along AB and ordinate as the value of I at that point. Values of I at numerous points on AB are obtained by linear interpolation from values at node points. If S equals the total area under the curve, then approximate location, s_n^a, of the nth node on AB is given by

$$\int_{s_{n-1}}^{s_n^a} I\,\mathrm{d}s = \frac{S}{N_{es}} \tag{4.65}$$

where N_{es} equals the number of elements on AB. We reposition the node to the interpolation point immediately to the left of its approximate location determined by Eq. (4.65). In Figure 4.8c, the position of a node as found by Eq. (4.65) is shown by a superimposed prime, and its relocation is shown in Figure 4.8d by superimposed two primes. Because end points on AB are kept fixed, the previously given procedure can be employed by starting from either node A or node B. Note that when nodes on an approximate horizontal curve are relocated, positions of nodes A and B will change.

Quadrilateral elements produced by the preceding technique are not always well shaped in the sense that an interior angle may be either too small or too large. It usually happens in regions where the element size varies noticeably. We use the mesh smoothing method of Zhu et al. [49] to improve upon shapes of quadrilateral elements. Each interior node is repositioned to the centroid of the polygon formed by all elements meeting at the node. As illustrated in Figure 4.8e, the internal node i is moved to i' with coordinates given by:

$$x_i' = \frac{1}{4M}\sum_{a=1}^{M}(x_j + 2x_k + x_l)_a, \quad y_i' = \frac{1}{4M}\sum_{a=1}^{M}(y_j + 2y_k + y_l)_a \tag{4.66}$$

where M is the number of elements sharing node i. After having relocated all internal nodes, element shapes are checked to see if all interior angles of every element are between 20° and 160°; these limiting values of interior angles are arbitrarily chosen. If not, nodes are repositioned according to Eq. (4.66) till such is the case. Because of smoothening of the mesh, the value of a_e defined by Eq. (4.60) is only approximately the same for all elements in the mesh.

4.4.3 Numerical Results

The two preceding adaptive mesh-refinement techniques have been used to delineate the initiation and development of ASBs in plane-strain compression of a square block. Because of the presumed symmetry of deformations about the horizontal and the vertical centroidal axes, deformations of only a quarter of the block are investigated. The constitutive relation for the material is taken to be

$$\sigma_{ij} = -B\left(\frac{\rho}{\rho_0} - 1\right)\delta_{ij} + \frac{\sigma_0}{\sqrt{3}I}(1 + bI)^m(1 - \upsilon\theta)D_{ij} \tag{4.67}$$

$$2I^2 = \overline{D}_{ij}\overline{D}_{ij}, \quad \overline{D}_{ij} = D_{ij} - \frac{1}{3}D_{kk}\delta_{ij} \tag{4.68}$$

Here, σ_{ij} is the Cauchy stress tensor, $\boldsymbol{D}$ is the strain-rate tensor and B may be thought of as the bulk modulus for the material of the block. Batra [50] proposed Eq. (4.67) as a generalization of Litonski's law [32] to 3D problems. Equation (4.67) has been used in Refs [51,52] to analyse asymptotic structure of propagating ASBs. All boundaries of the block are thermally insulated, its vertical edges traction free and the upper and the lower smooth horizontal surfaces are moved vertically in opposite directions to induce a nominal strain rate of 5000 s^{-1}.

4.4.4 The h-Refinement

Figure 4.9 depicts the initial coarse mesh at time $t = 0$ and the generated refined meshes at an average strain or non-dimensional time $t = 0.025$, 0.040 and 0.047. In the solution of the problem, the mesh was also adaptively refined at $t = 0.015$, 0.030 and 0.035; however, these are not shown for the sake of brevity. The times

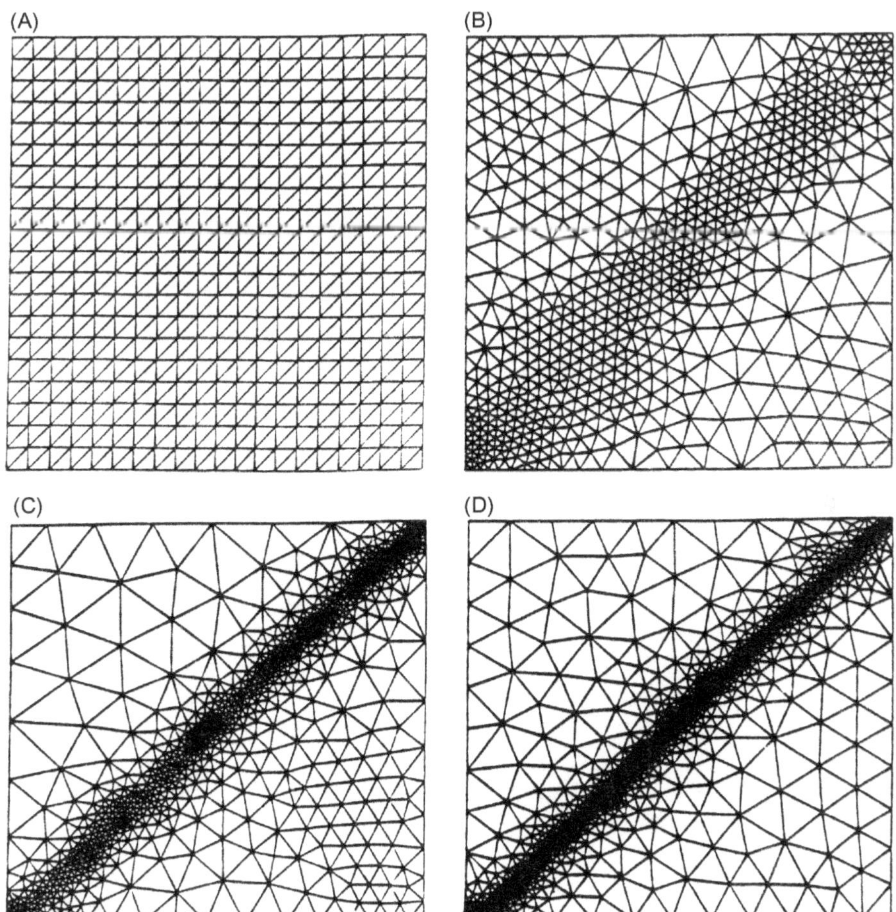

Figure 4.9 FE meshes at $t = 0.0$ (A), 0.025 (B), 0.040 (C) and 0.047 (D).
Source: From Ref. [42].

for mesh refinement were manually selected and are arbitrary. A possibility is to refine the mesh when either I or the temperature at the specimen centre has risen by a certain amount. Meshes shown in Figure 4.9 vividly reveal that the refinement technique described in Section 4.2 generates non-uniform meshes with small elements in the severely deforming region and large elements elsewhere. No restriction was imposed on the number of new nodes that can be introduced during the refinement process. The distribution of the velocity field in the deforming region at $t = 0.047$, shown in Figure 4.10A, supports Tresca's [11] and Massey's [12] assertions that the tangential velocity is discontinuous across an ASB. In our work, the velocity field is forced to be continuous throughout the domain. The sharp jumps in the velocity components across the ASB lend credence to the discontinuity of the tangential velocity across the ASB. The plot of the effective stress, s_e, defined as

$$s_e = \sqrt{\frac{2}{3}}(1 - \upsilon\theta)(1 + bI)^m$$

in Figure 4.10B reveals that s_e drops considerably within the ASB.

We now investigate the improvement, if any, in the approximate solution obtained by refining the mesh. Because the analytical solution of the problem is unknown, we compare the approximate solution with a higher-order approximate solution [53] obtained by smoothing out the computed solution. Let g be a solution variable to be smoothed. For the three-noded triangular element, we write

$$g(\xi, \eta) = a\xi + b\eta + c \tag{4.69}$$

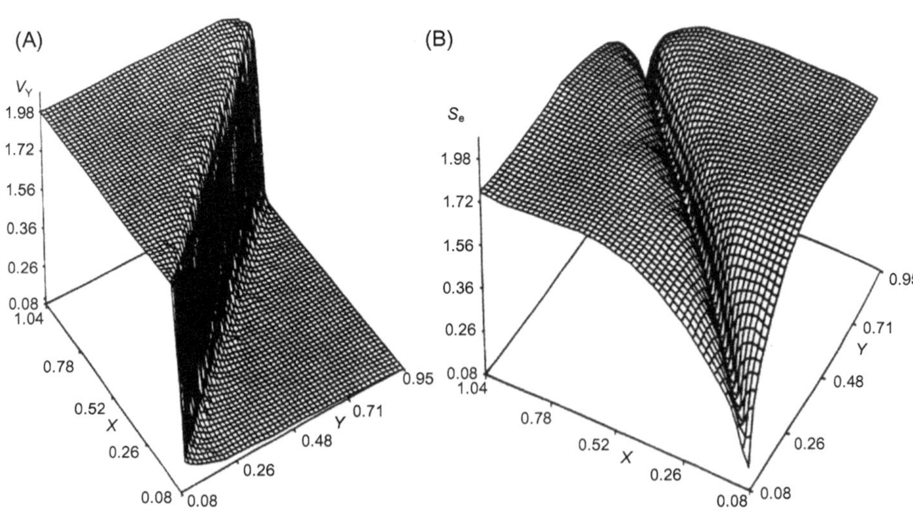

Figure 4.10 At $t = 0.047$, distribution of (A) velocity field and (B) effective stress.
Source: From Ref. [42].

where ξ and η are area coordinates of a point, and constants a, b and c are determined from values of g at three quadrature points located in the triangular element. From Eq. (4.69), we evaluate g at vertices of the triangle. Then the value g_n^* of the smoothed solution at node n is given by

$$g_n^* = \frac{1}{N_e}\sum_{n=1}^{N_e} g_n \tag{4.70}$$

where N_e equals the number of elements sharing the node n. Knowing g^* at each node, we interpolate its value at any other point by using the FE basis functions and define the percentage error η in $\overline{\mathbf{D}}$ by

$$\eta = \left(\frac{||\mathbf{e}||_0^2}{||\mathbf{e}||_0^2 + ||\overline{\mathbf{D}}||_0^2}\right)^{1/2} \times 100 \tag{4.71}$$

where

$$\mathbf{e} = \overline{\mathbf{D}} - \overline{\mathbf{D}}^*, \quad ||\mathbf{e}||_0^2 = \sum_{e=1}^{N_{el}} \int_{\Omega_e} \mathbf{e}^{\mathrm{T}}\mathbf{e}\, \mathrm{d}\Omega \tag{4.72}$$

and $\overline{\mathbf{D}}^*$ is computed from the smoothened velocity field. The plot of η in Figure 4.11 for the three meshes shows that the error is lower for the approximate solution obtained by using the adaptively refined mesh than that for the other two meshes. The

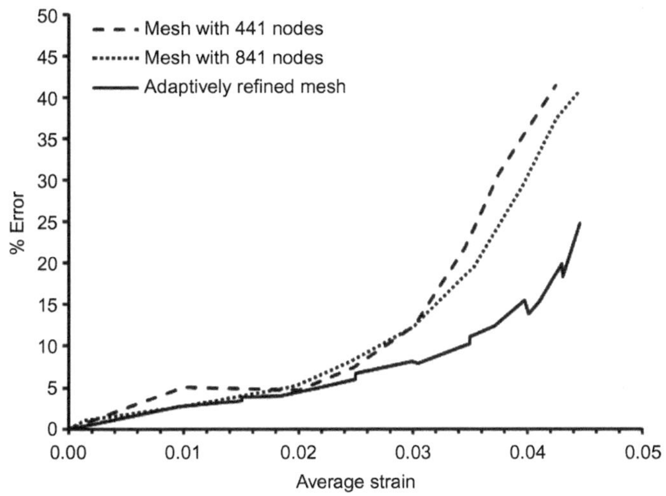

Figure 4.11 Comparison of the error in the computed approximate solution and the higher-order approximate solution for three different meshes.
Source: From [42].

error measure η is not robust as indicated by slightly larger errors obtained with a fixed mesh of 841 nodes than those with 441 nodes. It could be due to the larger errors caused by smoothing out the approximate solution with 841 nodes because the band in this case is more intense than that for the mesh with 441 nodes.

4.4.5 The r-Refinement

Figure 4.12A shows the initial mesh having 400 uniform elements and Figure 4.12B, C and D show the generated refined meshes when the non-dimensional temperature θ at the plate centroid equalled 0.25, 0.35 and 0.45. The mesh was also refined when θ at the plate centroid reached 0.30 and 0.40, but these are not depicted. We chose to refine the mesh for equal increments of the temperature rise. However, other criteria

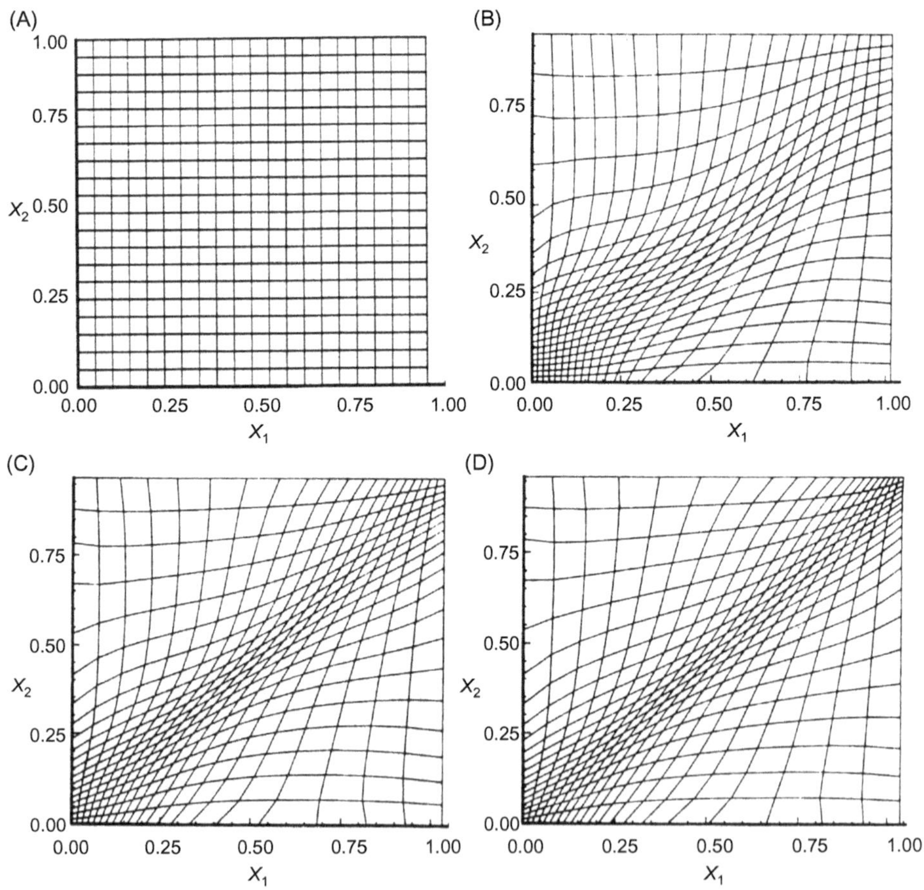

Figure 4.12 (A) Initial mesh of uniform 400 square elements and FE meshes (B,C,D) generated using the *r*-refinement technique when the temperature rise at the specimen equalled 0.25, 0.35 and 0.45.
Source: From Ref. [43].

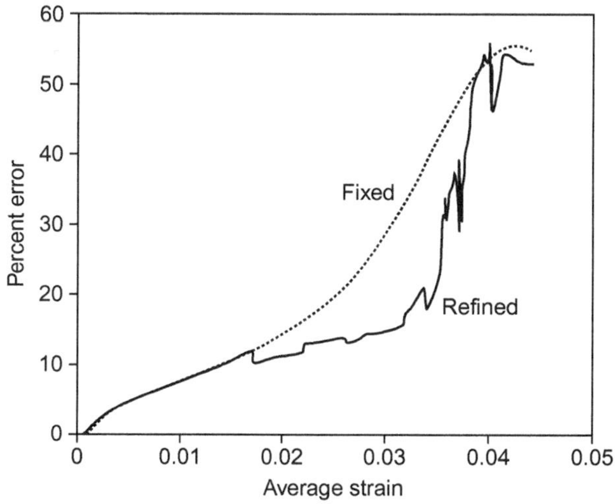

Figure 4.13 Comparison of the error in the approximate solution computed with a fixed mesh and an adaptively refined mesh.
Source: From Ref. [43].

such as I or θ attaining preassigned values would be equally as good. The meshes shown in Figure 4.12 vividly reveal that the r-refinement technique generated a nonuniform mesh with small elements in the severely deformed region and large elements elsewhere. The mesh smoothing criterion in Eq. (4.66) was applied at most three times to satisfy the requirement that the interior angles of every quadrilateral element be between 20° and 160°. We note that the mesh-generation scheme does not impose any restriction on the ratio of the area of the largest to that of the smallest element in the mesh. Even though this did not generate an unduly skewed mesh for the present problem, in other situations, such a restriction may be necessary. One could avoid this either by having more elements in the initial mesh or by adding elements at a few intermediate stages. The latter would necessitate the creation of a new mesh topology.

The percentage error in $\overline{\mathbf{D}}$ is plotted in Figure 4.13. The error for the solution obtained by using the adaptively refined mesh is lower until an average strain of 0.037, and it then suddenly increases and becomes larger than that obtained for the solution computed with a fixed mesh. It is probably due to large errors caused by smoothing out the approximate solution in the late stages of the band development when deformations within the band are very intense.

4.5 ASBs and Cracks in Microporous Thermo-Viscoplastic Solids

4.5.1 Problem Formulation

It is assumed that the material is microporous and exhibits thermal softening, strain and strain-rate hardening. In the referential description of motion, Eqs (4.1) and (4.2) are replaced by:

$$\text{Balance of mass:} \quad (\rho J(1-f))^{\bullet} = 0 \tag{4.73}$$

Balance of linear momentum: $\rho_0(I - f_0)\dot{\mathbf{v}} = Div\ \mathbf{T}$ (4.74)

where f_0 (f) equals the volume fraction of voids in the reference (current) configuration. We assume that the strain-rate tensor has additive decomposition into elastic $\mathbf{D}^{\mathrm{e}}$, plastic $\mathbf{D}^{\mathrm{p}}$ and thermal parts $\alpha\dot{\theta}\mathbf{1}$:

$$\mathbf{D} = \mathbf{D}^{\mathrm{e}} + \mathbf{D}^{\mathrm{p}} + \alpha\dot{\theta}\mathbf{1} \tag{4.75}$$

We postulate the constitutive relations

$$\overset{\nabla}{\boldsymbol{\sigma}} \equiv \dot{\boldsymbol{\sigma}} + \boldsymbol{\sigma}\mathbf{W} - \mathbf{W}\boldsymbol{\sigma} = \frac{E(1-f)}{1+v}\mathbf{D}^{\mathrm{e}} + \frac{Ev(1-f)}{(1+v)(1-2v)}\mathrm{tr}(\mathbf{D}^{\mathrm{e}})\mathbf{1} \tag{4.76}$$

$$\frac{3}{2}\frac{\mathrm{tr}(\mathbf{s}\mathbf{s}^{\mathrm{T}})}{\sigma_m^2} + 2f^*\beta_1 \cosh\left(\frac{\beta_2 \mathrm{tr}\,\boldsymbol{\sigma}}{2\sigma_m}\right) - 1 - \beta_1^2 f^{*2} = 0 \tag{4.77}$$

$$\mathbf{D}^{\mathrm{p}} = \frac{(1-f)\sigma_m\dot{\varepsilon}_m^{\mathrm{p}}}{\mathrm{tr}(\boldsymbol{\sigma}\mathbf{N}^T)}\mathbf{N}, \quad \mathbf{s} = \boldsymbol{\sigma} - \frac{1}{3}(\mathrm{tr}\,\boldsymbol{\sigma})\mathbf{1} \tag{4.78}$$

$$\mathbf{N} = \frac{3\mathbf{s}}{\sigma_m^2} + \frac{f^* q_1 q_2}{\sigma_m}\left[\sinh\left(\frac{q_2 \mathrm{tr}\,\boldsymbol{\sigma}}{2\sigma_m}\right)\right]\mathbf{1} \tag{4.79}$$

$$f^* = \begin{cases} f, & \text{if } f \le f_{\mathrm{c}} \\ f_{\mathrm{c}} + \left(\dfrac{f_{\mathrm{u}}^* - f_{\mathrm{c}}}{f_{\mathrm{f}} - f_{\mathrm{c}}}\right)(f - f_{\mathrm{c}}), & \text{otherwise} \end{cases} \tag{4.80}$$

$$\sigma_m = \sigma_0(1 + b\dot{\varepsilon}_m^{\mathrm{p}})^m\left(1 + \frac{\varepsilon_m^{\mathrm{p}}}{\varepsilon_{\mathrm{y}}}\right)^n(1 - \beta\theta) \tag{4.81}$$

$$\dot{f} = (1-f)\mathrm{tr}\,\mathbf{D}^{\mathrm{p}} + \frac{f_2\dot{\varepsilon}_m^{\mathrm{p}}}{s_2\sqrt{2\pi}}\exp\left(-\frac{1}{2}\left(\frac{\varepsilon_m^{\mathrm{p}} - e_N}{s_2}\right)^2\right)H(-\sigma_{kk} - 0) \tag{4.82}$$

$$\mathbf{q} = -k\left(1 - \frac{3}{2}f\right)\mathrm{grad}\,\theta \tag{4.83}$$

$$\dot{e} = (1 - f_0)c\dot{\theta} + \frac{1}{\rho}\mathrm{tr}(\boldsymbol{\sigma}(\mathbf{D} - \mathbf{D}^{\mathrm{p}})) \tag{4.84}$$

where

$$\mathbf{T} = J\sigma(\mathbf{F}^{-1})^{\mathrm{T}}, \quad \mathbf{Q} = J\mathbf{F}^{-1}\mathbf{q}, \quad \sigma_m^2 = \frac{3}{2}\mathrm{S}_{ij}\mathrm{S}_{ij} \tag{4.85}$$

$$2\mathbf{W} = \mathrm{grad}\ \mathbf{v} - (\mathrm{grad}\ \mathbf{v})^{\mathrm{T}}, \quad (\dot{\varepsilon}_m^{\mathrm{p}})^2 = \frac{2}{3}\mathrm{D}_{ij}^{\mathrm{p}}\mathrm{D}_{ij}^{\mathrm{p}} \tag{4.86}$$

Here, E, v and α, respectively, are Young's modulus, Poisson's ratio and the coefficient of thermal expansion for the matrix material; **1** is the unit tensor. We refer the reader to Refs [54–57] for detailed descriptions of these equations. Nearly all of the thermo-physical material parameters depend on the temperature; however, such dependencies have been neglected for the sake of simplicity. While homogenizing material properties of a particulate composite, the rule of mixtures is used.

4.5.2 Plane-Strain Shearing of an Inhomogeneous Plate

We discuss results for plane-strain shearing of a plate of thickness $2H$ with all sides thermally insulated, vertical edges traction free and equal and opposite tangential velocities applied to the top and the bottom surfaces as depicted in Figure 4.14. The prescribed speeds increase from zero to the steady value v_0 in time t_r, giving an eventual nominal shear strain rate of v_0/H.

An ASB and a crack are assumed to initiate at a point when the effective plastic strain there equals 1.5; the difference between simulating the two is that for an ASB, no separation between adjoining surfaces occurs, but for a crack, the two surfaces slide relative to each other. The latter is simulated by using the node-splitting technique. The FE mesh used to analyse the problem had elements of size 4.6 μm × 4.6 μm in the central and end portions but are larger in other regions. For inhomogeneous bodies, the composition varied from 100% W (tungsten) at the centroid to 0% W at the edges. The yield stress of the material of elements in the 0.2 mm long, 4.6 μm thick layers located symmetrically about the centroidal axis was reduced by 30% to nucleate an ASB there. The ASB and the crack were found to propagate horizontally to the left and to the right with equal and opposite velocities; thus, their propagation to the right is described here. Results taken from Refs [58,59] presented next are for $t_r = 1$ μs and $v_0 = 0.625$ m/s giving $\dot{\gamma}_{\mathrm{avg}} = 5000\ \mathrm{s}^{-1}$.

Figures 4.15 and 4.16 depicts variations of the crack and the ASB lengths as they advance in the two homogeneous materials, W and NiFe (nickel–iron), and the

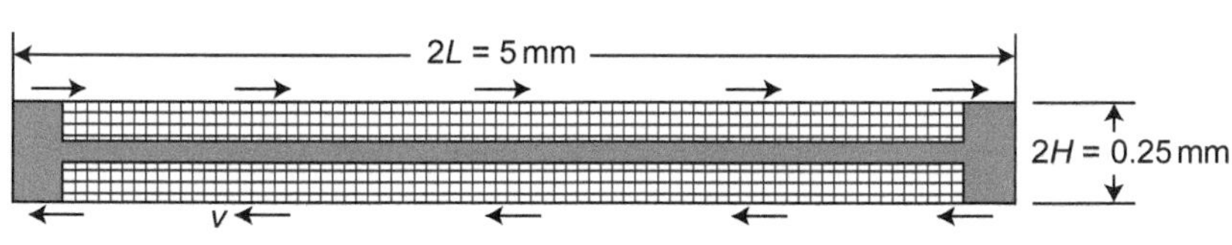

Figure 4.14 Schematics and FE mesh for plane-strain shearing of a plate.

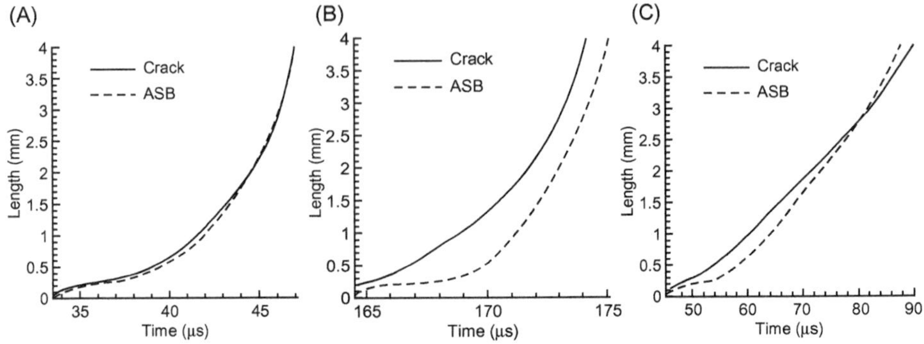

Figure 4.15 Time histories of evolution of a crack and an ASB in (A) W, (B) NiFe and (C) W2NiFe plates deformed in plane-strain shear.
Source: From Ref. [58].

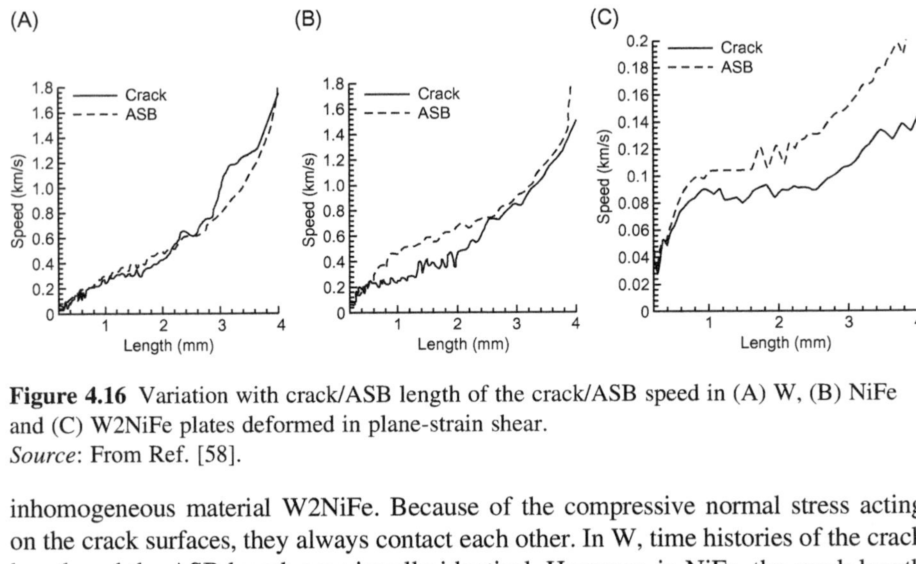

Figure 4.16 Variation with crack/ASB length of the crack/ASB speed in (A) W, (B) NiFe and (C) W2NiFe plates deformed in plane-strain shear.
Source: From Ref. [58].

inhomogeneous material W2NiFe. Because of the compressive normal stress acting on the crack surfaces, they always contact each other. In W, time histories of the crack length and the ASB length are virtually identical. However, in NiFe, the crack length is larger than the ASB length, and the crack propagation speed is higher than the ASB speed.

Except when the crack arrives at the edges, its speed increases from about 0.1 km/s at initiation to about 1.8 km/s when computations were terminated. The crack accelerates slowly in the beginning but quite rapidly towards the end. An ASB/crack initiates in NiFe at 165 μs but in W at 35 μs. Interestingly, plots of the crack propagation speed versus the crack length for W and NiFe are nearly coincident; cf., Figs 4.16A,B. We note that the Rayleigh wave speeds in W and NiFe equal 2.61 and 3.02 km/s, respectively. For the W2NiFe plate, the effective plastic strain at a point reaches 1.5 at about 46 μs. The plot of the crack speed versus the crack length (cf., Fig. 4.16C) is quite different from that for either W or NiFe, and

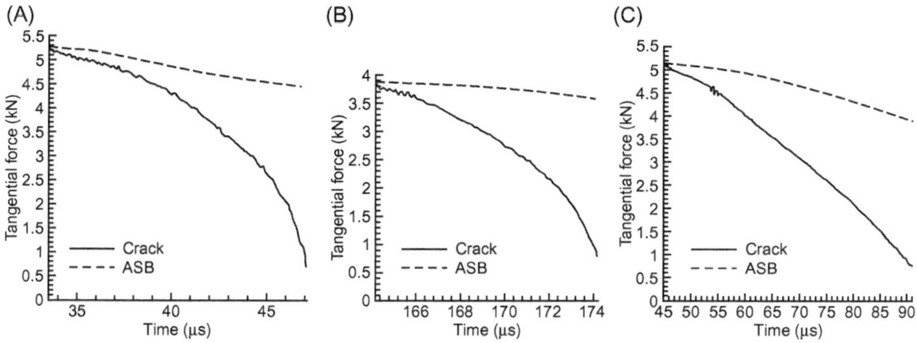

Figure 4.17 Time histories of the tangential force applied on the top and the bottom surfaces of (A) W, (B) NiFe and (C) W2NiFe plates deformed in plane-strain shear. *Source*: From Ref. [58].

the maximum crack speed achieved equals about 0.14 km/s, which is about one-tenth of that in W and NiFe. Also, except immediately after the crack and the ASB initiate, the ASB speed is less than that of the crack.

Time histories of the tangential force required to deform the specimen are exhibited in Figure 4.17. Recalling that the top and the bottom surfaces are constrained from moving in the vertical direction, the working of external forces is due to tangential traction and is proportional to it because the tangential velocity, except for the first 1 μs, is constant. The tangential force decreases with the opening of a crack and continues to decrease as the crack elongates. With the crack extension, a smaller surface area supports the external load. Also, with continued plastic deformations, the material softens, and its capacity to support external load diminishes. The rate of decrease of the applied tangential force is higher for W than that for NiFe because the strain and strain-rate hardening effects are higher in NiFe than those in W. With the extension of the crack, the driving force drops more rapidly for the W2NiFe plate than that for the W and NiFe plates.

The hydrostatic pressure in the material within the ASB, ahead of it and directly in front of the crack tip, was found to be compressive. Thus, no new voids nucleated in the material even though it had undergone enormous plastic deformations.

The homogenization of material properties for the W2NiFe plate suppressed sharp gradients in deformations likely to occur near interfaces between the matrix and the particulates. Furthermore, extensive plastic deformations during the ductile failure may debond particulates from the matrix, as discussed in the following section.

4.5.3 Plane-Strain Compression of a Particulate Composite Plate

We present results for plane-strain compression of a rectangular plate made of W circular cylindrical fibres immersed in NiFe matrix and explore effects of the interfacial bonding on the ASB initiation; a schematic sketch of the problem studied is shown in Fig. 4.18. The prescribed velocity increases linearly from zero

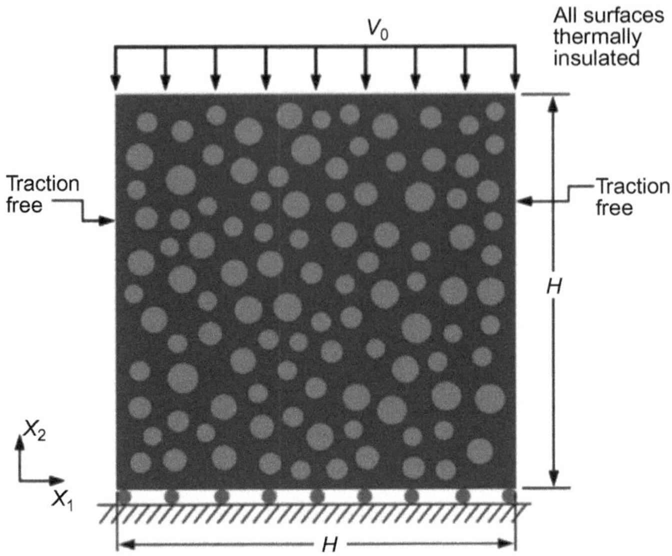

Figure 4.18 Schematic sketch of plane-strain compression of a particulate composite plate.

to its steady value in 1 μs so that the maximum nominal axial strain rate equals $10^4\ s^{-1}$. An ASB is assumed to initiate at a point when the energy dissipation rate there suddenly increases by nearly an order of magnitude, the material point is deforming plastically and deformations in its neighbourhood are inhomogeneous. For high strain-rate deformations of particulate composites, it was found [57] that the ASB initiation criterion stated in Section 4.3.2 is satisfied at a material point at time t_1 but not at time $t_2 > t_1$ due to load exchange between particulates and the surrounding matrix.

We use the cohesive zone model [60–63] to delineate interfacial failures between particulates and the matrix. After every time step, each interface between two adjoining FEs is checked for failure (i.e., separation and/or sliding) by examining the state variables of the two elements. The traction–separation relation is taken to be triangular with the area of the triangle representing fracture energy per unit surface area. The traction first linearly increases with an increase in the separation between the two faces of an interface to a pre-assigned maximum value and then affinely decreases to zero when the separation reaches a critical value. The mode mixity is considered by following the approach suggested by Ortiz and Pandolfi [63].

When interfaces are not allowed to separate, ASBs formed with plastic strains of 100% and a nearly discontinuous velocity field along surfaces inclined at about 45° to the loading axis. Figure 4.19 shows fringe plots of the effective plastic strain and the vertical component of the velocity. It is clear that one dominant ASB formed and divided the region into two parts: the lower essentially stationary pyramid and the upper one moving downward with the velocity prescribed on the top

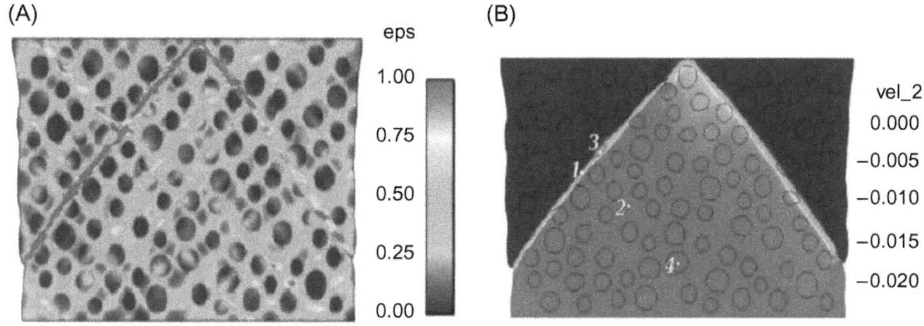

Figure 4.19 For plane-strain compression of a particulate composite plate: (A) fringe plots of the effective plastic strain at $t = 22.5$ μs and (B) plots of the vertical component of velocity.
Source: From Ref. [59].

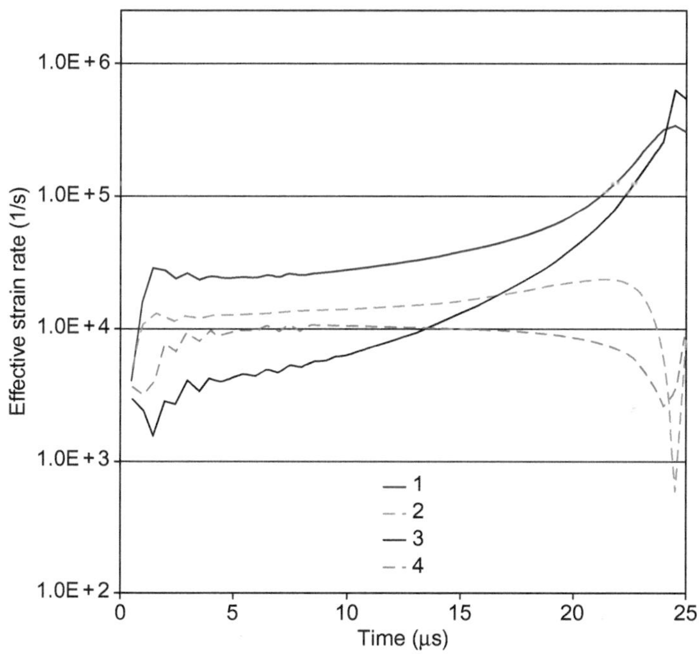

Figure 4.20 Time histories of the effective plastic strain rate at four points marked in Figure 4.19B.
Source: From Ref. [59].

surface. There is a sharp gradient in the velocity field between these two regions where strain rates, the effective plastic strain and the temperature are very high. In Figure 4.19B, points 1 and 3 are in the ASB, and points 2 and 4 are outside the ASB; points 1 and 2 are in the matrix, and points 3 and 4 are in the particulates. Time histories of the effective plastic strain rate exhibited in Figure 4.20 reveal

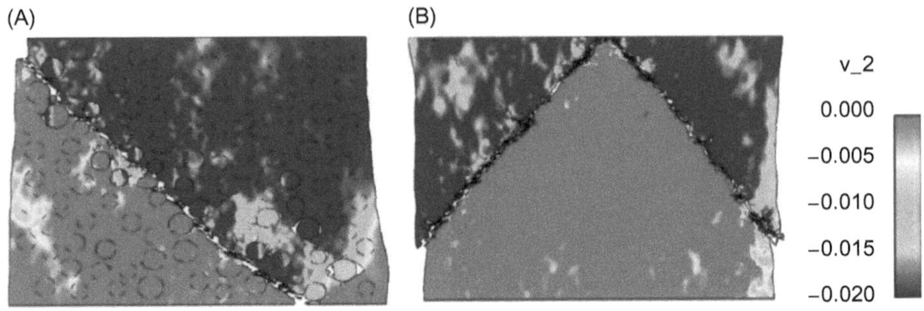

Figure 4.21 Fringe plots of the vertical component of velocity for mode-mixity parameters and times of (A) 0.6 and 13.2 μs and (B) 1.2 and 22.8 μs. Dark lines denote cracks and new interfaces.
Source: From Ref. [59].

that the effective plastic strain rates at these four points exceed $10^4\ s^{-1}$, except at late times when the ASB had formed. The strain rates at points 1 and 3 increase by a factor of 10 as the ASB develops.

Fringe plots of the x_2-velocity depicted in Figure 4.21 are for the simulations with failure allowed at interfaces. Dark circles denote regions where the matrix has separated from the particulates, and white regions denote voids and cracks. The deformed shapes and locations of ASBs strongly depend on the mode-mixity parameter, β. A smaller value of β makes the interface more susceptible to shear fracture; for $\beta = 0.6$, many particulate/matrix interfaces are fractured prior to the ASB development that result in totally different stress states from when there is no debonding. The ASB initiation time for $\beta = 0.6$ is nearly one-half of that for $\beta = 1.1$. Thus, the mode mixity and the cohesive failure criterion play dominant roles in not only the formation of ASBs but also in their spatial locations.

4.6 Concluding Remarks

As should be clear from results discussed earlier, 1D and 2D problems can be analysed with fine and properly graded meshes to obtain essentially mesh-independent results that are usually controlled by the thermal length scale. Very few 3D problems involving the localization of deformation into narrow bands have been scrutinized. Batra and Zhang [64] used the large-scale explicit FE code DYNA3D to analyse torsional deformations of a thin-walled tube and computed the speed of ASBs initiating from weak elements near the tube's centre and propagating outwards circumferentially. The speed strongly depended on the nominal strain rate; at an average strain rate of 5000 s^{-1}, it varied from 180 m/s at the instant of initiation to about 900 m/s by the time it reached the opposite end. Batra and Rattazzi [65] also used DYNA3D to study torsional deformations of a thick-walled tube and found that an ASB initiating from the root of a V-notch at the tube's centre propagated in the radial direction at about 100 m/s. However, when the thermal softening of the material was modelled by a different function that enhanced the effect, the speed of the shear band in the radial direction increased to about 1000 m/s. During the simulation of the Taylor impact test for a tungsten heavy alloy (WHA) rod, Batra and Stevens [66] found that the speed of an ASB (a contour of effective plastic strain of 1.0) initially equalled 800 m/s in the axial direction and 550 m/s in the radial direction and quickly dropped to 150 m/s in each direction.

Zhu and Batra [67] studied the possibility of phase transformation within an ASB. In the undeformed state, the specimen was assumed to be fully annealed and isotropic, and its microstructure was a mixture of coarse ferrite and cementite. A material point was assumed to transform into austenite after its temperature exceeded the transformation temperature with the rate of transformation governed by a simple kinetic equation. Proper account was taken of the latent heat required for the transformation, the associated volume change and the variation in the thermo-physical properties. It was found that the austenite is quenched rapidly enough by the surrounding material for it to be converted into martensite rather than a mixture of pearlite and martensite.

Wang and Batra [68] accounted for the texture development by using a theory with two internal variables, a scalar to account for the isotropic hardening of the material and a symmetric traceless second-order tensor to account for the kinematic hardening. They found that the consideration of kinematic hardening does not alter the qualitative nature of results.

It is hoped that this article has elucidated upon some aspects of ASBs and has raised enough questions in the reader's mind to warrant further enquiry into the subject. It will be interesting to find out if any new physics is revealed by studying 3D problems with adaptively refined meshes.

References

1. Bai, Y. L., and Dodd, B. (1992). "Adiabatic Shear Localization: Occurrence, Theories and Applications". Pergamon, Oxford.
2. Dodd, B., and Bai, Y. L. (1987). "Ductile Fracture and Ductility with Applications to Metal Working". Academic Press, London.
3. Wright, T. W. (2002). "The Physics and Mathematics of Adiabatic Shear Bands". Cambridge University Press, Cambridge.
4. Perzyna, P. (Ed.) (1998). "Localization and Fracture Phenomenon in Inelastic Solids". Pergamon Press, Wien, New York, NY.
5. Needleman, A., and Tvergaard, V. (1984). Finite element analysis of localization in plasticity. *In* "Finite Elements: Special Problems in Solid Mechanics" (J. T. Oden and G. F. Carey, Eds.), pp. 94–157. Prentice Hall, Englewood Cliffs, NJ.
6. Zbib, H. M., Shawki T., and Batra, R. C. (Eds.) (1992). Material instabilities. *Appl. Mech. Rev.* **45**, 1–173.
7. Tomita, Y. (1994). Simulations of plastic instabilities in solid mechanics. *Appl. Mech. Rev.* **47**, 171–205.

8. Hughes, T. J. R. (1987). "The Finite Element Method. Linear Static and Dynamic Finite Element Analysis". Prentice Hall, Englewood Cliffs, NJ.
9. Carey, G. F., and Oden, J. T. (1984). "Finite Elements. Computational Aspects". Vol. III. Prentice Hall, Englewood Cliffs, NJ.
10. Belytschko, T., Liu, W. K., and Moran, B. (2001). "Nonlinear Finite Elements for Continua and Structures". Wiley, New York, NY.
11. Tresca, H. (1878). On further application of the flow of solids. *Proc. Inst. Mech. Eng.* **30**, 301–345.
12. Massey, H. F. (1921). The flow of metal during forging. *Proc. Manchester Assoc. Eng.*, 21–26.
13. Kalthoff, J. F. (2000). Modes of dynamic shear failure in solids. *Int. J. Fract.* **101**, 1–31.
14. Kalthoff, J. F., and Winkler, S. (1988). Failure mode transition at high strain rates of loading. *In* Proceedings of the International Conference on Impact Loading and Dynamic Behavior of Materials (C. Y. Chiem, H. D. Kunze and L. W. Meyer, Eds.), pp. 185–196. Deutsche Gewsellschaft für Metallkundcle, DGM, Bremen.
15. Kalthoff, J. F. (1990). Transition in the failure behavior of dynamically shear loaded cracks. *Appl. Mech. Rev.* **43**, S247.
16. Zhou, M., Ravichandran, G., and Rosakis, A. (1996). Dynamically propagating shear bands in impact-loaded prenotched plates – I. *J. Mech. Phys. Solids* **44**, 981–1006.
17. Zhou, M., Ravichandran, G., and Rosakis, A. (2000). Dynamically propagating shear bands in impact-loaded prenotched plates – II. *J. Mech. Phys. Solids* **44**, 1007–1032.
18. Batra, R. C., and Gummalla, R. R. (2000). Effect on material and geometric parameters on deformations near the notch-tip of a dynamically loaded prenotched plate. *Int. J. Fract.* **101**, 99–140.
19. Batra, R. C., and Ravisankar, M. V. S. (2000). Three-dimensional numerical simulation of the Kalthoff experiment. *Int. J. Fract.* **105**, 161–186.
20. Batra, R. C., and Jaber, N. A. (2001). Failure mode transition speeds in an impact loaded prenotched plate with four thermoviscoplastic relations. *Int. J. Fract.* **110**, 47–71.
21. Batra, R. C., and Lear, M. H. (2004). Simulation of brittle and ductile fracture in an impact loaded prenotched plate. *Int. J. Fract.* **126**, 179–203.
22. Wang, S., and Liu, H. (2012). Failure transitions in metal plates with a two-dimensional meshfree continuum method. *Thin-Walled Struct.* **50**, 1–13.
23. Gear, C. W. (1971). "Numerical Initial Value Problems in Ordinary Differential Equations". Prentice Hall, Englewood Cliffs, NJ.
24. Zienkiewicz, O. C., and Zhu, J. Z. (1992). The superconvergent patach recovery and a posteriori error estimates, Part I, II. *Int. J. Numer. Methods Eng.* **33**, 1331–1382.
25. Oh, H.-S., and Batra, R. C. (1999). Application of Zienkiewicz–Zhu's error estimate with superconvergent patch-recovery to hierarchical *p*-refinement. *Finite Elem. Anal. Des.* **31**, 273–280.
26. Oh, H.-S., and Batra, R. C. (1999). Locations of optimal stress points in higher-order elements. *Commun. Numer. Methods Eng.* **15**, 127–136.
27. Marchand, A., and Duffy, J. (1988). An experimental study of the formation process of adiabatic shear bands in a structural steel. *J. Mech. Phys. Solids* **36**, 251–283.
28. Batra, R. C., and Kim, C. H. (1990). Effect of viscoplastic flow rules on the initiation and growth of shear bands at high strain rates. *J. Mech. Phys. Solids* **38**, 859–874.
29. Batra, R. C., and Kim, C. H. (1990). Adiabatic shear banding in elastic–viscoplastic nonpolar and dipolar materials. *Int. J. Plast.* **6**, 127–141.

30. Batra, R. C., and Chen, L. (2001). Effect of viscoplastic relations on the instability strain, shear band initiation strain, the strain corresponding to the minimum shear band spacing, and the band width in thermoviscoplastic materials. *Int. J. Plast.* **17**, 1465–1489.
31. Wright, T. W., and Batra, R. C. (1985). The initiation and growth of adiabatic shear bands. *Int. J. Plast.* **1**, 205–212.
32. Litonski, J. (1977). Plastic flow of a tube under adiabatic torsion. *Bull. Acad. Polon. Sci.* **25**, 7–17.
33. Bodner, S. R., and Partom, Y. (1975). Mechanical properties at high rate of strain. *Inst. Phys. Conf. Ser. No. 21*, 102.
34. Johnson, G. R., and Cook, W. H. (1983). A constitutive model and data for metals subjected to large strains, high strain rates and high temperatures. *In* "Proceedings of the 7th International Symposium on Ballistics", pp. 1–7. The Hague, The Netherlands.
35. Klopp, R. W., Clifton, R. J., and Shawki, T. G. (1985). Pressure-shear impact and the dynamic viscoplastic response of metals. *Mech. Mater.* **4**, 375–385.
36. Batra, R. C., and Kim, C. H. (1992). Analysis of shear bands in twelve materials. *Int. J. Plast.* **8**, 425–452.
37. Dilellio, J. A., and Olmstead, W. E. (1997). Temporal evolution of shear band thickness. *J. Mech. Phys. Solids* **45**, 345–359.
38. Batra, R. C., and Kim, C. H. (1991). Effect of thermal conductivity on the initiation, growth and band width of adiabatic shear bands. *Int. J. Eng. Sci.* **29**, 949–960.
39. Batra, R. C. (1988). Effect of nominal strain-rate on the initiation and growth of adiabatic shear bands in steels. *J. Appl. Mech.* **55**, 229–230.
40. Batra, R. C. (1987). Effect of material parameters on the initiation and growth of adiabatic shear bands. *Int. J. Solids Struct.* **23**, 1435–1446.
41. Kim, C. H., and Batra, R. C. (1992). Effect of initial temperature on the initiation and growth of shear bands in a plain carbon steel. *Int. J. Nonlinear Mech.* **27**, 279–291.
42. Batra, R. C., and Ko, K. I. (1992). An adaptive mesh refinement technique for the analysis of shear bands in plane strain compression of a thermoviscoplastic solid. *Comput. Mech.* **10**, 369–379.
43. Batra, R. C., and Hwang, J. (1993). An adaptive mesh refinement technique for two-dimensional shear band problems. *Comput. Mech.* **12**, 255–268.
44. Cescotto, S., and Zhou, D. W. (1989). A variable density mesh generation for planar domains. *Commun. Appl. Numer. Methods* **5**, 473–481.
45. Lo, S. H. (1985). A new mesh generation scheme for arbitrary planar domains. *Int. J. Numer. Methods Eng.* **21**, 1403–1426.
46. Peraire, J., Vahdati, M., Morgan, K., and Zienkiewicz, O. C. (1987). Adaptive remeshing for compressible flow computations. *J. Comput. Phys.* **72**, 449–466.
47. Habraken, A. M., and Cescotto, S. (1990). An automatic remeshing technique for finite element simulation of forming processes. *Int. J. Numer. Methods Eng.* **30**, 1503–1525.
48. Löhner, R. (1988). Useful data structures for the generation of unstructured grids. *Commun. Appl. Numer. Methods* **4**, 123–135.
49. Zhu, J. Z., Zienkiewicz, O. C., Hinton, O., and Wu, J. (1991). A new approach to the development of automatic quadrilateral mesh generation. *Int. J. Numer. Methods* **32**, 849–886.
50. Batra, R. C. (1988). Steady state penetration of thermoviscoplastic targets. *Comput. Mech.* **3**, 1–11.
51. Wright, T. W., and Walter, J. W. (1996). The asymptotic structure of an adiabatic shear band in antiplane motion. *J. Mech. Phys. Solids* **44**, 77–97.

52. Chen., L., and Batra, R. C. (1999). The asymptotic structure of a shear band in mode-II deformations. *Int. J. Eng. Sci.* **37**, 895–919.
53. Hinton, E., and Campbell, J. S. (1974). Local and global smoothing of discontinuous finite element functions using a least squares method. *Int. J. Numer. Methods Eng.* **8**, 461–480.
54. Gurson, A. L. (1977). Continuum theory of ductile rupture by void nucleation and growth, Part 1. Yield criteria and flow rules for porous ductile media. *J. Eng. Mater. Technol.* **99**, 2–15.
55. Pan, J., Saje, M., and Needleman, A. (1983). Localization of deformation in rate-sensitive porous plastic solids. *Int. J. Fract.* **21**, 261–278.
56. Saje, M., Pan, J., and Needleman, A. (1982). Void nucleation effects on shear localization in porous plastic solids. *Int. J. Fract.* **19**, 163–182.
57. Batra, R. C., and Love, B. M. (2004). Adiabatic shear bands in functionally graded materials. *J. Therm. Stresses* **27**, 1101–1123.
58. Batra, R. C., and Love, B. M. (2005). Crack propagation due to brittle and ductile failures in microporous thermoelastoviscoplastic functionally graded materials. *Eng. Fract. Mech.* **72**, 1954–1979.
59. Batra, R. C., and Love, B. M. (2010). Effect of particulate/matrix debonding on the formation of adiabatic shear bands. *Int. J. Mech. Sci.* **52**, 386–397.
60. Dugdale, D. (1960). Yielding of steel sheets containing slits. *J Mech. Phys. Solids* **8**, 100–104.
61. Barrenblatt, G. (1962). The mathematical theory of equilibrium of cracks in brittle fracture. *Adv. Appl. Mech.* **7**, 55–129.
62. Xu, X.-P., and Needleman, A. (1994). Numerical simulations of fast crack growth in brittle solids. *J. Mech. Phys. Solids* **42**, 1397–1434.
63. Ortiz, M., and Pandolfi, A. (1999). Finite-deformation irreversible cohesive elements for three-dimensional crack-propagation analysis. *Int. J. Numer. Meth. Eng.* **44**, 1267–1282.
64. Batra, R. C., and Zhang, X. (1994). On the propagation of a shear band in a steel tube. *J. Eng. Mater. Technol.* **116**, 155–161.
65. Batra, R. C., and Rattazzi, D. (1997). Adiabatic shear banding in a thick-walled steel tube. *Comput. Mech.* **20**, 412–426.
66. Batra, R. C., and Stevens, J. B. (1998). Adiabatic shear bands in axisymmetric impact and penetration problems. *Comput. Methods Appl. Mech. Eng.* **151**, 325–342.
67. Zhu, Z. G., and Batra, R. C. (1992). Consideration of phase transformations in the study of shear bands in a dynamically loaded steel block. *J. Eng. Mater. Technol.* **114**, 368–377.
68. Wang, Y., and Batra, R. C. (1994). Effect of kinematic hardening on the initiation and growth of shear bands in plane strain compression of a thermoviscoplastic solid. *Acta Mechanica* **102**, 217–233.

5 Theory of Adiabatic Shear Bands

Thomas W. Wright

Department of Mechanical Engineering, The Johns Hopkins University, Baltimore, MD

Nomenclature

α thermal softening parameter
$\hat{\mathbf{a}}$ relative thermal softening, eq. (13)
$\hat{\mathbf{a}}$ nondimensional softening, eq. (24)
$\boldsymbol{b}$ characteristic reciprocal strain rate
$\boldsymbol{c}$ heat capacity
D_p plastic strain rate tensor
E rate of dissipation per unit area
$\boldsymbol{g}(|s|,T,\kappa)$ Gibbs free energy
$\boldsymbol{g}(\vartheta)$ softening factor
$\boldsymbol{G}(\varphi)$ nondimensional thermal factor
$\boldsymbol{h}$ length scale
$\boldsymbol{k}$ thermal conductivity
$\hat{k}$ nondimensional thermal conductivity
$\boldsymbol{l}$ characteristic length
$\boldsymbol{L}$ guage length
$\boldsymbol{m}$ strain rate sensitivity
$\boldsymbol{m}_c$ rate sensitivity at center of shear band
$\boldsymbol{n}$ Fourier element number
$\boldsymbol{Q}$ energy stored from work hardening
$\hat{Q}, \hat{K}, \hat{\Gamma}$ functions, eq. (2)
$\boldsymbol{r}$ strain rate, eq. (12)
$\boldsymbol{r}$ radius from band tip
$\boldsymbol{R}(s,\vartheta)$ function equal to strain rate
s shear traction
s^+ stress ahead of propagating band
$\tilde{s}, \tilde{v}_y, \tilde{\vartheta}$ perturbations
$\boldsymbol{S}(t)$ homogeneous stress
$\boldsymbol{S}(\varphi)$ nondimensional stress
$\boldsymbol{S}(r,\vartheta)$ function equal to stress, eq. (12)
$\boldsymbol{S}$ deviatoric stress tensor
$\boldsymbol{t}$ time
t_{cr} a critical time
$\boldsymbol{T}$ absolute temperature

Adiabatic Shear Localization. DOI: http://dx.doi.org/10.1016/B978-0-08-097781-2.00005-8

T stress tensor
U speed of shear band
v velocity parallel to band
v_y velocity gradient
V driving speed on edge of slab
w relative displacement, eqs. (19,20)
w z-component of velocity, eq. (35)
$W(\varphi)$ nondimensional velocity
W^{-} nondimensional velocity at $\varphi \rightarrow \pi$
x distance parallel to shear band
y distance perpendicular to shear band
z change of variable for temperature
α temporal decay parameter, eq. (7)
α similarity exponent
β Taylor-Quinney fraction
$\dot{\gamma}_0$ reference strain rate
γ_{cr}, γ_{crit} critical strain
γ_{max} strain in center of shear band
$\dot{\gamma}_p$ plastic strain rate
$\Gamma(\varphi)$ nondimensional strain rate
δ_M mechanical band width, eq. (16)
δ_T thermal band width, eq. (17)
$\delta\vartheta$ magnitude of temperature defect
ϑ temperature above reference
ϑ_c temperature at center of shear band
$\tilde{\vartheta}_n$ Fourier component
$\Theta(t)$ homogeneous temperature
κ work hardening parameter
κ_0 initial work hardening parameter
λ strain rate at center of shear band
μ elastic shear modulus
ρ density
$\hat{\rho}$ nondimensional density, eq. (34)
σ_e effective stress
φ angle from band tip
φ, Ψ energy functions
$[v]$, $[w^{-}]$ velocity jump across shear band

5.1 Introduction

Over the past several decades, the theory of adiabatic shear bands has developed from the continuum point of view so that today a great deal is known about their general structure and the dynamic conditions that lead to their formation. Theoretical scaling laws have also been developed to describe many of these features, including timing or critical strain, thickness, spacing and speed of propagation, so that an experimentalist can attempt to optimize the design of a particular experiment in a rational way.

The technological importance of adiabatic shear bands lies in the fact that they are often sites for further damage or failure. In some applications, only the conditions for band formation need be known so that they may be avoided by design, e.g. through restrictions of loading rates and magnitudes. In the fields of impact protection or ballistics, however, failure may be expected as part of proper functioning, and, therefore, the kinetics and dynamics of band formation should be accounted for as part of the design. In these cases, the spatial localization of adiabatic shear bands and the strong coupling between mechanical and thermal effects during formation make it impractical to calculate their occurrence from first principles in large-scale numerical simulations.

Alternate means of predicting their behaviour must therefore be sought. This has been attempted through thorough study of the physics and mechanics of formation, by analysis of simple, yet physically representative, problems in continuum mechanics and by extracting from those problems algebraic (or equally simple) scaling laws that capture the essential physics of band formation. The resulting scaling laws may then be injected into large-scale calculations in a realistic way so that the macroscopic effect is adequately captured. The necessity and difficulty of this last step should not be underestimated.

This chapter outlines some of the advances made in the theoretical understanding of adiabatic shear bands since the publication of Bai and Dodd [1].

5.2 Theoretical Setting and Mathematical Preliminaries

Although adiabatic shear banding is the usual name given to the intense localization often seen in metals that have been subjected to high-rate shearing, the name itself is something of a historical misnomer. It is sometimes supposed that because heat conduction is a relatively slow process, it may be ignored in high-rate shearing, but, in fact, it is the eventual failure of the adiabatic approximation that leads to the intense localization. Small local variations in temperature or material strength tend to become magnified under rapid deformation in a viscoplastic material with softening due to heating from plastic work being offset by rate hardening as the plastic strain rate increases. In the early stages of deformation, these two tendencies cancel out more or less so that the stress evolves initially as if there were no early variations in environmental or material properties at all. Eventually, however, softening wins with a rapid and dramatic loss of the material's capability to carry shear stress at the place of the original local variation. As the stress collapses, the local temperature and strain rate become very high with enormous thermal gradients, and in this stage of formation, the adiabatic approximation fails completely. If there are multiple shear bands in some part of the material, thermal conductivity also plays a dominant role in determining their possible spacings from one another.

Because of the large deformations that finally develop in an adiabatic shear band, non-linear elasticity/viscoplasticity is the obvious theoretical setting for adiabatic shearing. Complete continuum equations with internal variables and a multiplicative elastic/plastic decomposition have been given by Anand [2], Rosakis et al. [3] and Wright [4], among others, for example, where the internal variables are meant to model plasticity or other damage mechanisms. Wright [4] does not assume plastic incompressibility a priori, so although not exploited in the reference, it does make provision for modelling complex interactions among multiple damage mechanisms.

When any material is subjected to extreme conditions, it will respond with whatever combination of deformation and damage mechanisms that may be required (elasticity, dislocations, stacking faults, twins, phase changes, dynamic recovery, voids, cleavage, etc.). Each mechanism will have its own threshold and kinetic requirements, so the easiest mechanism to be excited will be the first to appear, then another will appear if still required by the imposed conditions and so forth. If several mechanisms are excited, there will be a competition among them that depends on the kinetics associated with each one. In this chapter, only strain-rate hardening and thermal effects will be considered, although some authors have considered void nucleation and growth to be essential for rapid shear failure, e.g. Cowie et al. [5]. Voids should undoubtedly be considered if the maximum principal stress at the damage site is positive. In any case, shear band initiation is a process that takes time, so the evolution of the various mechanisms may influence one another as the process develops.

The conservation laws for mass, momentum, angular momentum and energy must always be satisfied, and the Clausius–Duhem inequality must also hold to enforce the second law of thermodynamics. (See Ericksen [6] for a discussion of the Clausius–Duhem inequality in non-linear elasticity.) In addition, constitutive relations must be specified to complete the full statement of the theory. Although amenable to numerical analysis, the 3D theory is much too complicated for simple analysis. Fortunately, experimental evidence shows that adiabatic shear bands tend to be extremely thin and quite extensive in other directions compared to the thickness. Therefore, the theory is usually reduced to a set of 1D equations but only at the cost of a number of approximations. These include constant specific heat; heat flux given by Fourier's law with constant thermal conductivity, thermo-elasticity ignored; laminar flow parallel to the shear band; plastic work $\mathbf{T}:\mathbf{D}^{\mathrm{p}}$ replaced by the simple expression $s\dot{\gamma}^{\mathrm{p}}$, where s is shear traction and $\dot{\gamma}^{\mathrm{p}}$ is plastic shear strain rate on planes parallel to the shear band; and thermo-plastic terms arising from changes in internal variables simply being replaced by reducing plastic work by a constant factor. These approximations are discussed in some detail in Wright [4]. Only the last approximation will be discussed further in this chapter.

The reduced 1D version with only a single internal variable, but still retaining the thermo-plastic term, is presented here. Constants ρ, c, k stand for density, heat capacity and thermal conductivity, respectively. Variables v, s, T, $\dot{\gamma}_{\mathrm{p}}$, Q, κ are functions only of the distance perpendicular to the band and time, denoted by (y,t), and stand for particle velocity, shear stress, temperature, plastic strain rate, a thermo-plastic variable and a work-hardening parameter, respectively.

$$
\begin{aligned}
&\text{momentum} &:\quad & \rho\dot{v} = s_{\mathrm{y}} \\
&\text{energy} &:\quad & \rho c\dot{T} = kT_{\mathrm{yy}} + s\dot{\gamma}_{\mathrm{p}} - \left(Q - T\frac{\partial Q}{\partial T}\right)\dot{\kappa} \\
&\text{entropy} &:\quad & s\dot{\gamma}_{\mathrm{p}} - Q\dot{\kappa} \geq 0
\end{aligned}
\tag{5.1}
$$

where $Q = -\rho(\partial g/\partial\kappa)$ and $g(|s|,T,\kappa)$ is the Gibbs free energy. The third line of Eq. (5.1) states that the rate of plastic work must be greater than the rate of storage of cold work, i.e. increasing internal energy associated with an increasing level of work hardening. The term $T\frac{\partial Q}{\partial T}\dot{\kappa}$ is a thermo-plastic term, which is completely analogous in its origin to the familiar thermo-elastic term. If the thermo-plastic term is thought to be small, then the energy is sometimes partitioned into a mechanical and a thermal term, $g = \varphi(\kappa,|s|) + \psi(T)$. To complete the system, constitutive relations must be specified for s, $\dot{\gamma}_p$, Q, $\dot{\kappa}$. In general, these may be written as:

$$\begin{aligned} \dot{s} &= \mu(v_y - \dot{\gamma}_p) \\ Q &= \hat{Q}(|s|, T, \kappa) \\ \dot{\gamma}_p &= \hat{\Gamma}(|s|, T, \kappa)\frac{s}{|s|} \\ \dot{\kappa} &= \hat{K}(|s|, T, \kappa) \end{aligned} \tag{5.2}$$

In Eq. (5.2), μ is the elastic shear modulus, $\hat{\Gamma} \geq 0$ and $\hat{\Gamma} = 0$ if $|s|$ is not greater than a specified and possibly evolving threshold. The following restriction must be placed on the constitutive functions $\hat{Q}$, $\hat{\Gamma}$, $\hat{K}$ to satisfy the Clausius–Duhem inequality:

$$|s|\hat{\Gamma} - \hat{Q}\hat{K} \geq 0 \tag{5.3}$$

Because work hardening and increasing numbers of dislocations are associated with the stored energy of cold work, it can be assumed that $\hat{Q} \geq 0$, so that when $\hat{Q} \neq 0$,

$$\hat{K} \leq |s|\hat{\Gamma}/\hat{Q} \tag{5.4}$$

Equation (5.4) does not forbid negative values of $\hat{K}$, so work hardening may actually be negative under some circumstances, as has been pointed out by Anand and Brown [7]. More complex models with additional internal variables could be used to model additional phenomena, such as dynamic recrystallization, by additional terms in the energy and entropy relations analogous to the ones containing Q in Eq. (5.1). Other treatments of recovery have been given by Hines and Vecchio [8].

More commonly, rather than the second line of Eq. (5.1), the balance of energy has been approximated by

$$\rho c\dot{T} = kT_{yy} + \beta s\dot{\gamma}_p \tag{5.5}$$

where β is a constant slightly less than 1. Experimental evidence indicates that β is not constant, according to Hodowany et al. [9]. At zero plastic strain, it starts at a value in a range at least as large as 0.3–0.8 and then tends to increase with plastic strain, becoming asymptotic to 1 for strains larger than 0.4 or so, at least for the alloys investigated. Suppose that $Q_T/Q \ll 1/T$ so that the thermo-plastic term may be ignored. Then, $0 \leq 1 - \beta = \hat{Q}\hat{K}/|s|\hat{\Gamma} \leq 1$, and β must lie between 0 and 1. If work hardening saturates, then $\hat{K} \rightarrow 0$ and $\beta \rightarrow 1$, so the theoretical limits certainly contain the observed limits. If the thermo-plastic term is retained, then β may actually be greater than 1 at times, indicating that dynamic recovery is occurring, but it still tends to 1 if work hardening saturates. These issues have been considered in some detail by Rosakis et al. [3], and others who have recognized the overly simplistic nature of assuming a constant β, e.g. Wright [4] and Longere and Dragon [10].

At this point, the general theory has been reduced to one spatial dimension by clear and consistent approximations. Before specific problems can be posed for analysis, it remains only to choose appropriate constitutive equations that are properly constrained by the Clausius–Duhem inequality and that are compatible with appropriate experimental data for constitutive response. Most constitutive models in use today appear to have a substantial phenomenological basis and a lesser theoretical basis, although the relative proportions of each vary widely. Some examples are power law, e.g. Molinari and Clifton [11]; Litonski [12]; Johnson-Cook [13]; Zerilli-Armstrong [14]; Bodner-Partom, e.g., see [15]; MTS (Material Threshold Stress), e.g. Follansbee and Kocks [16]; and Bammann-Johnson-Chiesa, see Bammann [17]. Many other forms have been used as well, and development of these and other constitutive models continues to be an active field today.

What seems apparent is that there is no generally agreed upon set of constitutive models that are adequate for viscoplasticity with several physically motivated and modelled internal variables, even though the general framework, at least for fully isotropic materials, is available. Therefore, in some sense, only qualitative results can be obtained with further analysis until more fully developed, physically based and experimentally verified models become available. A physically based model tends to have quantitative usefulness even far outside the database for which it was developed, at least until overtaken by other mechanisms of deformation and failure, whereas a purely phenomenological model cannot be expected to perform well outside its database. Even so, this author believes that great insight can be developed through the solution of simplified problems, even if quantitative accuracy is not yet fully available due to inadequate constitutive models. In the next several sections, various aspects of shear band formation will be explored. These will include the timing of stress collapse, the canonical structure of a fully formed shear band (including a discussion of the elusive concept of shear band width), the spacing between shear bands, the speed of a shear band and the energy rates and dissipation associated with adiabatic shear bands.

Stress collapse after adiabatic shearing is a matter of loss of stability of continued homogeneous deformation. The usual analytical approach is first to find the homogeneous solution; next to investigate the behaviour of small perturbations through linearized equations; and finally to make some judgement as to loss of stability from the behaviour of the perturbations. Because the homogeneous solution is generally time dependent, the linear perturbation equations will have time-dependent coefficients. Often in the literature, the assumption of 'frozen coefficients' has been made, i.e. because the coefficients in some sense change only slowly, they may be regarded as constants, and time dependence of the perturbations is taken to be exponential. Then a positive real part of the exponent is taken to mean instability and a negative real part to mean stability. The author is unaware of any theory that may guide a researcher as to when this assumption may be effective, although its use has often seemed to provide reasonable conclusions. However, there are known counterexamples in the literature of linear equations with time-dependent coefficients where the real part of the exponent, assuming frozen coefficients, is negative, supposedly indicating stability, but known exact solutions actually increase exponentially, e.g. Markus and Yamabe [18]. In the shear band literature with work hardening, there is an example where solutions to the perturbation equations show very strong boundary layers in time (unless initial conditions are specially chosen) with some variables decreasing rapidly and others increasing rapidly, but this is of no consequence in judging the ultimate stability, according to Wright [19] (see also Wright [4] for an outline of these two examples). With these warnings in mind, asymptotic solutions to linearized perturbation equations in this chapter will be constructed instead of appealing to frozen coefficients.

5.3 Timing of Stress Collapse or Critical Strain

The timing or critical strain when an adiabatic shear band first nucleates is not a fixed material property, as is sometimes supposed. Although timing does depend on material properties, it also depends on material or geometric defects and on the process of deformation itself, as influenced for example by loading, boundary conditions and the geometry of the material being deformed. It was predicted by Molinari and Clifton [11] that the nominal strain at localization in a torsional Kolsky bar experiment would depend logarithmically on the strength of a particular geometric defect. A further generalization of this result was given by Wright [20] to include small variations in initial temperature or in material strength, as well as defects in geometry, all of which are equivalent defects when properly scaled. Experimental verification of the logarithmic dependence was given by Duffy and Chi [21], who carefully introduced known defects into the wall thickness of thin-walled, torsional specimens. However, the experimental results systematically showed a larger strain at localization than the theoretical predictions. An explanation for the offset has been offered by Wright [4], who demonstrated that heat conduction alone is sufficient to produce the delay in localization (Figure 5.1). If heat conduction is not included, the estimate is a lower bound for localization.

A simple model for steady shearing of a short strip illustrates these features. Imagine an idealized material that deforms as a rigid perfectly plastic solid, i.e. $\mu \to \infty$, $v_y \to \dot{\gamma}_p$ in the first line of Eq. (5.2) and $\dot{\kappa} = 0$ in the second and fourth lines of Eq. (5.2). In addition, the material softens as it heats up and strengthens with increasing strain rate. Linear thermal softening and power law rate hardening are sufficient for illustrative purposes:

$$\begin{aligned} &\text{momentum}: && s_y = 0 \\ &\text{energy} \quad: && \rho c\dot{\vartheta} = k\vartheta_{yy} + s v_y \\ &\text{flow law} \quad: && s = \kappa_0(1 - a\vartheta)(b v_y)^m \end{aligned} \tag{5.6}$$

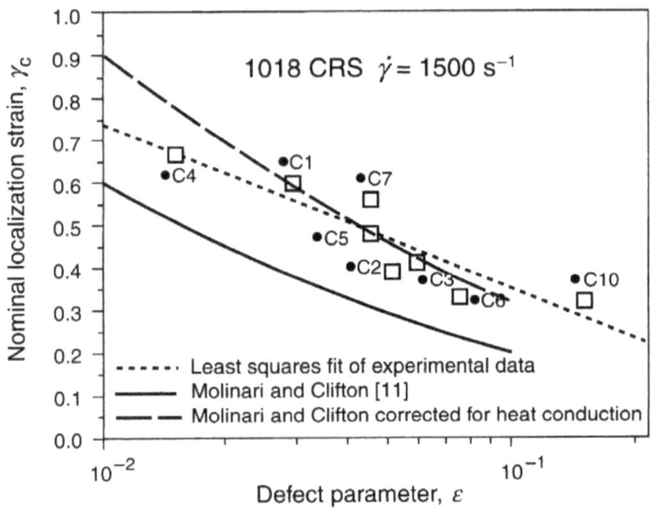

Figure 5.1 Localization strain versus defect parameter showing the theoretical logarithmic dependence, from the data of Duffy and Chi [21], and the correction for heat conduction, from Wright [4].

The strip is defined by $-h \leq y \leq +h$, with boundary conditions $\vartheta_y(-h,t) = \vartheta_y(h,t) = 0$ and $v(h,t) = -v(-h,t) = \dot{\gamma}_0 h$ and with initial conditions $v(y,0) = \dot{\gamma}_0 y$, $\vartheta(y,0) = \vartheta_0(y)$ and $s(y,0) = s_0 = \kappa_0(b\dot{\gamma}_0)^m$. These are adaptations from Eq. (5.1) where only the quasi-static case is considered, i.e. when $\rho(\dot{\gamma}_0 h)^2/s_0 \ll 1$, where $\beta = 1$ because there is no work hardening; the temperature difference from the initial average temperature, θ, is used rather than T; and a rudimentary flow law is given. New material constants are the initial quasi-static flow strength, κ_0; rate of thermal softening, a; reference time, b; and strain-rate sensitivity, m. Following the basic procedure outlined previously, we assume that the initial temperature distribution is almost uniform. In this particular case, the homogeneous response is given by

$$s = S(t) = s_0 e^{-\alpha t}, \quad a\theta = a\Theta(t) = 1 - e^{-\alpha t} \tag{5.7}$$

where $\alpha = \frac{as_0}{\rho c}\dot{\gamma}_0$. The stress decays exponentially from its initial maximum and the temperature increases exponentially toward an upper limit.

The perturbations are made relative to the homogeneous solution, and therefore the equations corresponding to Eq. (5.6) are

$$\begin{aligned}
&\text{momentum}: \quad \tilde{s}_y = 0 \\
&\text{energy} \quad : \quad \rho c \tilde{\theta}_t = k\tilde{\theta}_{yy} + S\tilde{v}_y + \tilde{s}\dot{\gamma}_0 \\
&\text{flow law} \quad : \quad \tilde{s} = -s_0 a\tilde{\theta} + mS\tilde{v}_y/\dot{\gamma}_0
\end{aligned} \tag{5.8}$$

where the tilde signifies perturbation quantities, $\tilde{s} = s - S(t)$, $\tilde{v}_y = v_y - \dot{\gamma}_0$, $\tilde{\vartheta} = \vartheta - \Theta(t)$. It is easy to demonstrate that $\tilde{s} = 0$ – see Wright [4] – so that the third line of Eq. (5.8) requires the perturbation in strain rate to be proportional to the perturbation in temperature, and the energy equation reduces to an equation for the perturbation of temperature only. The Fourier solution is

$$\tilde{\theta} = \exp\left(\frac{as_0}{\rho cm}\dot{\gamma}_0 t\right)\sum_{n=1}^{\infty}\tilde{\theta}_n \exp\left\{-\frac{k}{\rho c}\frac{n^2\pi^2}{h^2}t\right\}\cos\frac{n\pi y}{h} \tag{5.9}$$

where $\tilde{\theta}_n$ are the Fourier coefficients for the initial distribution of temperature.

In the absence of heat conduction, the initial distribution of temperature, and therefore the strain rate, is simply amplified by the leading exponential term. At early times, if the applied strain rate is small enough, $as_0\dot{\gamma}_0/m - \pi^2 k/h^2 < 0$, then heat conduction causes all terms to decrease, but at higher rates, a finite number of Fourier components will grow exponentially. To get an idea of the time of localization and the appropriate scaling parameters, simply set $\tilde{v}_y = \dot{\gamma}_0$ (when the perturbation has certainly lost its validity and rescaling would be required, as suggested by H. Ockendon, personal communication, 1990), and assume that the Fourier series only has one term. This procedure results in the estimate

$$\gamma_{cr} = \dot{\gamma}_0 t_{cr} \approx \frac{1}{\dfrac{as_0}{\rho c}\dfrac{1+m}{m} - \dfrac{k\pi^2}{\rho c\dot{\gamma}_0 h^2}}\ln\frac{m}{a\delta\theta} \tag{5.10}$$

where $\delta\theta$ is the first Fourier coefficient. In fact, although this estimate was made from a linear theory, a more thorough non-linear analysis shows that the formula is correct if m is replaced by $2m$ in the numerator of the logarithmic term, as shown by Wright [20] (Figure 5.2).

It is reasonably effective in summarizing the consequences of the model for a finite range of applied strain rates, $\frac{m}{1+m}\frac{k\pi^2}{as_0 h^2} \leq \dot{\gamma}_0 \leq \dot{\gamma}_{max}$, where $\dot{\gamma}_{max}$ is an upper bound for validity of Eq. (5.10) that corresponds to a strain rate where inertia begins to play a prominent role. An estimate for $\dot{\gamma}_{max}$ is given later in Eq. (5.28), and it gives an excellent result of approximately 2500 s^{-1} for the example shown in Figure 5.2.

Equation (5.10) represents analytical results extracted from a simple model. Establishing the validity of any model, and hence quantitative estimates, for a particular material is another matter entirely. Nevertheless, it is expected that the principal qualitative influences of physical and geometric properties on adiabatic shear are revealed by Eq. (5.10) and can easily be read off from it or from previous equations. For example:

- As the magnitude of the defect, $\delta\theta$, tends to zero, the critical strain increases without limit.
- As the strain-rate sensitivity tends to zero, $m \to 0$, the critical strain tends to zero, $\gamma_{cr} \to 0$.

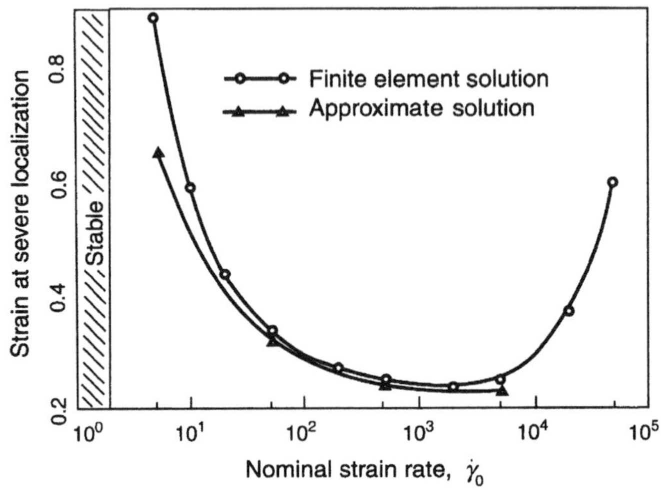

Figure 5.2 Nominal strain at localization for a given initial defect versus nominal strain rate, from Wright [4]. The upper curve was obtained by finite element calculations and shows a left branch dominated by heat conduction and a right branch by inertia. At the minimum, the two effects are in balance. The approximate solution is from Eq. (5.10) with the use of $2m$ rather than m, following Wright [20].

- In the early stages of deformation, stress follows the homogeneous path with zero perturbation (to first order) even as the perturbations in temperature and strain rate start to grow but offset each other in their effect on stress, as shown in the third line of Eq. (5.8).
- Ignoring heat conduction by setting $k = 0$ gives a lower bound estimate for the critical strain, but finite thermal conductivity increases the critical strain.
- The effect of finite thermal conductivity would be particularly strong in specimens with a short gage length, h, although thermal boundary conditions may need to be reconsidered for extremely short specimens.
- As the strain rate increases, heat conduction has a diminishing influence.
- Increased density or heat capacity increases the critical strain.
- There is a lower limit on strain rate below which it is impossible to form a shear band. This limit is estimated quantitatively by setting the denominator of the leading term to zero in the expression for critical strain.

The preceding results were developed for materials with no work hardening. When work hardening is included, the analysis is more complicated, but the end result is actually quite similar, according to Wright [19,22]. More realistic flow laws would undoubtedly change the quantitative results, but not the qualitative results.

Perhaps the most important result of all is the identification of three non-dimensional quantities whose interplay dictates the results, namely $a(\delta\vartheta)/m$, which is a measure of the size of the perturbation and should be small; $as_0/\rho cm$, which is a measure of an increment of flow stress per increment of plastic work relative to the strain-rate sensitivity; and $k\pi^2/\rho c\dot{\gamma}_0 h^2$, which compares the squares of the thermal length at a given nominal strain rate and a typical physical length in the material or a thermal rate to a strain rate.

5.4 Structure of a Shear Band and Width

In a photo-micrograph of an adiabatic shear band, taken perpendicularly to the plane of the band, the band may appear only as a narrow, almost featureless trace, but with higher magnification, a rich structure can appear. In certain steels, the original microstructure can be seen to be distorted smoothly from the exterior material right into the interior of the band. In other materials, such as the titanium alloy Ti6Al4V, there seems to be an abrupt change in microstructure between material exterior and interior to the band. In the first case, the material seems to be responding only to rapid viscoplastic deformation. In the second case, the abrupt change suggests that a second deformation mechanism may also come into play, or perhaps metallurgical changes have continued after the loading phase has ceased (see, for example, Chichili et al. [23]).

With viscoplasticity as the single deformation mechanism, and assuming that work hardening has saturated, a canonical description of thermo/mechanical response through the thickness of a shear band is possible. Begin by assuming that deformation is quasi-steady so that time dependence is neglected in the balance laws for momentum and energy:

$$s_y = 0\,, \quad (k\vartheta_y)_y + s v_y = 0 \tag{5.11}$$

These balance laws must be supplemented by a constitutive relation that connects stress, temperature and strain rate, which may be written in two ways:

$$v_y \equiv r = R(s, \vartheta), \quad s = S(r, \vartheta) \tag{5.12}$$

These in turn define two new auxiliary functions: the strain-rate sensitivity, m, and the relative softening, $\hat{a}$, which are defined as follows:

$$m \equiv \frac{r}{S}\frac{\partial S}{\partial r} = \frac{R}{s}\frac{1}{\partial R/\partial s}, \qquad \hat{a} \equiv -\frac{1}{S}\frac{\partial S}{\partial \vartheta} = \frac{1}{s}\frac{\partial R/\partial \vartheta}{\partial R/\partial s} \tag{5.13}$$

Equation (5.11) may be integrated to yield

$$\begin{aligned} &s = \text{const}, \quad k\vartheta_y + sv = \text{const} \\ &\frac{1}{2}(k\vartheta_y)^2 + s\int_{\vartheta_c}^{\vartheta} k(\vartheta)R(s,\vartheta)\mathrm{d}\vartheta = 0 \end{aligned} \tag{5.14}$$

where the subscript c denotes evaluation at the centre of the shear band. With a change of variable from ϑ to z, defined by $R(s,\vartheta) = \lambda(s,\vartheta_c)\mathrm{e}^{-z/m_c}$, the integral takes a form that can be evaluated by standard asymptotic methods for integrals with a large parameter, which in this case is $1/m_c$. From here on, asymptotic integration is straightforward, but it is necessary to keep track of the magnitude of error terms. Full details are given by Wright and Ravichandran [24] and an abbreviated account by Wright [4]. Less general analyses were given earlier by Wright and Ockendon [25], Glimm et al. [26] and also by Dinzart and Molinari [27].

A summary of the principal results follows. Provided that the strain-rate sensitivity at the centre is small, $m_c \ll 1$, and that $(k_\vartheta/k)_c \ll (\hat{a}/m)_c$ and $-[d(m/\hat{a})/d\vartheta]_c \ll 1$, then the asymptotic results for any flow law are given by

$$\begin{aligned} v &= \lambda\delta_M \tanh(y/\delta_M) \\ \dot{\gamma} &= v_y = \lambda\,\text{sech}^2(y/\delta_M) \\ \vartheta &= \vartheta_c - 2(m/\hat{a})_c \ln\cosh(y/\delta_M) \end{aligned} \tag{5.15}$$

In Eq. (5.15), each result is mathematically correct within only a small error. The characteristic mechanical length scale is

$$\delta_M = \left(\frac{2km}{s\lambda a}\right)_c^{1/2} \tag{5.16}$$

For a given model for flow stress, it is possible to calculate an explicit value for the central strain rate, λ_c, and hence for δ_M. For example, for the model given by the third line in Eq. (5.6) with $v_0 = \dot{\gamma}_0 h$, the values are

$$\lambda_c = \left(\frac{1-m}{m}\frac{as_0\dot{\gamma}_0}{k}\right)^{\frac{1}{1-m}} \dot{\gamma}_0 h^{\frac{2}{1-m}} \quad \text{and} \quad \delta_M = \left(\frac{1-m}{m}\frac{as_0\dot{\gamma}_0}{k}\right)^{-\frac{1}{1-m}} h^{-\frac{1+m}{1-m}} \tag{5.16a}$$

The third line of Eq. (5.15) suggests that there is a second length scale that refers to the extent of heat conduction from the centre of the band. At a distance not much greater than δ_M from the centre of the band, the temperature gradient rapidly tends to $\pm\frac{(2m/\hat{a})_c}{\delta_M}$ on either side of the band, so a measure of the half-distance in which the temperature in the band is elevated above ambient temperature, ϑ_0, is

$$\delta_T = (\vartheta_c - \vartheta_0)\frac{\hat{a}_c\delta_M}{2m_c} = \left(\frac{k\hat{a}}{2s\lambda m}\right)_c^{1/2}(\vartheta_c - \vartheta_0) \tag{5.17}$$

which can be substantially larger than δ_M. This might be a rough estimate of the heat-affected zone where changes in hardness are seen, but metallurgical changes are the result of a history of thermal and mechanical events and will not usually be well represented by a snapshot during deformation.

Note also that from the first line of Eq. (5.15) and the second line of Eq. (5.14), the jumps in velocity and temperature gradient across a shear band are given by

$$\begin{aligned} [v] &= 2\delta_M\lambda \\ k[\vartheta_y] + s[v] &= 0 \end{aligned} \tag{5.18}$$

A line embedded in the material and starting perpendicular to the plane of shearing will be distorted to a curve, which from approximate integration of the first line of Eq. (5.15) will be given by

$$w = (\lambda t)\delta \tanh(y/\delta) + w_0(y) \tag{5.19}$$

In a recovered specimen after deformation, the curve should look like

$$w = (\gamma_{\text{max}} - \gamma_{\text{crit}})\delta \tanh(y/\delta) + \gamma_{\text{crit}} y \tag{5.20}$$

where γ_{crit} is the asymptotic slope of the curve outside the shear band in the matrix material, and γ_{max} is the strain at the centre of the band. We could also write $w_0 = (\gamma_{\text{max}} - \gamma_{\text{crit}})\delta$ for the first coefficient in Eq. (5.20). In the form $w = \underbrace{(w_0 \tanh y/\delta)}_{\text{inner}} + \underbrace{(w_0 + \gamma_{\text{crit}} y)}_{\text{outer}} - \underbrace{(w_0)}_{\text{intermediate}}$, we recognize the inner, outer and intermediate solutions of a typical boundary layer, as given by Bender and Orszog [28]. The displacement of material on one side of the band relative to the other side after relative motion stops will be $2w_0$, and the shear strain in the material just before localization will be approximately γ_{crit}. As the stress falls below a level that can sustain further plastic deformation, the critical strain gets locked into material exterior to the band. The most commonly used material models, mentioned previously [11–16], appear to meet the criteria for validity stated just before Eq. (5.15); see Wright [4].

If the general validity of Eq. (5.20) is accepted for materials that meet the stated restrictions, then it should be possible to interpret certain observed, post-experiment microstructures. In the optical micrographs of an adiabatic shear band, formed by explosively driving a punch through a steel plate, lines of impurities, originally lying in the rolling planes of the plate, have been driven so far that they are nearly perpendicular to their original plane, yet the lines appear to be smooth and continuously curved from the interior of the plate all the way to the cut surface, according to Moss [29] (Figure 5.3).

Furthermore, the shape of the observed curve is closely matched by the theoretical curve, Eq. (5.20), and values for each of w_0, γ_{crit} and δ_{M} may be read directly from the micrograph. Twice the characteristic mechanical length, $2\delta_{\text{M}}$, is very close to the location of the maximum curvature of the trace, according to Wright and Ravichandran [24] (Figure 5.4).

Conversely, if a particular theoretical flow law is assumed, then detailed calculations can be made to predict the dependence of δ_{M} and λ on the nominal strain rate and the shear stress. As the stress falls, the strain rate in the centre of the band can be expected to increase, and if the jump in particle velocity across the band is constant, the value of δ_{M} must decrease, according to the first line of Eq. (5.18).

By now, you will have recognized that the question what is the width of an adiabatic shear band can only be answered subjectively. However defined, the 'width'

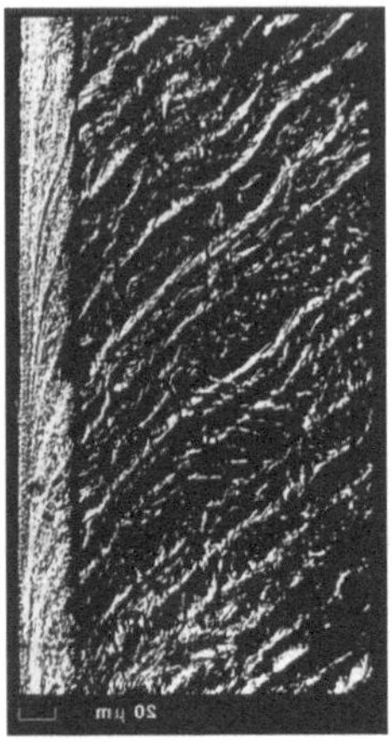

Figure 5.3 Adiabatic shear band, formed by an explosively driven punch, in a Ni–Cr steel alloy, from Moss [29]. Inhomogeneities, showing white, were horizontal before punching. (The image is flipped over for easy comparison to Figure 5.4.)

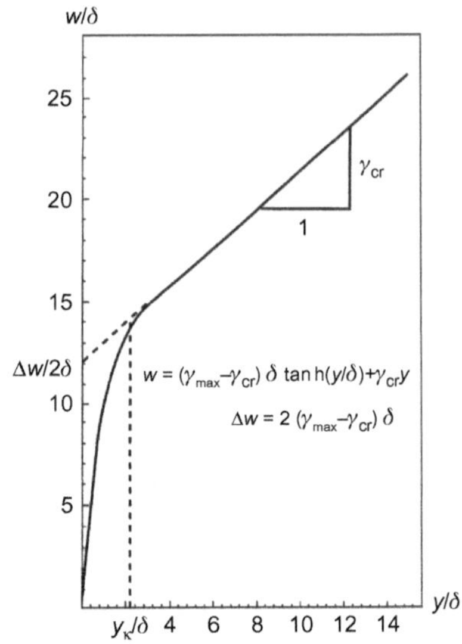

Figure 5.4 The theoretical curve, Eq. (5.20), matching Figure 5.3. The vertical dotted line to the point of maximum curvature shows that the horizontal distance is approximately $2\delta_M$. From the micrograph, therefore, $\delta_M \approx 9\ \mu m$, from [24].

of an adiabatic shear band is a matter of constitutive response, response history and judgement from various points of view, ranging from characteristic lengths found from mathematical modelling to visual inspection of experimental results. Although there can be no definitive answer, Eq. (5.16) indicates the elements that contribute most to determining the width. As a final note, however, the use of the first line of Eq. (5.18) and the second equation of Eq. (5.13) shows that alternate representations for the mechanical scale are

$$\delta_M = \frac{4}{[v]}\left(\frac{km}{s\hat{a}}\right)_c = \frac{4}{[v]}\left(\frac{km}{\partial S/\partial \vartheta}\right)_c \tag{5.21}$$

5.5 Spacing of Shear Bands

Multiple adiabatic shear bands can form in a small volume of material, so it is natural to ask what spacing between bands can be expected. This has been approached from three different perspectives: analysis, physical reasoning and specialized computations. As before, specific results must depend ultimately on the content of constitutive equations, but much insight can be gained from greatly simplified versions.

5.5.1 Analysis

Consider again the problem posed by Eq. (5.6) but this time with inertia retained. The homogeneous solution is the same as Eq. (5.7), but now, the perturbation equations become

$$\begin{aligned}&\tilde{s}_{\mathrm{y}} = \rho\tilde{v}_{\mathrm{t}}\\&\rho c\tilde{\vartheta}_{\mathrm{t}} = k\tilde{\vartheta}_{\mathrm{yy}} + s_0\,\mathrm{e}^{-\alpha t}\tilde{v}_{\mathrm{y}} + \tilde{s}\dot{\gamma}_0\\&\tilde{s} = -s_0 a\tilde{\vartheta} + m s_0\,\mathrm{e}^{-\alpha t}\tilde{v}_{\mathrm{y}}/\dot{\gamma}_0\end{aligned} \tag{5.22}$$

The last two lines of Eq. (5.22) may be solved for $\tilde{v}_{\mathrm{y}}$ and $\tilde{s}$ in terms of $\tilde{\vartheta}$ and then substituted into the first line of Eq. (5.22) after differentiating once with respect to y to give a single differential equation for $\tilde{\vartheta}$, which may now be expressed as a Fourier series with time-dependent coefficients. The resulting differential equations for the Fourier coefficients are linear but still contain the time-dependent term $\mathrm{e}^{\alpha t}$ so are not easily interpreted. (The coefficient α is defined following Eq. (5.7).) This problem was discussed by Wright and Ockendon [30] and (in less detail) by Wright [4]. Again the time-dependent coefficients must be treated carefully, but when standard asymptotic methods are used, an apparently simple result is found for the Fourier coefficients of strain rate, $\dot{\gamma}_n = (a\dot{\gamma}_0/m)\vartheta_n(0)\mathrm{e}^{\alpha_n t}$, where

$$\alpha_n = n^2\pi^2\dot{\gamma}_0\frac{m}{2\hat{\rho}}\left\{\left[\left(1-\frac{\hat{k}\hat{\rho}}{m}\right)^2+\frac{4+2m}{n^2\pi^2}\frac{\hat{a}\hat{\rho}}{m^2}\right]^{1/2}-1-\frac{\hat{k}\hat{\rho}}{m}+\frac{1}{n^2\pi^2}\frac{\hat{a}\hat{\rho}}{m^2}\right\} \tag{5.23}$$

Equation (5.23) contains the three non-dimensional coefficients:

$$\hat{k} = \frac{k}{\rho c\dot{\gamma}_0 h^2},\quad \hat{\rho} = \frac{\rho\dot{\gamma}_0^2 h^2}{s_0},\quad \hat{a} = \frac{a s_0}{\rho c} \tag{5.24}$$

where h is a characteristic length. Although the appearance is much the same, Eq. (5.23) does not correspond to an exponent with the assumption of frozen coefficients.

Equation (5.23) can be useful for estimating scaling laws for a variety of situations. By optimizing on various parameters in the equation, we find the following five results:

1. Reasoning that the largest number of shear bands that can grow in a given gauge length can be no greater than the number of Fourier components that can grow, and then

maximizing Eq. (5.23) on n, we find that the largest number of shear bands likely to nucleate in a gauge length $L = 2h$ is

$$n = \frac{L}{2\pi}\left(\frac{\dot{\gamma}_0^3 a^2 s_0}{m^3 kc}\right)^{1/4} \tag{5.25}$$

2. The cut-off strain rate below which a shear band cannot be formed is found from setting $\alpha_n = 0$ for $n = 1$ with the approximate result

$$\dot{\gamma}_0 = \frac{4\pi^2 mk}{aL^2 s_0} \tag{5.26}$$

3. If many bands can form, the least spacing to be expected between nucleating bands is found by minimizing Eq. (5.23) on L with $n = 1$, with the result

$$L = 2\pi\left(\frac{m^3 kc}{\dot{\gamma}_0^3 a^2 s_0}\right)^{1/4} \tag{5.27}$$

4. The strain rate at which heat conduction and inertia are balanced so that localization occurs at the least-nominal strain for a fixed gauge length is

$$\dot{\gamma}_0 = \frac{m}{2^{1/3}}\left[\left(\frac{2\pi}{L}\right)^4 \frac{kc}{a^2 s_0}\right]^{1/3} + \frac{m^2}{3\cdot 2^{2/3}}\left[\left(\frac{2\pi}{L}\right)^2 \frac{c^2 s_0}{ka}\right]^{1/3} \tag{5.28}$$

   In this result, the strain-rate dependence of s_0 has been accounted for by noting that $\frac{d}{d\dot{\gamma}_0}\left[\kappa_0(b\dot{\gamma}_0)^m\right] = ms_0\dot{\gamma}_0^{-1}$, and an asymptotic argument has been used to estimate a root of a complicated algebraic equation.

5. The inverse of the previous formula, interpreted as the band spacing corresponding to a given strain rate that minimizes the nucleating strain for localization, is

$$L = \sqrt{\frac{2}{2+m}}\pi m^{3/2}\left(\frac{c}{a\dot{\gamma}_0^2}\right)^{1/2}\left[1 + \left(1 + 8\frac{k\dot{\gamma}_0}{m^3 c s_0}\right)^{1/2}\right]^{1/2} \tag{5.29}$$

5.5.2 Physical Reasoning

In a classic paper, Grady and Kipp [31] argued that the rapid decrease of flow stress as a shear band forms will cause unloading in adjacent material. As the unloading propagates away from the forming shear band, it will quench other potential bands as the available driving stress drops below current flow stress at those points. Then, using a two-temperature model for a shear band and its ambient surroundings and considering the energy being dissipated as the band forms for a given width of shear band, an estimate is made of the amount of material quenched and the time of quenching. The optimum width of band for maximum rate of dissipation of its energy is then calculated where dissipation mechanisms include both momentum diffusion and thermal diffusion. Then it is argued that to reach full growth, two neighbouring shear bands would have to be separated by twice the distance calculated as being unloaded during the time of unloading. The result is

$$L = 2\left[\frac{9kc}{a^2\kappa_0\dot{\gamma}_0^3}\right]^{1/4} \tag{5.30}$$

Strain-rate sensitivity is not considered, so κ_0 is the initial yield stress. In the linear thermal-softening model used, the thermal rate of softening would be $a = (T_m - T_0)^{-1}$, where T_m is the melting temperature of the metal and T_0 is the initial temperature. Furthermore, the width of the band is predicted to be

$$2\delta_M = 2^{3/4}3^{1/2}\left[\frac{k^3}{ca^2\kappa_0^3\dot{\gamma}_0}\right]^{1/4} \tag{5.31}$$

5.5.3 Specialized Computations

To check the analytical results for accuracy and to correlate some of the results with the Grady–Kipp model, a specialized 1D code, based on the method of characteristics, was constructed by Zhou et al. [32]. The system of equations used includes linear elasticity and viscoplasticity, but assumes shearing motions only, and as a result is only semi-linear for the mechanical part of the calculation. As a consequence, no adjustment of the spatial discretization is required during a calculation run. The code has second-order accuracy and is very fast, thus permitting many high-quality parametric calculations to be made even with fine gridding. Thermal diffusion is also included in the calculation. A symmetric temperature variation in the middle of an interval was assumed for the initial perturbing condition in a benchmark calculation (Figure 5.5).

The code was judged to be reliable after comparison with previous work that used the completely different computational method of lines in Walter [33]. Results from the two approaches were essentially identical in every particular. These calculations on a finite domain also showed that elastic waves, reflecting from boundaries in this case, can interrupt growth, which resumes when favourable conditions return.

In a second paper of Zhou et al. [34], the code was used to calculate the response of a 50 mm segment when the initial temperature contained many small perturbations with random amplitudes around the initial average temperature. The calculations showed that at first there was no communication between perturbation sites where increases in strain rate and temperature offset each other so that there was almost no variation in stress, as predicted by the linear perturbation analysis. Furthermore, the average growth rates of the perturbation spectrum also agreed well with linear analysis. Later, as localization occurred with an accompanying loss

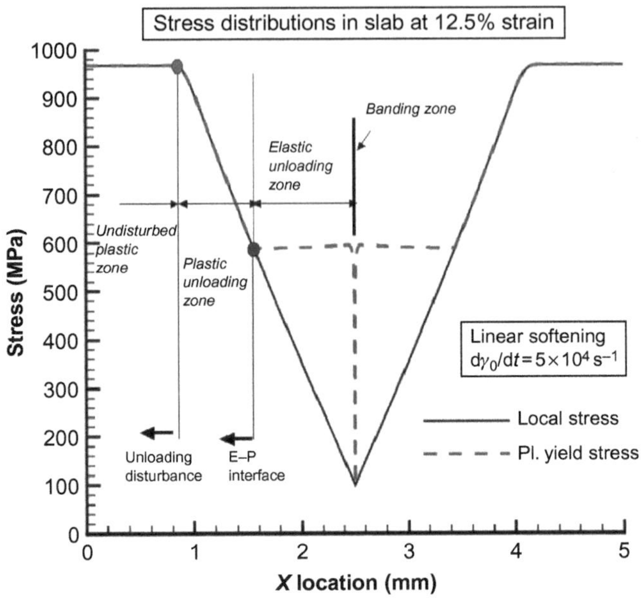

Figure 5.5 Unloading from the shear band as material loses shear strength. Sloping lines are unloading waves propagating away from the shear band in the centre, from Zhou et al. [32].

of flow stress, the higher spatial harmonics grew rapidly so that the spectrum became essentially flat above the initially fastest-growing harmonic, indicating a tendency towards a spiked spatial response in temperature and strain rate but with some structure still indicated by the lower harmonics.

As fully formed bands appeared, the spacing tended to be much larger than that previously found for initial growth of shear bands. For fully formed shear bands, the Grady–Kipp result for band spacing, Eq. (5.30), was found to be quite accurate for the dependence on strain rate, but the prediction of the dependence on k, c, a, and s_0 was sometimes weaker and sometimes stronger than indicated by the exponential powers in Eq. (5.30). In Zhou et al. [34], Eq. (5.29) was also noted but not fully exploited. In Eq. (5.16a), explicit formulas are given for the peak strain rate in the centre of a shear band, $\lambda_c = \dot{\gamma}_c$, and for the characteristic length scale (half-width) of a fully formed shear band, δ_M. Each is given in terms of a half-width, h, of a slab within which the shear band is embedded. Combining Eq. (5.16a) for δ_M and maximum strain rate with Eq. (5.29) for the band spacing, even though the latter only applies strictly to the spacing at nucleation, we obtain intriguing results when compared first with Eqs (5.30) and (5.31) and then with the computational results of Zhou et al. [34]. First note that Eq. (5.29) gives different results for various values of the non-dimensional parameter, $8k\dot{\gamma}_0/m^3cs_0 = 8k\hat{p}/m^3$ (see Eq. (5.24)), so let us consider only the limiting values when it is much larger or much smaller than 1. In addition, for simplicity, we ignore the strain-rate sensitivity when it is added or subtracted from 1. Then the half-spacing between shear bands is predicted to have the following behaviour:

$$\frac{L}{2} = h \to \begin{cases} 2^{3/4}\pi m^{3/4}\left(\dfrac{kc}{s_0 a^2 \dot{\gamma}_0^3}\right)^{1/4} & : \quad 8k\dot{\gamma}_0/m^3 c s_0 \gg 1 \\ 2^{1/2}\pi m^{3/2}\left(\dfrac{c}{a\dot{\gamma}_0^2}\right)^{1/2} & : \quad 8k\dot{\gamma}_0/m^3 c s_0 \ll 1 \end{cases} \tag{5.32}$$

The peak strain rate in the centre of the band is predicted to have the following behaviour:

$$\dot{\gamma}_c \to \begin{cases} 2^{3/2}\pi^2 m^{3/2}\left(\dfrac{s_0 c \dot{\gamma}_0}{k}\right)^{1/2} & : \quad 8k\dot{\gamma}_0/m^3 c s_0 \gg 1 \\ 2\pi^2 m^3 \dfrac{s_0 c}{k} & : \quad 8k\dot{\gamma}_0/m^3 c s_0 \ll 1 \end{cases} \tag{5.33}$$

The characteristic mechanical half-width of the band is predicted to have the following behaviour:

$$\delta_{\mathrm{M}} \to \begin{cases} \dfrac{m}{2^{3/4}\pi}\left(\dfrac{k}{a s_0 \dot{\gamma}_0}\right)^{3/4}\left(\dfrac{a\dot{\gamma}_0^2}{c}\right)^{1/4} & : \quad 8k\dot{\gamma}_0/m^3 c s_0 \gg 1 \\ \dfrac{1}{2^{1/2}\pi m^{1/2}}\dfrac{k}{(ac)^{1/2} s_0} & : \quad 8k\dot{\gamma}_0/m^3 c s_0 \ll 1 \end{cases} \tag{5.34}$$

The leading numerical coefficients in each formula are probably not correct, but the dependences on the physical parameters and strain rate appear to be reliable. Equations (5.32) and (5.34) give exactly the same power law scaling for the effect of physical parameters and strain rate on band spacing and width as do the formulas of Grady and Kipp, shown in Eqs (5.29) and (5.31). Furthermore, they suggest that the formula $8k\dot{\gamma}_0/m^3 c s_0 \gg 1$ is the required limiting condition for Eqs (5.30) and (5.31) to be valid. Clearly, the smallness of m has a large influence on the size of this parameter.

The actual effects found by Zhou et al. [34] are all in the direction of Eqs (5.32) and (5.34). For example, in the spacing formula, they found a $(k/s_0)^{1/8}$ dependence rather than $(k/s_0)^{1/4}$, and the limiting dependence is predicted to be $(k/s_0)^0$, i.e. no dependence at all. All other scaling exponents similarly seem to move approximately halfway from the rate-independent Grady–Kipp limit to the other extreme case. Numerical values used by Zhou et al. [34] were centred on the values $k = 50$ W/m s K, $c = 500$ J/kg K, $m = 0.025$, $s_0 = 0.6$ GPa and $\dot{\gamma}_0 = 5 \times 10^4 \text{s}^{-1}$. For these values, the parameter that controls the scaling has the value $8k\dot{\gamma}_0/m^3 c s_0 = 4.3$. This is an intermediate value, neither large nor small compared to 1, so it is not too surprising that the scalings found empirically are more or less midway between the possible extremes.

5.6 Structure and Speed of a Propagating Shear Band

In the material covered so far, shear bands have for the most part been considered to originate spontaneously in bulk material due to the growth of small perturbations that have their origin in the fabric of the material itself. At another extreme, shear bands can originate from a surface that has been loaded very rapidly and has experienced an intense stress concentration due to the loading. A sharp stress concentration may be thought of as an extremely large perturbation, as opposed to the small perturbations considered so far. Experiments performed by Kalthoff [35] are examples of this kind of loading. These experiments consist of impact by a flat-ended projectile upon the edge of a plate with a thin slit precut into it. The projectile centreline and the midplane of the plate are aligned, and the flank of the projectile strikes immediately to one side of the slit. The result is to force the material on one side of the slit to move with respect to the other side in rapid mode II motion. If struck with enough velocity, a shear band rather than a crack propagates away from the end of the slit. The situation described raises theoretical questions concerning the propagation of the shear band, such as the nature of deformation and stresses around the tip of the band where there must be a transition from relatively undisturbed material to material in a rapidly deforming state, and the speed at which the tip of the band must move.

In an attempt to gain some insight into these questions, several simplified approaches have been used, perhaps the simplest one being by Wright and Walter [36]. They chose to examine the advance of a shear band in deformation similar to mode III crack motion. They assumed steady motion with the band moving in the x-direction, material velocity only in the z-direction and no variation at all in the z-direction. Thus, the velocity field is given by $\mathbf{v} = (0,0,w)$. The deviatoric traction is written as $\boldsymbol{S}\boldsymbol{e}_3 = s\boldsymbol{e}_w$, where $\boldsymbol{S}$ is the deviatoric stress tensor, s is the magnitude of the shear traction and $\mathbf{e}_w = \nabla w/|\nabla w|$ is a unit vector in the direction of the velocity gradient, ∇w. The balance equations and an assumed flow law may now be written as

$$\begin{aligned} \rho w_t &= \nabla \cdot (s\mathbf{e}_w) \\ \rho c \vartheta_t &= k\nabla^2 \vartheta + s|\nabla w| \\ s &= \kappa_0 g(\vartheta)(b|\nabla w|)^m \end{aligned} \tag{5.35}$$

In Eq. (5.35), the thermal factor is assumed to be $g(\vartheta) = 1 - a\vartheta$ as previously. If the band propagates edgewise with constant speed, U, and if the resulting fields, as seen from the tip of the band, are steady, then in the coordinate system $x' = x - Ut$, $y' = y$, $t' = t$, Eq. (5.35) transforms into

$$\begin{aligned} -\rho U w_x &= \nabla\cdot(s\mathbf{e}_w) \\ \rho c U g_x &= -k\nabla^2 g + as|\nabla w| \\ s &= \kappa_0 g(b|\nabla w|)^m \end{aligned} \tag{5.36}$$

In Eq. (5.36), the prime on x' has been dropped for convenience.

If it is assumed that there is a similarity solution in the form $w = Cr^{\alpha}W(\varphi)$, etc., where $r = \sqrt{x^2 + y^2}$ is the radial distance from the tip, and $\varphi = \tan^{-1} x/y$ is the polar angle measured from straight ahead, and where $r_1 < r < r_2$ and $-\pi < \varphi < +\pi$ limits the range of the variables to a partial ring ahead of the tip, then after finding the necessarily consistent exponents, the similarity solution takes the form

$$\begin{aligned} w &= U\left\{\frac{2m}{1+m}\frac{\rho c}{as_0}\right\}^{1/(1+m)}\left(\frac{r\dot{\gamma}_0}{U}\right)^{m/(1+m)} W(\varphi) \\ g &= \frac{\rho U^2}{s_0}\left\{\frac{2m}{1+m}\frac{\rho c}{as_0}\right\}^{(1-m)/(1+m)}\left(\frac{r\dot{\gamma}_0}{U}\right)^{2m/(1+m)} G(\varphi) \\ \dot{\gamma} &= \dot{\gamma}_0\left\{\frac{2m}{1+m}\frac{\rho c}{as_0}\right\}^{1/(1+m)}\left(\frac{r\dot{\gamma}_0}{U}\right)^{-1/(1+m)} \Gamma(\varphi) \\ s &= \rho U^2\left\{\frac{2m}{1+m}\frac{\rho c}{as_0}\right\}^{1/(1+m)}\left(\frac{r\dot{\gamma}_0}{U}\right)^{m/(1+m)} S(\varphi) \end{aligned} \tag{5.37}$$

Equation (5.37) appears to be different from the equations given in Wright and Walter [36], but they are only rearranged so that the parameter b does not appear explicitly anywhere. The strain rate has been found from $\dot{\gamma} = |\nabla w| = (w_r^2 + w_\varphi^2/r^2)^{1/2}$. The functions W, G, Γ, S are all non-dimensional, and the leading scale factors have been chosen so that $S = G\Gamma^m$ and $\Gamma^2 = W_\varphi^2 + \{[m/(1+m)]W\}^2$. One further change of variables defines Ψ to be the angle between $\boldsymbol{e}_w$ and $\boldsymbol{e}_\varphi$ so that $(1+m)\Gamma \sin\Psi = mW$ and $\Gamma\cos\Psi = W'$, where now the prime signifies the derivative with respect to φ. Finally, with W and W' eliminated in favour of Γ and Ψ, Eq. (5.36) reduces to a system of ordinary differential equations with the angle φ as independent variable because r^{α} factors out of every equation. The equations become

$$
\begin{aligned}
&(S\cos\Psi)' + \frac{1+2m}{1+m} S\sin\Psi + \Gamma\sin(\Psi-\varphi) = 0 \\
&\sin\varphi G' + \frac{2m}{1+m}(\Gamma^{1+m} - \cos\varphi) G = 0 \\
&(\Gamma\sin\Psi)' - \frac{m}{1+m}\Gamma\cos\Psi = 0 \\
&S' - (G\Gamma^m)' = 0
\end{aligned}
\tag{5.38}
$$

Initial conditions at $\varphi = 0$ are chosen as follows:

- $\Psi(0) = 0$ from its definition because W must be odd in φ for anti-plane motion, and therefore $W(0) = 0$.
- The second line of Eq. (5.38) requires that $(\Gamma^{1+m} - 1)G(0) = 0$ because G must be even in φ and $G(0) \neq 0$. There will be no solution unless $\Gamma(0) = 1$.
- $S(0) = G(0) = S_0$, where S_0 is arbitrary, but $S_0 = O(1)$ is consistent with Eq. (5.37) if initial evaluations are made where $r\dot{\gamma}_0/U = O(1)$.

It is now straightforward to integrate the system of equations numerically. The results are strongly dependent on the strain-rate sensitivity, m. There is a discontinuity in the solution at $\varphi = \pm\pi$, which must be bridged by a fully formed shear band. An approximate matching procedure verified that the thickness of the band is small compared to its length and that the matching depends only on a characteristic length, $l = k/\rho c U$, but not at all on the parameters s_0, a or b. Figures 5.6 and 5.7 illustrate some of the features of the calculation.

Chen and Batra [37] applied the same technique to mode II band motion. The details are quite a bit more complex, but there are many strong similarities between the mode I and II similarity solutions. The system also reduces to a set of ordinary differential equations in φ, and boundary conditions at $\varphi = 0$ are easy to choose. Again there is one equation that has no solution unless a specific value for the rate function is chosen at $\varphi = 0$. One marked difference between mode I and II is that the similarity solution for mode II predicts that there will be a small transverse material velocity at the tip of the shear band. This seems to be indicating that the band will have a tendency to deviate slightly towards the compressive side of the propagating band, which would be in accord with the experimental evidence [35]. There is also a discussion of this problem in Wright [4].

A third treatment of a propagating shear band was given by Gioia and Ortiz [38]. This paper is quite complex and develops the complete structure of a 2D boundary layer emanating from the tip of a notched plate that has been struck by a projectile in the manner described previously for the Kalthoff experiment. All constitutive quantities are described in terms of power laws, which enable a similarity solution to be constructed and matched to the plastic flow on each side of the slit. As in the previous two examples of steady flows, there is evidence of a strong forward-facing plume of higher temperature and plastic work extending deep into the plate ahead of the notch.

The edgewise speed of propagation of an adiabatic shear band has been considered computationally by Bonnet–LeBouvier et al. [39] and theoretically by

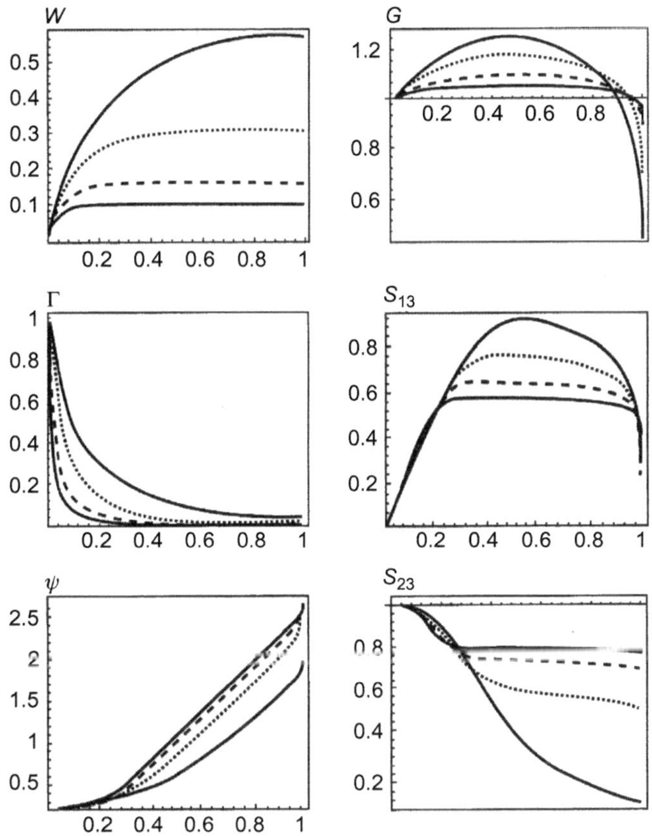

Figure 5.6 Non-dimensional functions $W(\varphi)$, $\Gamma(\varphi)$, $G(\varphi)$, and shear components $S_{13}(\varphi)$, $S_{23}(\varphi)$ plotted versus φ/π for strain-rate sensitivities $m = 0.01$, 0.02, 0.05 and 0.1, from Wright and Walter [36]. The driving traction at the shear band is found from the scale factor $S(\pi - \varepsilon) \approx S_{23}(\pi - \varepsilon)$.

Gioia and Ortiz [38] and Wright [40]. The first estimate of speed, made by Gioia and Ortiz [38], tracked the speed of a particular level of plastic work in the leading part of the plume propagating away from the notch at the impacted edge of the plate. They found that the ratio of this speed to the speed of the plate on the impact edge could be expressed as:

$$\frac{U}{V} = f\left(\frac{\rho V^2}{\sigma_0}, \frac{w_c}{\sigma_0 \gamma_0}\right) \tag{5.39}$$

where U is the speed of the shear band, V is the driving speed at the impacted edge, w_c is a critical level of plastic work, σ_0 is a characteristic value of flow stress

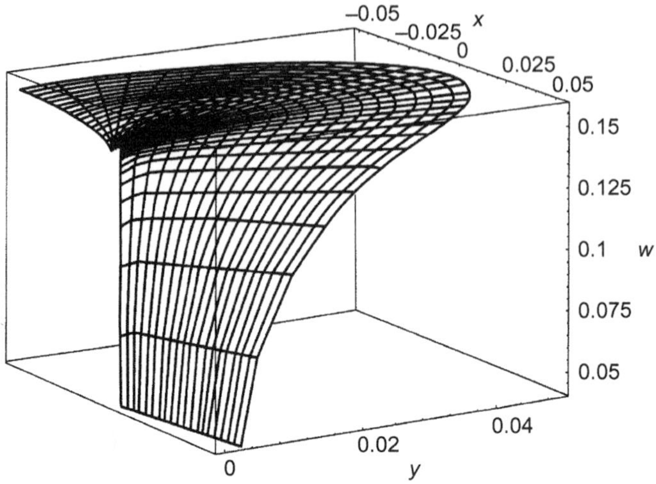

Figure 5.7 Surface of velocity, w, in arbitrary scales. Shear band extends to the left along the x-axis from $x = 0$, from Wright and Walter [36].

and $\sigma_0\gamma_0$ is a characteristic level of plastic work. For values of the Reynolds number, $\rho V^2/\sigma_0$, greater than 10 or 12, the ratio U/V becomes nearly constant.

Bonnet-LeBouvier et al. [39] considered a strip of material 2.5 mm by 200 mm with a notch at one end and driven by constant velocities, $\pm V$, on the long upper and lower surfaces. With the constitutive law $\sigma_e = K(\varepsilon + \varepsilon_0)^n T^{-v}(D + D_0)^m$, where σ_e and ε are effective stress and strain, T is temperature, D is plastic strain rate, and all other parameters are constants; they made careful parametric calculations, including heat conduction, and tracked the point where the fully developed shear band transitioned into a forward-facing plume of heated material. As the driving velocities increased above a small initial critical velocity, they found that the ratio U/V was nearly constant:

$$\frac{U}{V} = \alpha \frac{K\beta}{\rho c T_0 m}(-An + B) \tag{5.40}$$

where T_0 is the initial temperature, β is the Taylor–Quinney factor and α, A and B are empirical constants. As the driving velocity increased further, it passed through a set of values that could not be characterized, and finally there was a region where U became constant:

$$U = \eta \frac{K}{\rho}\sqrt{\frac{\beta}{cT_0 m}}(-A'n + B') \tag{5.41}$$

where η depends on v, and A' and B' are empirical constants.

Wright [40] re-examined the results from Wright and Walter [36] and Chen and Batra [37] and was able to show that for a perfectly plastic material (work-hardening exponent $n = 0$), the speed of a shear band in either mode II or III motion could be theoretically expressed as:

$$U = \sqrt{\frac{s^+}{\rho}}\sqrt{\frac{-s_\vartheta^+}{\rho c}}\sqrt{\frac{1+m}{2m}}\frac{1}{\sqrt{S_0}} \tag{5.42}$$

where s^+ and s_ϑ^+ mean that the stress and its thermal derivative are to be evaluated immediately ahead of the fully formed band, and $S_0 = S(0) = O(1)$ is the initial value of the function introduced in the fourth line of Eq. (5.37). Setting $S_0 = 1$ is equivalent to calibrating the flow law, third line of Eq. (5.35), at a point in the forward flow where $r = \frac{2m}{1+m}\frac{\rho c}{as^+}\frac{U}{\dot{\gamma}_0}$, and then setting $\dot{\gamma} = \dot{\gamma}_0$, $\vartheta = 0$ and $s^+ = \kappa_0(b\dot{\gamma}_0)^m$. In essence, the necessity of setting $\Gamma(0) = 1$ requires this particular calibration to be consistent throughout. For strain rates of 10^4, the calibration distance is typically only a few millimetres.

Furthermore, the jump in the tangential velocity (either mode II or III) across the fully developed shear band can now be worked out to be

$$\frac{1}{2}[w^-] = W^- \sqrt{\frac{2m}{1+m}}\sqrt{\frac{s^+ c}{-s_\vartheta^+}} \tag{5.43}$$

and therefore,

$$\frac{U}{[w^-]/2} = \frac{1}{W^-}\frac{1+m}{m}\frac{-s_\vartheta^+}{\rho c} \tag{5.44}$$

where $[w^-]/2$ may be regarded as the driving velocity. These results are unconstrained by nearby boundaries.

It is difficult to make a direct comparison between these formulas with those of Bonnet-LeBouvier et al. [39] because of the difference in assumptions, one being the result of similarity solutions and the other arising from a specific boundary value problem. Nevertheless, after noting that the effective stress and the shear flow stress are related by $\sigma_e = \sqrt{3}s$ and that their flow rule preceding Eq. (5.40) gives $-s_\vartheta = vs/T$, it is striking to see that Eqs (5.42) and (5.44) now become

$$U = \frac{\sigma_e^+}{\sqrt{3}\rho}\sqrt{\frac{v}{cT^+}}\sqrt{\frac{1+m}{2m}} \quad \text{and} \quad \frac{U}{[w^-]/2} = \frac{1}{W^-}\frac{1+m}{2m}\frac{v\sigma_e^+}{\sqrt{3}\rho c T^+} \tag{5.45}$$

For the case of no work hardening, both parts of Eq. (5.45) bear a close resemblance to Eqs (5.41) and (5.40), respectively. However, in the unconstrained case, both equations in Eq. (5.45) hold simultaneously, but Eqs (5.41) and (5.40) represent different ranges of the driving velocities on the upper and lower boundaries of a strip. You might speculate that the distance to the boundary in the calculations of Bonnet-LeBouvier et al. [39] interferes with development of the full width of the leading thermal plume as the driving velocity increases and so loses some aspects of the similarity solution.

Equation (5.42) suggests that the steady velocity of a shear band, if it can be reached, is proportional to the product of a plastic wave speed, $\sqrt{s/\rho}$; the square root of a thermal sensitivity, $-s_\vartheta/\rho c$; and the square root of the reciprocal of strain-rate sensitivity, m. The non-dimensional thermal sensitivity is a measure of the decrease in flow stress for an adiabatic increment of plastic work because $\rho c d\vartheta = s d\gamma_p$. The ratio of thermal sensitivity to strain-rate sensitivity may be called shear band susceptibility. Estimates of characteristic steady speeds of adiabatic shear bands, given in Wright [40], range roughly from 200 to 1200 m/s. The ratio of largest to smallest plastic wave speeds is only about a factor of 2, but the ratio of the square root of susceptibility is in the range of 4–5 and has a greater influence on possible shear band speed, with strain-rate sensitivity having the greatest individual influence.

There is only a little direct evidence concerning shear band speed. Guduru et al. [41] were able to measure both the speed and the temperature field attached to a propagating adiabatic shear band in a Maraging C300 steel plate. The plate was struck on its edge adjacent to a crack, which had been introduced ahead of time, as in the Kalthoff experiment. The authors point out that the motion is mixed, containing both mode II and III elements. Although the tip of the shear band was always rather diffuse, it was identified and tracked approximately in five experiments. Its motion in each case was somewhat intermittent, sometimes accelerating, sometimes nearly stopping, but never reaching a steady speed, most likely because of various elastic waves that moved rapidly and reflected from a variety of boundaries. Such intermittent behaviour would be in rough accord with the calculations in an elastic viscoplastic material, as reported by Walter [33], DeLellio and Olmstead [42], and Zhou et al. [32], where reflections of elastic waves from boundaries cause unloading at the shear band.

The temperature measurements of Guduru et al. [41] captured the tip and a fully developed segment of the shear band following the tip. These measurements not only supported the idea of a diffuse tip but also revealed hot spots distributed along the fully developed shear band. Highly resolved, full-field calculations by Li et al. [43] and Teng et al. [44] also found hot spots along fully developed shear bands. Both experimental and computational results seem to be indicating that a secondary bifurcation from laminar flow may occur after the initial bifurcation from spatially homogeneous flow.

5.7 Energy Rates and Dissipation

The results reported in Eqs (5.32)–(5.34) may also be used to estimate the rate of energy dissipation per unit area of the shear band, the average rate of dissipation per unit volume and the energy dissipated per unit area of shear band per unit of advance of the tip. First, consider the areal rate of dissipation. A fully developed shear band has a half-width, δ_M, within which plastic work is converted to heat, and a half-spacing, h, to an adjacent fully developed shear band. Outside the shear band and between neighbouring bands, heat conduction redistributes the thermal energy. The rate of plastic work in half a shear band just behind the tip is $s_-\lambda_c\delta_M = s_-\dot{\gamma}_0 h$, which can be seen from Eq. (5.16a). The heat flux away from one side of the shear band is given by the expression $k\vartheta_y = \pm 2(km/\hat{a}\delta_M) = \pm 2^{1/2}(s_-\lambda_c m/k\hat{a})$, as is easily obtained from Eqs (5.15) and (5.16). The driving shear stress on the shear band behind the leading tip is s_-. This driving stress may also be expressed in terms of the stress immediately ahead of the tip, termed s^+, by the relationship, $s_- = s_+ S_-/S_0$. $S(\varphi)$ is the angular function for shear stress developed in Eqs (5.37) and (5.38), and $S_- = S(\pi - \varepsilon)$, $S_0 = S(0)$ for some small ε. With these preliminaries established, the rate of dissipation within a shear band and the adjacent material halfway to the next band may be worked out to be

$$\frac{1}{2}\dot{E} = s_-\lambda_c\delta_M = s_-\dot{\gamma}_0 h$$
$$\rightarrow \begin{cases} 2^{3/4}\pi m^{3/4}(kcs_+^3\dot{\gamma}_0/\hat{a}^2)^{1/4}(S_-/S_0)^{3/4} \\ 2^{1/2}\pi m^{3/2}(cs_+/\hat{a})^{1/2}(S_-/S_0)^{1/2} \end{cases} \tag{5.46}$$

where the limits of Eqs (5.32) and (5.33) have been used for the second and third line of Eq. (5.46). The ratio S_-/S_0 depends only on the strain-rate sensitivity, decreasing monotonically from 1.0 to approximately 0.2 as m increases from 0.0 to 0.1 (Figure 5.8).

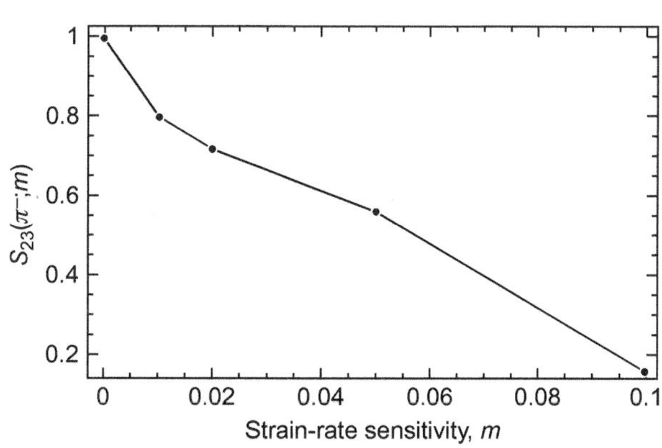

Figure 5.8 Non-dimensional shear traction S^-/S_0 as a function of strain-rate sensitivity, m. Recall that $S^- = S(\pi - \varepsilon) \approx S_{23}(\pi - \varepsilon)$ and that $S_0 = 1$ for calibration a short distance ahead of the band tip. From Wright [40].

The units of $\dot{E}$ are watts per square metre in the MKS system, and it accounts for the dissipation in the narrow layer of the shear band. But because there is essentially no dissipation between fully developed bands (little plastic work), the average rate of dissipation in the volume containing one band is $\dot{E}/2h$, which turns out to be approximately

$$\dot{E}/2h \approx s_{-}\dot{\gamma}_0 = s^{+}\dot{\gamma}_0 S^{-}/S_0 \tag{5.47}$$

With the average stress having fallen to s_{-} throughout the local region, Eq. (5.47) states that multiple shear bands dissipate the same amount of energy as would be dissipated by a uniform strain rate of $\dot{\gamma}_0$ at the reduced stress s_{-}. All of these estimates are only approximate, of course, and furthermore, the stress will continue to fall and the temperature to rise behind the tip, continually changing the balance between strain-rate hardening and thermal softening.

Finally, consider the extra dissipation per unit advance of the shear band tip. Because the radial variation of no field is as sharp as r^{-1} – see Eq. (5.37) where the sharpest variation is $\dot{\gamma} \sim r^{-1/(1+m)}$ – all contour integrals around the tip must reduce to zero with vanishing radius. Nevertheless, the extra dissipation may be estimated by computing $\dot{E}/U$, which in the two limits works out to be

$$\frac{\dot{E}}{2U} = \begin{cases} 2\cdot 2^{1/4}\pi m^{5/4}\dfrac{\rho c^{3/4}}{a}\left(\dfrac{k\dot{\gamma}_0}{s_{+}}\right)^{1/4}\left(\dfrac{S_{-}}{S_0}\right)^{3/4} \\ 2\pi m^2\dfrac{\rho c}{a}\dfrac{S_{-}}{S_0} \end{cases} \tag{5.48}$$

The dimensions are joules per square metre of shear band per metre of advance in the MKS system.

5.8 Discussion and Conclusions

All of the calculations in this chapter have been made with a simplified constitutive model for a rigid, perfectly plastic material with linear thermal softening and power law rate hardening. The purpose is not so much to make predictions for real materials, which invariably have a more complicated constitutive response, but rather to get a sense of the interplay of various physical properties. In each problem investigated, this interplay or competition between various physical processes shows up as non-dimensional parameters, which always display the relative intensity of those processes. The one that shows up in virtually every problem is $s_0 a/\rho c m$, or more generally, $m^{-1}\delta s/\delta W_{\mathrm{p}}$, which represents the ratio of thermal sensitivity (decrease of stress per unit of plastic work, $\delta s/\delta W_p = \delta s/s\ \delta\gamma = \delta s/\rho c\ \delta\vartheta$) to strain-rate sensitivity, $m = \delta \ln s/\delta \ln \dot{\gamma}$. This represents the fundamental competition between thermal softening of flow stress and strain-rate increase of flow stress. With any kind of small background perturbation (strength, structure, temperature, dimensions), these two processes will tend to amplify the initial perturbation until either the driving forces are released or the material is forced into a rate-dependent bifurcation and drops to a substantially lower flow stress. As the material loses strength at a shear band, the adjacent material is forced to unload.

The ratio $k/\rho c\dot{\gamma}_0 h^2$ compares the square of a local thermal length to the square of a characteristic length in the material. Alternatively, it compares the rate at which thermal conductivity can carry enough heat away to reduce the temperature one degree to the rate at which enough heat is supplied to raise the temperature one degree. However phrased, the ratio expresses the thermal competition between heat removal and supply.

Many other ratios involving physical properties as well as characteristic lengths, rates, stresses and so on occur naturally as approximate solutions are constructed, and often it seems possible to find the important non-dimensional parameters only after first finding an approximate solution to the balance laws. An example of this property is the parameter $8\hat{k}\hat{\rho}/m^3$ in Eqs (5.32)–(5.34) that is imagined to be either very large or very small. This parameter is a composite of the competitions between heat supply and removal and between kinetic energy and strength, as modulated by strain-rate sensitivity. It would be difficult to find the upper and lower limits of the various expressions without guidance from the balance laws.

The scaling relationships for propagating shear bands are found after an extra layer of complexity, namely finding similarity solutions (intermediate asymptotics according to Barenblatt [45]), which require solving a non-linear set of ordinary differential equations, Eq. (5.38). Fortunately, with the software now available, the solutions themselves are not difficult to find, but equally fortunate is the fact that the scalings introduced in Eq. (5.37) have eliminated all parameters from Eq. (5.38), except strain-rate sensitivity, thus greatly reducing the need for parametric runs. The similarity solutions have permitted the steady shear band speed to be estimated, but the limited amount of available experimental data suggests that this is an upper bound only, as it seems to be easy to interrupt the band motion as elastic unloading waves arrive.

Various new estimates of energy rates have been suggested, at least for mode III motion. Although the mode II steady speed is the same as for mode III, the various energy rates have not been checked for mode II. Energy dissipation per band is considered to consist only of plastic work within the narrow band itself, which converts completely to heat as work hardening is not considered, and, therefore, there is no storage of cold work. Heat flux, which leaves the hot shear band towards the halfway points between the band and an adjacent band or towards distant points where the thermal gradient vanishes, is considered as redistributing the thermal energy, and thus perhaps making it unrecoverable, but has not been included in the estimation of dissipation. It has been found that, on average, the volumetric rate of dissipation at a given stress is the same as it would be if the material had the average strain rate throughout but with a reduced stress. Ahead of the shear band, the rate of dissipation per unit volume is $s^+\dot{\gamma}_0$, but just behind the tip of the band, the average volumetric rate of dissipation is $s_-\dot{\gamma}_0$ and is therefore less. Considering Eqs (5.28) and (5.29), it would seem that the material attempts to localize as rapidly as possible, thus lowering its overall rate of dissipation near the tip and perhaps permitting the shear band to continue advancing as a coherent entity. This discussion has only accounted for the rate of dissipation at the tip, and it must be remembered that dissipation occurs all along the shear band and may vary in intensity all along its length. This must be especially true if the ultimate driving force is located remotely from the tip, as in the Kalthoff experiment, because momentum balance will require variable stress fields along the length. Total dissipation along the length of a shear band cannot exceed the energy delivered at the boundary, and, therefore, it seems reasonable to expect that the motion of the band must adjust itself to the changing power delivered to the external boundary.

At least two issues remain to be investigated. First is the question of hot spots along a shear band. Are they secondary instabilities, as suggested previously, or are they related to the necessary changes along the length due to the requirements of momentum balance and local energetics, or both? Or should the hot spots be understood in some completely different way? The second issue has been pointed out by Teng et al. [44]. Adiabatic shear bands do not in themselves represent ultimate material failure with separation of surfaces, as in cracks or spall. Other mechanisms must come into play. It is known that shear bands often are regions where cracking may occur, and that often there is an interaction with void nucleation and growth. But how these interactions should be described and understood is still largely open.

References

1. Bai, Y., and Dodd, B. (1991). "Adiabatic Shear Localization". Pergamon Press, Oxford.
2. Anand, L. (1985). Constitutive equations for hot-working of metals. *Int. J. Plast.* **1**, 213–231.
3. Rosakis, P., Rosakis, A. J., Ravichandran, G., and Hodowany, J. (2000). A thermodynamic internal variable model for the partition of plastic work into heat and stored energy in metals. *J. Mech. Phys. Solids* **48**, 581–607.
4. Wright, T. W. (2002). "The Physics and Mathematics of Adiabatic Shear Bands". Cambridge University Press, Cambridge.
5. Cowie, J. G., Azrin, M., and Olson, G. B. (1989). Microvoid formation during shear deformation of ultrahigh strength steels. *Metall. Trans. A* **20**, 143–153.
6. Ericksen, J. L. (1998). "Introduction to the Thermodynamics of Solids", Revised ed. Springer, New York, NY.
7. Anand, L., and Brown, S. (1987). Constitutive equations for large deformations of metals at high temperatures. *In* "Constitutive Models of Deformation" (J. Chandra and R. Srivastav, Eds.), pp. 1–26. SIAM, Philadelphia, PA.
8. Hines, J. A., and Vecchio, K. S. (1997). Recrystallization kinetics within adiabatic shear bands. *Acta Mater.* **45**, 635–649.

9. Hodowany, J., Ravichandran, G., Rosakis, A. J., and Rosakis, P. (2000). Partition of plastic work into heat and stored energy in metals. *Exp. Mech.* **40**, 113–123.
10. Longere, P., and Dragon, A. (2007). Adiabatic heat evaluation for dynamic plastic localization. *J. Theor. Appl. Mech.* **45**, 203–223.
11. Molinari, A., and Clifton, R. J. (1987). Analytical characterization of shear localization in thermoviscoplastic materials. *J. Appl. Mech.* **54**, 806–812.
12. Litonski, J. (1977). Plastic flow of a tube under adiabatic torsion. *Bull. Acad. Pol. Sci. – Ser. Sci. Tech.* **25**, 7–14.
13. Johnson, G. R., and Cook, W. H. (1983). A constitutive model and data for metals subjected to large strains, high strain rates, and high temperatures. *In* "Proceedings of the 7th International Symposium on Ballistics", pp. 541–547.
14. Zerilli, F. J., and Armstrong, R. W. (1987). Dislocation-mechanics-based constitutive relations for material dynamics calculations. *J. Appl. Phys.* **61**, 1816–1825.
15. Bodner, S. R. (1987). Review of a unified elastic–viscoplastic theory. *In* "Unified Constitutive Theories for Creep and Plasticity" (A. K. Miller, Ed.), pp. 273–301. Elsevier, New York, NY.
16. Follansbee, P. S., and Kocks, U. F. (1988). A constitutive description of copper based on the use of the mechanical threshold stress as an internal state variable. *Acta Metall.* **36**, 81–93.
17. Bammann, D. J. (1990). Modeling the temperature and strain rate dependent large deformation of metals. *Appl. Mech. Rev.* **43**, S312–S319.
18. Markus, L., and Yamabe, H. (1960). Global stability criteria for differential systems. *Osaka Math. J.* **12**, 305–317.
19. Wright, T. W. (1992). Shear band susceptibility: work hardening materials. *Int. J. Plast.* **8**, 583–602.
20. Wright, T. W. (1990). Approximate analysis for the formation of adiabatic shear bands. *J. Mech. Phys. Solids* **38**, 515–530.
21. Duffy, J., and Chi, Y. C. (1992). On the measurement of local strain and temperature during the formation of adiabatic shear bands. *Mat. Sci. Eng.* **A157**, 195–210.
22. Wright, T. W. (1994). Toward a defect invariant basis for the susceptibility to adiabatic shear bands. *Mech Mater.* **17**, 215–222.
23. Chichili, D. R., Ramesh, K. T., and Hemker, K. J. (2004). Adiabatic shear localization in alpha-titanium: experiments, modeling and microstructural evolution. *J. Mech. Phys. Solids* **52**, 1889–1909.
24. Wright, T. W., and Ravichandran, G. (1997). Canonical aspects of adiabatic shear bands. *Int. J. Plast.* **13**, 309–325.
25. Wright, T. W., and Ockendon, H. (1992). A model for fully formed shear bands. *J. Mech. Phys. Solids* **40**, 1217–1226.
26. Glimm, J., Plohr, B. J., and Sharp, D. H. (1995). "Tracking of Shear Bands. I. The One-Dimensional Case". SUNY, Stony Brook, NY, SUNYSB-AMS-95-04.
27. Dinzart, F., and Molinari, A. (1998). Structure of adiabatic shear bands in thermoviscoplastic materials. *Eur. J. Mech. A. Solids* **17**, 923–938.
28. Bender, C. M., and Orszog, S. A. (1978). "Advanced Mathematical Methods for Scientists and Engineers". McGraw-Hill, New York.
29. Moss, G. L. (1980). Shear strains, strain rates and temperature changes in adiabatic shear bands. *In* "Shock Waves and High-Strain-Rate Phenomena in Metals" (M. A. Meyers and L. E. Murr, Eds.), pp. 299–312. Plenum Press, New York.
30. Wright, T. W., and Ockendon, H. (1996). A scaling law for the effect of inertia on the formation of adiabatic shear bands. *Int. J. Plast.* **12**, 927–934.

31. Grady, D. E., and Kipp, M. E. (1987). The growth of unstable thermoplastic shear with application to steady-wave shock compression in solids. *J. Mech. Phys. Solids* **35**, 95–120.
32. Zhou, F., Wright, T. W., and Ramesh, K. T. (2006). A numerical methodology for investigating the formation of adiabatic shear bands. *J. Mech. Phys. Solids* **54**, 904–926.
33. Walter, J. W. (1992). Numerical experiments on adiabatic shear band formation in one dimension. *Int. J. Plast.* **8**, 657–693.
34. Zhou, F., Wright, T. W., and Ramesh, K. T. (2006). The formation of multiple adiabatic shear bands. *J. Mech. Phys. Solids* **54**, 1376–1400.
35. Kalthoff, J. F. (2000). Modes of dynamic shear failure in solids. *Int. J. Fract.* **101**, 1–31.
36. Wright, T. W., and Walter, J. W. (1996). The asymptotic structure of an adiabatic shear band in antiplane motion. *J. Mech. Phys. Solids* **44**, 77–97.
37. Chen, L., and Batra, R. C. (1999). The asymptotic structure of a shear band in mode-II deformation. *Int. J. Eng. Sci.* **37**, 895–919.
38. Gioia, G., and Ortiz, M. (1996). The two-dimensional structure of dynamic boundary layers and shear bands in thermoviscoplastic solids. *J. Mech. Phys. Solids* **44**, 251–292.
39. Bonnet-LeBouvier, A.-S., Molinari, A., and Lipinski, P. (2002). Analysis of the dynamic propagation of adiabatic shear bands. *Int. J. Solids Struct.* **39**, 4249–4269.
40. Wright, T. W. (2003). On the speed of an unconstrained shear band in a perfectly plastic material. *Int. J. Solids Struct.* **40**, 871–879.
41. Guduru, P. R., Rosakis, A. J., and Ravichandran, G. (2001). Dynamic shear bands: an investigation using high speed optical and infrared diagnostics. *Mech. Mater.* **33**, 371–402.
42. DiLellio, J. A., and Olmstead, W. E. (1998). Numerical solutions of shear localization in a finite slab. *Mech. Mater.* **29**, 71–80.
43. Li, S. F., Liu, W. K., Qian, D., Guduru, P. R., and Rosakis, A. J. (2001). Dynamic shear band propagation and micro-structure of adiabatic shear band. *Comput. Methods Appl. Mech. Eng.* **191**, 73–92.
44. Teng, X., Wierzbicki, T., and Couque, H. (2007). On the transition from adiabatic shear banding to fracture. *Mech. Mater.* **39**, 107–125.
45. Barenblatt, G. I. (1979). "Scaling, Self-Similarity, and Intermediate Asymptotics". Consultants Bureau, New York.

6 Adiabatic Shear Bands in Penetrators and Targets

Hervé Couque

Nexter Munitions, Bourges, France

Nomenclature

C_p specific heat
E quasi-static tensile ductility
K thermal conductivity
K_0 material constant of a constitutive law
L length of a kinetic penetrator
m strain rate coefficient
n hardening coefficient
p pressure
P penetration
UTS quasi-static ultimate tensile strength
V_{bl} ballistic limit velocity
v_c rigid penetrator critical speed
ε_c dynamic compression shear banding failure strain
ν thermal softening coefficient
σ_y *0.2%* quasi-static conventional yield strength at a plastic deformation of 0.002
v_{rp} speed of the rear of the penetrator
v_i penetration speed of the penetrator
R_t material strength of the target
W mechanical work
Y_p material strength of the penetrator
α plastic zone extent coefficient of the target
α **phase** crystallographic phase of the tungsten and titanium alloys
α' **phase** martensitic crystallographic phase of the titanium alloys
β Quinney's coefficient
β **phase** crystallographic phase of the depleted uranium and titanium alloys
γ shear strain
$\dot{\gamma}$ shear strain rate
γ **phase** crystallographic phase of the depleted uranium, tungsten and titanium alloys
θ temperature (usually absolute)
θ_m melting temperature (usually absolute)
ρ_p density of the penetrator
ρ_t density of the target

Adiabatic Shear Localization. DOI: http://dx.doi.org/10.1016/B978-0-08-097781-2.00006-X

σ stress at the interface target/penetrator
σ_{yp} quasi-static yield strength of the penetrator
σ_{yt} quasi-static yield strength of the target
τ shear stress

Adiabatic shear banding is one of the main failure mechanisms occurring under dynamic compression for metallic materials. This failure process involves the initiation and propagation of a shear band followed by a crack going through the shear band. This type of failure is generated when metallic materials are subjected to compression loads over a short duration of less than 1 ms. Such loading conditions are encountered not only during impact and blast events but also during catastrophic failure of structures which contain large stored energy.

An intensive research activity related to this problem has been conducted around the world focusing on the perforation processes associated with these highly dynamic events. This research is conducted not only by defence agencies with regard to armour and penetrator developments but also by civilian researchers dealing with shielding to contain failing pressure vessels or high-speed rotating machines. There are two distinct fundamental researches associated with the development and evaluation of materials, one with regard to the increase of the material integrity used in shielding structures, and the other with regard to the increase of the penetration performance of armour-piercing materials. These researches with opposite goals are known as the sword and armour race.

6.1 Adiabatic Shear Bands in Penetrators

For penetrators, the focus is the interaction of a metallic penetrator against metallic materials employed in armoured vehicles or against concrete used in bunkers and shelters. These penetrators consist of either rods made of dense alloys providing large kinetic energy per unit area, such as depleted uranium or tungsten alloys, or projectiles generated by an explosive charge placed against a liner made of ductile metallic materials such as copper, iron, tantalum and nickel.

For impact conditions in the range of 500–1800 m/s, adiabatic shear banding is the main mode of degradation of the penetrator, which determines the rate of consumption. Rapid initiation of the shear bands implies a reduced consumption of the penetrator per unit length, resulting in a higher penetration performance. As of today, depleted uranium alloys are the best-performing penetrator materials because they easily initiate localized shear bands, as detailed in the following paragraphs.

To identify the importance of adiabatic shear banding with regard to the penetrator, it is important to be clear that the interaction of the penetrator with the target is highly dependent on the impact speed as shown in Figure 6.1, as well as the material strengths. At low impact velocities, usually less than 500 m/s, and when penetrator and target strengths are of the same order, the penetrator will penetrate practically without degradation. Under these conditions, the penetrator is usually

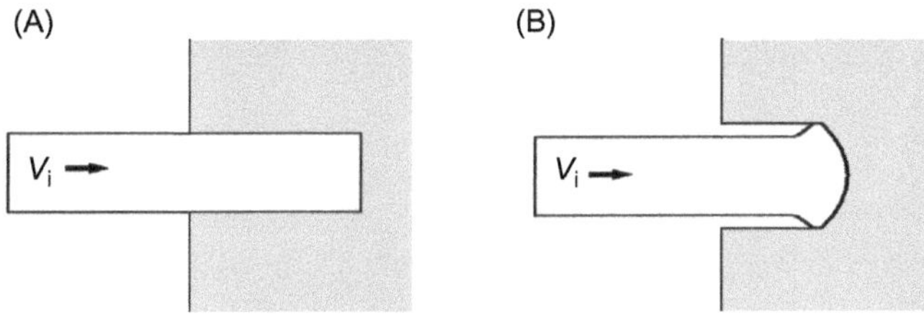

Figure 6.1 Interaction of a metallic penetrator at an impact speed v_i on a metallic target based on the critical speed v_c of Eq. (6.3). (A) Penetration with rigid penetrator $v_i < v_c$ and (B) penetration with deforming penetrator $v_i > v_c$.

described as a 'rigid penetrator'. At higher impact speeds, above a critical impact speed v_c defined below, the penetrator starts to erode following the plastic deformation of the penetrator head.

The transition in the penetrator behaviour is analysed by examining the stress at the interface between the penetrator and the target using 1D models proposed by Alekseevskii [1] and Tate [2]. These models are based on the modification of the Bernouilli equation through the conservation of energy and provide the stress, σ, at the interface,

$$\sigma = ½\, \rho_p (v_{rp} - v_i)^2 + Y_p = ½\, \rho_t v_i^2 + R_t \tag{6.1}$$

where v_i is the penetration speed, v_{rp} is the speed of the rear of the penetrator, ρ_p and ρ_t are the density of the penetrator and the target and Y_p and R_t are proportional to the quasi-static yield strength σ_{yp} and σ_{yt} of the penetrator and target, respectively. Through comparison with experimental data, Tate [3,4] estimated Y_p to be equal to $1.7\sigma_{yp}$. This stress can be considered as a yield stress representative of the highly shocked material at the tip of the penetrator. Based on numerical simulations, Walker and Anderson [5] have expressed R_t with regard to the quasi-static yield stress of the target material, σ_{yt},

$$R_t = 7/3\, \ln(\alpha)\, \sigma_{yt} \tag{6.2}$$

with α describing the extent of the plastic zone in the target. Depending on the impact speed, the value of α ranges from 4.94 to 11.2 resulting in the ratio R_t/σ_{yt} varying from 3.7 to 5.6. It is important to note that the estimates of R_t and Y_p are mainly based on fitting penetration data with the 1D models. Further material experimental investigations are required to confirm their values.

For R_t less than Y_p, the penetrator behaves as a 'rigid penetrator' when the penetration speed is less than the critical speed v_c:

$$v_c = \{2(Y_p - R_t)/\rho_t\}^{½} \tag{6.3}$$

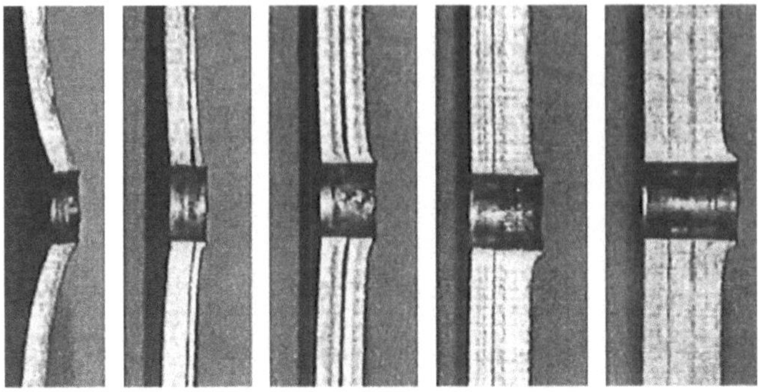

Figure 6.2 Rigid perforation of a blunt-nosed projectile 20 mm in diameter and 80 mm in length made of tool steel against a steel plate of thickness 6, 8, 12, 16 and 20 mm at impact speeds ranging from 150 to 300 m/s [6].

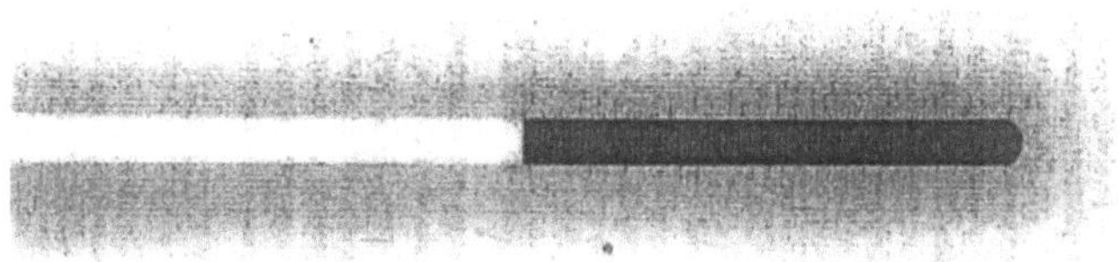

Figure 6.3 Rigid perforation of a blunt-nosed projectile 7 mm in diameter and 71 mm in length made of maraging steel against an aluminium target at an impact speed of 1120 m/s [7].

When dealing with penetrator and target materials of similar strengths, rigid penetration occurs up to an impact speed of 250 m/s as illustrated in Figure 6.2 based on the work of Borwick et al. [6]. Such rigid penetration can prevail at an impact speed of 1120 m/s when the target material strength is much lower than the penetrator material strength as is the case in the work of Forrestal et al. [7] (Figure 6.3).

When impact velocities exceed the critical speed, the head of the penetrator starts to deform plastically. This deformation process is associated with very high pressures exceeding 5 GPa as shown in Figure 6.4 [8].

Associated with these high pressures, large shear strains are encountered by the penetrator [9], exceeding 100% for tungsten alloy penetrators. The rate of displacement is very high, inducing highly shocked materials at the front penetrator target interface followed by plastic deformation occurring at strain rates exceeding $10^4\ s^{-1}$ [9].

As suggested by Magness [10,11], the consumption of the penetrator occurs through adiabatic shear banding, which appears to be delayed for tungsten alloys when compared to depleted uranium alloys, as schematically represented in Figure 6.5. No experimental data are available to describe shear banding at these high pressures. Instead, experimental evidences have been generated using dynamic compression data generated with Kolsky–Hopkinson pressure bars at a pressure of about

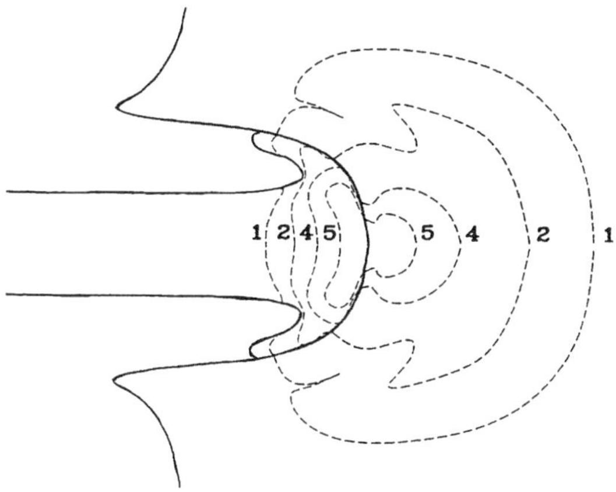

Figure 6.4 Pressure in GPa in the vicinity of a penetrator made of a depleted uranium alloy penetrating a steel target at 1650 m/s based on numerical simulations of Mescall [8]. *Source*: Published with the kind permission from Springer Science +Business Media B.V.

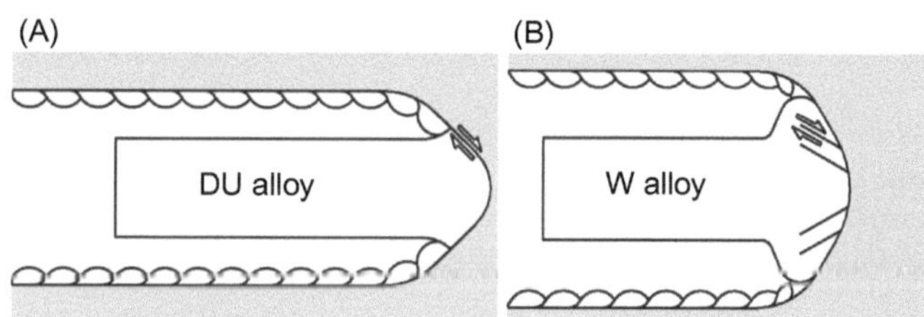

Figure 6.5 Schematic of the penetrator consumption for depleted uranium alloy (A) and tungsten alloy (B) kinetic penetrators.

0.4 GPa [12,13] (Figure 6.6). These data reveal that the dynamic compression failure strain associated with shear banding decreases with the perforation impact speed or ballistic limit of 15 mm thick steel plates for a series of uranium materials alloyed with titanium and/or molybdenum. For a given penetrator, the decrease of this perforation impact speed is equivalent to an increase of the penetration depth in a semi-infinite target. The same dependence has been observed for a series of tungsten alloys, as shown in Figure 6.7, evaluated with small kinetic penetrators of length over a diameter ratio of 24, and symmetric impact Taylor tests [14]. As for uranium alloys, the penetration performance of the tungsten alloy increases with the decrease of the dynamic compression failure strain.

When considering Kolsky–Hopkinson pressure bar data, shear banding occurs at a low compressive failure strain less than 0.35 for depleted uranium materials as compared to 0.40 for tungsten alloys. Although no data are available on the influence of pressure on failure strain for depleted uranium alloys, some data are available for tungsten alloys as shown in Figure 6.8A. For pressures down to 0.1 GPa, as is the case during shear band formation in the symmetric impact Taylor test [15], the failure strain drops to 0.24. With the increase of pressure to 1.5 GPa, shear banding occurs at twice this failure strain [16]. It is not clear for pressures reaching 5 GPa, as is the case at the head of the penetrator, whether the strain will continue

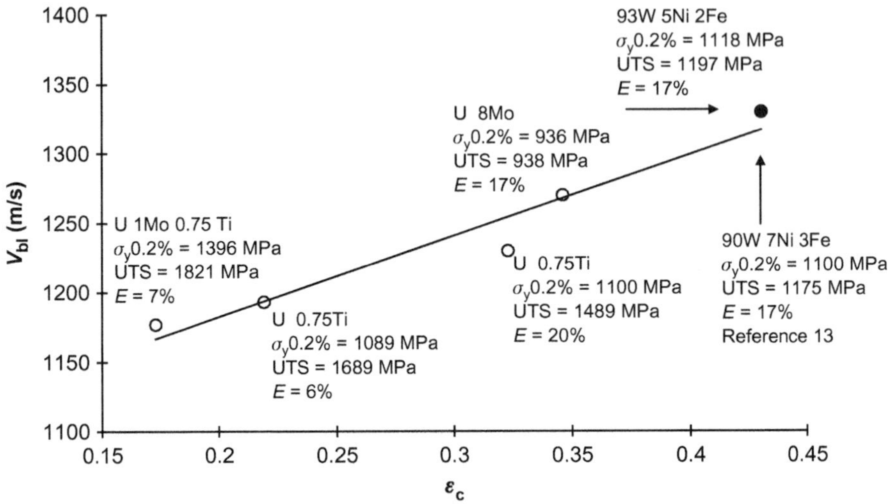

Figure 6.6 Ballistic limit velocity function of dynamic compression shear banding failure strain for uranium and tungsten alloys [12,13]. The failure data were generated with Kolsky–Hopkinson pressure bar tests for which shear banding occurs at pressures ranging from 0.3 to 0.5 GPa depending on the material strength.

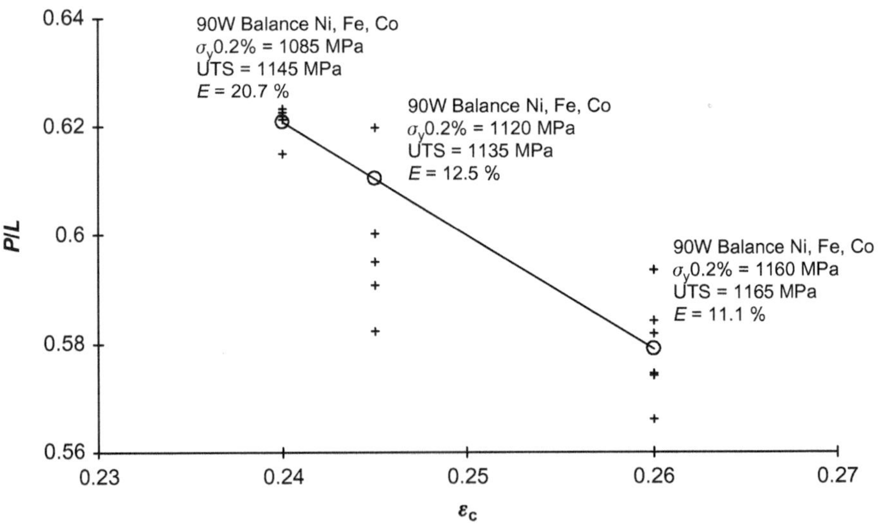

Figure 6.7 Relative penetration P/L, with P the penetration and L the length of the kinetic penetrator, against steel targets at 1400 m/s function of the dynamic compression shear banding failure strain for tungsten alloys [14]. The failure data were generated with symmetric impact Taylor tests for which shear banding occurs at pressures of the order of 0.1 GPa [15].

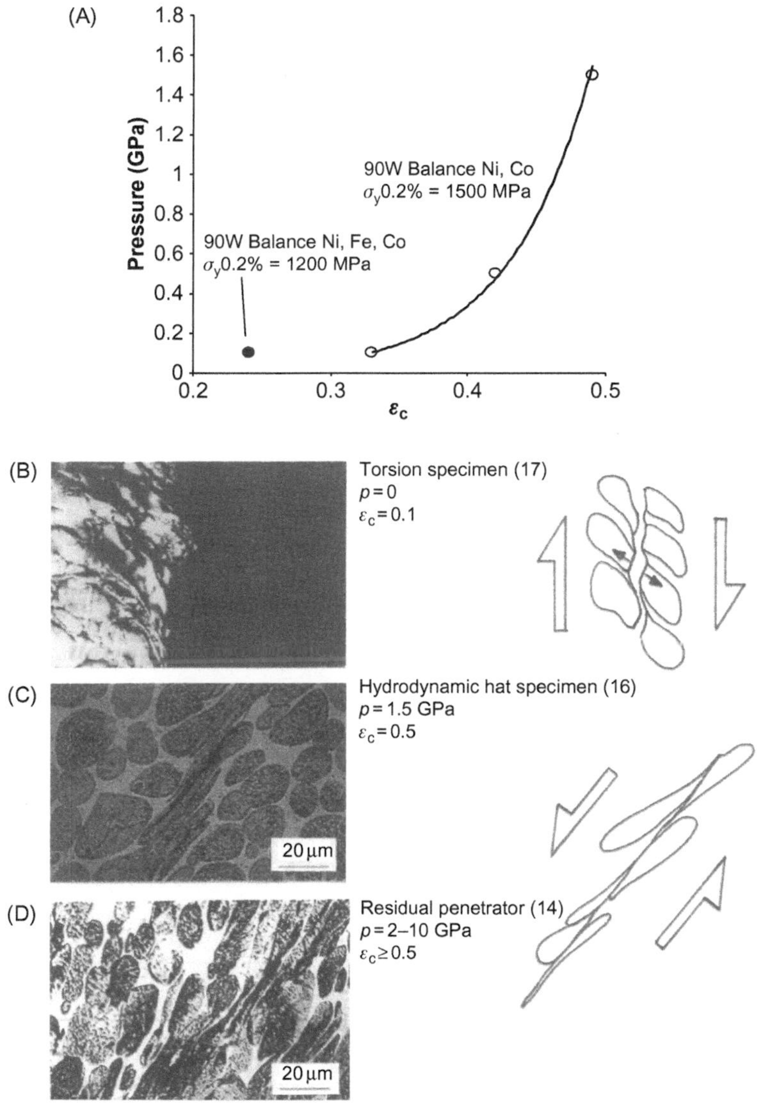

Figure 6.8 Influence of the pressure on shear band formation in tungsten alloys: (A) pressure function of the dynamic failure strain in compression; failure characteristics at (B) p = 0, (C) p = 1.5 GPa and (D) p = 2–10 GPa.

to increase. It is important to note that when no pressure is applied to tungsten alloys as in the torsion test, the tungsten alloys fail at a low strain of the order of 0.10 via shear without localizing [17] (Figure 6.8B). This is due to the triggering of a failure mechanism observed in tension consisting in the opening of the weak bonding between the pure tungsten grains known as the α phase [18].

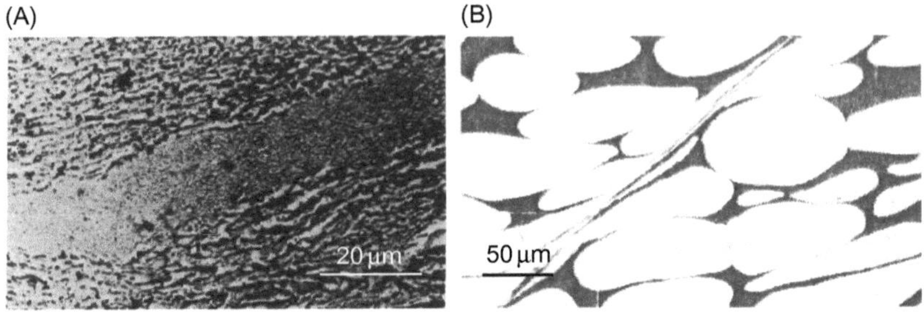

Figure 6.9 Shear band morphologies of a U 1.5 Mo (A) and a 90% W alloy (B) generated in dynamically loaded compression specimens using Kolsky–Hopkinson pressure bars [19,20].

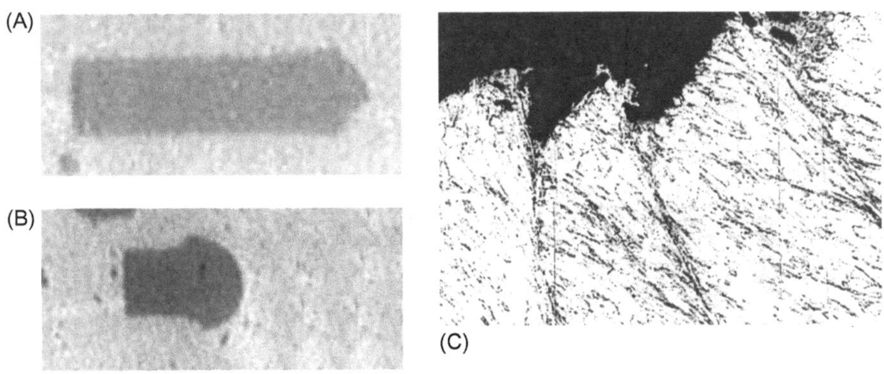

Figure 6.10 Residual penetrator made of (A) U 0.75Ti and (B) 90% W alloy after passing a metallic target [10,22]. (C) Shoulder of a residual penetrator made of 90% W alloy revealing a deformation process involving multiple shear bands [9].

The characteristics of the shear bands were obtained from metallographic observations of dynamic compression specimens for depleted uranium and tungsten alloys [19,20] (Figure 6.9). For depleted uranium alloys, narrow shear bands are generated of width less than 20 μm with undeformed surrounding material. In the case of tungsten alloys, narrow shear bands are also observed of width 20 μm, but they are generated after a large shearing of the surrounding material occurring over a width of 50–100 μm. In fact, localized shear occurred after a macroscopic path had been created in the nickel base matrix material known as the face-centered cubic γ phase (The dark phase of Figure 6.9B). The key event in the process is the transformation of the body-centered cubic tungsten α grains to lenses by combined compression and shear. Analytical modelling of the instability conditions for the γ-like materials reveals that low hardening and high thermal softening facilitate adiabatic shearing [21]. These results reveal that a heavily worked γ phase will provide conventional tungsten alloy penetrators with high penetration performances.

As shown in Figure 6.10A and B, X-ray observations of penetrators after passing through a metallic target reinforce the hypothesis with regard to chiselling via

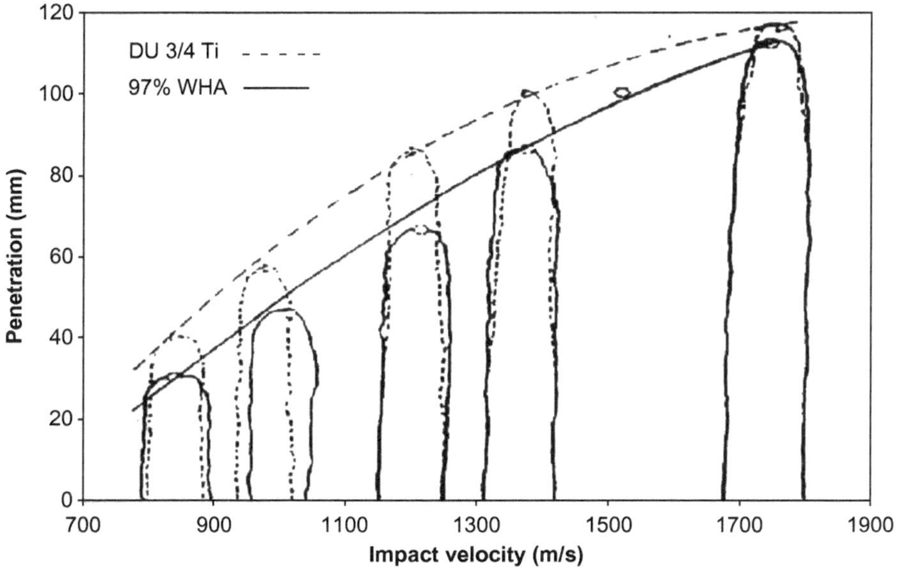

Figure 6.11 Comparison of penetration tunnels displaced by U 0.75Ti and 97% W alloy into mild steel [10].

localized shear bands of the nose of a depleted uranium alloy penetrator [10] and to the mushrooming deformation of the tungsten alloy penetrator head [22]. Post-mortem observations of tungsten alloy residual penetrators confirm the hypothesis of erosion via multiple localized shear banding of tungsten alloys [14,22] (Figures 6.8B and 6.10C). Larger penetration tunnels have been observed by Magness [10] with penetrators made of 97% W alloy, which when compared to U 0.75Ti reinforce the view that chiselling via adiabatic shear banding is the eroding process for depleted uranium alloy penetrators (Figure 6.11).

There is an ongoing debate about the shear-failure process in depleted uranium penetrators. Because of the adiabatic conditions, the temperature rise at the head of the penetrator can lead to a phase transition from the tetragonal β phase to the body-centered cubic γ phase, which may favour strain localization [23]. However, occurrence of this phase transition is not obvious as the high pressure at the penetrator target interface may delay the γ phase domain to higher temperatures.

Several researches have been undertaken with the objective of reducing the shear banding failure strain of penetrator materials. These researches focus mainly on tungsten-based materials, which are less hazardous to handle. As revealed previously, small-scale dynamic testing using Hopkinson pressure bars and Taylor impact tests are a first step to provide some insights in shear banding failure strain of novel materials usually available in small quantities. Because such laboratory tests cannot generate the high pressure encountered at the tip of the penetrator, small-scale ballistic testing remains the second step towards the validation of the penetrator materials. Before performing such studies, it is of interest to consider an intermediate step by evaluating materials previously shocked to the intensity encountered at the head of the material.

The novel tungsten-based materials currently being investigated for penetrators are highly worked, nanostructured and metallic glassy materials. Because the material systems for kinetic penetrators are mainly tungsten based, the hardening and high-strain-rate sensitivity are the fundamental mechanical parameters, which can be used towards designing a penetrator material favouring shear banding. On the other hand, as detailed in Section 6.1.2, metallic target materials can be chosen by considering other physical and mechanical parameters that influence shear banding occurrences such as density, specific heat and softening.

Using high-hydrostatic extrusion (HHE), Xiaoqing et al. [24] showed that highly working tungsten alloys will favour self-sharpening with improved ballistic penetration performance of 4% with respect to conventional swaged tungsten alloys. Using equal channel angular pressing (ECAP) techniques, Wei et al. [25] have generated ultra-fine pure tungsten. In this work, shear banding under dynamic compression using the Kolsky–Hopkinson pressure bars technique has been generated at a low failure strain of 0.20, which shows promise towards favouring self-sharpening during penetration (Figure 6.12A). Kecskes et al. [26] have demonstrated that alloys with a metallic glass phase provide the necessary self-sharpening behaviour for alloys containing 60% W with a penetration performance comparable to the depleted uranium alloy performance (Figure 6.12B). However, such penetration performance does not prevail when the amount of tungsten is increased.

Nanostructured and metallic glass alloys are very attractive because the plastic deformation and failure processes involve new mechanisms. A particular result from Schuster et al. [27] is obtained with nanostructure tantalum elaborated with a high-pressure torsion (HPT) technique, as they reveal localized shear banding under quasi-static conditions (Figure 6.12C).

These novel materials are promising. However, today, they provide only research directions as they do not generate the necessary tensile ductility to preserve the integrity of the penetrator during the launching and impact phases.

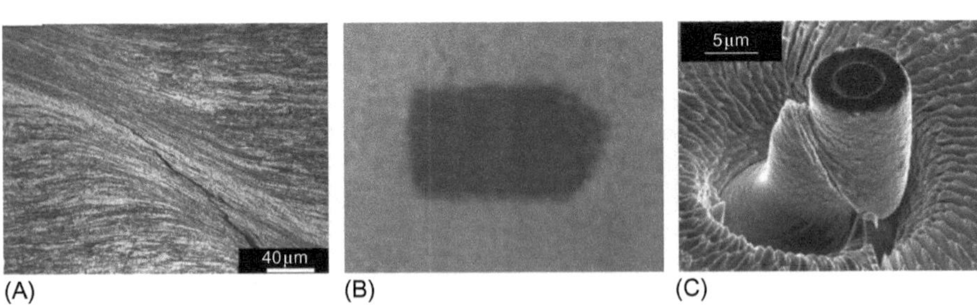

Figure 6.12 Novel materials for penetrator: (A) dynamic shear localization in a nanostructure tungsten generated with the Kolsky–Hopkinson pressure bars technique [25], (B) residual penetrator made of bulk metallic glass (BMG) alloy after passing a metallic target [26] and (C) quasi-static shear localization of a nanostructure tantalum [27].

6.2 Adiabatic Shear Bands in Targets

For targets, shear localization phenomena are observed both for conventional metallic armours and for lightweight metallic armours. Metallic materials cover a wide range of properties from low-strength to high-hardness steels, aluminium, titanium and nickel alloys. These metals are intensively used in the conception of armours and shielding protections. They are usually combined with laminate composites, ceramics and eventually metallic glasses and metallic foams to increase their efficiency.

Research into target materials focuses, on the one hand, on degradation of the penetrator and, on the other hand, on the evolution of damage in the target. The different shear banding failures observed in metallic targets are schematized in Figure 6.13.

With regard to the penetration process, the failure mechanism of the target involves cracks propagating through shear bands. This failure process is known as plugging (Figure 6.13A). For low-strength targets made of aluminium alloys or low-strength steels, this failure mechanism is the only one observed.

When a stronger interaction occurs between the target and the penetrator, which can occur for intermediate-strength metallic materials, secondary shear bands are observed along the perforated channel (Figure 6.13B).

Also observed with intermediate-strength steels is non-negligible bending at the back face of the target, which generates failed layers through initiation and propagation of shear bands. This failure process is known as discing (Figure 6.13C).

When intense stress waves are generated during impact in the target, these waves reflected as tensile waves generate spall cracks. These cracks create plates that separate through shear banding (Figure 6.13D). This failure process is known as spalling or scabing.

With the exception of the plugging failure, the pressures associated with the other localized shear banding failures are less than 1 GPa, making conventional dynamic mechanical testing a means to establish the material sensitivity to localized shear banding.

With the objective to have a target material delaying adiabatic shear banding, the ability to generate heat from mechanical work has to be minimized. By neglecting the heat conduction as a first approximation, the heat equation relates the mechanical work, W, to the increase of the temperature, $\Delta\theta$,

$$W = \beta \rho_t C_p \Delta\theta \tag{6.4}$$

with ρ_t the density of the target material, C_p the specific heat and β the Quinney coefficient describing the amount of work converted in heat, usually taken equal to 0.9. To reduce the amount of heat generated, metallic materials with a high $\rho_t C_p$ are favourable. Among candidate metals for targets, steels, nickel and copper alloys exhibit the highest $\rho_t C_p$ of about $3.4-3.9 \times 10^6$ J/(m^3 kg). Other candidates such as

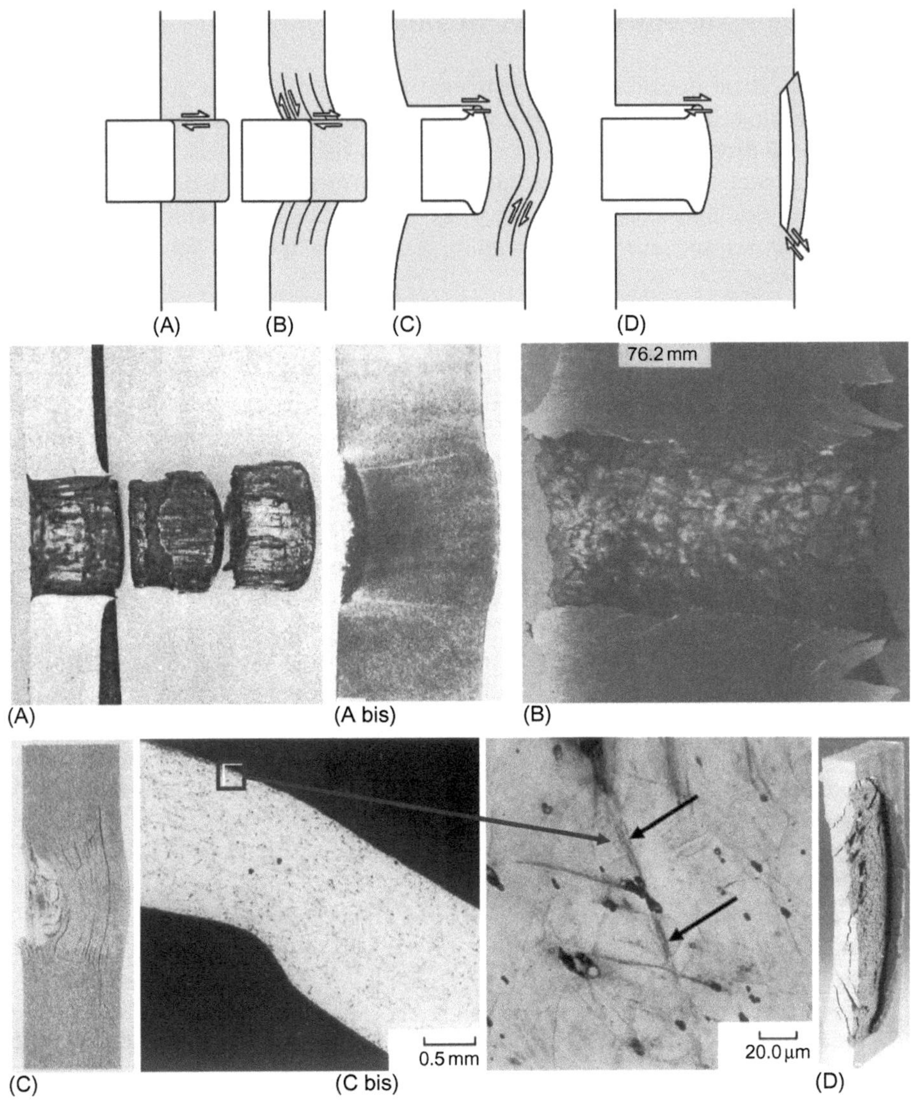

Figure 6.13 Shear localization occurring in metallic targets: during the perforation process (A) with the main shear bands during plugging for a 15HRC steel [28] and (A bis) for a 52HRC steel [29], and (B) with secondary shear bands along the penetration channel for an armour 42HRC steel [30]; during the bending of the back face (C) for an aluminium alloy [31] and (C bis) for a nickel alloy [32]; and after spalling of the back face for an aluminium alloy (D) [33].

(A bis) Source: Published with the kind permission from Springer Science+Business Media B.V. *(B and C) Source*: Published with the kind permission of the International Ballistics Society.

titanium and aluminium alloys favour heat generation as $\rho_t C_p$ is much lower of about 2.3×10^6 J/(m^3 kg).

To generate a better metallic target, the mechanical parameters to be primarily considered are the hardening (n), strain rate (m) and thermal softening (v) coefficients, which express the stress (τ), as a function of the strain (γ), strain rate ($\dot{\gamma}$) and the temperature (θ) for metallic materials through the constitutive law,

$$\tau = K_0 \gamma^n \dot{\gamma}^m \theta^v \tag{6.5}$$

with K_0 as a constant.

Molinari and Clifton [34] proposed a criterion with regard to shear localization 'if the sum of the materials coefficients, m, n, v, is positive, then the material is less prone to localized shear banding'. Because hardening and strain-rate sensitivity are positive, these parameters have to be maximized. With regard to the softening, which is negative, a low absolute value should be looked for.

High hardening, n, is required to make plastic deformation difficult to generate. Such behaviour is usually obtained in annealed conditions when metallic materials have not been work hardened. It is associated with low yield strength and a difference between the ultimate tensile failure stress and the yield strength of the order of the yield strength. However, such target materials will be easily defeated as large amounts of plastic deformation will be generated favouring failure of the target.

High strain-rate sensitivity, m, is required to provide an elevated elastic–plastic threshold when high strain rates are applied. Metallic materials with body-centered cubic and hexagonal closed-packed crystallographic structures provide such high sensitivity (such as steels and titanium alloys).

Low absolute value of the softening parameter, v, will delay plastic deformation when heat is generated. Such characteristic is dependent on the melting temperature of the metal. With a high melting temperature, softening is delayed to higher temperatures, making steels and titanium alloys the most appropriate candidates.

In developing and selecting better target materials, cost and mass efficiency have to be considered [35]. For army vehicles, either aluminium alloys or intermediate-strength steels are candidates depending on the threats. Because of the high cost of titanium alloys, their use will be limited to applications where a high level of protection is required. For aeronautic applications, where cost is not of the first importance, nickel alloys are considered because they provide a high resistance to localized shear bands while having enough strength to meet structural requirements.

Critical failure strain data describing the initiation of localized shear banding of metals have been gathered in the book by Bai and Dodd, constituting a reference work for the different families of metals used in targets [36]. However, care has to be taken when interpreting these data because within one metal family, different heat treatments and processing roots can generate significant variations of this critical failure strain. The following paragraphs summarize the trends observed with localized shear banding in steels, and aluminium, titanium and nickel alloys.

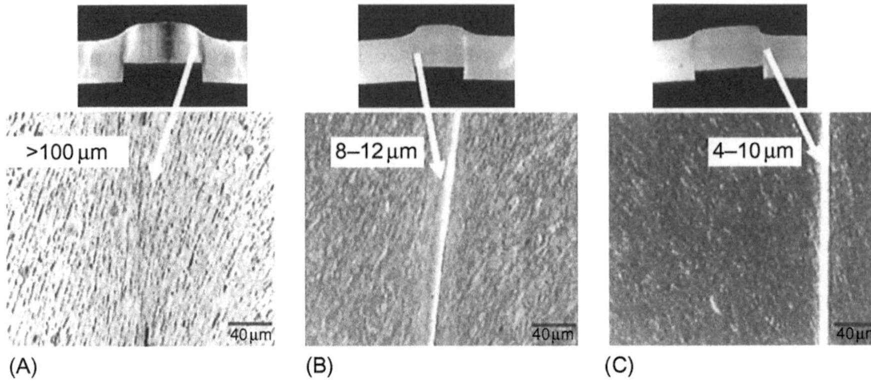

Figure 6.14 Shear band morphologies in target steels of yield strength (A) 500, (B) 850 and (C) 1000 MPa impacted by a blunt-nosed projectile made of tool steel [37,38].
Source: Published with the kind permission from Springer Science+Business Media B.V.

The influence of mechanical properties on shear band morphology and on plugging failure is provided by Dey et al. [37] for a series of steels. A steel with a low yield strength of 500 MPa and with non-negligible hardening capacities generates large shear bands of widths greater than 100 μm (Figure 6.14). This characteristic is associated with a critical impact speed of 190 m/s for perforation of a 12 mm thick steel target with a blunt-nosed projectile. The projectile made of tool steel has a length of 80 mm and a diameter of 20 mm.

With the increase of the yield strength to 850 and 1000 MPa and the reduction of the hardening characteristics, the shear band width reduces to about 10 μm (Figure 6.14). For these two steels, the critical impact speed drops to 150 and 130 m/s, which confirms that highly worked steels favour localized shear bands resulting in less efficient target materials.

It is important to note that in this work, the use of conical-nosed and ogival-nosed shapes reveals an opposite result. Such reverse behaviour has been attributed to the presence of another ductile failure mechanism ahead of the projectile involving void coalescence and growth, which is known to exhibit a failure energy increasing with the increase of the yield strength [39].

As revealed in Figures 6.14 and 6.15, white bands are observed within the localized shear bands of tempered martensite steels coming from a rapid phase transformation during cooling from austenitic to martensitic structures [40,41]. These bands, often referred to as transformed bands, suggest that a temperature in excess of 720°C corresponding to the austenitic domain of steels is reached.

Because of the low strength of aluminium alloys, large shear bands greater than 80 μm are generated [43] (Figure 6.16A). As in steels, secondary shear banding can be observed during plugging [35] (Figure 6.16B).

As to generate improved aluminium alloys for targets, high-strength alloys have been developed with the objective to maintain a non-negligible resistance to localized shear banding. Owolabi et al. [44] have studied reinforced 6061-T6 aluminium

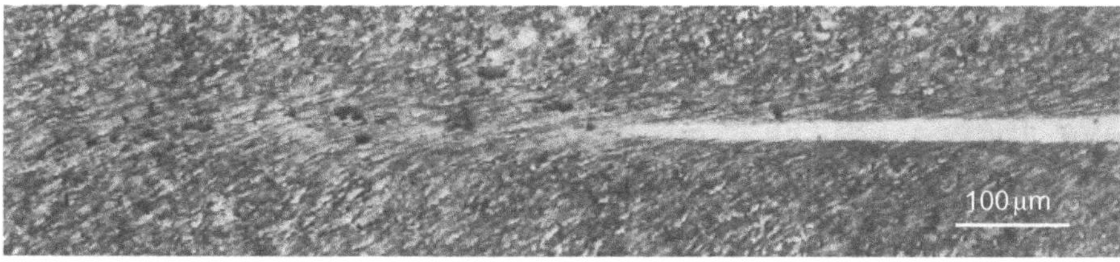

Figure 6.15 Shear band partially transformed in a 1040 steel after impact with a blunt-nosed projectile made of tool steel [42].
Source: Published with the kind permission from Springer Science+Business Media B.V.

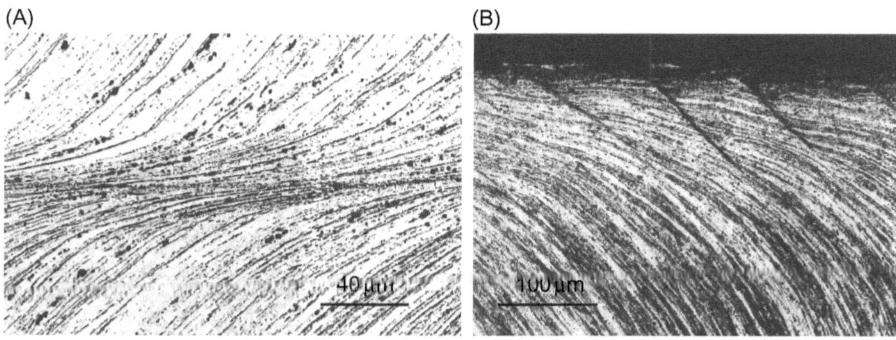

Figure 6.16 Localized shear bands in 7039 aluminium alloys generated from the impact of a blunt-nosed projectile: (A) Ref. [43] and (B) Ref. [35].
Source: Published with the kind permission from Springer Science+Business Media B.V.

alloy with 10% and 20% volume fraction of alumina presenting very high yield strength in excess of 800 MPa [44].

This work reveals that large shear bands about 100 μm in width are generated along with a failure strain, for the initiation of shear bands, reduced by 25% when compared to the monolithic material.

Titanium alloys for target applications are usually of the type Ti−6Al−4V composed of the hexagonal closed-packed α phase and of the body-centered cubic β phase. These alloys are sensitive to localized shear banding as they present a low $\rho_t C_p$. Because advantage is taken in the capability to generate high strengths, titanium alloys present a low hardening rate, which is another reason that they are favouring localized shear banding. Among metallic materials, titanium alloys as well as depleted uranium alloys present the lowest thermal conductivity K, which seem to be another reason, notwithstanding rapid loading, to favour localized shear bands. Associated to the high strength of titanium alloys, narrow shear bands of width 10−20 μm are generated [45,46] (Figure 6.17).

In spite of their high sensitivity to localized shear banding, titanium alloys with their low density and non-negligible strength are excellent candidates for target

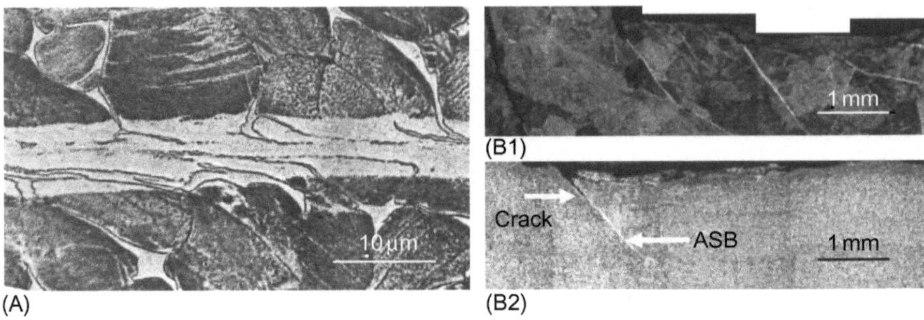

Figure 6.17 Localized shear bands in Ti–6Al–4V titanium alloys generated from impact with (A) a hard steel sphere [45] and (B) with a 7.62 mm penetrator [47]. The titanium alloys present a α and β phase structure (A and B1) and a β and martensitic α' phase structure (B2).

materials [45,46]. In fact, high-strength titanium alloys with α and β phases compared to β and martensitic α phases are found to be more resistant for localized shear banding [47] (Figure 6.17B), when a 7.62 mm projectile was arrested against a 15 mm target. This higher resistance goes along with ballistic data generated with a 12 mm target, for which only the β and martensitic α phases target material was perforated.

As for steels, localized shear bands revealed white bands (Figure 6.17). The white bands in titanium alloys are composed of fine α grains generated through dynamic recrystallization [48]. Dynamic recrystallization indicates that the homologous temperature, θ/θ_m, is about 0.55, with θ the temperature after the impact, and θ_m the melting temperature of the alloy, corresponding to temperatures above 790°C [49].

When high efficiency against penetration is required, such as for containing the failure of rotating machines or gas tanks, nickel-based alloys are employed. These materials are characterized by high resistance to the two main failure modes, which are tensile failure (mode I) and in-plane shear banding (mode II).

For the case of severe containment requirements, such as the failure of blades rotating at 50,000 rpm in auxiliary power units used in commercial airplanes, nickel based alloys such as the Inconel 718 are used. The resistance to shear has been revealed to be extremely high by DeMange et al. [32] for an annealed Inconel 718. They reveal localized shear bands initiating at a strain in excess of 150% using hat-shaped specimens dynamically loaded with the Kolsky–Hopkinson bars technique (Figure 6.18A).

Such high resistance to shear has been observed in nickel-iron alloys representative of the γ phase of tungsten alloys with symmetric impact Taylor specimens [31]. This high resistance to localized shear is associated with a low-yield stress of 400 MPa, a high ratio of ultimate tensile yield strength to yield stress of about 3 and large shear bands of widths ranging from 50 to 100 μm (Figure 6.18B).

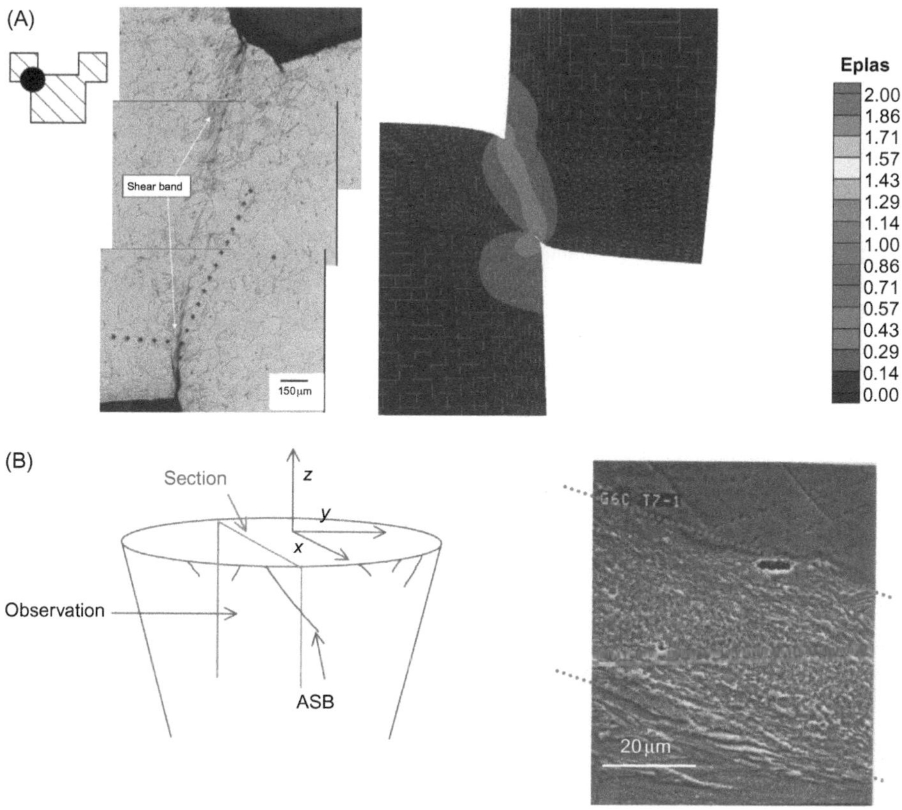

Figure 6.18 Shear band morphologies in (A) an Inconel 718 [32] and (B) a nickel–iron alloy [21].

References

1. Alekseevskii V. P. (1966), Penetration of a rod into a target at high velocity, *Combust. Explosion Shock Waves* **2**(2), 63–66.
2. Tate A. (1967), A theory for the decelaration of long rods after impact, *J. Mech. Phys. Solids* **15**, 387–399.
3. Tate A. (1986), Long rod penetration models – Part I. A flow field model for high speed long rod penetration, *Int. J. Mech. Sci.* **28**, 535–548.
4. Tate A. (1986), Long rod penetration models – Part I. Extensions to the hydrodynamic theory of penetration, *Int. J. Mech. Sci.* **28**, 599–612.
5. Walker J. D. and Anderson C. E. (1995), A time-dependent model for long-rod penetration, *Int. J. Impact Eng.* **16**, 19–48.
6. Borwick T., Hopperstad O. S., Langseth M. and Malo K. A. (2003), Effect of target thickness in blunt projectile penetration of Weldox 460 E steel plates, *Int. J. Impact Eng.* **28**, 413–464.

7. Forrestal M. J., Tzou D. Y., Askari E. and Longcope D. B. (1995), Penetration into ductile metal targets with rigid spherical nose rods, *Int. J. Impact Eng.* **16**, 699–710.
8. Mescall J. (1983), Materials issues in computer simulations of penetration performances, in Computational Aspects of Penetration Mechanics, eds. J. Chandra and J. E. Flaherty, pp. 47–62, Springer, Berlin.
9. Magness L. S., Scheffer D., Weerasooriya T. and Chung M. (2000), Deformation and failure of a tungsten alloy penetrator as a function of stress state, in Proceedings of the Tungsten and Tungsten Alloys 2000 Conference, eds. M. S. Greenfield and J. J. Oakes, pp. 35–42, Metal Powders Industries Federation, Princeton, NJ.
10. Magness L. S. (1992), A Phenomenological Investigation of the Behavior of High-Density Materials Under the High Pressure, High Strain Rate Loading, Ph.D. Thesis, Johns Hopkins University.
11. Magness L. S. (1993), Properties and performance of KE penetrators materials, in Proceedings of the Tungsten and Tungsten Alloys 1992 Conference, eds. A. Bose and R. J. Dowding, pp. 15–22, Metal Powders Industries Federation, Princeton, NJ.
12. Weerasooriya T. and Magness L. S. (1995), High rate testing of kinetic energy penetrator materials, in Metallurgical and Materials Applications of Shock-Wave and High-Strain-Rate Phenomena, eds. L. E. Murr, K. P. Staudhammer and M. A. Meyers, pp. 893–899, Elsevier, Amsterdam.
13. Couque H., Lankford J. and Bose A. (1992), Tensile fracture and shear localization under high loading rate in tungsten alloys, *J. Phys. III* **2**, 2225–2238.
14. Couque H., Nicolas G. and Altmayer C. (2006), Relation between shear banding and penetration characteristics of conventional tungsten alloys, *Int. J. Impact Eng.* **34**, 412–423.
15. Couque H., Nicolas G. and Eches N. (1998), Pressure effects on shear band formation in as sintered and swaged nickel cobalt tungsten alloys, in Proceedings of the Tungsten and Tungsten Alloys 1997 Conference, eds. A. Bose and R. J. Dowding, pp. 103–110, Metal Powders Industries Federation, Princeton, NJ.
16. Couque H. (2003), A hydrodynamic hat specimen to investigate pressure and strain rate dependence on adiabatic shear band formation, *J. Phys. IV* **110**, 423–428.
17. Ramesh K. T., Yadav S. and Davis J. (1993), Shear localization in tungsten heavy alloys, in Proceedings of the Tungsten and Tungsten Alloys 1992 Conference, eds. A. Bose and R. J. Dowding, pp. 299–306, Metal Powders Industries Federation, Princeton, NJ.
18. Lankford J., Couque H., Bose A. and German R. (1991), Dynamic deformation and failure of tungsten heavy alloys, in Tungsten and Tungsten Alloys, eds. A. Crowson and E. S. Chen, pp. 151–159, The Minerals, Metals and Materials Society, Warrendale, PA.
19. Stelly M. and Dormeval R. (1986), Adiabatic shearing, in Metallurgical Applications of Shock-Wave and High-Strain-Rate Phenomena, eds. L. E. Murr, K. P. Staudhammer and M. A. Meyers, pp. 607–632, Marcel Dekker, New York, NY.
20. Bose A., Couque H. and Lankford J. (1993), Influence of microstructure on shear localization in tungsten heavy alloys, in Proceedings of the Tungsten and Tungsten Alloys 1992 Conference, eds. A. Bose and R. J. Dowding, pp. 291–298, Metal Powders Industries Federation, Princeton, NJ.
21. Couque H., Molinari A., Salesse B. and Nicolas G. (2006), Adiabatic shear banding characteristics of conventional tungsten alloys, in Proceedings of the Tungsten and Tungsten Alloys 2006 Conference, eds. A. Bose and R. J. Dowding, pp. 385–394, Metal Powders Industries Federation, Princeton, NJ.

22. Magness L. S. (2002), An overview of the penetration performances of tungsten and depleted uranium alloy penetrators: ballistic performances and metallographic examinations, in Proceedings of the 20th International Symposium on Ballistics, eds. J. Carleone and D. Orphal, pp. 1104–1111, National Defense Industrial Association, Lancaster, PA.
23. Zurek A. K. and Follansbee P. S. (1993), Shear localization in U 0.75Ti and comparison to W–Ni–Fe, in Proceedings of the Tungsten and Tungsten Alloys 1992 Conference, eds. A. Bose and R. J. Dowding, pp. 439–445, Metal Powders Industries Federation, Princeton, NJ.
24. Xiaoqing Z., Shukui L., Jinxu L., Yingchun W. and Xing W. (2010), Self-sharpening behavior during ballistic impact of the tungsten heavy alloy rod penetrators processed by hot-hydrostatic extrusion and hot torsion, *Mater. Sci. Eng. A* **527**, 4881–4886.
25. Wei Q., Zhang H. T., Schuster B. E., Ramesh K. T., Valiev R. Z., Kecskes L. J., Dowding R., Magness L. S. and Cho K. (2006), Microstructure and mechanical properties of super-strong nanocrystalline tungsten processed by high-pressure torsion, *Acta Mater.* **54**, 4079–4089.
26. Kecskes L. J., Magness L. S. and Schuster B. E. (2008), Impact, flow and deformation behavior of high density alloys during penetration into armor steel, in Proceedings of the Tungsten and Tungsten Alloys 2008 Conference, Vol. 6, eds. A. Bose, R. J. Dowding and J. A. Shields, pp. 39–52, Metal Powders Industries Federation, Princeton, NJ.
27. Schuster B. E., Ligda J. P., Pan Z. L. and Wei Q. (2011), Nanocrystalline refractory metals for extreme condition applications, *JOM* **63**, 27–31.
28. Zukas J. (1990), High Velocity Impact Dynamics, Wiley, New York, NY.
29. Mescall J. and Papirno R. (1974), Spallation in cylinder-plate impact, *Exp. Mech.* **14**, 257–266.
30. Raftenberg M. N. and Kennedy E. W. (1995), Steel plate perforation by tungsten rods of small L/D, in eds. M. Mayseless and S. R. Bodner, Vol. 1, pp. 315–321, Jerusalem. ISBN:0-961-8156-0-4.
31. Crouch I. (1992), Discing failures in both traditional and composite armour materials, in Proceedings of the 13th International Symposium on Ballistics, eds. E. B. Bjorck, Vol. 3, pp. 163–170, National Defence Research Establishment, Sundbyberg, Sweden.
32. DeMange J. J., Prakash V. and Pereira J. M. (2009), Effects of material microstructure on blunt projectile penetration of a nickel-based super alloy, *Int. J. Impact Eng.* **36**, 1027–1043.
33. Gailly, B. (1996), Etude du Comportement Dynamique et de la Rupture de Trois Aciers à Blindage, Ph.D. Thesis, Ecole Nationale Supérieure des Mines de Paris.
34. Molinari A. and Clifton R. (1987), Analytical characterization of shear localization in thermoviscoplastic materials, *J. Appl. Mech.* **54**, 806–812.
35. Yellup J. M. and Woodward R. L. (1980), Investigations into the prevention of adiabatic shear failure in high strength armour steels, *Res. Mech.* **1**, 41–57.
36. Bai Y. and Dodd B. (1992), Adiabatic Shear Localization, Pergamon Press, Oxford.
37. Dey S., Borvik T., Hopperstad O. S., Leinum J. R. and Langseth M. (2004), The effect of target strength on the perforation of steel plates using three different projectile nose shapes, *Int. J. Impact Eng.* **30**, 1005–1038.
38. Borvik T., Dey S., Hopperstad O. S. and Langseth M. (2009), On the main mechanisms in ballistic perforation of steel plates at sub-ordnance impact velocities, in Predictive Modeling of Dynamic Processes, ed. S. Hiermaier, Chapter 11, pp. 189–219, Springer, New York, NY.

39. Couque H., Asaro R. J., Duffy J. and Lee S. (1988), Correlations of microstructure with dynamic and quasi-static fracture in plain carbon steel, *Met. Trans. A* **19A**, 2179–2206.
40. Cho K., Chi Y. C. and Duffy J. (1990), Microscopic observations of adiabatic shear bands in three different steels, *Metall. Trans. A* **21A**, 1161–1175.
41. Dormeval R. (1988), Adiabatic shear phenomena, in Proceedings of the Impact Loading and Dynamic Behavior of Materials Conference, eds. C. Y. Chiem, H. D. Kunze and L. W. Meyer, pp. 43–56, DGM Verlag, Germany.
42. Rogers H. C. and Shastry C. V. (1981), Material factors in adiabatic shearing in steels, in Proceedings of the Shock Waves and High Strain Rate Phenomena Conference in Materials: Concepts and Applications, eds. M. A. Meyers and L. E. Murr, pp. 285–298, Plenum Press, New York, NY.
43. Leech P. W. (1985), Observation of adiabatic shear band formation in 7039 aluminum alloy, *Metall. Trans. A* **16**, 1900–1903.
44. Owolabi G. M., Odeshi A. G., Singh M. N. K. and Bassim M. N. (2007), Dynamic shear band formation in aluminum 6061-T6 and aluminum 6061-T6/Al_2O_3 composites, *Mater. Sci. Eng. A* **457**, 114–119.
45. Timothy S. P. and Hutchings I. M. (1985), The structure of adiabatic shear bands in a titanium alloy, *Acta Metall.* **33**, 667–676.
46. Gooch W. A., Burkins M. S., Ernst H. J. and Wolf T. (1995), Ballistic performance of titanium against laboratory penetrators with aspect-ratios of 10 or greater, in Proceedings of the 15th International Symposium on Ballistics, eds. M. Mayseless and S. R. Bodner, pp. 207–214, Jerusalem, ISBN:0-961-8156-0-4.
47. Bhav Singh B., Sukumar G., Bhattacharjee A., Siva Kumar K., Balakrishna T. and Gogia A. K. (2011), Effect of heat treatment on ballistic impact behavior of Ti–6Al–4V against 7.62 mm deformable projectile, *Mater. Des.* **36**, 640–649.
48. Shahan A. R. and Karimi Taheri A. (1993), Adiabatic shear bands in titanium and titanium alloys: a critical review, *Mater. Des.* **14**, 243–250.
49. Murr L. E. and Esquivel E. V. (2004), Observations of common microstructural issues associated with dynamic crystallization, *J. Mater. Sci.* **39**, 1153–1168.

7 Strain Localization in Energetic and Inert Granular Materials

Stephen M. Walley

SMF Fracture and Shock Physics Group, Cavendish Laboratory, University of Cambridge, Cambridge, UK

Nomenclature

U_s shock velocity
U_p particle velocity
C low-amplitude elastic wave velocity
S Hugoniot material-specific constant
P hydrostatic pressure
ρ density
σ_y yield stress
τ shear stress
γ shear strain
θ shear band inclination angle
ϕ friction angle
Ψ dilatancy angle

7.1 Granular Materials

Before considering how and why strain may localize when a granular material is deformed, it is necessary to discuss how stress propagates through grains. Powders, sand and soil differ from polycrystalline materials (such as cast metals) in one important respect: in a granular material, the grains are in contact at just a few points rather than over their whole surface. One effect of this is that in a vessel or silo, the load of a granular material is partly supported by the walls [1] through a phenomenon known as 'arching'. This can lead to problems in emptying a silo [2–4]. Another is that the load under a sandpile is not uniform but is concentrated at a few places [5–7]. These observations may be explained through the way loads are transmitted via so-called force chains. The first observation of this phenomenon I have been able to find was made by Seelig and Wulff in 1946 [8] (Figure 7.1) (they made no comment about it in the text of their paper), although Housel had predicted their existence (along with arching) analytically in 1936 [5].

Adiabatic Shear Localization. DOI: http://dx.doi.org/10.1016/B978-0-08-097781-2.00007-1

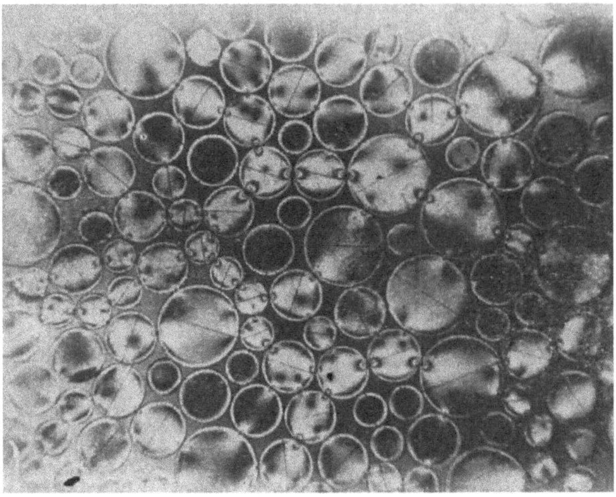

Figure 7.1 First known observation of force chains using photoelastic polymer discs.
Source: From Ref. [8].

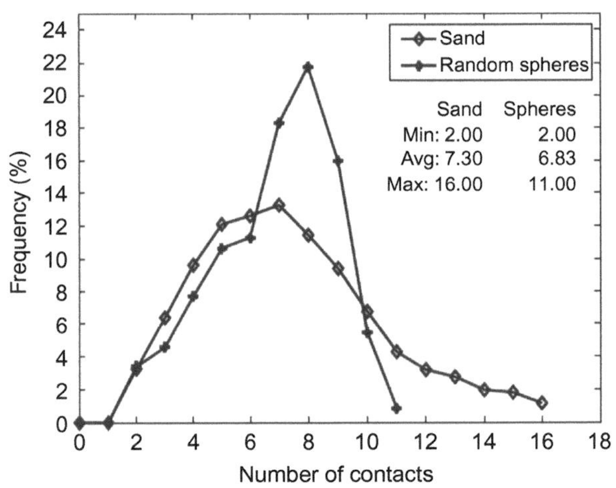

Figure 7.2 Distribution of the average number of particles in contact with a given particle in sand and random spheres systems.
Source: From Ref. [12].

The number of contact points is called the coordination number. This was long ago calculated for random packings of spheres [9,10] but was only relatively recently measured and simulated for more irregular particles such as sand [11,12]. A comparison of the frequency distribution of coordination numbers for sand and a random packing of spheres is given in Figure 7.2. Note that the maximum number of contacts can be significantly greater for sand grains than for spheres. An additional complication is that some studies have shown that two load-bearing networks can exist where a minority of the particles (the 'strong network') bear most of the load [13,14]. The strong network is not the same as the network of force chains [14].

Force chains are still an active area of research both quasi-statically [13–18] and dynamically [19].

Zhu et al. [19] used the 2D photoelastic technique pioneered (quasi-statically) by Seelig and Wulff [8] to investigate how stresses propagate dynamically through

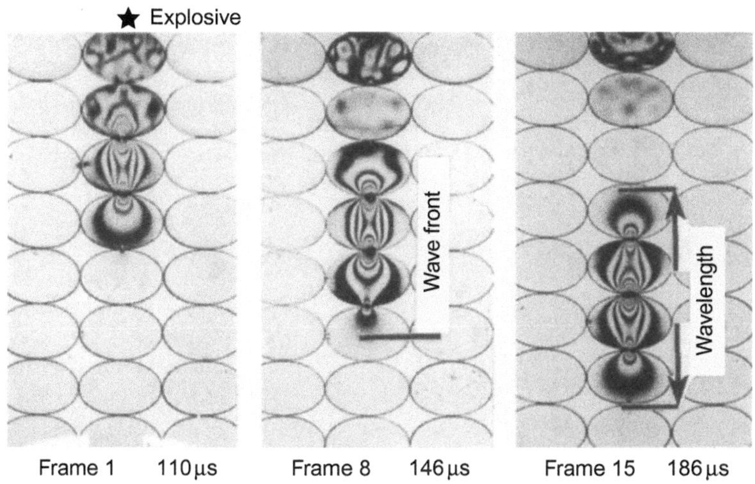

Figure 7.3 Selected frames from a high-speed photographic sequence of the propagation of a shock wave propagating down one chain of a bcc array of Homalite-100 elliptical discs whose long axes are perpendicular to the direction of propagation of the explosive stimulus. Long axes: 38 mm; short axes: 25 mm.
Source: From Ref. [19].

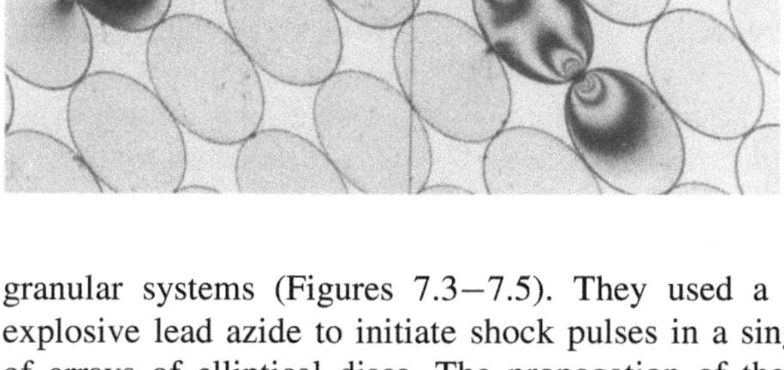

Figure 7.4 Selected frame from a high-speed photographic sequence of the propagation of a shock wave propagating down two chains of a bcc of Homalite-100 elliptical discs whose long axes are at 45° to the direction of propagation of the explosive stimulus. Long axes: 38 mm; short axes: 25 mm.
Source: From Ref. [19].

granular systems (Figures 7.3–7.5). They used a small charge of the primary explosive lead azide to initiate shock pulses in a single disc located at the surface of arrays of elliptical discs. The propagation of these pulses was recorded using high-speed photography. The pulse lengths were typically three to four times the size of the discs.

The way the pulse spread out through the array was found to depend strongly on the 'crystallography' of the array. If the array is square (bcc), the pulse (at least initially) spreads away from the loaded disc down a single chain if the long axis is either perpendicular (shown in Figure 7.3) or parallel to the shock. If the long axis is at 45° to the shock, then the pulse propagates down two chains (Figure 7.4).

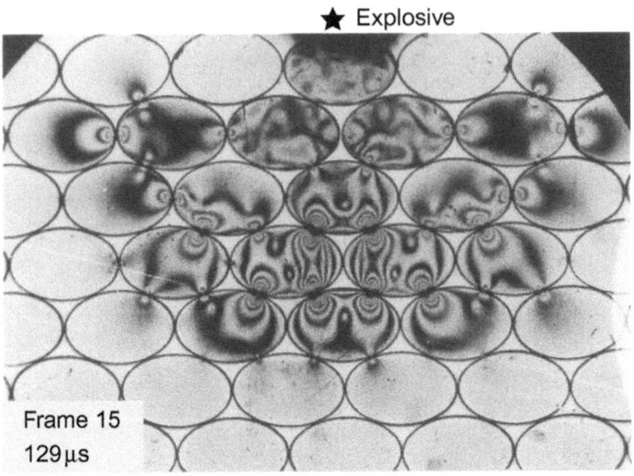

Figure 7.5 Selected frame from a high-speed photographic sequence of the propagation of a shock wave propagating through an hcp array of Homalite-100 elliptical discs whose long axes are perpendicular to the direction of propagation of the explosive stimulus. Long axes: 38 mm; short axes: 25 mm.
Source: From Ref. [19].

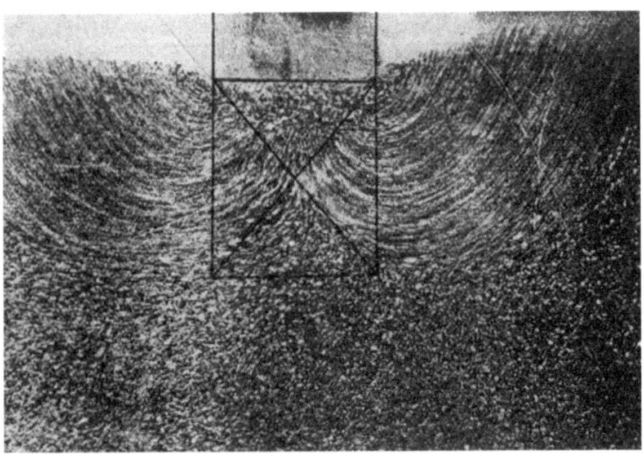

Figure 7.6 Slip lines and shear localization observed in a granular material (ballast) produced by loading with a punch.
Source: From Ref. [5] (originally published without lines showing slip planes in Ref. [23]).

If, however, the discs are arranged in a triangular (hcp) array, the pulse spreads out through the array (Figure 7.5). Roessig et al. [20] used this technique to study how stresses might propagate dynamically through a polymer-bonded explosive (PBX) with a bimodal particle distribution. One of the main differences to the real problem that they identified is that real explosive crystals not only are irregular in shape but can also have quite sharp corners. These can in turn strongly indent their neighbours causing fracture of grains [21], although rubbing of explosive crystals against each other is a more likely hot-spot mechanism [22].

Despite the major differences in the way loads are transmitted through granular and polycrystalline materials, they exhibit some of the same behaviour when deformed, e.g. slip lines (Figure 7.6) and strain localization (Figure 7.7). A substantial literature exists on the mechanics of granular materials, which has been summarized and reviewed in a number of authored and edited books, e.g. Refs [24–28]. The book by Nesterenko [27] is particularly relevant to the dynamic loading of energetic powders and that by Tejchman [28] gives an overview of the theory and simulation of shear localization in inert granular materials. Previous review papers on shear localization in granular materials have been written by Bardet [29]

Figure 7.7 Shear localization in polymer-bonded sand produced by loading with a 20 mm diameter cylindrical punch.
Source: Courtesy of Dr D.M. Williamson, Cavendish Laboratory, Cambridge.

(who concentrated on soil mechanics and the description of the phenomenon by the Mohr–Coulomb theory), Lade [30] (who concentrated on laboratory experimental methods) and Schall and van Hecke [31] (who, while including some observations on sand, concentrated on softer, colloidal suspensions).

The emphasis of the present review will be on experimental results and observations in energetic materials and sand.

7.2 Localization of Deformation in Explosives

It has long been known that explosives can be ignited by raising their temperature. However, during the 1930s, some researchers found that during dynamic loading, solid explosives can ignite even when the mechanical work dissipated within them is much less than that required to raise the bulk temperature to that required for self-sustaining reaction [32,33]. A major increase in the understanding of the microscopic processes leading up to detonation was produced by an investigation in Australia during the Second World War into an industrial accident involving the liquid explosive nitroglycerine (NG) [34,35].

Subsequent work on the ignition of explosive solids by friction led to the concept of 'hot spots' as the necessary condition for explosion to occur [36]. The critical temperature of these hot spots was determined by some classic work involving the use of various inert grits having a range of melting points [37]. By this method, it was found that pentaerythritol tetranitrate (PETN) does not ignite if it is mixed with powders having a melting point less than around 400°C. Theoretical work on heat flow showed that critical hot spots (i.e. those that produce ignition) had to be of micron dimensions or greater, last for at least a microsecond and have a minimum temperature of around 500°C [38].

Bowden and Yoffe [39] later summarized their understanding of hot-spot ignition mechanisms in powdered and granular explosives due to mechanical stimuli as being due to three causes: (i) adiabatic compression of gas spaces; (ii) frictional rubbing of explosive crystals with each other, with a containing surface, or with inert grits; or (iii) viscous heating due to rapid flow. The last mechanism is now no longer believed to be a hot-spot mechanism as such but by raising the background temperature it increases the probability of propagation of hot spots [40,41]. Recent surveys of microscopic mechanisms that lead to detonation have been published by Leiber [42] and Walley et al. [22].

One very important addition to the mechanisms originally identified by Bowden and co-workers was proposed by Afanasev et al. [43], namely shear banding, although in some sense it could be considered a combination of mechanisms (ii) and (iii). Shear banding occurs when the mechanical resistance of a material or structure decreases with increasing deformation or strain [44]. This softening occurs because either (i) damage accumulates (this is the only softening mechanism that can operate under low-rate loading) or (ii) the temperature increases due to mechanical work dissipated as heat. The latter mechanism often goes under the name 'adiabatic shear banding' (ASB).

A study was performed by Winter and Field [45] on the impact of small (80 μm) aluminium and glass spheres on crystals of primary explosives (silver and lead azide). They found that for any given particle size, there was a critical impact speed to cause initiation. Although they did not observe ASBs in the crystals (primary explosives are so sensitive that if an initiation mechanism operates, the crystals explode destroying the evidence), they were the first to propose ASBs as a possible hot-spot mechanism. The motivation for their study was that when such crystals explode, millimetre-sized fragments are sent out at high velocity, which can initiate nearby energetic crystals [46].

Later studies by Field and co-workers [47–51] on ASBs in explosives concentrated on PBXs and powders of secondary explosives such as cyclotetramethylene tetranitramine (HMX) and PETN [49,50] (Figures 7.8 and 7.9).

Two techniques that have proved very useful in investigating hot-spot initiation mechanisms in energetic materials are (i) high-speed photography [47,51–57] and (ii) heat-sensitive film [49,51,58]. An example of the use of high-speed photography to observe the development of ASBs and associated ignition in powdered PETN is given in Figure 7.10. Not only does deflagration begin in the shear bands,

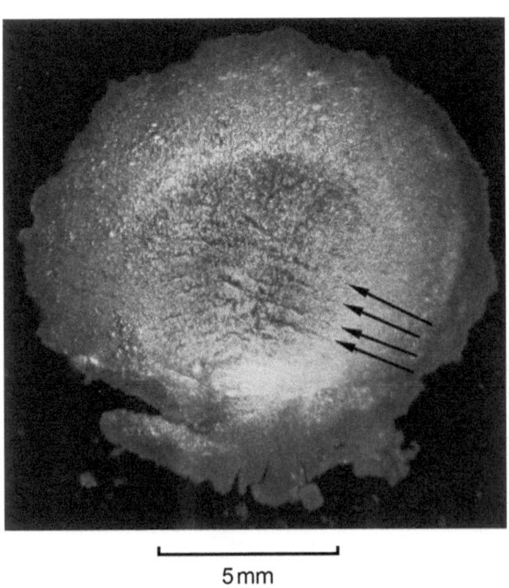

Figure 7.8 ASBs (arrowed) formed in a PBX due to drop-weight impact.
Source: From Ref. [48].

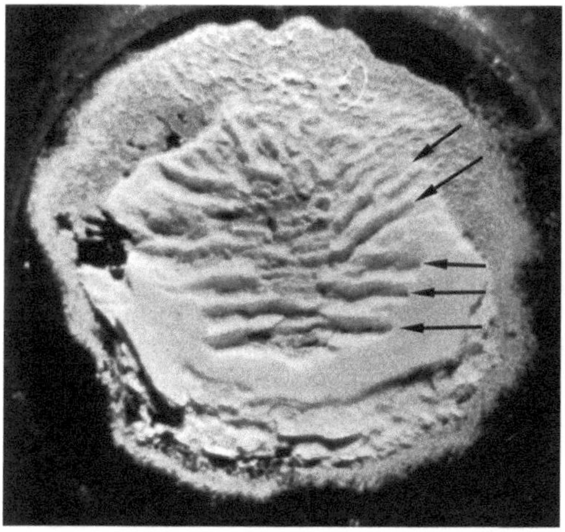

Figure 7.9 Shear bands (arrowed) in powdered HMX deformed by drop-weight impact between 10 mm diameter hardened steel rollers.
Source: From Ref. [49].

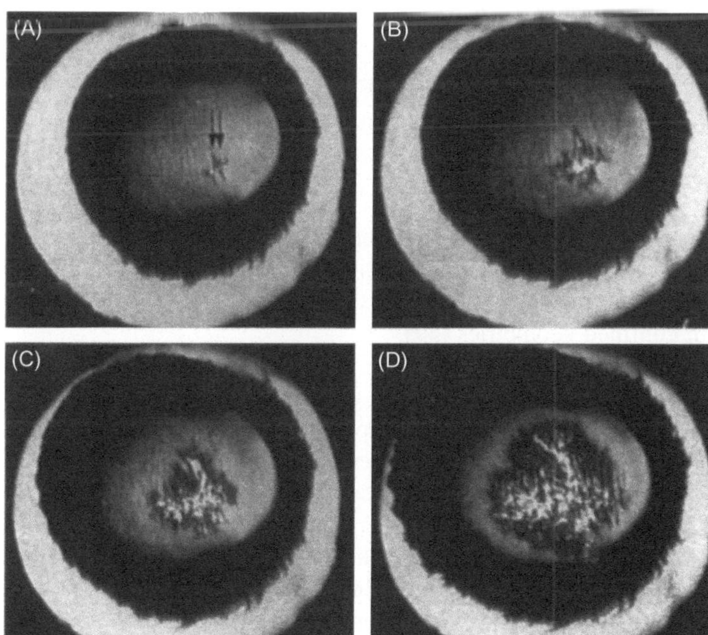

Figure 7.10 High-speed photographic sequence of the formation of ASBs (arrowed) in powdered PETN deforming between toughened float-glass anvils in a drop-weight machine. Interframe time 7 μs; field of view 20 mm.
Source: From Ref. [51].

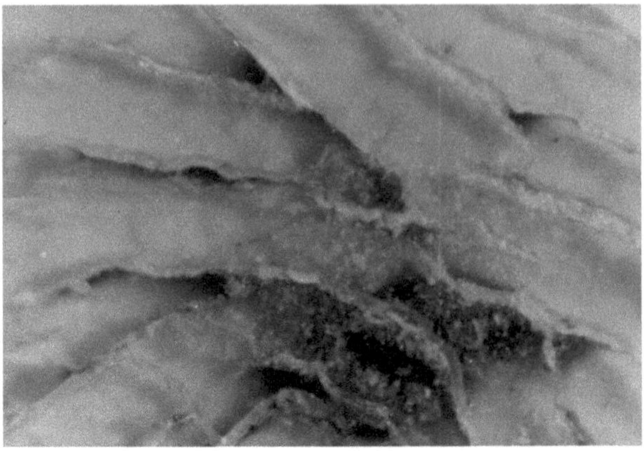

Figure 7.11 Heat-sensitive film showing hot spot formed at the intersection of shear bands formed in powdered HMX by drop-weight impact. Width of figure 6 mm.
Source: From Ref. [51].

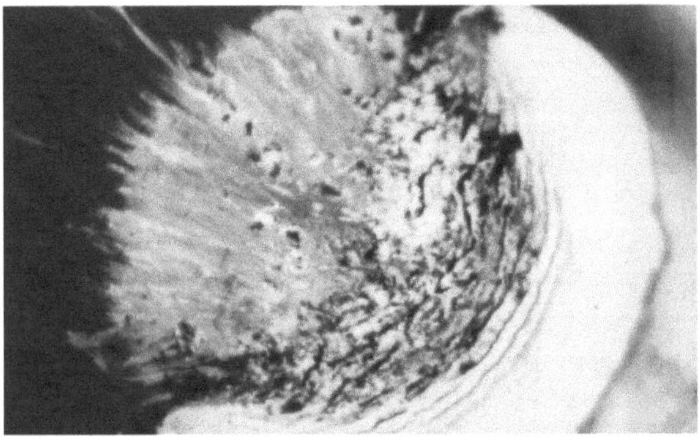

Figure 7.12 Heat-sensitive film record of an impact on powdered PETN. Width of figure 14 mm.
Source: From Ref. [51].

but it also spreads preferentially along them as they form planes of weakness in the compressed powder. A recent study has shown that ASB formation is a likely cause of the seemingly anomalously high sensitivity of PBXs to low-velocity impact, especially in confined geometries [59].

Examples of the use of heat-sensitive film in giving spatial information about shear bands and associated hot spots are given in Figures 7.11 and 7.12. The film used was calibrated for both time and temperature by Swallowe et al. [58] using various polymers. If combined with high-speed photography (to give the timescale of the process that caused darkening), heat-sensitive film can be used to give an estimate of the temperature reached at various places in the specimen during the impact. In Figure 7.11, the HMX did not ignite, whereas in Figure 7.12, the PETN did ignite, scouring away the surface of the film. However, the flame can clearly be seen to have begun in the vicinity of the shear bands to the right of the photograph.

Obvious questions that arise from Figures 7.8–7.12 are the following:

- What is the thickness of the shear bands?
- What determines the thickness of the shear bands?
- What governs the spacing of the shear bands (0.24 mm in Figure 7.8 and 0.4 mm in Figure 7.12)?

To answer these questions, we need to turn to the extensive literature on the mechanics of soils and other granular materials as well as that on the shocking of reactive metallic powders.

7.3 Shear Localization Due to Shock Loading of Reactive Powders

Shear has been known to enhance reactions at high static pressures since at least the time of Bridgman, sometimes markedly and destructively [60]. Reviews of research on this topic may be found in Refs [55,61].

Shock loading also produces high pressures combined with shear. The main difference between the outcome of shocking reactive as opposed to inert powders is that chemical reaction can be initiated by the passage of the shock. It has been found that the shock can act in two distinct ways [62]: (i) if reaction only starts sometime after the shock has passed through, the reaction is called 'shock-assisted'; (ii) if, on the other hand, the reaction starts at the location of the shock, the reaction is called 'shock-induced'. In the second case, the reaction may, of course, still go on to completion after the shock has passed. Insight into the processes that occur may be found from the large amounts of literature on self-propagating high-temperature synthesis, a technique that was both initiated and taken forward in Russia [63–65], though there is now interest in the method in other countries, particularly the United States [66].

Often in a reactive powder mix, one of the components is a reactive metal. Hence a barrier to chemical reaction will exist in the form of an oxide layer. This will need to be continuously disrupted for reaction to occur quickly and efficiently. Even though compression of powders does disrupt brittle surface layers (Figure 7.13),

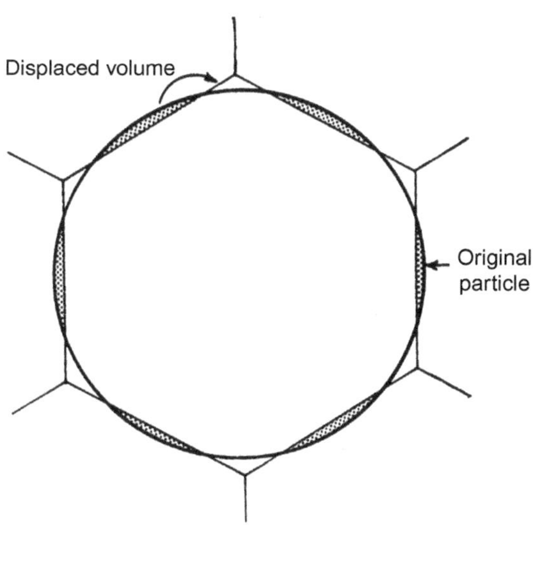

Figure 7.13 Schematic diagram of the idealized deformation of a spherical particle during shock loading.
Source: From Ref. [67].

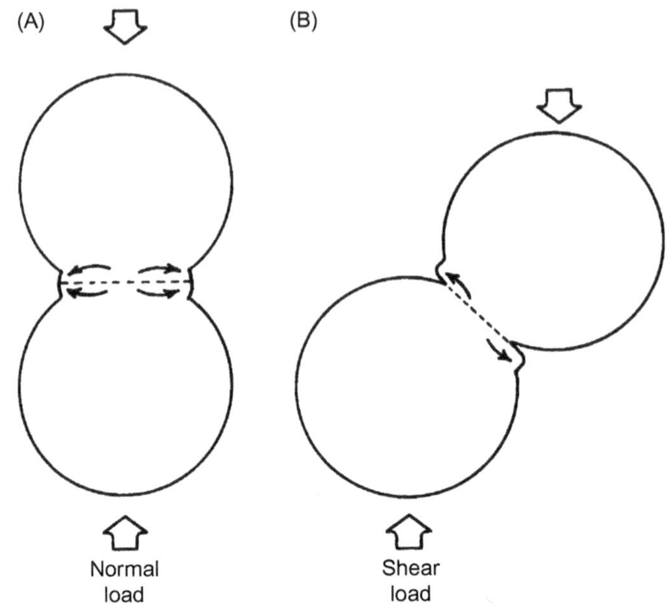

Figure 7.14 Schematic diagram of two main modes of more realistic particle interactions in shock loading. Particle interaction where the contact is (A) normal and (B) at 45° to the direction of passage of the compaction front.
Source: From Ref. [67].

shear and/or rotation of particles is desirable (Figure 7.14) to bring fresh surfaces from the two components into contact. If the particles are large enough [68–70], fresh reactive surfaces can also be formed by fragmentation [71], but these new surfaces will still need to be brought rapidly into contact with material from the other component(s) of the mixtures. So shear and rotation are still necessary.

Fracture introduces another level of complexity into how a granular material responds to shock loading (Figure 7.15). Effectively, it changes the state of the material and produces an effect similar to a phase change; i.e. during loading, the material does not follow a straight (or Rayleigh) line from its initial (p_0, V_0) to its (p_f, V_f) final state (described here by pressure p and volume V). The shock response of a granular/powder mixture also strongly depends on the shape and size of the particles (Figure 7.16), which are often far from spherical (Figure 7.17). Also most powder mixtures are made from at least two different sizes of particles (bimodal) to maximize the packing density [74–77].

The passage of a shock induces different particle velocities in the grains of the two materials [78]. This causes collisions between the grains generating compressive stresses at their points of contact. Tensile fracture can then occur normal to these stresses in a process akin to the Brazilian test [79]. The

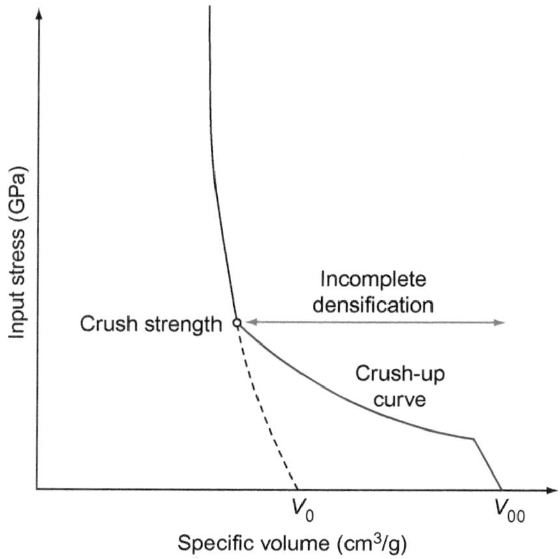

Figure 7.15 Schematic diagram illustrating the crush-up curve for a powder and how it connects to the Hugoniot for the fully dense material.
Source: From Ref. [66].

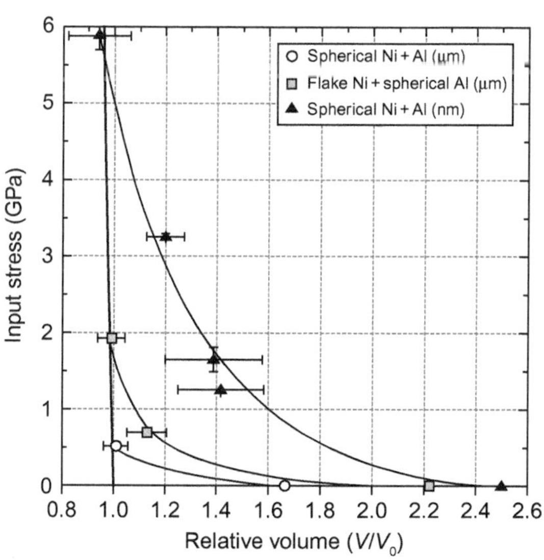

Figure 7.16 Experimental data on the crush-up behaviour of three different nickel and aluminium powder mixes showing the strong effect of both particle size and morphology on the shock stress required to reach full density.
Source: From Ref. [72].

standard equations from shock wave theory for the Hugoniot and the shock pressure are

$$U_s = C + SU_p \quad (7.1)$$

$$P = \rho U_s U_p \quad (7.2)$$

where U_s is the shock velocity, U_p is the particle velocity, C is the low-amplitude elastic wave velocity, S is material-specific constant, P is the hydrostatic pressure and ρ is the density.

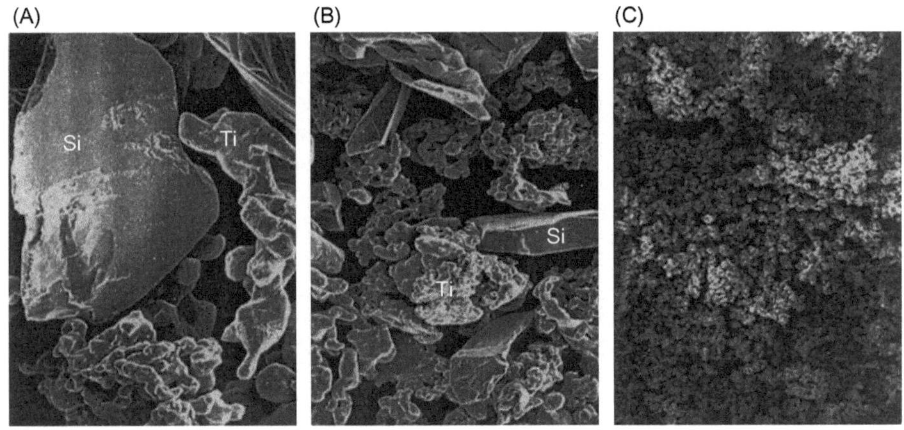

Figure 7.17 SEM images of real titanium–silicon powder mixes: (A) coarse; (B) medium; (C) fine. The silicon particles are blocky single crystals, whereas the titanium particles are rounded polycrystalline aggregates.
Source: From Ref. [73].

Solving Eqs (7.1) and (7.2) for U_p gives

$$U_p = \left[\left(\frac{C}{2S}\right)^2 + \frac{P}{S\rho}\right]^{0.5} - \frac{C}{2S} \tag{7.3}$$

Batsanov [78] used this equation to calculate the differences in particle velocities between silicon and various refractory metals to make shock synthesis possible. The only additional data needed was the size of the particles (44 μm) and the time needed for reaction (0.1 μs). This gave a threshold particle velocity difference of 880 m/s.

To bring fresh reactive surfaces continuously into contact during shock loading, many researchers have developed more complex loading geometries than the standard plate impact or explosive lens arrangements to increase the shear component [80–83] (see, for example, Figures 7.18–7.20), particularly ASBs (Figures 7.19, 7.21–7.23).

As well as producing localized frictional heating by causing particles to rub against each other, ASBs can also induce void collapse [87], which is well known as a hot-spot mechanism for the initiation of energetic materials [40]. The effect of void collapse on the overall densification of powdered materials has been analysed in a number of different ways. The first is the empirical P-α pore collapse model due to Carroll and Holt [88,89]. Being an empirical model, the parameters that describe it have to be fitted to the data. The second is a more analytic void collapse

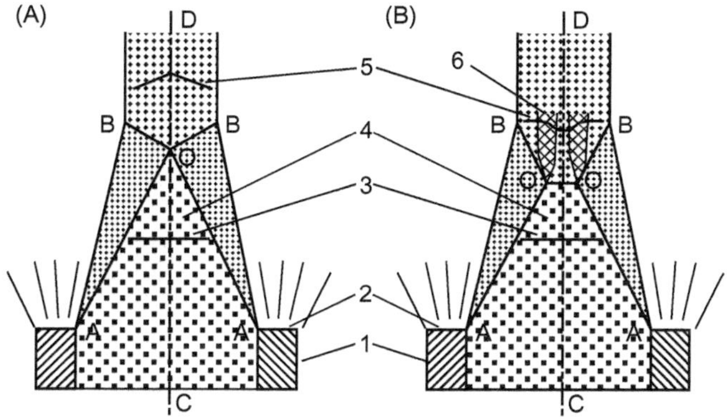

Figure 7.18 Schematic diagram illustrating the flow (and hence mixing) of powders under oblique shock loading (A) without and (B) with a Mach stem between the two shock reflections. (Labelling: 1. Explosive charge; 2. Detonation front; 3. Witness foil in unshocked region; 4. Unshocked powder; 5. Witness foil in shocked region; 6. Boundary zone. AO: incident shock front; OB: reflected shock front; OO: Mach stem; CD: shock collision axis.)
Source: From Ref. [84].

model due to Fischmeister and Arzt [90], which takes into account the known strength of the material of which the powder is made (see Eq. (7.4)),

$$P = 2.97\rho^2 \frac{(\rho - \rho_0)}{(1 - \rho_0)} \sigma_y \tag{7.4}$$

where P is the applied pressure, ρ is the relative density (volume fraction of metal phase), ρ_0 is the initial relative (fill) density and σ_y is the material yield stress.

In modelling the shock properties of reactive powders, the effect of reaction must be taken into account both on the Rayleigh line and on the Hugoniot (Figure 7.24), remembering from the earlier discussion of shocking non-reacting powders that these lines are not as well defined for powders as for fully dense solids (see Figures 7.15 and 7.16).

Both the ambient and shock temperature are important variables, on which there is little data. For example, Figure 7.25 shows that the yield of the various nickel aluminides from shocking nickel–aluminium powder mixes depends on both the original (ambient) temperature and the shock pressure. Shock temperature measurement is more difficult for a powder than for a solid due to the heterogeneous nature of the specimen [93]. Also it is not at all clear what the emissivity of any material is (let alone a powder) during shock loading, making optical methods of temperature measurement problematic [94–97]. Indirect evidence (such as partial melting of the constituents) and calculation [98] implies local temperature rises of thousands of degrees, but the temperature field will be very heterogeneous (like the

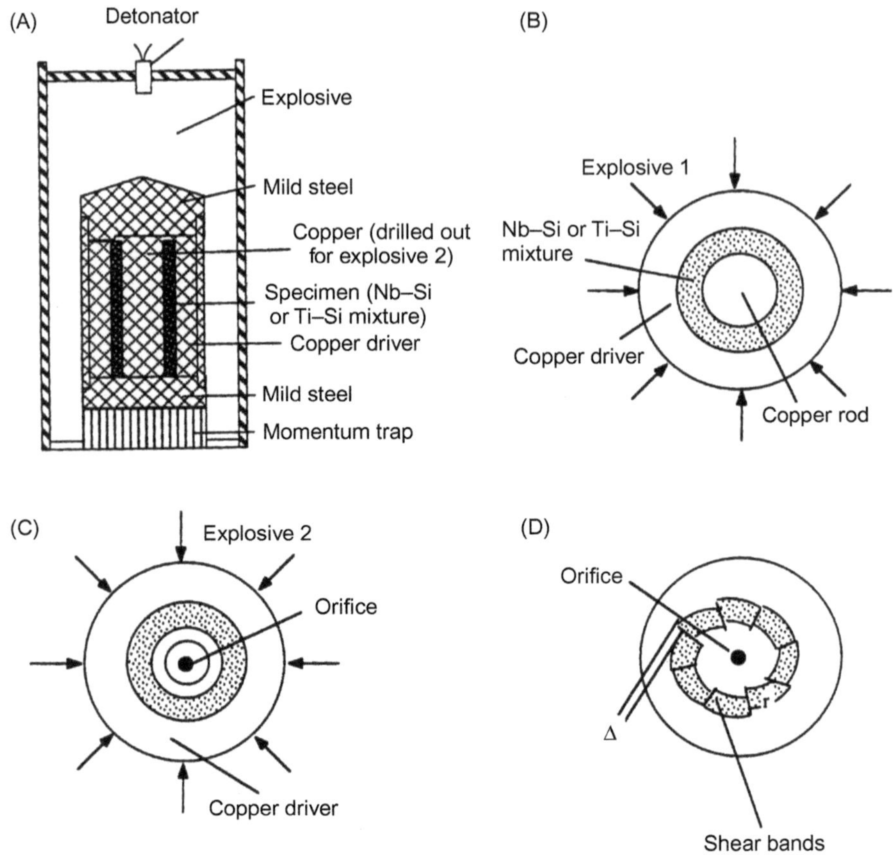

Figure 7.19 (A) Schematic diagram of thick-walled cylinder (TWC) method for producing shear bands in a powder tube by external shock loading; (B–D) sequence of events leading to the formation of shear bands.
Source: From Ref. [82].

powder mix itself) in both space (Figure 7.25) and time [99]. The heterogeneity of the conditions within a powder specimen leads to large variations in reaction rates and hence to a very heterogeneous product [100] (Figure 7.26).

The temperature reached also depends on the particle size [99]. This is because large particles have to deform more to fill the voids between them than small particles. This can be seen qualitatively in the micrographs presented in Figure 7.27.

The reaction kinetics depend on the processes occurring at the interfaces between the reacting powders (Figure 7.28). For example, if melting occurs, diffusion of reactants is speeded up considerably [102]. Nabatov et al. [103] state that melting of at least one component (silicon in their studies) is essential if reactions are to proceed on the timescales typical of shock loading.

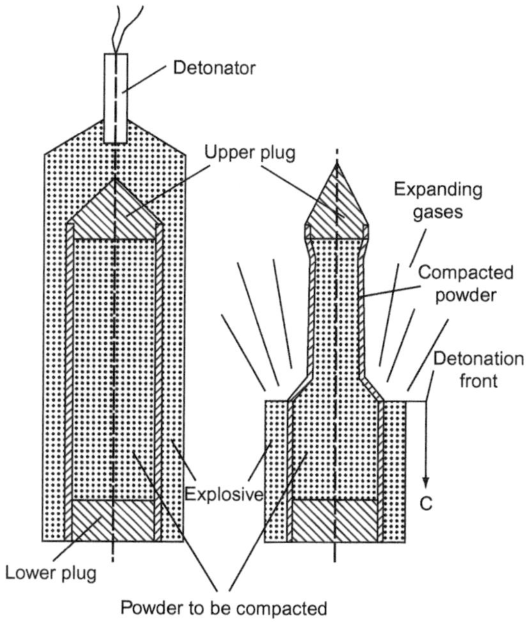

Figure 7.20 Schematic diagram of the shock collapse of a tube containing a powder in order to generate shear.
Source: From Ref. [85].

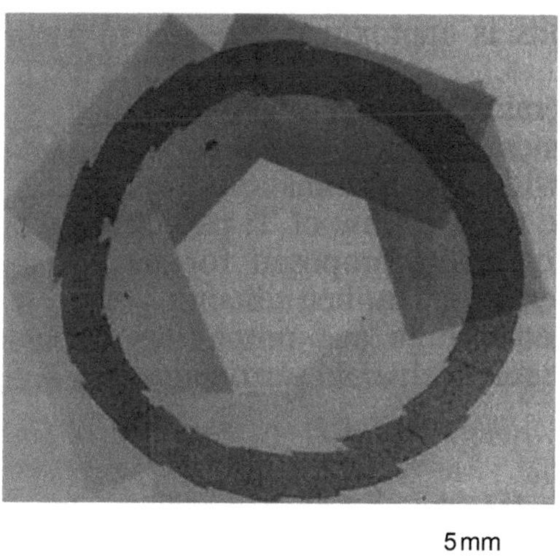

Figure 7.21 Global view of ASBs in a shocked TWC specimen (shown schematically in Figure 7.19) of a mixture of titanium and silicon powders.
Source: From Ref. [86].

The final outcome of a shock experiment also depends on the ductility of the material. For example, a numerical modelling study showed that if the same arrangement (Figure 7.29) of particles of a metallic alloy (Inconel) and a ceramic (silicon carbide) is subjected to a range of different shock strengths, the final outcome for the two materials is very different (compare Figures 7.30 and 7.31). For sufficiently strong shocks in ductile materials, some particles will form jets (Figures 7.30 and 7.32).

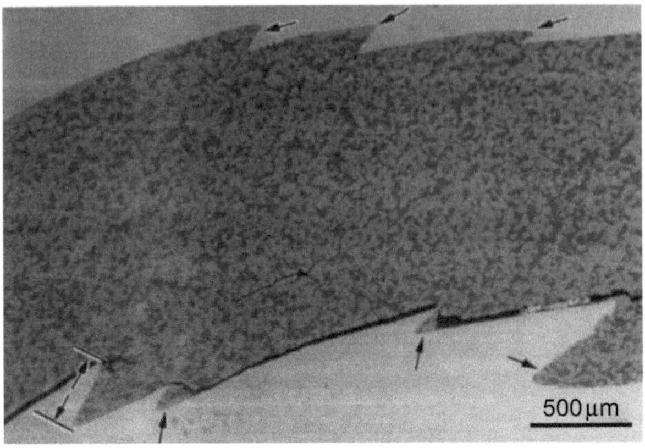

Figure 7.22 Overall view of shear bands in a mixture of titanium and silicon powders formed using the technique shown in Figure 7.19.
Source: From Ref. [82].

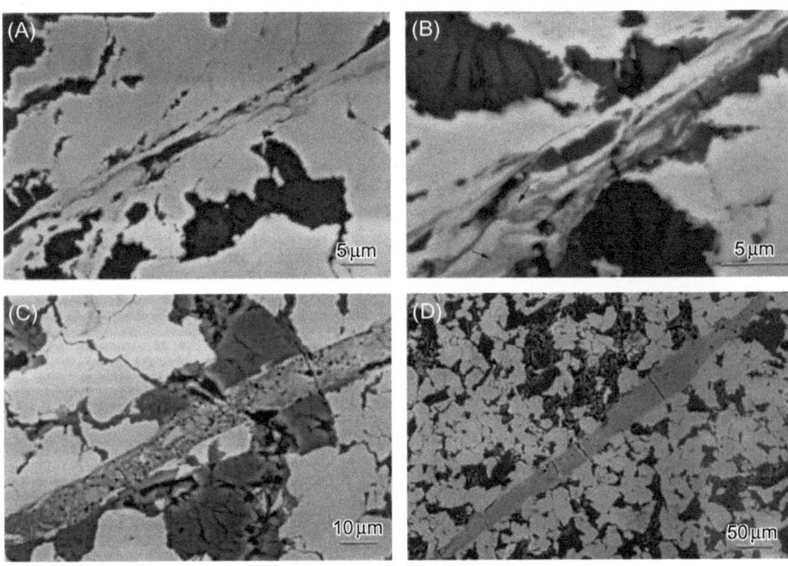

Figure 7.23 Stages in shear localization and chemical reaction in a titanium–silicon powder mixture: (A) mechanical deformation and fracture; (B) initiation of reaction (where arrows pointing); (C) partial reaction; (D) full reaction.
Source: From Ref. [82].

One of the most thorough studies of shock-loading powder mixtures was conducted by Meyers et al. [106]. They found that simple Arrhenius kinetics substantially underestimated the reaction rates even allowing for shock energies. To explain this, they postulated that liquid globules form at the interface between the two powders they were studying (niobium and silicon) (Figure 7.33) leading to a substantial increase in reaction rate. The main feature of the mechanism they

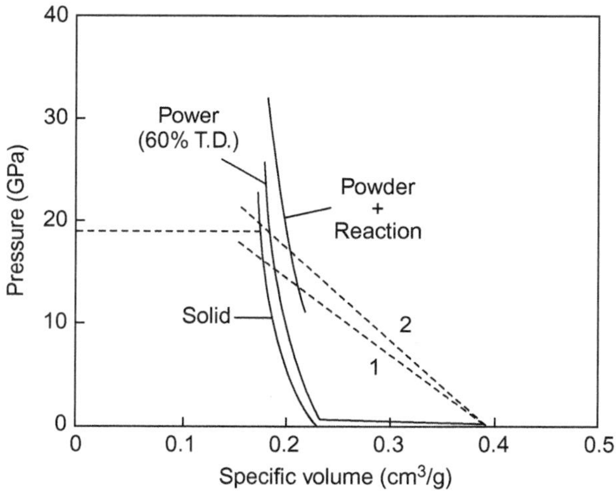

Figure 7.24 Schematic diagram showing the effect of reaction on the Rayleigh line and Hugoniots for a reactive powder mix of nickel and silicon. *Source*: From Ref. [66]; adapted from Ref. [91].

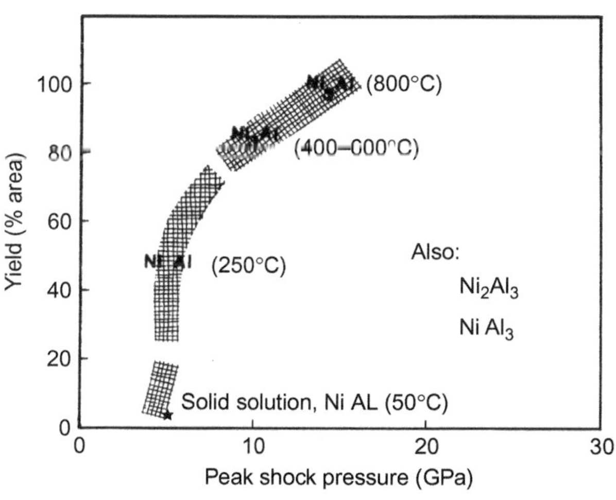

Figure 7.25 Yield of various nickel aluminides as a function of pressure for various temperatures. *Source*: From Ref. [92].

proposed is that first shock heating melts the silicon (Figure 7.33A). This then leads to rapid reaction with the refractory metal (niobium) producing niobium silicide in its molten state (Figure 7.33B and C). Surface tension then pulls this into a globule (Figure 7.33D). As the contact area with the niobium is now reduced, the rate of reaction falls sufficiently that heat loss to the niobium becomes much greater than the heat released by reaction, and hence the globule begins to freeze from the interface (Figure 7.33E). Thus fresh niobium surfaces are exposed to the silicon at this point through the mechanical forces exerted on the globule by neighbouring solid globules (shown schematically in Figure 7.33F–H). Another possible mechanism they proposed was turbulent flow of the silicon leading to detachment of the globules. It is not at all clear how to analyse these processes quantitatively in terms of kinetic equations recognizable to a chemist, and the authors did not do so. They restricted themselves to performing heat transfer and nucleation calculations.

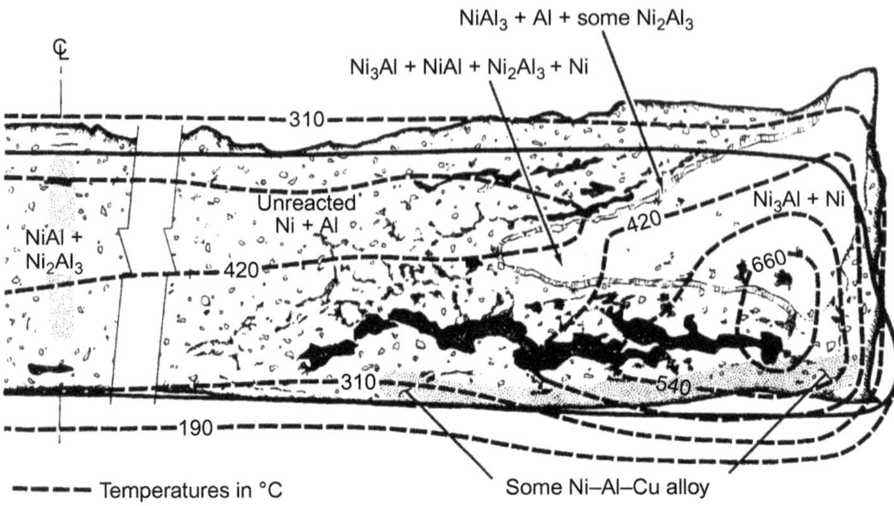

Figure 7.26 Calculated temperature distribution and comparison with the regions of the various reaction products for a nickel–aluminium powder specimen shock loaded in a recovery experiment within a copper vessel.
Source: From Ref. [101].

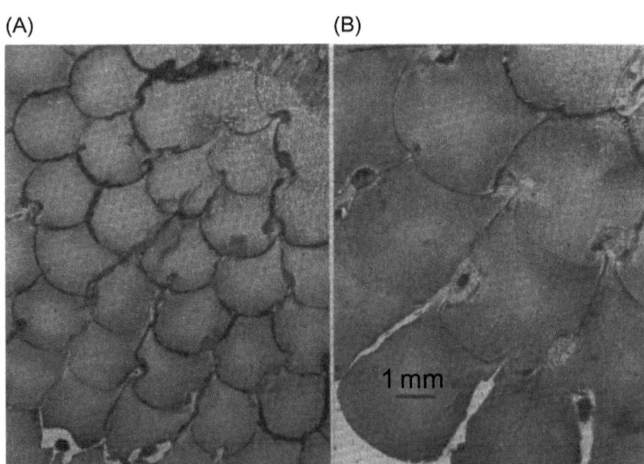

Figure 7.27 Optical micrographs of the effect of shock deformation of arrays of different-sized tungsten rods (same scale): (A) 0.8 mm diameter; (B) 3.2 mm diameter.
Source: From Ref. [99].

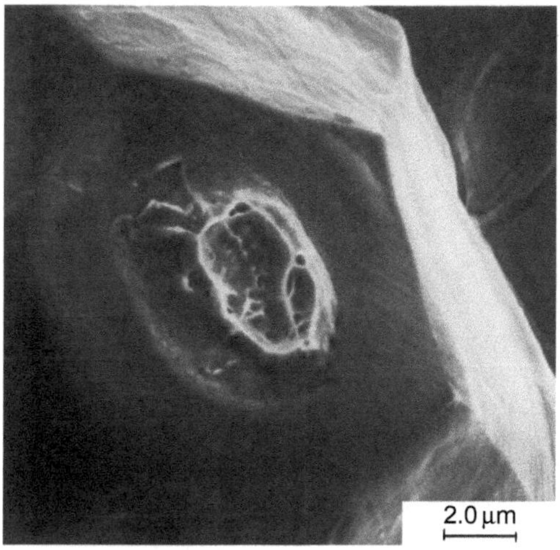

Figure 7.28 SEM micrograph showing disruption of the surface layer due to interparticle bonding.
Source: From Ref. [67].

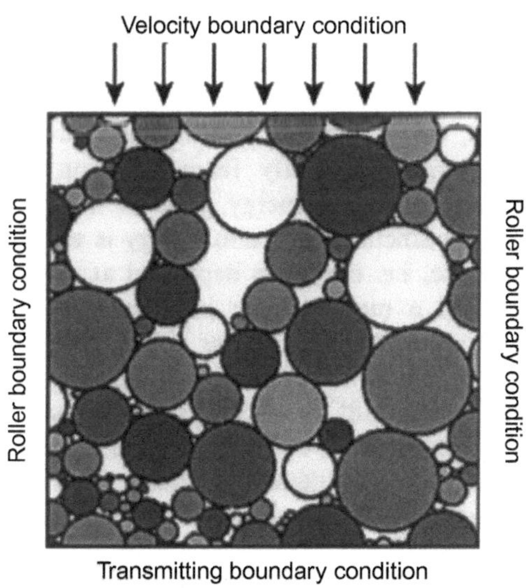

Figure 7.29 Schematic diagram showing the original arrangement of particles and the boundary conditions for modelling the shock compaction of various powder mixtures.
Source: From Ref. [104].

7.4 Modern Methods of Study

Up until recently, studies of the initiation and growth of ASBs have been performed using optically transparent ductile materials, i.e. polymers [45,53,58,107,108]. Granular materials are optically opaque, but many that are of interest are X-ray transparent [109,110]. This opens up the possibility of using tomography to study shear banding within specimens composed of particulates [111–119] (Figure 7.34).

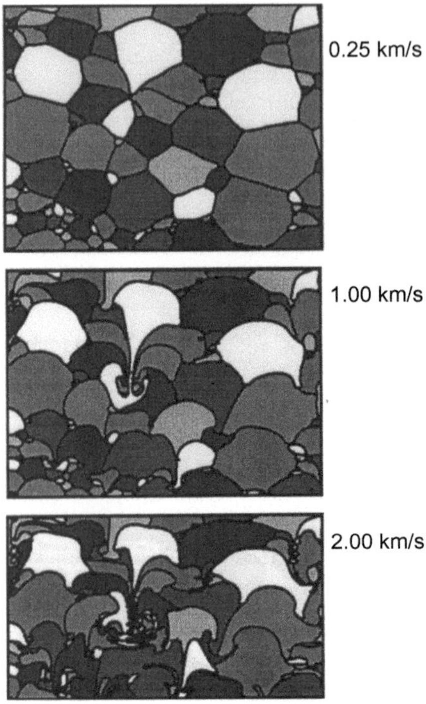

Figure 7.30 Modelling results for final deformation of Inconel particles whose original arrangement is shown in Figure 7.29 for three different strength shocks (indicated on the right by the particle velocity).
Source: From Ref. [104].

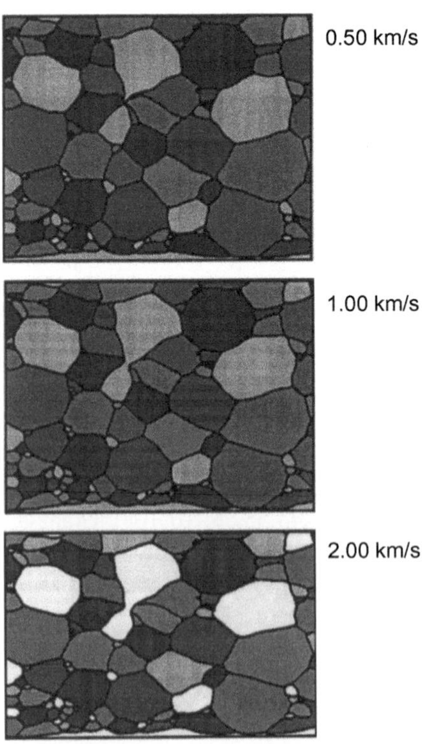

Figure 7.31 Modelling results for final deformation of SiC particles whose original arrangement is shown in Figure 7.29 for three different strength shocks (indicated on the right by the particle velocity).
Source: From Ref. [104].

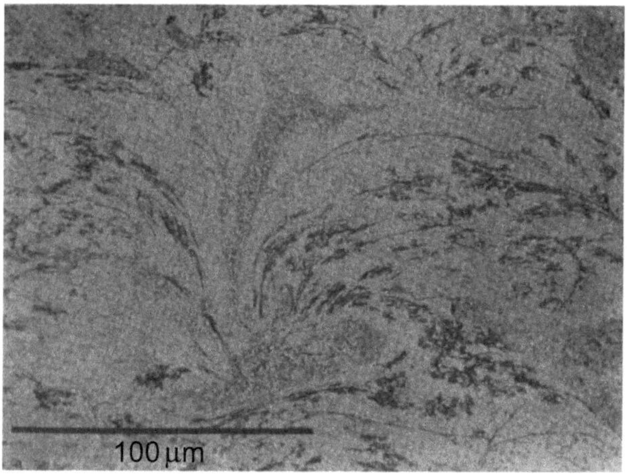

Figure 7.32 Optical micrograph of dynamically compacted aluminium powder shocked at 2 GPa. The shock propagation direction was from top to bottom. Note the similarity of the deformation of the grain to one of the grains in the simulation for the 1 km/s particle velocity in Inconel powder in Figure 7.30.
Source: From Ref. [105].

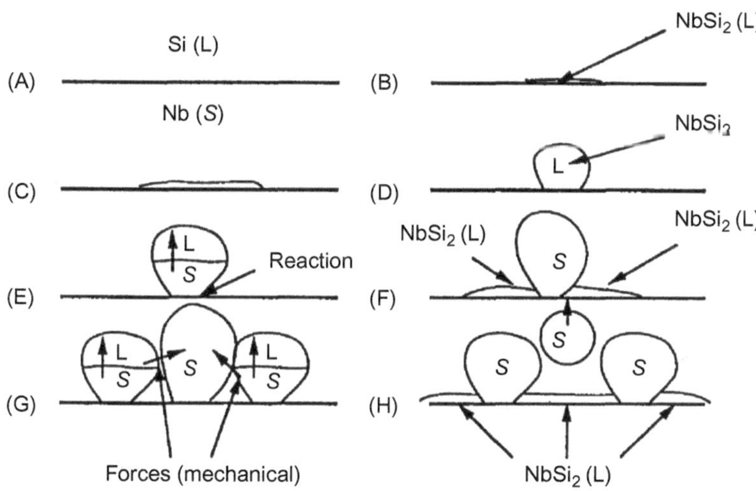

Figure 7.33 Schematic diagram of the sequence of events that are believed to take place at the interface between niobium and silicon particles under shock loading.
Source: From Ref. [106].

Tomography has proved particularly useful in quantifying the effect that a shear band in sand has on coordination number (Figure 7.35) and grain orientation [119].

Other techniques that are proving or may prove useful in the future include (i) magnetic resonance imaging (MRI) [120–122] and (ii) digital image correlation (DIC), both optical [123–132] and X-ray based [3,118,133–135]. Rechenmacher and Finno [129] used DIC to measure the development and the thickness of shear bands in sand (Figure 7.36). They found that the shear-band thickness depended on the type of sand used but lay between 10 and 20 mean grain diameters (D_{50}). This is

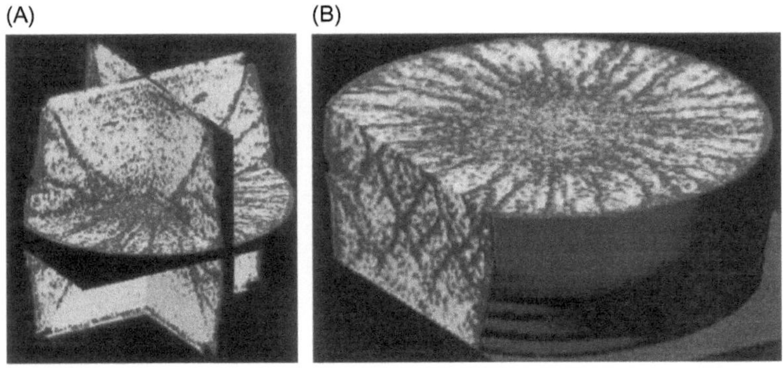

Figure 7.34 (A) Lengthwise and (B) cross-section X-ray tomographs of a confined sand cylinder after 25% axial strain.
Source: From Ref. [113].

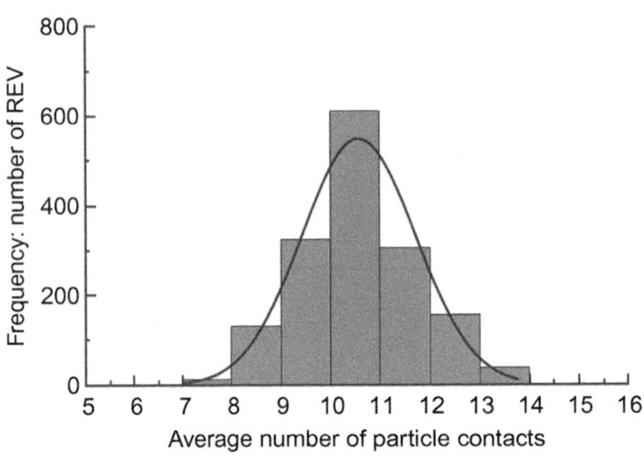

Figure 7.35 Frequency histogram of the average number of particle contacts in a shear band in sand.
Source: From Ref. [119].

larger than previous measurements by Roscoe [136] and Mühlhaus and Vardoulakis [137] who reported thicknesses of between 8 and 10 mean grain diameters.

Rechenmacher and Finno were also able to follow the dilation (volume increase) within and nearby the shear bands as a function of global strain (Figure 7.37). Yoshida and Tatsuoka summarized the data of three research groups (Figure 7.38) showing that the trend is for the shear-band thickness to decrease with grain size.

Bésuelle and Rudnicki [140] analysed the shear-band thickness problem for granular materials from which they derive the (deceptively) simple formula that the thickness t is given by:

$$t = 2\pi\sqrt{\frac{-c}{H}} \tag{7.5}$$

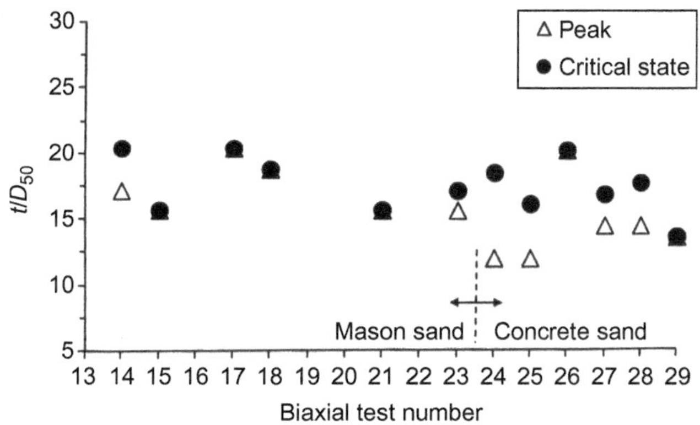

Figure 7.36 Normalized shear-band thickness at peak stress for two different sands. *Source*: From Ref. [129].

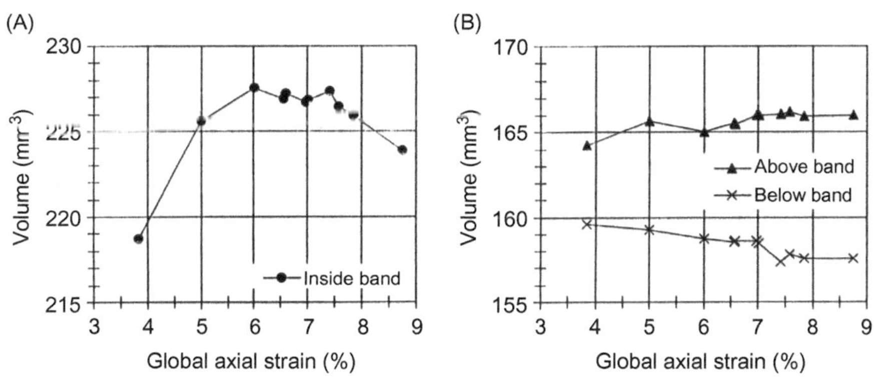

Figure 7.37 (A) Volume change as a function of axial strain inside a shear band in sand. (B) Volume change as a function of axial strain in the vicinity of a shear band in sand. *Source*: From Ref. [129].

where c (a negative number) and H are plastic moduli defined by the following equation:

$$d\tau = H\, d\gamma^{p} - c\frac{\partial^2(d\gamma^{p})}{\partial y^2} \tag{7.6}$$

where τ is the shear stress and γ is the shear strain (superscript p denotes 'plastic'). However, they did not compare this prediction with observation. Recent modelling has shown that the shear-band thickness increases with confining stress [141,142].

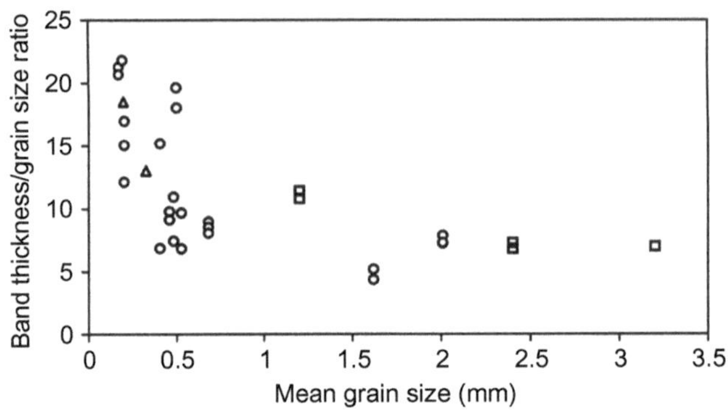

Figure 7.38 Summary of data on effect of mean sand grain size on shear-band thickness. Plot compiled by Yoshida and Tatsuoka [138]. (Key: circle, data obtained by Yoshida and Tatsuoka [138]; square, data obtained by Mokni [139]; triangle, data obtained by Mühlhaus and Vardoulakis [137].)

It is now understood that for fully dense materials, no analytic solutions exist for either the shear-band thickness or the shear-band separation as both parameters depend on the diffusion of heat and inertia [143,144], making it a coupled-variable, non-analytic problem that can only be solved by numerical methods. The same will be true for granular materials (although attempts have recently been made to tackle this problem analytically [145,146]). Analysis is made more difficult by poor knowledge of the heat and inertia diffusion coefficients.

Even for shear bands formed under globally quasi-static deformation, the material in the vicinity of the shear band will be subject to dynamic rates of strain ($>10^2\ s^{-1}$) [147] requiring acoustic techniques to study them [148–152].

Rechenmacher and Finno also found (as others had before them [123,136,153–159]) that the inclination of the shear band to a line normal to the side in an axially loaded cylinder of sand tends to be larger than for fully dense metal cylinders, being between 55° and 65° (depending on the type of sand and whether it is wet or dry) (Figure 7.39) as opposed to around 50° for metal cylinders [144,160]. Experimental evidence summarized by Lade et al. [157] suggests that the inclination angle θ that shear bands make with the compressive principal stress direction lies between that predicted by the classic analysis of Coulomb [161] on the basis of force equilibrium (including friction) (see Eq. (7.7)) and the angle predicted by Roscoe's theory [136] (see Eq. (7.8)) from considerations of dilation (Figure 7.40),

$$\theta_C = \pi/4 + \phi/2 \tag{7.7}$$

$$\theta_R = \pi/4 + \Psi/2 \tag{7.8}$$

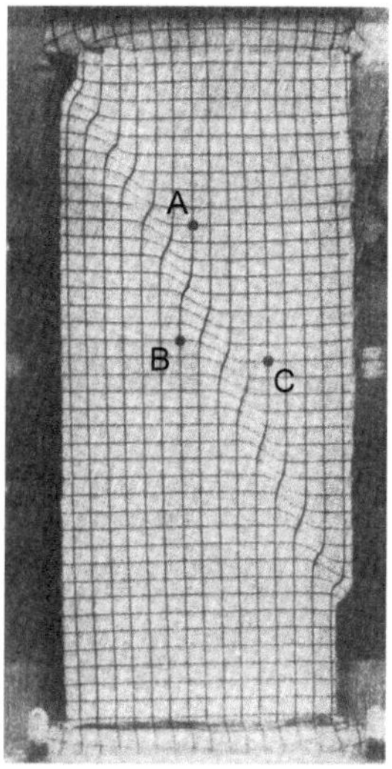

Figure 7.39 Distortion of a square grid on a half-cylinder of sand in which a shear band has formed.
Source: From Ref. [156].

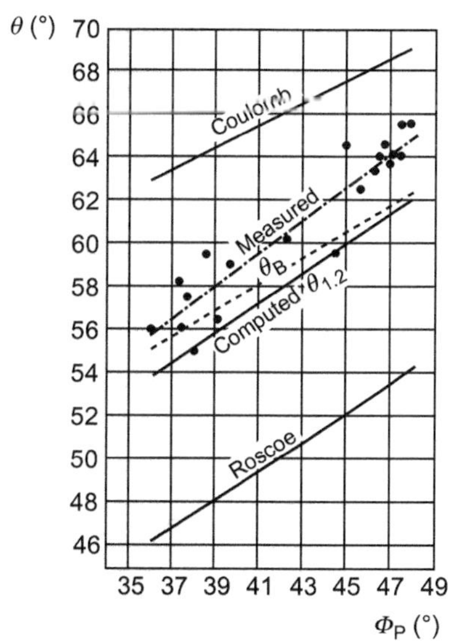

Figure 7.40 Comparison of the shear-band inclination θ against the obliquity at failure ϕ_p for sand with Coulomb and Roscoe theories.
Source: From Ref. [154].

where the subscripts C and R refer to Coulomb and Roscoe, ϕ is the friction angle and Ψ is the dilatancy angle. The Roscoe angle is usually smaller than the Coulomb angle by about 15° [140] (see Figure 7.40). The inclination angle is also found to depend on grain size (with coarse sands following the Roscoe theory and fine sands following the Coulomb theory [29]) and density with dense sands more closely following the Coulomb theory [158].

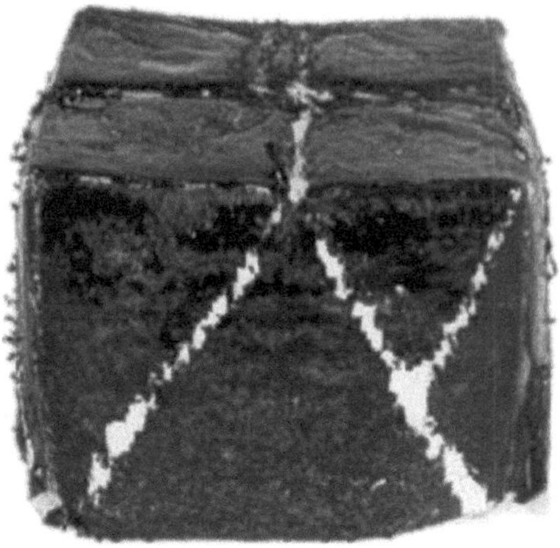

Figure 7.41 A dry dense sand specimen ($100 \times 100 \times 100\ mm^3$) after deformation in a tri-axial test apparatus.
Source: From Ref. [164].

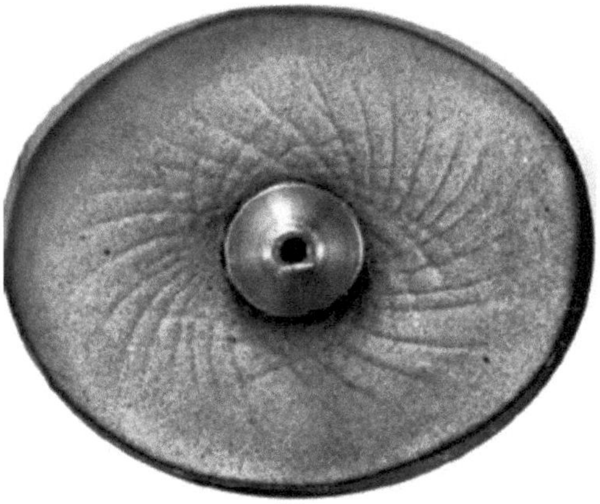

Figure 7.42 Intersecting families of slip lines in an elliptical sand cylinder after being subjected to slow twist motion at the centre. Dimensions: diameter of circular core 31 mm, major axis of ellipse 63 mm, minor axis of ellipse 53 mm.
Source: From Ref. [165].

All this may appear to imply that only one shear band forms, but as in fully dense metals [160] and polymers [107], families of shear bands may form [123,128,162–165] (Figures 7.41 and 7.42) (roughly following trajectories typical of slip line field solutions [143,166]) of which one suddenly becomes dominant. The closer to the peak stress that this occurs, the less pronounced the evidence for the other shear bands.

At present, neither tomography nor MRI operates fast enough to study dynamic events while they are happening. For example, the data shown in Figure 7.34 will probably have taken tens of minutes to obtain. However, both techniques are proving very useful in examining the interiors of specimens after deformation. They also have the advantage that mechanical sectioning is not required. Digital speckle radiography (DSR), the X-ray equivalent of DIC, is fast enough even for ballistic studies [134] but cannot at present be used to study the development of deformation in a single event. If, however, the phenomenon of interest is sufficiently repeatable, a sequence of images can be obtained by triggering the X-ray flash at different times during repeat experiments [133,135]. We await the development of a CCD X-ray framing camera cheap enough for use by people other than fusion researchers [167]!

In granular materials, strain localization can also take place normal to the principal compressive stress-forming so-called compaction [140,168–186] or dilation bands [187,188]. These deformation modes are not seen in fully dense materials such as cast metals. Compaction bands are usually denser than the surrounding rock and are believed to act as permeability barriers to the movement of fluids in the Earth's crust [170,178,189–191]. Field evidence shows that compaction bands are usually more erosion resistant than the surrounding rock (Figure 7.43).

One problem in replicating localization bands in rocks in the lab is that the stress conditions, both in magnitude and in tri-axiality, are hard to replicate [179,192]. For example, Figure 7.44 presents experimental data obtained from laboratory and field observations. The length scales of the data obtained from the two sources are radically different.

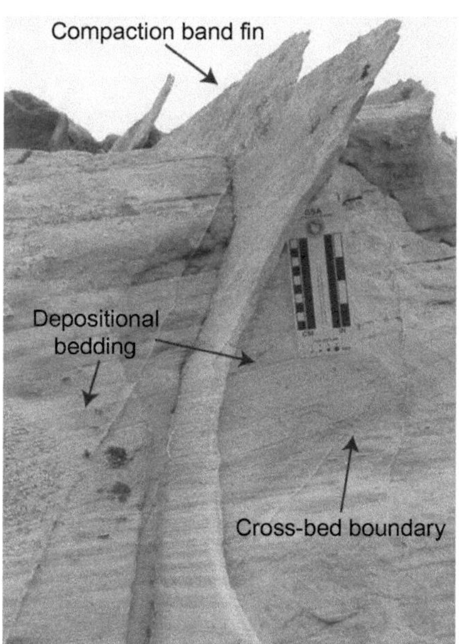

Figure 7.43 Photograph of a compaction band in a sandstone.
Source: From Ref. [175].

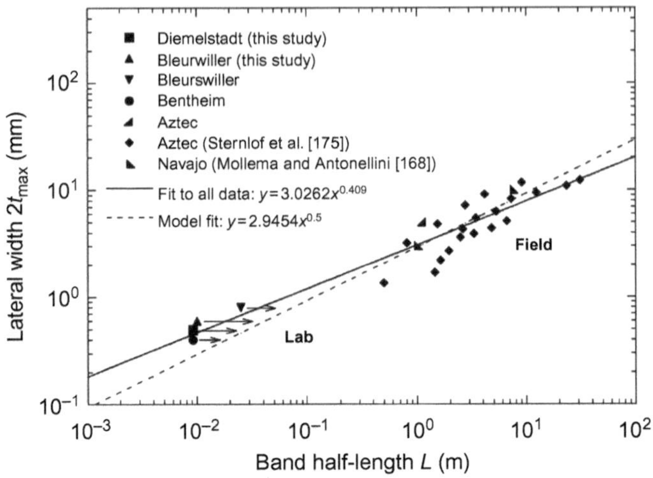

Figure 7.44 Comparison of laboratory and field observations of the thickness and half-length of compaction bands.
Source: From Ref. [179].

However, Issen and Challa [188] found experimentally that compaction bands are favoured if

$$\sigma_1 > \sigma_2 = \sigma_3 \tag{7.9}$$

whereas dilation bands are favoured if

$$\sigma_1 = \sigma_2 > \sigma_3 \tag{7.10}$$

Shear bands are favoured if

$$\sigma_1 < \sigma_2 < \sigma_3 \tag{7.11}$$

where the stresses σ are defined in Figure 7.45 so long as σ_2 is 'not too close to either'. But as Issen and Challa point out [188] in the field, it is not usually possible to tell what the stress conditions were in the rocks that led to the formation of localization bands.

Unlike in fully dense materials, which require deformation softening [143,193], shear localization can take place under certain circumstances while a granular material hardens [158,194,195] (Figure 7.46).

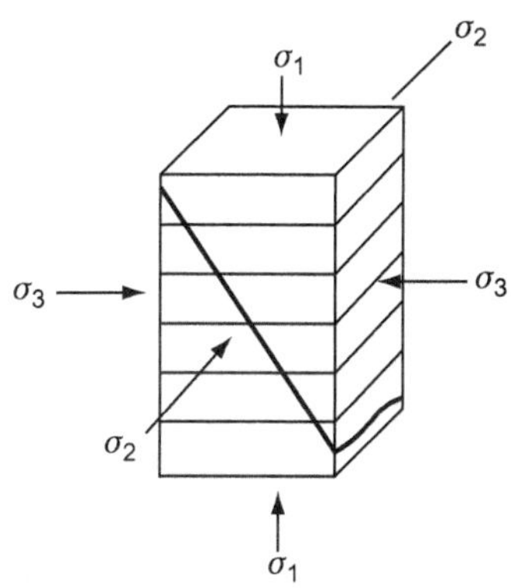

Figure 7.45 Definition of the stresses for Eqs (7.9)–(7.11).
Source: From Ref. [158].

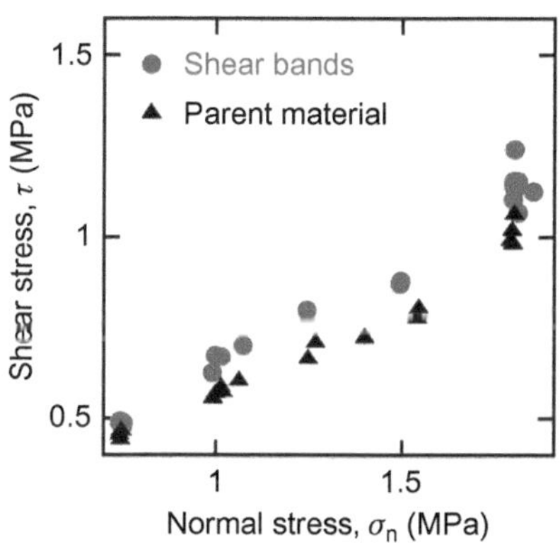

Figure 7.46 Coulomb–Mohr failure envelope for shear bands in sand compared to parent material.
Source: From Ref. [195].

Lade and Wang [158] give the following condition on the stress system for this to occur:

$$0.18 \le b = (\sigma_2 - \sigma_3)/(\sigma_1 - \sigma_3) \le 0.85 \tag{7.12}$$

Note the limits given in Eq. (7.12) are only approximate. Abelev and Lade [196,197] soon afterwards reported a more detailed study into the effect of the parameter b on localization behaviour of sand under tri-axial loading.

Kaproth et al. [195] say it is still not understood how strain localization can take place when the material is strengthening, but the explanation may be that weakening takes place locally even though this does not feed through into global stress–strain measurements. Analysis of the acoustic energy emitted during this process may in due course give insight into their initiation and propagation [182,198]. Acoustic waves themselves are also one possible trigger of strain localization [199].

Figure 7.47 gives some impression of what happens to those grains that are incorporated into a shear or compaction band. It can be seen in this case that not only have the grains fragmented (crushing) but also the void fraction has decreased. Similar observations were made by Shih et al. [200] in the study of the rapid collapse of cylinders of silicon carbide under explosive shock. Void collapse does not always happen when shear bands form in granular media. For example, in initially dense sand, dilatation can be positive when localization occurs [140,201]. The change in particle size distribution during localization has been quantified by Kaproth et al. [195] (Figure 7.48).

Other matters that may have to be taken into account when attempting to model this phenomenon include particle rotation [192], particle shape [202], interparticle friction [203], anisotropy [204,205], presence of fluids [206,207], vorticity and buckling of force chains [18,131,208–211] (Figure 7.49), although up to the present there have not been many direct experimental investigations of these phenomena [182,212–214].

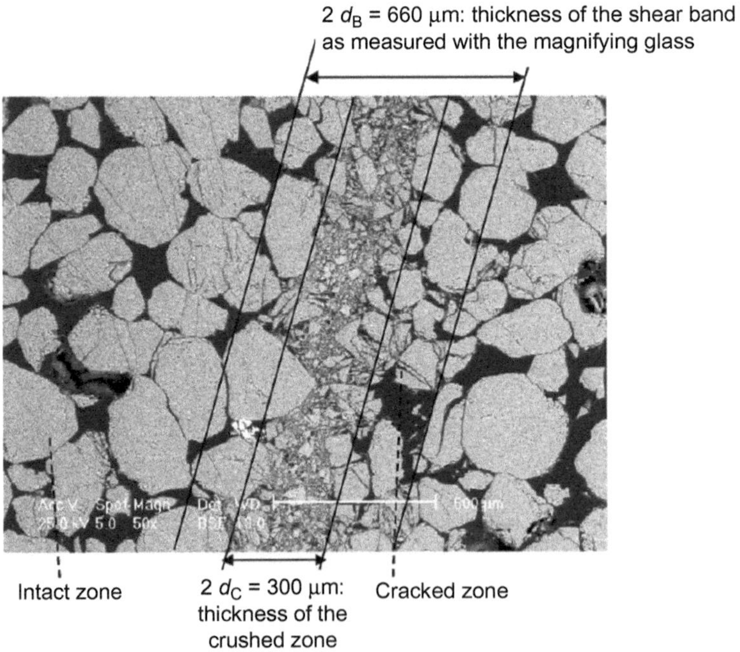

Figure 7.47 Micrograph of a highly compacted shear band in a sandstone for a specimen quasi-statically deformed under 28 MPa confining pressure showing a fracture of grains incorporated into the shear band.
Source: From Ref. [159].

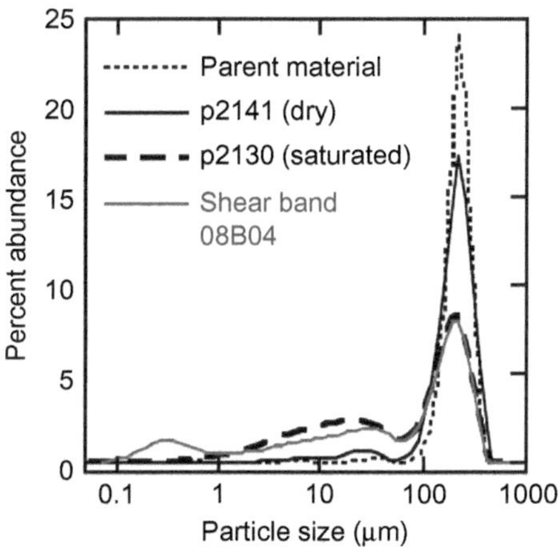

Figure 7.48 Plots showing the particle size distribution in undeformed sand (parent material) and shear bands formed in the wet and dry states.
Source: From Ref. [195].

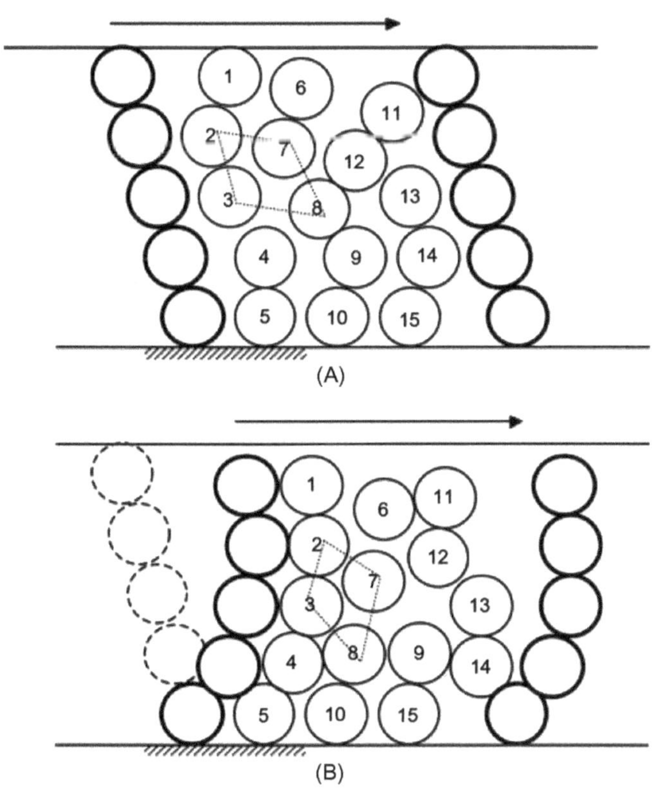

Figure 7.49 Hypothesized force-chain behaviour during shear-band formation: (A) before; (B) during.
Source: From Ref. [131].

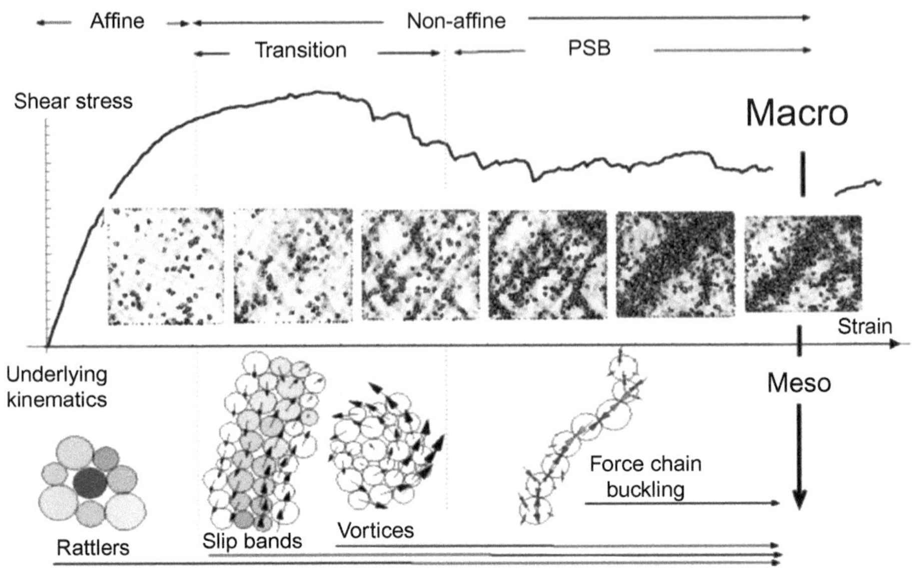

Figure 7.50 Summary of hypothesized mesoscale phenomena associated with localization of deformation in granular materials.
Source: From Ref. [210].

7.5 Summary

Strain localization is now well established as one of the hot-spot mechanisms that can lead to ignition of secondary explosives. Insights into the processes involved may be found in two other bodies of the scientific literature: (i) shock loading of intermetallic reactive mixtures and (ii) soil and granular mechanics. Extra modes of strain localization are possible for granular materials that are not open to fully dense materials, namely compaction and dilatation bands. These have been observed in granular rocks, such as sandstones, but have not (so far as I know) been observed in explosives. One recommendation of this review is that those who study granular explosives should familiarize themselves with the soil mechanics literature. A visual summary of the mesoscale phenomena believed to be associated with strain localization in granular materials is given in Figure 7.50.

Acknowledgements

The author would like to thank Dr R. Blumenfeld, Dr A.L. Collins, Dr B. Dodd, Prof. J.E. Field and Prof. M.A. Meyers for their comments on this review paper.

References

1. Ristow, G.H. and Herrmann, H.J. (1995). Forces on the walls and stagnation zones in a hopper filled with granular material. *Physica A* **213**, 474–481.
2. Wolf, D.E. and Grassberger, P. (Eds.) (1997). "Friction, Arching, Contact Dynamics." World Scientific, Singapore.
3. Grantham, S.G. and Forsberg, F. (2004). Measurement of granular flow in a silo using Digital Speckle Radiography. *Powder Technol.* **146**, 56–65.
4. Wojcik, M. and Tejchman, J. (2009). Investigations of shear localization during granular silo flow with nonlocal hypoplastic constitutive model. In "Powders and Grains 2009" (M. Nakagawa and S. Luding, Eds.), pp. 641–644. American Institute of Physics, Melville, NY.
5. Housel, W.S. (1936). Internal stability of granular materials. *Proc. Am. Soc. Test. Mater.* **36**(2), 426–468.
6. Brockbank, R., Huntley, J.M., and Ball, R.C. (1997). Contact force distribution beneath a three-dimensional granular pile. *J. Phys. II* **7**, 1521–1532.
7. Huntley, J.M. (1999). Force distribution in an inhomogeneous sandpile. *Eur. Phys. J. B* **8**, 389–397.
8. Seelig, R.P. and Wulff, J. (1946). The pressing operation in the fabrication of articles by powder metallurgy. *Trans. AIME* **166**, 492–505.
9. Graton, L.C. and Fraser, H.J. (1935). Systematic packing of spheres with particular relation to porosity and permeability. *J. Geol.* **43**, 785–909.
10. Bernal, J.D. and Mason, J. (1960). Coordination of randomly packed spheres. *Nature* **188**, 910–911.
11. Al-Raoush, R. and Alsaleh, M. (2007). Simulation of random packing of polydisperse particles. *Powder Technol.* **176**, 47–55.
12. Al-Raoush, R. (2007). Microstructure characterization of granular materials. *Physica A* **377**, 545–558.
13. Radjai, F., Wolf, D., Jean, M., and Moreau, J.J. (1998). Bimodal character of stress transmission in granular packing. *Phys. Rev. Lett.* **90**, 61–64.
14. Peters, J.F., Muthuswamy, M., Wibowo, J., and Tordesillas, A. (2005). Characterization of force chains in granular material. *Phys. Rev. E* **72**, 041307.
15. Hidalgo, R.C., Grosse, C.U., Kin, F., Reinhardt, H.W., and Herrmann, H.J. (2002). Evolution of percolating force chains in compressed granular media. *Phys. Rev. Lett.* **89**, 205501.
16. Otto, M., Bouchaud, J.P., Claudin, P., and Socolar, J.E.S. (2003). Anisotropy in granular media: Classical elasticity and directed force chain network. *Phys. Rev. E* **67**, 031302.
17. Blumenfeld, R. (2004). Stresses in isostatic granular systems and emergence of force chains. *Phys. Rev. Lett.* **93**, 108301.
18. Tordesillas, A. and Muthuswamy, M. (2009). On the modeling of confined buckling of force chains. *J. Mech. Phys. Solids* **57**, 706–727.
19. Zhu, Y., Shukla, A., and Sadd, M.H. (1996). The effect of microstructural fabric on dynamic load transfer in two dimensional assemblies of elliptical particles. *J. Mech. Phys. Solids* **44**, 1283–1303.
20. Roessig, K.M., Foster J.C., Jr., Bardenhagen, S.G. (2002). Dynamic stress chain formation in a two-dimensional particle bed. *Exp. Mech.* **42**, 329–337.

21. Rae, P.J., Palmer, S.J.P., Goldrein, H.T., Field, J.E., and Lewis, A.L. (2002). Quasi-static studies of the deformation and failure of PBX 9501. *Proc. R. Soc. London A* **458**, 2227–2242.
22. Walley, S.M., Field, J.E., and Greenaway, M.W. (2006). Crystal sensitivities of energetic materials. *Mater. Sci. Technol.* **22**, 402–413.
23. Enger, M.L. (1922). Discussion of the paper by G. Paaswell entitled "Transmission of pressure through solids and soils and the related engineering phenomena". *Trans. Am. Soc. Civ. Eng.* **85**, 1579–1589.
24. Chambon, R., Desrues, J., and Vardoulakis, I. (Eds.) (1994). "Localisation and Bifurcation Theory for Soils and Rocks." Balkema, Rotterdam, The Netherlands.
25. Duran, J. (2000). "Sands, Powders and Grains: An Introduction to the Physics of Granular Materials." Springer, New York, NY.
26. Labuz, J.F. and Drescher, A. (Eds.) (2003). "Bifurcations and Instabilities in Geomechanics." Balkema, Lisse, The Netherlands.
27. Nesterenko, V.F. (2001). "Dynamics of Heterogeneous Materials." Springer, Berlin.
28. Tejchman, J. (2008). "Shear Localization in Granular Bodies with Micro-Polar Hypoplasticity." Springer, Berlin.
29. Bardet, J.P. (1990). A comprehensive review of strain localization in elastoplastic soils. *Comput. Geotech.* **10**, 163–188.
30. Lade, P.V. (2002). Instability, shear banding, and failure in granular materials. *Int. J. Solids Struct.* **39**, 3337–3357.
31. Schall, P. and van Hecke, M. (2010). Shear bands in matter with granularity. *Annu. Rev. Fluid Mech* **42**, 67–88.
32. Taylor, W. and Weale, A. (1932). The mechanism of the initiation and propagation of detonation of solid explosives. *Proc. R. Soc. London A* **138**, 92–116.
33. Andreev, K.K. and Chariton, J.B. (1935). The mechanism of self propagating chain reactions. *Trans. Faraday Soc.* **31**, 797–804.
34. Bowden, F.P., Eirich, F., Ferguson, A.E., and Yoffe, A. (1943). "The Detonation of Nitroglycerine by Impact." (*CSIR Bulletin* **167**), Council for Scientific and Industrial Research, Melbourne, Australia.
35. Bowden, F.P., Eirich, F., Mulcahy, M.F.R., Vines, R.G., and Yoffe, A.D. (1943). "The Detonation of High Explosives by Impact: The Sensitivity and the Propagation of the Explosion in Liquids." (*CSIR Bulletin* **173**), Council for Scientific and Industrial Research, Melbourne, Australia.
36. Bowden, F.P. and Yoffe, A.D. (1952). "Initiation and Growth of Explosion in Liquids and Solids." **(republ. 1985)**, Cambridge University Press, Cambridge.
37. Bowden, F.P. and Gurton, O.A. (1949). Initiation of solid explosives by impact and friction: The influence of grit. *Proc. R. Soc. London A* **198**, 337–349.
38. Rideal, E.K. and Robertson, A.J.B. (1948). The sensitiveness of solid high explosives to impact. *Proc. R. Soc. London A* **195**, 135–150.
39. Bowden, F.P. and Yoffe, A.D. (1958). "Fast Reactions in Solids." Butterworth, London.
40. Field, J.E. (1992). Hot spot ignition mechanisms for explosives. *Acc. Chem. Res.* **25**, 489–496.
41. Yuan, H. and Prosperetti, A. (1996). Transient viscous heating of liquid layers under intense shear. *Proc. R. Soc. London A* **452**, 2767–2783.
42. Leiber, C.O. (2003). "Assessment of Safety and Risk with a Microscopic Model of Detonation." Elsevier, Amsterdam.

43. Afanasev, G.T., Bobolev, V.K., Kazarova, Y.A., and Karabanov, Y.F. (1972). Formation of local hot spots during the fracture of thin layers under shock. *Combust. Explosion Shock Waves* **8**, 241–246.
44. Zener, C. and Hollomon, J.H. (1944). Effect of strain rate upon plastic flow of steel. *J. Appl. Phys.* **15**, 22–32.
45. Winter, R.E. and Field, J.E. (1975). The role of localized plastic flow in the impact initiation of explosives. *Proc. R. Soc. London A* **343**, 399–413.
46. Courtney-Pratt, J.S. and Rogers, G.T. (1955). Conditions of explosion of azides: Initiation of explosion by light and by flying fragments. *Nature* **175**, 632–633.
47. Field, J.E., Swallowe, G.M., and Heavens, S.N. (1982). Ignition mechanisms of explosives during mechanical deformation. *Proc. R. Soc. London A* **382**, 231–244.
48. Field, J.E., Palmer, S.J.P., Pope, P.H., Sundararajan, R., and Swallowe, G.M. (1985). Mechanical properties of PBXs and their behaviour during drop-weight impact. In "Proc. Eighth Symposium (Int.) on Detonation" (J.M. Short, Ed.), pp. 635–644. Naval Surface Weapons Center, White Oak, MD.
49. Field, J.E., Parry, M.A., Palmer, S.J.P., and Huntley, J.M. (1989). Deformation and explosive properties of HMX powders and polymer bonded explosives. In "Proc. Ninth Symposium (Int.) on Detonation," pp. 886–896. Office of the Chief of Naval Research, Arlington, VA.
50. Krishna Mohan, V., Jyothi Bhasu, V.C., and Field, J.E. (1989). Role of adiabatic shear bands in initiation of explosives by drop-weight impact. In "Proc. Ninth Symposium (Int.) on Detonation," pp. 1276–1283. Office of the Chief of Naval Research, Arlington, VA.
51. Field, J.E., Bourne, N.K., Palmer, S.J.P., and Walley, S.M. (1992). Hot-spot ignition mechanisms for explosives and propellants. *Philos. Trans. R. Soc. London A* **339**, 269–283.
52. Heavens, S.N. and Field, J.E. (1974). The ignition of a thin layer of explosive by impact. *Proc. R. Soc. London A* **338**, 77–93.
53. Swallowe, G.M. and Field, J.E. (1982). The ignition of a thin layer of explosive by impact: The effect of polymer particles. *Proc. R. Soc. London A* **379**, 389–408.
54. Walley, S.M., Field, J.E., and Palmer, S.J.P. (1992). Impact sensitivity of propellants. *Proc. R. Soc. London A* **438**, 571–583.
55. Walley, S.M., Balzer, J.E., Proud, W.G., and Field, J.E. (2000). Response of thermites to dynamic high pressure and shear. *Proc. R. Soc. London A* **456**, 1483–1503.
56. Balzer, J.E., Field, J.E., Gifford, M.J., Proud, W.G., and Walley, S.M. (2002). High-speed photographic study of the drop-weight impact response of ultrafine and conventional PETN and RDX. *Combust. Flame* **130**, 298–306.
57. Balzer, J.E., Siviour, C.R., Walley, S.M., Proud, W.G., and Field, J.E. (2004). Behaviour of ammonium perchlorate-based propellants and a polymer-bonded explosive under impact loading. *Proc. R. Soc. London A* **460**, 781–806.
58. Swallowe, G.M., Field, J.E., and Horn, L.A. (1986). Measurements of transient high temperatures during the deformation of polymers. *J. Mater. Sci.* **21**, 4089–4096.
59. Picart, D., Delamire-Sizes, F., Gruau, C., and Trumel, H. (2009). Ignition of HMX-based PBX submitted to impact: Strain localization and boundary conditions. In "Shock Compression of Condensed Matter – 2009" (M.L. Elert, W.T. Buttler, M.D. Furnish, W.W. Anderson, and W.G. Proud, Eds.), pp. 41–44. American Institute of Physics, Melville, NY.
60. Bridgman, P.W. (1935). Effects of high shearing stress combined with high hydrostatic pressure. *Phys. Rev.* **48**, 825–847.

61. Enikolopov, N.S. (1991). Solid phase chemical reactions and new technologies. *Russ. Chem. Rev.* **60**, 283–287.
62. Thadhani, N.N. (1994). Shock-induced and shock-assisted solid-state chemical reactions in powder mixtures. *J. Appl. Phys.* **76**, 2129–2138.
63. Merzhanov, A.G. and Borovinskaya, I.P. (1972). Self-propagating high-temperature synthesis of inorganic compounds (in Russian). *Dokl. Akad. Nauk SSSR* **204**, 366–369.
64. Merzhanov, A.G. and Khaikin, B.I. (1988). Theory of combustion waves in homogeneous media. *Prog. Energy Combust. Sci.* **14**, 1–98.
65. Merzhanov, A.G. and Borovinskaya, I.P. (2008). Historical retrospective of SHS: An autoreview. *Int. J. Self Propag. High Temp. Synth.* **17**, 242–265.
66. Eakins, D.E. and Thadhani, N.N. (2009). Shock compression of reactive powder mixtures. *Int. Mater. Rev.* **54**, 181–213.
67. Linse, V.D. (1986). The particulate nature of dynamic compaction. In "Metallurgical Applications of Shock-Wave and High Strain Rate Phenomena" (L.E. Murr, K.P. Staudhammer, and M.A. Meyers, Eds.), pp. 29–55. Marcel Dekker, New York, NY.
68. Harris, C.C. (1967). On the limit of comminution. *Trans. AIME* **238**, 17–30.
69. Kendall, K. (1978). The impossibility of comminuting small particles by compression. *Nature* **272**, 710–711.
70. Hagan, J.T. (1981). Impossibility of fragmenting small particles: Brittle-ductile transition. *J. Mater. Sci.* **16**, 2909–2911.
71. Duvall, G.E. (1986). The physics connection. In "Metallurgical Applications of Shock-Wave and High-Strain-Rate Phenomena" (L.E. Murr, K.P. Staudhammer, and M.A. Meyers, Eds.), pp. 3–25. Marcel Dekker, New York, NY.
72. Eakins, D.E. and Thadhani, N.N. (2008). The shock-densification behavior of three distinct Ni + Al powder mixtures. *Appl. Phys. Lett.* **92**, 111903.
73. Thadhani, N.N., Dunbar, E., and Graham, R.A. (1994). Characteristics of shock-compressed configuration of Ti and Si powder mixtures. In "High Pressure Science and Technology 1993" (S.C. Schmidt, J.W. Shaner, G.A. Samara, and M. Ross, Eds.), pp. 1307–1310. American Institute of Physics, New York, NY.
74. Liniger, E. and Raj, R. (1987). Packing and sintering of two-dimensional structures made from bimodal particle size distributions. *J. Am. Ceram. Soc.* **70**, 843–849.
75. Velamakanni, B.V. and Lange, F.F. (1991). Effect of interparticle potentials and sedimentation on particle packing density of bimodal particle distributions during pressure filtration. *J. Am. Ceram. Soc.* **74**, 166–172.
76. Milne, S.J., Patel, M., and Dickinson, E. (1993). Experimental studies of particle packing and sintering behaviour of monosize and bimodal spherical silica powders. *J. Eur. Ceram. Soc.* **11**, 1–7.
77. Smith, L.N. and Midha, P.S. (1997). A computer model for relating powder density to composition employing simulations of dense random packings of monosized and bimodal spherical particles. *J. Mater. Process. Technol.* **72**, 277–282.
78. Batsanov, S.S. (1995). Shock-induced chemical reactions in metal-silicon systems. In "Metallurgical and Materials Applications of Shock-Wave and High-Strain-Rate Phenomena" (L.E. Murr, K.P. Staudhammer, and M.A. Meyers, Eds.), pp. 707–714. Elsevier, Amsterdam.
79. Hutchings, I.M. (1996). Deformation and fracture of brittle particles. In "Solid–Solid Interactions" (M.J. Adams, S.K. Biswas, and B.J. Briscoe, Eds.), pp. 211–224. Imperial College Press, London.
80. Nesterenko, V.F., Meyers, M.A., Chen, H.C., and LaSalvia, J.C. (1996). Chemical reactions in controlled high-strain-rate shear bands. In "Shock Compression of Condensed

Matter 1995" (S.C. Schmidt and W.C. Tao, Eds.), pp. 713–716. American Institute of Physics, Woodbury, NY.

81. Chen, H.C., Lasalvia, J.C., Nesterenko, V.F., and Meyers, M.A. (1998). Shear localization and chemical reaction in high strain, high strain-rate deformation of Ti–Si powder mixtures. *Acta Mater.* **46**, 3033–3046.
82. Levitas, V.I., Nesterenko, V.F., and Meyers, M.A. (1998). Strain-induced structural changes and chemical reactions. 1: Thermomechanical and kinetic models. *Acta Mater.* **46**, 5929–5945.
83. Levitas, V.I., Nesterenko, V.F., and Meyers, M.A. (1998). Strain-induced structural changes and chemical reactions. 2: Modelling of reactions in shear band. *Acta Mater.* **46**, 5947–5963.
84. Kostyukov, N.A. and Kuzmin, G.E. (1995). Fine structure of powder material flow in the region of shock wave branching. In "Metallurgical and Materials Applications of Shock-Wave and High-Strain-Rate Phenomena" (L.E. Murr, K.P. Staudhammer, and M. A. Meyers, Eds.), pp. 869–875. Elsevier, Amsterdam.
85. Kochsiek, D., Prummer, R., and Brunold, A. (1995). Synthesis of intermetallic aluminides by explosive reaction pressing. *Metall* **49**, 168–172.
86. Chen, H.C., Nesterenko, V.F., LaSalvia, J.C., and Meyers, M.A. (1997). Shear-induced exothermic chemical reactions. *J. Phys. IV* **7**(C3), 27–32.
87. Nemat-Nasser, S., Okinaka, T., Nesterenko, V., and Liu, M. (1998). Dynamic void collapse in crystals: Computational modelling and experiments. *Philos. Mag. A* **78**, 1151–1174.
88. Carroll, M.M. and Holt, A.C. (1972). Static and dynamic pore collapse relations for ductile porous materials. *J. Appl. Phys.* **43**, 1626–1635.
89. Carroll, M. and Holt, A.C. (1972). Suggested modification of the P-α model for porous materials. *J. Appl. Phys.* **43**, 759–761.
90. Fischmeister, H.F. and Arzt, E. (1983). Densification of powders by particle deformation. *Powder Metall.* **26**, 82–88.
91. Yu, L.H. and Meyers, M.A. (1991). Shock synthesis and synthesis-assisted shock consolidation of silicides. *J. Mater. Sci.* **26**, 601–611.
92. Graham, R.A., Morosin, B., Horie, Y., Venturini, E.L., Boslough, M., Carr, M.J., and Williamson, D.L. (1986). Chemical synthesis under high pressure shock loading. In "Shock Waves in Condensed Matter – 1985" (Y.M. Gupta, Ed.), pp. 693–711. Plenum Press, New York, NY.
93. Boslough, M.B. (1992). Postshock spectral radiance measurements in nickel and nickel/aluminum powders. In "Shock Compression of Condensed Matter – 1991" (S.C. Schmidt, R.D. Dick, J.W. Forbes, and D.G. Tasker, Eds.), pp. 617–620. Elsevier, Amsterdam.
94. Blanco, E. (1997). "Température et Émissivité des Matériaux Sous Choc: Etude Expérimentale par Pyrométrie Optique à Travers un Matériau Fenêtre," Ph.D. Thesis, University of Paris X.
95. Partouche-Sebban, D. and Pelissier, J.L. (2003). Emissivity and temperature measurements under shock loading along the melting curve of bismuth. *Shock Waves* **13**, 69–81.
96. Seifter, A., Boboridis, K., and Obst, A.W. (2004). Emissivity measurements on metallic surfaces with various degrees of roughness: A comparison of laser polarimetry and integrating sphere reflectometry. *Int. J. Thermophys.* **25**, 547–560.
97. Hallo, L., Riou, O., Stenz, C., and Tikhonchuk, V.T. (2006). Infrared emissivity studies of melting thresholds and structural changes of aluminium and copper samples heated by femtosecond laser pulses. *J. Phys. D: Appl. Phys.* **39**, 5272–5279.

98. Bennett, L.S. and Horie, Y. (1994). Shock-induced inorganic reactions and condensed phase detonations. *Shock Waves* **4**, 127–136.
99. Staudhammer, K.P. (1999). Fundamental temperature aspects in shock consolidation of metal powders. In "Powder Materials: Current Research and Industrial Practices" (F.D. S. Marquis, Ed.), pp. 317–325. The Minerals, Metals and Materials Society, Warrendale, PA.
100. Yu, L.H. and Meyers, M.A. (1992). Shock synthesis of silicides. In "Shock-Wave and High-Strain-Rate Phenomena in Materials" (M.A. Meyers, L.E. Murr, and K. P. Staudhammer, Eds.), pp. 303–309. Marcel Dekker, New York, NY.
101. Horie, Y. (1993). Shock-induced chemical reactions in inorganic powder mixtures. In "Shock Waves in Materials Science" (A.B. Sawaoka, Ed.), pp. 67–100. Springer, Berlin.
102. Yu, L.H., Nellis, W.J., Meyers, M.A., and Vecchio, K.S. (1994). Shock synthesis of niobium silicides. In "High Pressure Science and Technology 1993" (S.C. Schmidt, J. W. Shaner, G.A. Samara, and M. Ross, Eds.), pp. 1291–1294. American Institute of Physics, New York, NY.
103. Nabatov, S.S., Breusov, O.N., Lebedev, A.V., Pershin, S.V., and Tatsiy, V.F. (1997). Synthesis of titanium disilicide induced by the shock wave effect. *Phys. Chem. Mater. Treat.* **31**, 399–402.
104. Meyers, M.A., Benson, D.J., and Olevsky, E.A. (1999). Shock consolidation: Microstructurally-based analysis and computational modeling. *Acta Mater.* **47**, 2089–2108.
105. Morris, D.G. (1985). Dynamic powder compaction as a thermomechanical process. In "Rapidly Solidified Materials" (P.W. Lee and R.S. Carbonara, Eds.), pp. 67–75. American Society for Metals, Metals Park, OH.
106. Meyers, M.A., Yu, L.-H., and Vecchio, K.S. (1994). Shock synthesis of silicides. 2: Thermodynamics and kinetics. *Acta Metall. Mater.* **42**, 715–729.
107. Walley, S.M., Xing, D., and Field, J.E. (1995). Mechanical properties of three transparent polymers in compression at a very high rate of strain. In "Impact and Dynamic Fracture of Polymers and Composites" (J.G. Williams and A. Pavan, Eds.), pp. 289–303. Mechanical Engineering Publications, London.
108. Walley, S.M. (1999). Adiabatic shear instability: Observations and experimental techniques. In "Mechanical Properties and Testing of Polymers" (G.M. Swallowe, Ed.), pp. 10–14. Kluwer, Dordrecht, The Netherlands.
109. Sopp, O.I. (1964). X-ray radiography and soil mechanics: Localization of shear planes in soil samples. *Nature* **202**, 832.
110. Vardoulakis, I. and Graf, B. (1985). Calibration of constitutive models for granular materials using data from biaxial experiments. *Géotechnique* **35**, 299–317.
111. Desrues, J., Chambon, R., Mokni, M., and Mazerolle, F. (1996). Void ratio evolution inside shear bands in triaxial sand specimens studied by computer tomography. *Géotechnique* **46**, 529–546.
112. Alshibli, K.A., Sture, S., Costes, N.C., Frank, M.L., Lankton, M.R., Batiste, S.N., and Swanson, R.A. (2000). Assessment of localized deformations in sand using X-ray computed tomography. *Geotech. Test. J.* **23**, 274–299.
113. Alshibli, K.A., Batiste, S.N., and Sture, S. (2003). Strain localization in sand: Plane strain versus triaxial compression. *ASCE J. Geotech. Geoenviron. Eng.* **129**, 483–494.
114. Oda, M., Takemura, T., and Takahashi, M. (2004). Microstructure in shear band observed by microfocus X-ray computed tomography. *Géotechnique* **54**, 539–542.

115. Viggiani, G., Lenoir, N., Besuelle, P., Di Michiel, M., Marello, S., Desrues, J., and Kretzschmer, M. (2004). X-ray microtomography for studying localized deformation in fine-grained geomaterials under triaxial compression. *C. R. Méc.* **332**, 819–826.
116. Louis, L., Wong, T.-F., and Tembe, S. (2006). Imaging strain localization by X-ray computed tomography. *J. Struct. Geol.* **28**, 762–775.
117. Alshibli, K.A. and Hasan, A. (2008). Spatial variation of void ratio and shear band thickness in sand using X-ray computed tomography. *Géotechnique* **58**, 249–257.
118. Hall, S.A., Bornert, M., Desrues, J., Pannier, Y., Lenoir, N., Viggiani, G., and Bésuelle, P. (2010). Discrete and continuum analysis of localised deformation in sand using X-ray μCT and volumetric digital image correlation. *Géotechnique* **60**, 315–322.
119. Hasan, A. and Alshibli, K.A. (2010). Experimental assessment of 3D particle-to-particle interaction within sheared sand using synchrotron microtomography. *Géotechnique* **60**, 369–379.
120. Merwin, L.H., Nissan, R.A., Stephens, T.S., and Wallner, A.S. (1996). Magnetic resonance imaging study of solid rocket propellants and liners. *J. Appl. Polym. Sci.* **62**, 341–348.
121. Motz, J., Eisenreich, N., Geissler, A., and Geissler, E. (2002). Characterization of the structure in highly filled composite materials by means of MRI. *Propell. Explos. Pyrotech.* **27**, 179–184.
122. Fall, A., Bertrand, F., Ovarlez, G., and Bonn, D. (2009). Yield stress and shear banding in granular suspensions. *Phys. Rev. Lett.* **103**, 178301.
123. Harris, W.W., Viggiani, G., Mooney, M.A., and Finno, R.J. (1995). Use of stereophotogrammetry to analyze the development of shear bands in sand. *Geotech. Test. J.* **18**, 405–420.
124. Finno, R.J., Harris, W.W., Mooney, M.A., and Viggiani, G. (1996). Strain localization and undrained steady state of sand. *ASCE J. Geotech. Eng.* **122**, 462–473.
125. Mooney, M.A., Viggiani, G., and Finno, R.J. (1997). Undrained shear band deformation in granular media. *ASCE J. Eng. Mech.* **123**, 577–585.
126. Alshibli, K.A. and Sture, S. (1999). Sand shear band thickness measurements by digital image techniques. *J. Comput. Civ. Eng.* **13**, 103–109.
127. Alshibli, K.A. and Sture, S. (2000). Shear band formation in plane strain experiments of sand. *ASCE J. Geotech. Geoenviron. Eng.* **126**, 495–503.
128. Desrues, J. and Viggiani, G. (2004). Strain localization in sand: An overview of the experimental results obtained in Grenoble using stereophotogrammetry. *Int. J. Numer. Anal. Methods Geomech.* **28**, 279–321.
129. Rechenmacher, A.L. and Finno, R.J. (2004). Digital image correlation to evaluate shear banding in dilative sands. *Geotech. Test. J.* **27**, 13–22.
130. Wolf, H., König, D., and Triantafyllidis, T. (2006). The influence of the stress–strain behavior of non-cohesive soils on the geometry of shear band systems under extensional strain. *Eng. Struct.* **28**, 1760–1773.
131. Rechenmacher, A., Abedi, A., and Chupin, O. (2010). Evolution of force chains in shear bands in sands. *Géotechnique* **60**, 343–351.
132. Röchter, L., König, D., Schanz, T., and Triantafyllidis, T. (2010). Shear banding and strain softening in plane strain extension: Physical modelling. *Granular Matter* **12**, 287–301.
133. Addiss, J.W. (2009). "Measurement of Displacements in Granular Systems in Response to Penetration and Compaction," Ph.D. Thesis, University of Cambridge.
134. Field, J.E., Proud, W.G., and Walley, S.M. (2009). Review of optical and X-ray techniques used at the Cavendish Laboratory. *Imaging Sci. J.* **57**, 317–325.

135. Collins, A.L., Addiss, J.W., Walley, S.M., Promratana, K., Bobaru, F., Proud, W.G., and Williamson, D.M. (2011). The effect of rod nose shape on the internal flow fields during the ballistic penetration of sand. *Int. J. Impact Eng.* **38**, 951–963.
136. Roscoe, K.H. (1970). The influence of strains in soil mechanics. *Géotechnique* **20**, 129–169.
137. Mühlhaus, H.-B. and Vardoulakis, I. (1987). The thickness of shear bands in granular materials. *Géotechnique* **37**, 271–283.
138. Yoshida, T. and Tatsuoka, F. (1997). Deformation property of shear band in sand subjected to plane strain compression and its relation to particle characteristics. In "Proc. 14th Int. Conf. on Soil Mechanics and Foundation Engineering. Vol. 1," pp. 237–240, Balkema, Rotterdam, The Netherlands.
139. Mokni, M. (1992). "Relations Entre Déformations en masse et Déformations Localisées dans les Matériaux Granulaires," Ph.D. Thesis, University of Grenoble, France.
140. Bésuelle, P. and Rudnicki, J. (2004). Localization: Shear bands and compaction bands. In "Mechanics of Fluid-Saturated Rocks" (Y. Guéguen and M. Boutéca, Eds.), pp. 219–321. Elsevier, London.
141. Estrada, N., Lizcano, A., and Taboada, A. (2010). Simulation of cemented granular materials. 1: Macroscopic stress–strain response and strain localization. *Phys. Rev. E* **82**, 011303.
142. Estrada, N., Lizcano, A., and Taboada, A. (2010). Simulation of cemented granular materials. 2: Micromechanical description and strength mobilization at the onset of macroscopic yielding. *Phys. Rev. E* **82**, 011304.
143. Walley, S.M. (2007). Shear localization: A historical overview. *Metall. Mater. Trans. A* **38**, 2629–2654.
144. Xu, Y., Zhang, J., Bai, Y.L., and Meyers, M.A. (2008). Shear localization in dynamic deformation: Microstructural evolution. *Metall. Mater. Trans. A* **39**, 811–843.
145. Lu, X.B., Wang, S.Y., Wang, Y.H., and Peng, C. (2004). An approximate method for evaluating the shear band thickness in saturated soils. *Int. J. Numer. Anal. Methods Geomech.* **28**, 1533–1541.
146. Triantafyllidis, T., Röchter, L., Niemunis, A., and Prada-Sarmiento, L.F. (2011). Shear banding in geomaterials under extensional plane strain conditions: Physical and analytical model. *Acta Geotech.* **6**, 93–103.
147. Key, W.R.O. and Schultz, R.A. (2011). Fault formation in porous sedimentary rocks at high strain rates: First results from the Upheaval Dome impact structure, Utah, USA. *Geol. Soc. Am. Bull.* **123**, 1161–1170.
148. Lockner, D.A. and Byerlee, J.D. (1992). Fault growth and acoustic emissions in confined granite. *Appl. Mech. Rev.* **45**, S165–S173.
149. Olsson, W.A. and Holcomb, D.J. (2000). Compaction localization in porous rocks. *Geophys. Res. Lett.* **27**, 3537–3540.
150. Jouniaux, L., Masuda, K., Lei, X., Nishizawa, O., Kusunose, K., Liu, L., and Ma, W. (2001). Comparison of the microfracture localization in granite between fracturation and slip of a pre-existing macroscopic healed joint by acoustic emission measurements. *J. Geophys. Res.* **106**, 8687–8698.
151. Caroli, C. and Velicky, B. (2003). Anomalous acoustic reflection on a sliding interface or a shear band. *Phys. Rev. E* **67**, 061301.
152. Khidas, Y. and Jia, X. (2009). Acoustic probing of the contact network dynamics in a granular medium during shear band formation. In "Powders and Grains 2009" (M.

Nakagawa and S. Luding, Eds.), pp. 259–262. American Institute of Physics, Melville, NY.
153. Vardoulakis, I., Goldschneider, M., and Gudehus, G. (1978). Formation of shear bands in sand bodies as a bifurcation problem. *Int. J. Numer. Anal. Methods Geomech.* **2**, 99–128.
154. Vardoulakis, I. (1980). Shear band inclination and shear modulus of sand in biaxial tests. *Int. J. Numer. Anal. Methods Geomech.* **4**, 103–119.
155. Tatsuoka, F., Nakamura, S., Huang, C., and Tani, K. (1990). Strength anisotropy and shear band direction in plane strain tests of sand. *Soils Found.* **30**(1), 35–54.
156. Yoshida, T., Tatsuoka, F., and Siddiquee, M.S.A. (1994). Shear banding in sands observed in plane strain compression. In "Localisation and Bifurcation Theory for Soils and Rocks" (R. Chambon, J. Desrues, and I. Vardoulakis, Eds.), pp. 165–181. Balkema, Rotterdam, The Netherlands.
157. Lade, P.V., Yamamuro, J.A., and Skyers, B.D. (1996). Effects of shear band formation in triaxial extension tests. *Geotech. Test. J.* **19**, 398–410.
158. Lade, P.V. and Wang, Q. (2001). Analysis of shear banding in true triaxial tests on sand. *ASCE J. Eng. Mech.* **127**, 762–768.
159. El Bied, A., Sulem, J., and Martineau, F. (2002). Microstructure of shear zones in Fontainebleau sandstone. *Int. J. Rock Mech. Min. Sci.* **39**, 917–932.
160. Walley, S.M., Radford, D.D., and Chapman, D.J. (2006). The effect of aspect ratio on the compressive high rate deformation of three metallic alloys. *J. Phys. IV* **134**, 851–856.
161. Coulomb, C.A. (1776). Essai sur une application des règles des maximis et minimis à quelques problèmes de statique, relatifs à l'architecture. *Mém. Math. Phys. Acad. Sci. Paris* **7**, 343–387.
162. Desrues, J., Lanier, J., and Stutz, P. (1985). Localization of the deformation in tests on sand sample. *Eng. Fract. Mech.* **21**, 909–921.
163. Chemenda, A.I. (2007). The formation of shear-band/fracture networks from a constitutive instability: Theory and numerical experiment. *J. Geophys. Res.* **112**, B11404.
164. Ikeda, K., Yamakawa, Y., Desrues, J., and Murota, K. (2008). Bifurcations to diversify geometrical patterns of shear bands on granular material. *Phys. Rev. Lett.* **100**, 198001.
165. Bobryakov, A.P. and Revuzhenko, A.F. (2009). Experimental simulation of spiral slip lines on granular materials. *J. Min. Sci.* **45**, 99–104.
166. Backman, M.E. and Finnegan, S.A. (1973). The propagation of adiabatic shear. In "Metallurgical Effects at High Strain Rates" (R.W. Rohde, B.M. Butcher, J.R. Holland, and C.H. Karnes, Eds.), pp. 531–544. Plenum Press, New York, NY.
167. Kimbrough, J.R., Bell, P.M., Bradley, B.K., Holder, J.P., Kalantar, D.K., MacPhee, A. G., and Telford, S. (2010). Standard design for National Ignition Facility X-ray streak and framing cameras. *Rev. Sci. Instrum.* **81**, 10E530.
168. Mollema, P. and Antonellini, M.A. (1996). Compaction bands: A structural analog for anti-mode I cracks in aeolian sandstone. *Tectonophysics* **267**, 209–228.
169. Olsson, W.A. (1999). Theoretical and experimental investigation of compaction bands in porous rock. *J. Geophys. Res.* **104**, 7219–7228.
170. Issen, K.A. and Rudnicki, J.W. (2000). Conditions for compaction bands in porous rock. *J. Geophys. Res.* **105**, 21529–21536.
171. Issen, K.A. and Rudnicki, J.W. (2001). Theory of compaction bands in porous rock. *Phys. Chem. Earth A* **26**, 95–100.

172. Bésuelle, P. (2001). Compacting and dilating shear bands in porous rock: Theoretical and experimental conditions. *J. Geophys. Res.* **106**, 13435–13442.
173. Rudnicki, J.W. (2002). Conditions for compaction and shear bands in a transversely isotropic material. *Int. J. Solids Struct.* **39**, 3741–3756.
174. Rudnicki, J.W. and Sternlof, K.R. (2005). Energy release model of compaction band propagation. *Geophys. Res. Lett.* **32**, L16303.
175. Sternlof, K.R., Rudnicki, J.W., and Pollard, D.D. (2005). Anticrack inclusion model for compaction bands in sandstone. *J. Geophys. Res.* **110**, B11403.
176. Baxevanis, T., Papamichos, E., Flornes, O., and Larsen, I. (2006). Compaction bands and induced permeability reduction in Tuffeau de Maastricht calcarenite. *Acta Geotech.* **1**, 123–135.
177. Borja, R.I. (2006). Conditions for instabilities in collapsible solids including volume implosion and compaction banding. *Acta Geotech.* **1**, 107–122.
178. Holcomb, D., Rudnicki, J.W., Issen, K.A., and Sternlof, K. (2007). Compaction localization in the Earth and the laboratory: State of the research and research directions. *Acta Geotech.* **2**, 1–15.
179. Tembe, S., Baud, P., and Wong, T.-F. (2008). Stress conditions for the propagation of discrete compaction bands in porous sandstone. *J. Geophys. Res.* **113**, B09409.
180. Aydin, A. and Ahmadov, R. (2009). Bed-parallel compaction bands in aeolian sandstone: Their identification, characterization and implications. *Tectonophysics* **479**, 277–284.
181. Chemenda, A.I. (2009). The formation of tabular compaction-band arrays: Theoretical and numerical analysis. *J. Mech. Phys. Solids* **57**, 851–868.
182. Stanchits, S., Fortin, J., Gueguen, Y., and Dresen, G. (2009). Initiation and propagation of compaction bands in dry and wet Bentheim sandstone. *Pure Appl. Geophys.* **166**, 843–868.
183. Sulem, J. (2010). Bifurcation theory and localization phenomena. *Eur. J. Environ. Civ. Eng.* **14**, 989–1009.
184. Charalampidou, E.-M., Hall, S.A., Stanchits, S., Lewis, H., and Viggiani, G. (2011). Characterization of shear and compaction bands in a porous sandstone deformed under triaxial compression. *Tectonophysics* **503**, 8–17.
185. Chemenda, A.I. (2011). Origin of compaction bands: Anti-cracking or constitutive instability? *Tectonophysics* **499**, 156–164.
186. Das, A., Nguyen, G.D., and Einav, I. (2011). Compaction bands due to grain crushing in porous rocks: A theoretical approach based on breakage mechanics. *J. Geophys. Res.* **116**, B08203.
187. Du Bernard, X., Eichbuhl, P., and Aydin, A. (2002). Dilation bands: A new form of localized failure in granular media. *Geophys. Res. Lett.* **29**(24), 2176. doi:10.1029/2002GL015966.
188. Issen, K.A. and Challa, V. (2008). Influence of the intermediate principal stress on the strain localization mode in porous sandstone. *J. Geophys. Res.* **113**, B02103.
189. Wong, T.-F., David, C., and Menéndez, B. (2004). Mechanical compaction. In "Mechanics of Fluid-Saturated Rocks" (Y. Guéguen and M. Boutéca, Eds.), pp. 55–114. Elsevier, London.
190. Katz, Y. and Weinberger, R. (2005). Strain localization in sandstone during embryonic stages of shear-zone evolution. *Terra Nova* **17**, 311–316.
191. Rath, A., Exner, U., Tschegg, C., Grasemann, B., Laner, R., and Draganits, E. (2011). Diagenetic control of deformation mechanisms in deformation bands in a carbonate grainstone. *Am. Assoc. Pet. Geol. Bull.* **95**, 1369–1381.

192. Sengupta, A. (2010). Strain localization in geomaterials in nature, laboratory tests and numerical analyses. *Curr. Sci.* **98**, 1195–1201.
193. Culver, R.S. (1973). Thermal instability strain in dynamic plastic deformation. In "Metallurgical Effects at High Strain Rates" (R.W. Rohde, B.M. Butcher, J.R. Holland, and C.H. Karnes, Eds.), pp. 519–530. Plenum Press, New York, NY.
194. Rudnicki, J.W. and Rice, J.R. (1975). Conditions for the localization of deformation in pressure-sensitive dilatant materials. *J. Mech. Phys. Solids* **23**, 371–393.
195. Kaproth, B.M., Cashman, S.M., and Marone, C. (2010). Deformation band formation and strength evolution in unlithified sand: The role of grain breakage. *J. Geophys. Res.* **115**, B12103.
196. Abelev, A.V. and Lade, P.V. (2003). Effects of cross anisotropy on three-dimensional behavior of sand. 1: Stress–strain behavior and shear banding. *ASCE J. Eng. Mech.* **129**, 160–166.
197. Lade, P.V. and Abelev, A.V. (2003). Effects of cross anisotropy on three-dimensional behavior of sand. 2: Volume change behavior and failure. *ASCE J. Eng. Mech.* **129**, 167–174.
198. Labuz, J.F., Riedel, J.J., and Dai, S.T. (2006). Shear fracture in sandstone under plane-strain compression. *Eng. Fract. Mech.* **73**, 820–828.
199. Johnson, P.A., Savage, H., Knuth, M., Gomberg, J., and Marone, C. (2008). Effects of acoustic waves on stick-slip in granular media and implications for earthquakes. *Nature* **451**, 57–60.
200. Shih, C.J., Nesterenko, V.F., and Meyers, M.A. (1998). High strain rate deformation and comminution of silicon carbide. *J. Appl. Phys.* **83**, 4660–4671.
201. Tejchman, J. and Gorski, J. (2010). Finite element study of patterns of shear zones in granular bodies during plane strain compression. *Acta Geotech.* **5**, 95–112.
202. Börzsönyi, T., Unger, T., and Szabó, B. (2009). Shear zone refraction and deflection in layered granular materials. *Phys. Rev. E* **80**, 060302.
203. Huang, W., Nübel, K., and Bauer, E. (2002). Polar extension of a hypoplastic model for granular materials with shear localization. *Mech. Mater.* **34**, 563–576.
204. Lade, P.V. (2007). Modeling failure in cross-anisotropic frictional materials. *Int. J. Solids Struct.* **44**, 5146–5162.
205. Lade, P.V., Nam, J., and Hong, W.P. (2008). Shear banding and cross-anisotropic behavior observed in laboratory sand test with stress rotation. *Can. Geotech. J.* **45**, 74–84.
206. Pietruszczak, S. (1995). Undrained response of granular soil involving localized deformation. *ASCE J. Eng. Mech.* **121**, 1292–1297.
207. Pietruszczak, S. (1999). On homogeneous and localized deformation in water-infiltrated soils. *Int. J. Damage Mech.* **8**, 233–253.
208. Williams, J.R. and Rege, N. (1997). Coherent vortex structures in deforming granular materials. *Mech. Cohesive Frictional Mater.* **2**, 223–236.
209. Oda, M., Kazama, H., and Konishi, J. (1998). Effects of induced anisotropy on the development of shear bands in granular materials. *Mech. Mater.* **28**, 103–111.
210. Tordesillas, A., Muthuswamy, M., and Walsh, S.D.C. (2008). Mesoscale measures of nonaffine deformation in dense granular assemblies. *ASCE J. Eng. Mech.* **134**, 1095–1113.
211. Tordesillas, A., Hunt, G., and Shi, J.Y. (2011). A characteristic length scale in confined elastic buckling of a force chain. *Granular Matter* **13**, 215–218.
212. Vajdova, V. and Wong, T.-F. (2003). Incremental propagation of discrete compaction bands: Acoustic emission and microstructural observations on circumferentially

notched samples of Bentheim sandstone. *Geophys. Res. Lett.* **30**(14), 1775. doi:10.1029/2003GL017750.

213. Utter, B. and Behringer, R.P. (2008). Experimental measures of affine and nonaffine deformation in granular shear. *Phys. Rev. Lett.* **100**, 208302.
214. Rathbun, A.P. and Marone, C. (2010). Effect of strain localization on frictional behavior of sheared granular materials. *J. Geophys. Res.* **115**, B01204.

8 Shear Banding in Bulk Metallic Glasses

Lan Hong Dai

State Key Laboratory of Non-linear Mechanics (LNM), Institute of Mechanics, Chinese Academy of Sciences, Beijing, China

Nomenclature

$\dot{\gamma}$ shear-strain rate
τ shear stress
$\boldsymbol{T}$ temperature
ξ free-volume concentration
$\boldsymbol{D}$ diffusion coefficient of ξ
$\boldsymbol{g}$ function is the net creation rate of ξ
$\boldsymbol{k}$ wave number
$\boldsymbol{l}$ free-volume diffusion length
h subscript, the homogeneous solutions
G^{ξ} $\partial g/\partial \xi$
$\boldsymbol{A}$ a constant related to the Taylor–Quinney coefficient
G^{θ} free-volume creation rate due to temperature rise
$\boldsymbol{F}$ free-volume softening
κ thermal diffusivity
$\boldsymbol{Q}$ strain hardening
$\boldsymbol{R}$ strain-rate hardening
$\boldsymbol{P}$ thermal softening
Γ, Γ_c energy dissipation in shear band and its critical value
$\boldsymbol{w}$ shear-band thickness
Ω thermal-effect coefficient
K_s shear-band toughness
$\boldsymbol{G}$ shear modulus
$\boldsymbol{d}$ STZ size
γ_c average shear yield strain
ν Poisson's ratio
ℓ dilatation factor
$\boldsymbol{m}$ pressure-sensitivity coefficient
S, S_{cr} shear-band instability index and its critical value
κ_M stiffness of the testing machine
$\boldsymbol{E}$ Young's modulus

Adiabatic Shear Localization. DOI: http://dx.doi.org/10.1016/B978-0-08-097781-2.00008-3

d_s sample diameter
l_s sample length
L_{ext} an extrinsic length scale
L_{int} the internal resistance of a BMG to unstable shear banding
ι a dimensionless parameter accounting for the effect of machine stiffness

8.1 Introduction

Bulk metallic glasses (BMGs), due to the lack of long-range order (LRO) and the absence of traditional defects such as dislocations and grain boundaries [1–5], have a series of intriguing mechanical, physical and chemical properties [6–11]. They have shown widespread potential applications in many fields as structural and functional materials [12–16]. However, plastic flow of BMGs at room temperature (RT) is prone to be highly localized into nanoscale shear bands [17–21]. The initiation and rapid propagation of a shear band can induce catastrophic fracture with very limited ductility [22,23], impeding the further applications of BMGs. Essentially, the shear banding in BMGs is a multiple temporal–spatial and trans-scale process controlled by different physical mechanisms, during which rate-dependent processes such as viscosity/momentum diffusion, thermal/energy diffusion, free volume/mass diffusion, instability nucleation and development should be involved [19,20,24–30]. The key question is how these non-linear and coupled processes with respective characteristic time and length scales govern the shear-band formation and evolution in BMGs.

In this chapter, we present an overview of the inhomogeneous deformation behaviour of BMGs, focusing specifically on the origin (nucleation) and evolution (propagation) of shear banding. The development of BMGs as well as their atomic structure is briefly summarized, followed by an introduction to the general features of plastic flow and the underlying flow mechanisms. Then we proceed to review the results of recent research about shear banding, including experiments, continuum, atomistic modelling and theoretical developments. This chapter concludes with a summary and a view of important unresolved questions.

8.2 Development and Structure of BMGs

The existence of natural glassy materials can be traced back to around 5000 BC. However, metallic glasses or amorphous metallic alloys represent newcomers to glassy materials, having been first reported by Duwez at Caltech in 1960 [31]. Duwez's group made this discovery by rapidly quenching an Au–Si alloy at very high rates: 10^5–10^6 K/s. The work confirmed Turnbull's prediction [32,33] that glass formation in liquid metals is possible if cooling is sufficiently fast and crystallization does not occur. Since then, greater efforts have been made to explore metallic glass systems, which exhibit a high glass-forming ability (GFA) [34–38].

As for early alloys, the timescale for crystallization was usually in the hundreds of microseconds to millisecond range. Metallic glasses were therefore formed only under very rapid solidification conditions (10^3-10^6 K/s), and they were confined to very thin sheets, ribbons or even wires with a characteristic size of less than 50 μm [39]. From the Pd-based glasses studied by Chen [40] and the Turnbull group [35] to a series of new glasses developed by the Inoue group [36,38,41], their crystallization timescales decrease significantly to the range of 1–10 s. In particular, the Vitreloy family developed by Peker and Johnson [37] exhibits a very low critical cooling rate for glass formation of about 1 K/s. Now, a very wide range of multi-component alloys can form BMGs [7–9,14,42,43], including Pa-, Zr-, Cu-, rate earth-, Mg-, Fe-, Ti-, Ni-, Pt-based, etc. The advent of more and larger BMGs arouses a revival of interest in the basic science of glass forming, glass structure and their absorbing and potentially valuable properties [7,9–11,42,44,45].

Examination of BMGs by X-ray or electron diffraction shows the diffuse diffraction halos that may be taken as characteristic of amorphicity [46]. However, this characteristic alone is not sufficient to describe the atomic arrangements within the solid. Historically, a popular structural model for metallic glasses is that of Bernal's dense random packing of hard spheres [2,47]. The hard-sphere mode successfully describes the monatomic systems. As is well known, all BMGs are now multi-compositional, and hence their elements tend to form characteristic local atomic clusters [48]. Following the principle of the efficient filling of space, Miracle [1] proposed face-centred cubic (fcc) packing of solute-centred clusters or short-range order (SRO) as the building scheme for metallic glass structures. Such a packing mode within a medium-range order (MRO) has been confirmed by Ma and co-workers [3] using atomic simulations and directly observed by Hirata et al. [5]. Recent studies [49] suggest that over the MRO, the clusters are connected via a fractal network with the dimension of 2.31, although there are some puzzles regarding this [50]. Regions between clusters are loosely packed clusters with large free volumes [51].

Although the precise description of atomic structures for metallic glasses is still open even now, their mechanical, physical and chemical properties due to such unique structures have attracted a lasting attention. For example, they have high RT strengths much closer to the theoretical limit than their crystalline counterparts, also showing high hardness, large elastic deflection, relatively high RT fracture toughness, good wear resistance and corrosion resistance and so on [7,12]. Some Fe-based systems have excellent soft magnetic properties [52,53], and others have the capability of superconductivity [54] or hydrogen storage [55]. These strong points make them attractive candidates for potential applications ranging from defence and aerospace projects, biomedical devices and Micro Electro Mechanical systems apparatus, to communication equipment and sporting goods [7,12,13, 45,56–59]. However, BMGs have an intrinsic defect, i.e. macroscopic brittleness with very negligible RT ductility. For instance, a Zr-based BMG with substantial fracture toughness $K_{IIC} \sim 75$ MPa/m$^{1/2}$ still exhibits near-zero ductility in uni-axial tension at RT [22], as shown in Figure 8.1. Such low ductility results from the formation and rapid propagation of shear bands within samples, severely impeding

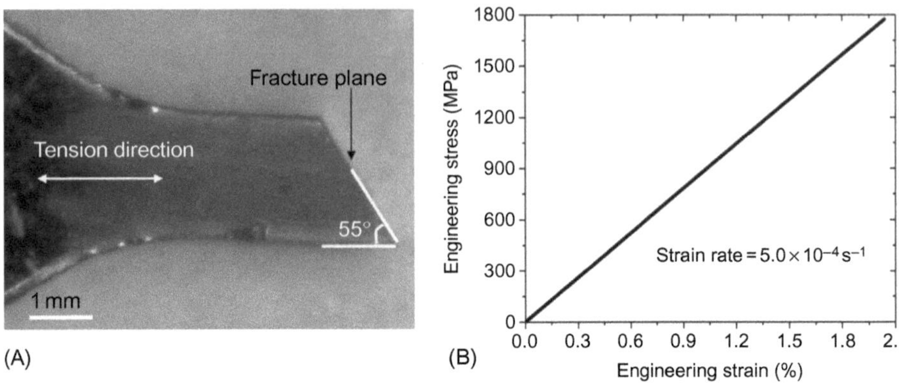

(A) (B)

Figure 8.1 (A) Side view of the fractured sample, a Zr-based BMG, in quasi-static uni-axial tension. (B) Corresponding engineering stress–strain curve for the material.

further exploitation of this class of advanced materials. Therefore, clarifying the mechanism of initiation and propagation of shear bands is of central importance in practice. The questions of why and how a shear band with a characteristic thickness (~10 nm) forms in an atomic-disordered medium are also interesting to scientists. Considerable efforts [10,11,19,44] have been made to examine this aspect during the past decades.

8.3 General Features of Deformation

Based on systematic examination of the deformation and fracture in metallic glasses, Spaepen [17] in his seminal paper constructed a deformation map, which draws the shear-strain rate $\dot{\gamma}$ contours with axes of shear stress τ and temperature T. The map distinguishes the plastic deformation behaviour into two basic modes: homogeneous, where each volume element of the sample contributes to the macroscopic strain, and inhomogeneous flow, where the strain is highly localized into a few very thin shear bands. The former, due to its potential application in fabrication of micro-devices and nano-devices, has been widely studied [45,60–62]. In this chapter, the latter is our interest. The inhomogeneous flow occurs at high stresses and low temperatures. Thin shear bands are the inhomogeneous flow mode for BMGs. Figure 8.2 shows the shear-band type flow in BMGs under different loads, such as compression, shear, bending and indentation. It can be seen from this picture that in unconstrained loading geometries, e.g. tension and shear, only a few shear bands form and dominate the final failure. However, multiple shear bands can form in constrained loading modes, contributing to global plasticity of materials.

Deformation maps have recently been developed by some other researchers. Based on instrumented nano-indentation experiments, Schuh et al. [63] found that

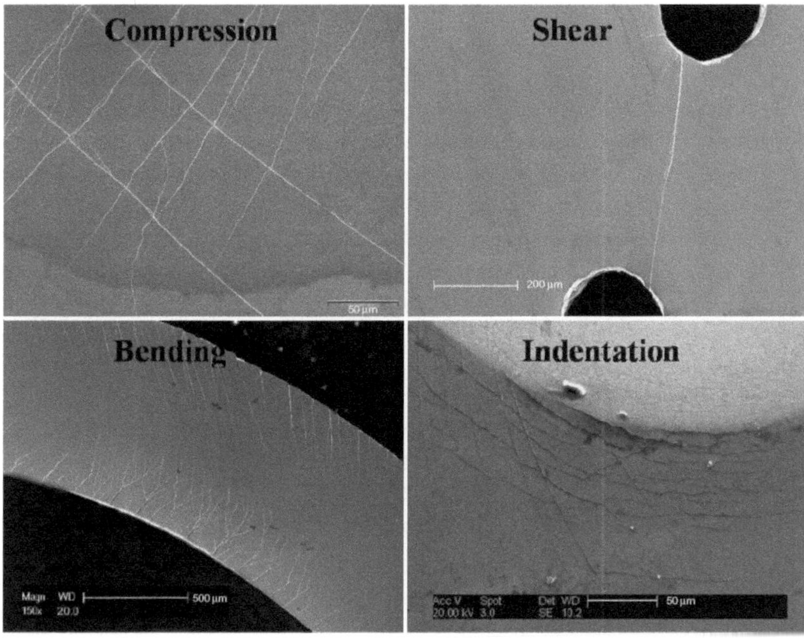

Figure 8.2 Shear bands in BMGs under different loads, including compression, shear, bending and indentation.

there exists a second homogeneous region at high-deformation rates even well below the glass transition (lying in the top-right corner of the Spaepen map). They explained that 'Homogeneous II' with no flow serrations occurs when deformation rates exceed the characteristic rate for shear-band nucleation. In fact, the disappearance of macroscopic flow serrations is mainly due to the synchronous and continued formation of many fine shear bands [64–68]. With increasing strain rates, the plastic deformation inhomogeneity tends to diminish only in time but not in space. In addition, Lee et al. [69] and Park et al. [70] recently found that BMGs can undergo homogenous deformation when they are subjected to a stress below yield at RT, held at the stress for 12 h. Such a deformation region is essentially creep flow. In particular, they observed that BMGs exhibiting more plastic strain during such homogeneous creep deformation tend to show lower global plasticity during inhomogeneous deformation. Such paradoxical phenomenon could be understood by the atomic packing density. Very recently, Furukawa and Tanaka [71] developed a novel rheological model of fracture to describe the fracture of both liquids and glasses in a unified manner. They constructed a dynamic phase diagram on the $T \sim \dot{\gamma}$ plane for a sheared, compressible viscoelastic liquid or glassy material that further divides the inhomogeneous deformation into three regions: solid-type instability, viscoelastic-type instability and liquid-type instability. This is actually consistent with the previous classification by Spaepen [17].

In general, the elastic deformation behaviour of materials is determined by the interaction of atoms inside. The plastic flow, however, should take into account the details of microstructure, more specifically defects. The flow in crystalline alloys can be described by the classical theory of plasticity underpinned by defects such as dislocations, twins or grain boundaries. These crystal defects do not exist in metallic glasses without LRO, and therefore the flow mechanism in metallic glasses must be different from that in crystalline alloys. Although the precise picture of how local atoms respond in deforming metallic glasses is not fully resolved, there is general consensus in the metallic glass community that the fundamental unit of the process underlying plastic flow must be a local *structural-rearrangement* type 'flow event' that can accommodate shear strain [17,18,72,73]. Whereas crystal dislocations allow changes in the atomic neighbourhood at low energies in crystals, the local flow event in metallic glasses requires relative high energy or high stress. Thus, how to describe such a local flow event and build a bridge between it and macroscopic deformation becomes the key to the flow mechanism.

A breakthrough has been made by Spaepen. For the first time, he applied the classical free-volume model developed by Turnbull and co-workers [74–76] to the deformation of metallic glasses [17]. Spaepen's model essentially views the flow event as an individual atomic jump. At a sufficiently high stress, an atom with a hard-sphere volume can be squeezed into a neighbouring hole with a smaller volume. This makes the neighbours of the new position move outwards and creates a certain amount of free volume. A series of discrete atomic jumps finally results in macroscopic plastic flow. The free-volume model presents a relatively systematic theoretical framework that introduces a simple state variable to the problem of glass flow. It opens a window into the understanding of the glass-flow mechanism through atomistic defects. Inspired by the shearing deformation of 2D bubble glasses [77], Argon proposed that the number of atoms involved in the flow event should be between a few to approximately 100. The flow event can be described by a local plastic or inelastic rearrangement of atoms, commonly termed the shear transformation zone (STZ). The first quantitative model of STZ behaviour was developed by Argon himself [18], who treated the problem in the context of an Eshelby-type inclusion [78,79]. He argued that an isolated STZ is not free but confined to the surrounding elastic matrix. The STZ is essentially a local cluster of atoms that undergoes an inelastic shear distortion from one relatively low energy configuration to a second such configuration, crossing an activated configuration of higher energy and volume. In a manner, an STZ in metallic glasses has an analogy with dislocation motion in crystals but at SRO length scales. Although an STZ is a flow event, not a defect, it is strongly affected by local atomic structures such as free volume, short-range chemical or topological order. For example, availability of free volume is important for the STZs to operate in a given volume of BMGs. In fact, STZ operations occur preferentially in those regions of higher free volume because relatively less dilatation is required. Spaepen [80] has envisioned that only a high free volume would contribute to deformation. He [73,80] introduced a concept of flow defect that is analogous to an STZ in terms of free volume, and he succeeded in extending an individual atomic jump to atomic cluster rearrangement. Since Argon proposed the STZ model, scientists have made great efforts to capture such flow events via experiment or simulation [72,81,82].

Either the free-volume or the STZ model is a mean-field theory. Nevertheless, the two models provide a basis for further analysis of the strain localization process in BMGs. Following the STZ model, Johnson and Samwer [83] made the first attempt to deal with such an aspect. They proposed a cooperative shear model (CSM) to understand the yield strength when inhomogeneous deformation occurs, although the spatial features of inhomogeneous deformation were still precluded in their analysis. As mentioned earlier, an isolated STZ is always confined by the surrounding elastic medium. Such elastic confinement would lead to reversible elastic energy storage in the STZ-matrix system, implying that transformed STZs have a memory of their original untransformed state. A question naturally arises: How do these STZs with memory induce memoryless plastic flow? Johnson and Samwer [83], Mayr [84] and Harmon et al. [85] argued that this process is analogous to a relaxation mechanism of glasses, which can be explained from a potential energy landscape (PEL) perspective [86–88]. The individual STZ operations can be viewed as fast β-relaxation processes that are sub-IS hopping events. The avalanche percolation of these STZs is associated with the slower α relaxation process, an intra-IS hopping event. The plastic irreversible α hopping event will contribute to macroscopically perceptive plasticity. The CSM model thus suggested that inhomogeneous yielding occurs when a critical fraction of 'minimum' barrier STZs becomes unstable and results in global instability. Merging this picture with the Frenkel model [89] of shear strengths in dislocation free solids, Johnson and Samwer [83] derived a universal power law of two-thirds of temperature for the yield strength of metallic glasses. Based on the CSM, the activation energy and size of STZs have been determined by simulations [84] or experiments [81]. Very recently, Yu et al. [90] provided possible evidence that the activation of STZs and β-relaxations are directly related by examining their activation energies.

8.4 Physical Origin of Shear-Banding Instability

The flow theories reviewed in the preceding material successfully explain how a metallic glass deforms plastically; however, these theories do not give a clear answer to inhomogeneous deformation or shear banding. Why does the flow in a metallic glass localize into such extremely thin shear-band regions? What initiates this localization process? What is the onset condition of shear-banding instability? These questions continue to plague and challenge scientists. Historically, there are two hypotheses as to why shear bands occur in metallic glasses. The first suggests that shear-induced dilatation inherent in the flow event causes the generation and coalescence of free volume, leading to a precipitous drop in viscosity within the shear band. This idea originated from the work of Spaepen [17]. Subsequent works from Argon [18], Steif et al. [91], Vakes [92], Falk and Langer [72], Huang et al. [24] and Wright et al. [93] have also shown the importance of free-volume dynamics to shear instability. Here, the free-volume formation is indeed a stress-drive structural change; hence, this hypothesis contends that shear banding in metallic glasses has a structural origin. Alternatively, Leamy et al. [94] proposed that shear-banding events are thermally initiated, similar to adiabatic shear bands (ASBs) in crystalline alloys. Local adiabatic heating occurs, decreasing the viscosity by several orders of magnitude within the shear band. Although the two concepts seem contradictory, in both cases, a change in viscosity localizes the deformation and incurs shear banding. Actually, based on systematic experimental and theoretical investigations [25,26,28,30,95–97], Dai and co-workers [19,20,27] proposed that the free-volume creation joining with the temperature rise induces the shear-band initiation in metallic glasses. In this coupled softening process, however, the free-volume softening plays the dominant role, while the thermal softening is the assistant effect. To clarify the third hypothesis, it is necessary to reinspect the shear bands in metallic glasses, especially focusing on variables that affect their origin.

8.4.1 Variables Relevant to Shear Banding

The most noticeable feature of shear bands in metallic glasses is that they can form at low (usually quasi-static) strain rates, whereas ASBs in crystalline alloys usually occur in dynamic cases. In addition, shear bands in metallic glasses also exhibit strain-rate dependence. To investigate whether the strain rate exerts a role in the shear-band formation in BMGs, Dai and co-workers [25,28,96] performed different strain-rate levels of plate shear, shear-punch and compressive testing on BMG specimens. Figure 8.3 shows the shear-band patterns (arrow point) at the notch tip during plate shear. The number of shear bands is smaller at quasi-static strain rates (Figure 8.3A) than that at dynamic strain rates (Figure 8.3B). The spatially positive strain-rate dependence of shear-banding formation is also clearly observed in the shear-punch tests, as shown in Figure 8.4. This figure shows that the number density of shear bands (arrow point) around the circular deformation region at dynamic strain rates (Figure 8.4B) is far larger than that at quasi-static cases (Figure 8.4A). This trend seems to be consistent with a series of otherwise experimental observations, such as tensile testing by Mukaia

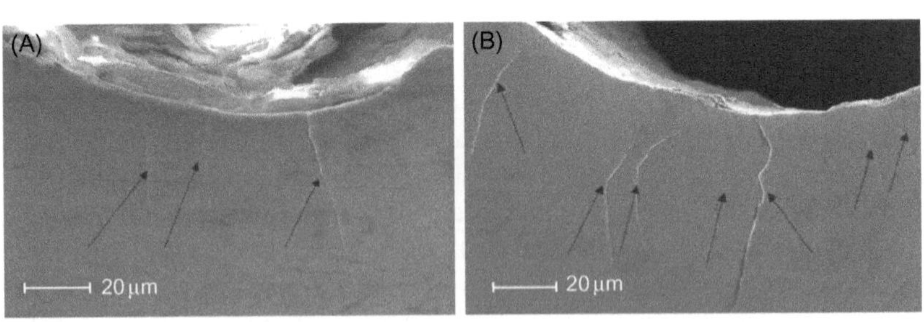

Figure 8.3 Shear-band patterns at the notch tip in Vit 1 BMG under (A) quasi-static and (B) dynamic plate shear.

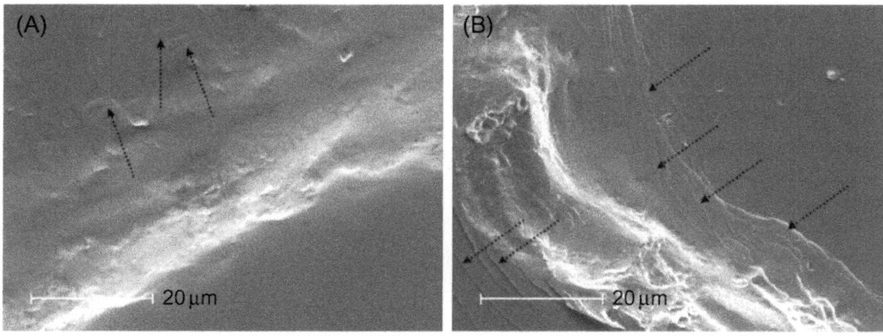

Figure 8.4 Shear-band patterns around the circular deformation region in Vit 1 BMG during shear punch under (A) quasi-static strain and (B) dynamic rates.

et al. [98], compressive testing by Jiang et al. [22] and indentation by Dai et al. [95] and Subhash and Zhang [65]. Recently, Jiang et al. [68] performed compression tests on Zr-based BMGs at various strain rates. They indeed observed that, with increasing strain rate, shear bands tend to form continuously in time. In other words, a higher strain rate promotes shear-band formation. However, the resultant spatial patterns can be of positive or negative effects, depending on material components and stress states. The strain-rate dependence on spatio-temporal features of shear banding implies that momentum dissipation plays an important role on the shear-banding process.

The characteristic thickness of mature shear bands (MSBs) in metallic glasses also attracts much attention by providing an important clue to their origin. Masumoto and Maddin [99], perhaps for the first time, reported deformation lines (shear bands) with a thickness of 20 nm on the tension side of the sample, during bending of a filament of $Pd_{80}Si_{20}$ metallic glass. Later, precise measurements showed a narrow range of thicknesses, listed in Table 8.1, identifying 10 nm as the characteristic thickness of shear bands in metallic glasses. It is impossible for such extremely localized flow to be thermal induced (discussed shortly) because atomic/cluster-scale structural changes should be responsible for it. For example, Pampillo [107] found that the shear bands were preferentially etched. Similar preferential attack phenomenon was also noticed by Donovan and Stobbs [100] in the deformation of an Fe-based metallic glass ribbon, and recently by Dai and co-workers [26] in the pre-etching sample surfaces, as shown in Figure 8.5. The preferential etching susceptibility of these shear bands clearly means that the chemical potential within the bands has been changed with respect to the rest of the materials. These researchers believed that it is the structural change that leads to this effect by destroying the SRO structure existent in metallic glasses to a more disordered structure. In fact, such disordered structures correspond to the free-volume defect, which was confirmed by work carried out by Li and Li [108,109], Gu et al.[104], Chen et al. [105], Cao et al. [110] and Flores et al. [111]. In particular, Donovan and Stobbs observed enhanced small-angle scattering in the shear bands formed in tension, which they attributed to the presence of sub-nanometre-scale voids. They speculated that excess free volume, stabilized by shear stress during deformation,

Table 8.1 Summary of Measurements of Shear-Band Thickness in Metallic Glasses

Metallic Glass (in at%)	Method	Thickness (nm)	Observers
$Pd_{80}Si_{20}$	TEM of replica	20	Masumoto and Maddin [99]
$Fe_{40}Ni_{40}B_{20}$	TEM	10–20	Donovan and Stobbs [100]
$Zr_{56.3}Ti_{13.8}Cu_{6.9}Ni_{5.6}Nb_{5.0}Be_{12.5}$	TEM	≤10	Pekarskaya et al. [101]
$Zr_{57}Ti_{5}Cu_{20}Ni_{8}Al_{10}$	HRTEM	≤20	Li et al. [102]
$Al_{90}Fe_{5}Gd_{5}$; $Al_{86.8}Ni_{3.7}Y_{9.5}$	HRTEM	10–15	Jiang and Atzmon [103]
$Zr_{52.5}Ti_{17.9}Cu_{20}Ni_{8}Al_{10}$	HRTEM	~15	Gu et al. [104]
$Zr_{65}Al_{7.5}Ni_{10}Cu_{7.5}Ag_{10}$	HRTEM	~5	Chen et al. [105]
$Cu_{47.5}Zr_{47.5}Al_{5}$	HRTEM	~10	Kim et al. [106]
$Zr_{41.2}Ti_{13.8}Cu_{10}Ni_{12.5}Be_{22.5}$	AFM	~20	Jiang and Dai [30]

TEM, transmission electron microscopy; HRTEM, high-resolution TEM; AFM, atomic force microscopy.

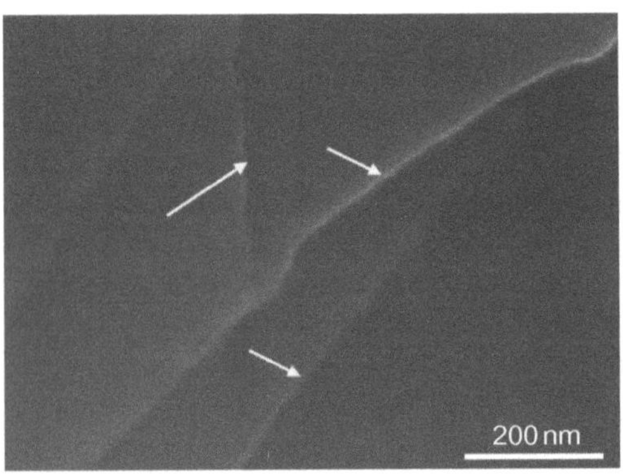

Figure 8.5 Preferential etching at shear bands in a Vit 1 BMG.

coalesced into voids after the deformation stopped. Dai and co-workers [26] did observe some nanovoids (Figure 8.6) formed at the intersection sites of the shear bands in a post-deformed Vit 1 BMG. These experimental phenomena – including the preferential etching of shear bands, excess free volume within shear bands and nanovoids at the intersection sites of shear bands after deformation – provide useful information that local structural changes occupy a unique niche during the initiation of shear bands in metallic glasses.

Shear bands, as a form of localized plastic flow, are not completely free from the thermal effect or local heating. Shear bands in metallic glasses are no exception either. Indirect evidence for local heating comes from melted droplets on the fracture surface (Figure 8.7) or a sparking phenomenon at the moment of fracture. However, it is incorrect to deduce that shear bands in metallic glasses have a thermal origin or are essentially an adiabatic phenomenon, especially at their initiation stage. One way to approach this question is to attempt to calculate the temperature

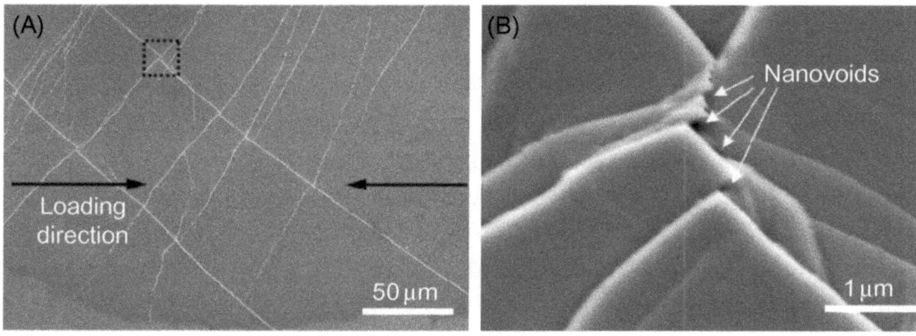

Figure 8.6 (A) Multiple shear bands formed during quasi-static compression of a Vit 1 BMG and (B) nanovoids exist at the intersection site of shear bands indicated in (A).

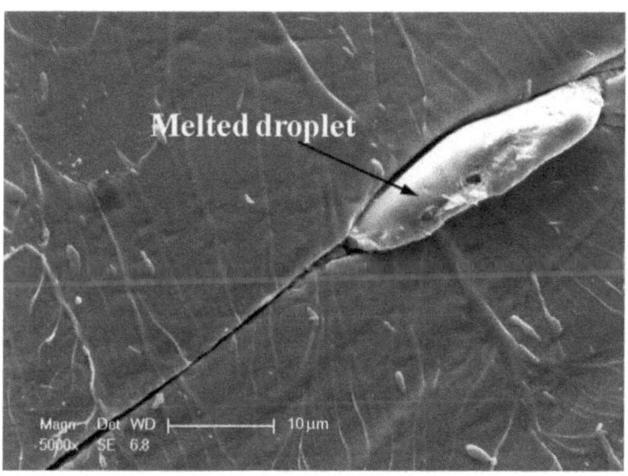

Figure 8.7 Melted droplet on tension fracture surfaces of a Zr-based BMG.

rise within a shear band. The reasonable calculation requires a precondition that we must know the mode of shear-band propagation or, more specifically, the shear-band velocity. It can be deduced that as the shear-band velocity increases, the predicted maximum temperature increases because less time is available for thermal dissipation away from the shear band. The shear-band velocity in metallic glasses has values ranging from 10^{-5} m/s to a few km/s. The calculated maximum temperature in shear band therefore varies from tens of kelvin to several thousand degrees kelvin [25,39,112–116], leading to confusions. It is well known that the most powerful evidence of whether the local heating occurs during shear banding or not is gained by directly measuring temperature rise in shear bands, which is hindered by the extremely thin thickness and short duration of shear bands, however [117–121]. Recently, Lewandowski and Greer [115] invented a clever, simple way to improve the resolution of measurements. They coated a Vit 1 BMG with a thin layer of tin and observed it after double-notched, four-point bending. This method has remarkable resolution: 30 ps (the thermal diffusion time through the coating) and 100 nm (the scale of the melting bead of the coating), respectively. Based on the observations of many melted hemispherical beads of a thin deposit at the notch that underwent shear banding, and further treating shear bands as planar sources of heat, they reported that the temperature rise in the shear band may be several thousand degrees. However, it is noted that their estimation strongly depends on the duration of shear banding (they adopted 90% of the shear-wave speed). In addition, shear bands were observed by Zhang et al. [122], which did not incur melting of the tin. Recently, by revising the shear time and stress for an operating shear band, Miracle et al. [116] recalculated the thermal profiles around the band and concluded that the temperature rises were insignificant. In fact, according to the ex situ observation, we do not exclude such a possibility that the melting of the tin occurs during the final fracture event. Notwithstanding a lot of controversy, the local heating, maybe as an accompanying effect, remains important to shear banding in metallic glasses [123].

As we know, shear banding is a dissipation system. ASBs in crystalline solids are determined by the competition between the momentum dissipation and the energy (heating) dissipation. The preceding review of experimental observations demonstrates that strain rate, free volume and local heating are all possible factors affecting the shear-banding instability in metallic glasses.

8.4.2 Continuum Modelling and Analysis

Argon [18] modelled the flow localization (shear band) as a consequence of strain softening from the free-volume clustering. He considered a shear band located in the centre of a 1D planar layer initially shearing at a constant strain rate. Then, he introduced the free-volume dynamics into the constitutive equations that are described within the STZ formulism; he further derived a bifurcation equation describing the divergence of strain rate in and out of the band. His numerical solutions show the strain acceleration in the band with increasing applied shear strain and the concomitant decrease of shear-strain rate in the surrounding matrix. A similar analysis was performed by Steif et al. [91], via developing Spaepen's free-volume model [17]. He derived the constitutive and free-volume evolution equations for the band and the matrix, respectively. In this case, the free-volume perturbation in the band was introduced directly. Calculations showed that such perturbations can indeed cause the shear-strain localization in the perturbation zone.

Later, Spaepen's free-volume mode was generalized by Steif et al. [124], and Huang et al. [24] added multi-axial stress states through the use of the Mises stress τ_e or the effective stress. In particular, Huang et al. [24] introduced the free-volume diffusion into the free-volume evolution process. The diffusion of the free volume is analogous to the diffusion of vacancies in crystalline materials. Thus, the free-volume evolution equation becomes

$$\frac{\partial \xi}{\partial t} = D\frac{\partial^2 \xi}{\partial y^2} + g(\xi, \tau_e) \tag{8.1}$$

where the g function is the net creation rate of free-volume concentration ξ, including the free-volume generation and annihilation [17], and D is the diffusion coefficient of the free-volume concentration. By linear stability analysis for homogeneous flow, Huang et al. revealed that the homogeneous flow will become inhomogeneous if the net generation rate of free volume is greater than its diffusion rate, i.e.

$$\frac{\partial g}{\partial \xi} > k^2 l^2 \tag{8.2}$$

where k is the wave number of the perturbations and l is the free-volume diffusion length scale. Further calculation indicated that, with the applied shear strain under simple shear, there is a peak in $\partial g/\partial \xi$ that corresponds to a stability point due to the free-volume coalescence. Such a physical picture was recently re-examined by Gao [125] who developed an implicit finite-element method (FEM).

All the results just mentioned lend insight into the free-volume origin of shear bands in metallic glasses. However, these analyses exclude the thermal effect. In other words, the shear-band formation is considered as an isothermal process. Recently, Gao et al. [29] performed a thermo-mechanical instability analysis of a shear band in metallic glasses, through the introduction of the thermal transport equation into the free-volume-based constitutive law. In contrast to the result of Huang et al. [24], they revealed that an increase in temperature perturbation brings about the bifurcation from homogeneous to inhomogeneous deformation modes. In fact, the thermal instability only plays a secondary role in the shear-band instability, which will be discussed shortly.

Since we proposed in 2005 for the first time [27] the coupled effect of free-volume softening and thermal softening on shear-banding instability, it has been identified by more and more work [126–130]. However, some critical questions have not been answered well. During such a coupled process, is either free-volume softening or thermal softening responsible for the onset of shear instability? How do the free volume and temperature interplay with each other, and how do they act? Recently, Dai and co-workers [19,20,27] presented a theoretical description of coupled thermo-mechanical deformation of BMG undergoing a 1D simple shear, focusing on the physical origin of shear-band instability. Considering that the strain rates, the free volume and the temperature rise are possible factors in initiating a shear band, their model takes into account the momentum balance, the energy balance and the dynamics of the free volume, during which the constitutive law can be the same as Steif et al. [91] and Huang et al. [24]. Firstly, they examined the homogeneous deformation case. By defining the thermal instability index and the free-volume instability index, they revealed that the free-volume production facilitates the sudden increase in the temperature before instability and vice versa. Through a rigorous linear perturbation analysis, they obtain the onset condition for a shear-band instability (subscript 'h' denotes the homogeneous solutions) [20],

$$G_{\mathrm{h}}^{\xi} + \frac{A\tau_{\mathrm{h}} G_{\mathrm{h}}^{\theta} F_{\mathrm{h}} - \kappa Q_{\mathrm{h}} k^2}{\kappa R_{\mathrm{h}} k^2 + Q_{\mathrm{h}} - A\tau_{\mathrm{h}} P_{\mathrm{h}}} > Dk^2 \tag{8.3}$$

where the left side denotes the net generation rate of free volume that is temperature dependent and the right side is its diffusion rate. Obviously, this criterion, Eq. (8.3), is physically analogous to the condition in Eq. (8.2), but the difference is that the former involves the temperature effect. Actually, if the free-volume softening dominates, the criterion in Eq. (8.3) can totally reduce to the condition in Eq. (8.2) proposed by Huang et al. [24]. If the shear instability is dominated only by the thermal softening under the adiabatic limit, the criterion in Eq. (8.3) is identical to the onset condition for the conventional thermoplastic shear instability revealed by Bai [131]. According to the instability condition in Eq. (8.3), the dynamic balance between the stabilizing and destabilizing effects determines a critical wavelength. By examining the dominant instability mode, we can obtain the internal timescale in the present coupling-softening instability. According to the relative importance of the free-volume softening and thermal softening, the internal scales can be converted into the internal free volume or thermal scales, respectively.

The internal length scale measures whether the instability occurs easily or not, while the internal timescale characterizes how fast the instability initiates. The internal length and timescales are plotted against the applied strain rates in Figure 8.8. Clearly, the internal scales of instability for the three cases decrease with increasing strain rate. This might be the main reason that the shear instability due to either thermal softening or free-volume creation is more probable at higher strain rates. This numerical result agrees well with the available experimental observations [25,28,95]. The internal free-volume length and timescales are remarkably smaller than those in the thermal-softening case, indicating that shear instability resulting from free-volume creation occurs easier and faster than thermal instability. At low strain rates, the thermal internal length and timescales are very large. This implies that thermal softening occurs with great difficulty at low strain rates. However, at this strain rate, the free-volume softening still appears. It is well known that shear banding instability in BMGs occurs not only at high strain rates but also at quasi-static loading, as described in Section 8.4.1. So, it is probable that shear banding or strain localization is started by free-volume softening. This weak influence of thermal softening on shear instability at low strain rates results in, the coupling shear-instability behaviour are more like that due to free-volume creation. In such cases, shear instability in BMGs approaches an isothermal process, during which the internal length scale is approximately tens of nanometres, and the internal timescale is roughly the inverse of strain rate. As the strain rate increases to the dynamic range, such as $10^3\ s^{-1}$, the internal length scales due to both free-volume softening and couple softening decrease to nanometres or sub-nanometres, while the thermal length scale is of the order of 10–100 μm. Also, the coupling of softening with the internal timescale (~μs) occurs much faster than the sole thermal-softening timescale (~ms). Therefore, under a dynamic strain rate, the thermal softening favours the shear instability originating from free-volume softening, leading to lower values of internal length and timescales in the coupled softening case.

Furthermore, a shear-band analysis similar to those by Argon [18] and Steif et al. [91] was performed with inclusion of temperature. After numerical calculations,

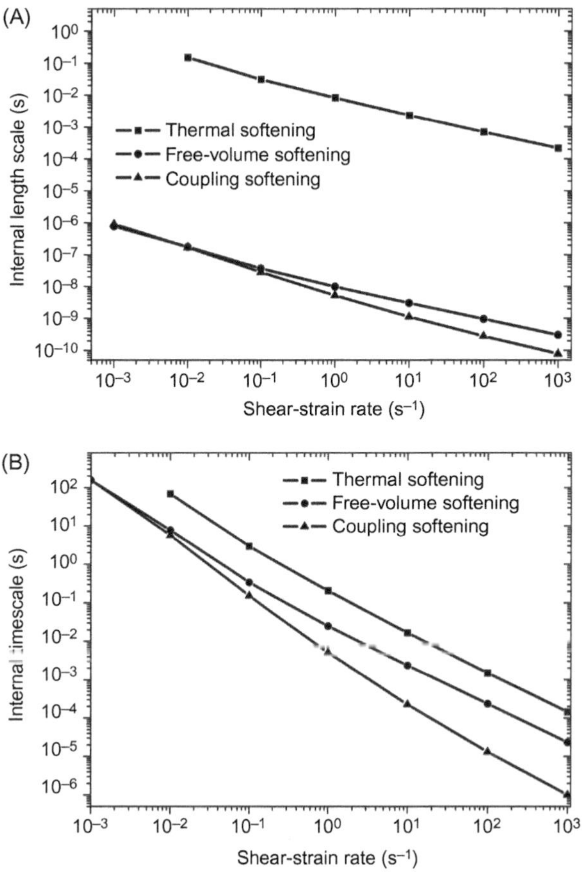

Figure 8.8 Stability analysis of the homogeneous deformation: the dependence of shear-strain rate on (A) the internal length scale and (B) the internal timescale.

they found that the catastrophic character of shear banding or strain localization is evident. Figure 8.9A illustrates the remarkable acceleration of strain development in the shear band after a peak stress has been reached and the corresponding drop of strain rate in the matrix. In addition, the great increase of strain rate in the band results in the rapid rise of the inner shear strain (Figure 8.9B), giving rise to a shear band. During such a process, the viscosity in the band drastically decreases to a value of approximately 10^{-5} poise, much smaller than that ($\sim 10^{11}$ poise) outside the band [20].

It is well known that the initiation of shear banding hinges strongly on the catastrophic drop of local viscosity. To ferret out the main reason for this material weakening, they calculated both the free-volume concentration and the temperature increase in the shear band during the shear deformation, as presented in Figure 8.10. Interestingly, it is noted that the sharp bend up in the curve of free-volume concentration is prior to that in the temperature rise. The result provides much clearer evidence that the local material softening or instability due to free-volume creation is earlier and faster than that due to temperature rise. This is consistent with that given by the

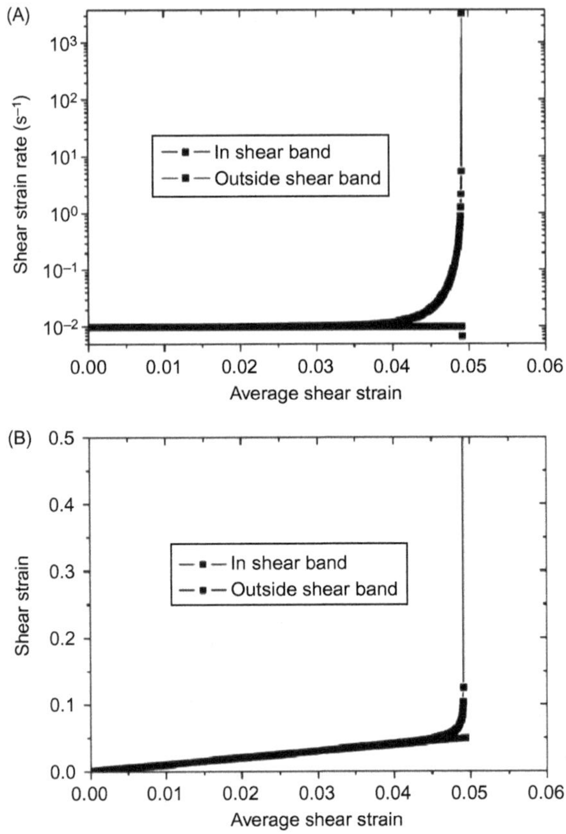

Figure 8.9 The shear-banding process. A history of (A) strain rate and (B) strain is shown for both the forming shear banding and the surrounding matrix.

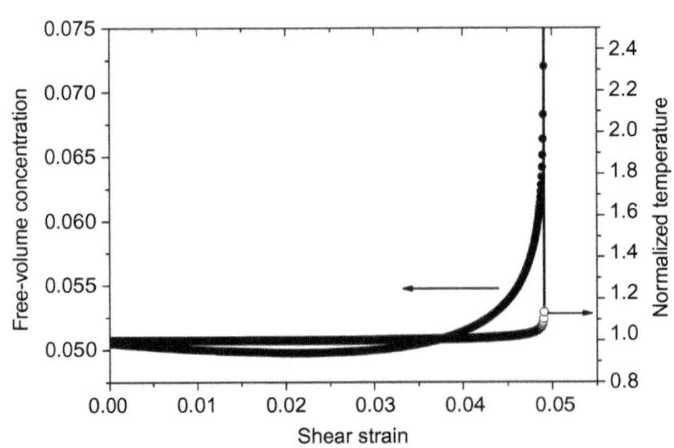

Figure 8.10 The evolution of free-volume concentration and normalized temperature in the shear band as the applied shear strain increases.

linear perturbation analysis. In addition, the temperature rise is the consequence of free-volume-induced shear localization, not its cause; this agrees well with the deduction of many researchers based on experimental observations. In the next section, atomistic simulations also show that local heating occurs after the onset of localized flow in amorphous alloys. In particular, these results provide a powerful theoretical expatiation on the puzzle about temperature rise at shear bands and thickness of shear bands in metallic glasses. The local free-volume creation, as the origin of shear localization, controls shear-band thickness (see Section 8.5 for more details), while local temperature rise, as a secondary effect, depends strongly on the development of this shear localization [128,129].

Both experiments and modelling show that large shear strain can be achieved in the shear band. Considering such large deformation, some finite-deformation coupled thermal–mechanical constitutive models of BMGs have been recently developed [126,127,132,133]. For example, the finite-deformation model proposed by Yang et al. [126] can well capture the quasi-static compression behaviour of Vit 1 BMG at RT. Further, they calculated the stress–strain curves with adiabatic heating at strain rates of 1.0×10^{-4}, 1 and $1.0 \times 10^{4}\ s^{-1}$. In contrast to the isothermal case, a significant decrease of the stress, or strong strain softening, can be predicted. Thus they conclude that although the free-volume increase initiates the shear band, the subsequent temperature rise accelerates the localization of deformation. This result is satisfactorily consistent with the previous conclusion [20]. Very recently, Jiang et al. [134] performed a series of three-point bending experiments on as-cast and annealed samples of Vit 105 BMGs over a wide range of temperatures varying from RT to liquid nitrogen temperature (77 K). They found that the significant decrease of free volume within samples, corresponding to the decease of STZ volume, can result in the disappearance of shear banding. However, the shear banding can still occur even at 77 K. This confirms the dominated role of free volume on the shear-banding instability in BMGs.

8.4.3 Atomistic Modelling

The formation mechanism of a shear band presented previously captures many experimental observations in metallic glasses. However, current experimental techniques, including positron annihilation, ultrafast infrared imaging, TEM and SEM and AFM, have difficulties to study such shear localization processes that have extremely short timescales and very small length scales. Moreover, most tests have to be performed post-mortem after sample deformation or fracture. The conclusions drawn from these tests are, therefore, often partial and sometimes contradict each other. The molecular dynamics (MD) simulation is generally believed to be an effective way, providing an in situ and real-time observation of the shear localization at the atomic level. In particular, the characteristic sizes of shear bands in metallic glasses just fall into the reach of direct MD simulations. In the following, we present some recent works pertinent to the initiation of shear banding by using the MD method.

Physically, a shear band is a narrow region with the plastic shear strain larger than that in the rest of the sample. During MD simulations, we can record the position of each atom at each run-step and then identify the localized region undergoing larger shear deformation using the atomic positions recorded. Currently, there are three popular methods: 'stripe-painting', proposed by Bailey et al. [135]; D^2_{min}, developed by Falk and Langer [72] and the atomic local shear strain η_i^{Mises}, proposed by Shimizu et al. [136].

In the framework of an MD simulation, the Voronoi method is usually used to characterize the glass structure. If a shear band is developing in a sample, this method still has a capability of capturing the structural change. There are two ways to indicate the structural change during deformation. The first is to directly look at a specific type of coordinate polyhedron and its volume fraction in a sample [110]. The second way is to estimate the free volume based on the Voronoi volume v_{voro} of different types of polyhedral [109].

Very recently, Ma and co-workers [110] combined the two methods to monitor the local structural evolution in the early stage of shear banding. They performed the uni-axial compression on a Cu−Zr metallic glass, during which the shear band is indicated by η_i^{Mises}. Furthermore, they used the fraction of the Cu-centred full icosahedra (FI) as the key indicator of the glass structure. They observed that a band is developing, in which the FI motifs are lost preferentially relative to other regions in the sample. This decrease in the fraction of stable and shear-resistant FI causes structural softening. Such local structural softening corresponds to the generation of a Voronoi or free volume in the shear band. Similar results were obtained recently by Li and Li [108,109]; in addition, they analysed the distribution of the NN atomic bond lengths (ABLs) at various strains [108]. It was found that the ABLs increase with increasing strains in the shear-band region, while the ABLs outside remain small and uniform. These results indicate that the structural disordering or free volume in the shear band is much more significant than that outside. In other words, the initiation of the shear band is always accompanied by the local structural softening, which is consistent with the experimental observation [26,39,100] and the previous continuum mechanical analyses.

To check whether a shear band originates from the thermal softening or not, we must determine the temperature rise within the shear band. Ma and co-workers [110] investigated the evolution of the temperature rise in the sample during deformation. Note that in this initial stage of localization, the temperature rise in the band is less than 150 K; the band remains cold. Their results also show that the temperature rise lags behind the structural softening in the shear band. By investigating the D^2_{ave} and temperature rise during localization, Bailey et al. [135] have obtained a similar result. The D^2_{ave} localization occurs over the interval 5−10% strain, whereas the temperature rise is somewhat delayed; it does not start until just before 10% strain. The simulation results agree well with our shear-band analysis (see Figure 8.10); i.e. the temperature rise is not a cause of shear banding but a consequence. Based on small-scale MD simulations and thermo-mechanical analysis, Shimizu et al. [136,137] have proposed an aged-rejuvenation-glue-liquid (ARGL) model for an MSB, during which a critical length scale of the order for STZs to develop into an MSB is predicted. It is found that, at the later stage of shearing, the local temperature at the centre of the shear band has reached the glass transition temperature, which is consistent with the result of Yang et al. [119].

8.5 The Shear-Band Evolution Process

After it nucleates, a shear band would propagate forward driven by far-field loading. This post-instability process is inextricably linked to final fracture, which has attracted growing attention. In this section, we review recent advances in this area, notwithstanding that there is a great deal unknown about the phenomenon.

8.5.1 *Shear-Band Propagation Dynamics*

The first attempt to clarify the detailed process of shear banding in metallic glasses was performed by Masumoto and Murata in 1976 [138]. Using a very hard tensile machine, they indirectly inferred that shear-band propagation is rapid and intermittent. Shortly afterwards, Neuhäuser [139] made observations by high-speed cinematography on the development of shear bands during tension of $Pd_{80}Si_{20}$ ribbon and during bending of an Fe-based ribbon, and observed that the bands propagated across the sample in several rapid bursts with time intervals of several seconds between the bursts. However, due to the insufficient temporal resolution (up to 1.7 ms), the single burst of the shear band was not captured. Interestingly, he noted that a pre-existing shear band can slow down the growth of a neighbouring band. In addition, some bands have been observed to disappear totally or partially, similar to crack healing. By recording acoustic emission (AE) signals in tension-strained Zr-based BMGs, Vinogradov and Khonik [140] revealed that the microscopic features of the shear banding in BMGs are very nearly the same as those found for ribbon metallic glasses. The shear band propagates in a jump-like mode as reflected by numerous AE bursts. Based on the AE signal, the initiation of shear bands was probed by Klaumünzer et al. [141]. Recently, the infrared camera with 1000 Hz frames was used to observe in situ the dynamic shear-banding process in compression of Zr-based BMGs by Jiang et al. [68], considering that shear banding causes an increase in temperature. Such spatial resolution is still not able to seize a full shear-banding process. However, based on the successive video frames, they demonstrated the spatio-temporality of shear bands in metallic glasses, as illustrated schematically in Figure 8.11. With decreasing strain rates, the plastic flow tends to be inhomogeneous with time and homogeneous in space; while with increasing rates, the plastic flow tends to be homogeneous with time and inhomogeneous in space. The temporal feature of shear bands has been confirmed by Dai and coworkers [66,67]. They conducted MD nano-indentation of Cu−Zr metallic glasses at different loading rates. The simulation result showed that at higher strain rates, the shear-banding events operate successively, whereas at lower rates, the shear events occur intermittently, exhibiting inhomogeneity in time, as shown in Figure 8.12. In addition to the strain rate, the shear-band temperature can also affect its propagation behaviour. Cheng et al. [142] pointed out that if it can remain cold, a shear band can slide in a stick-slip manner. However, a hot shear band will directly develop into a runaway catastrophic failure.

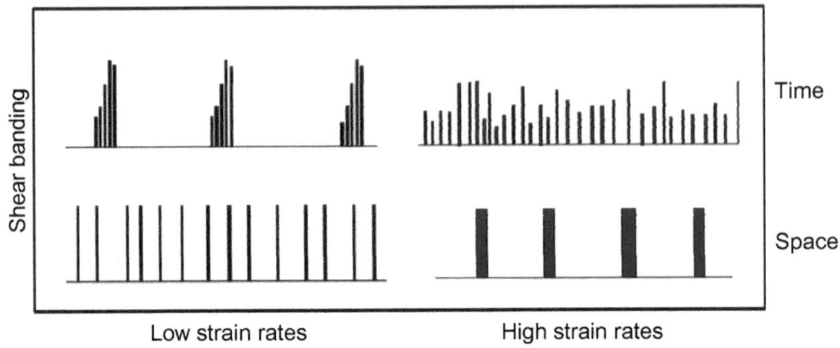

Figure 8.11 Schematic illustrations of temporal and spatial distributions of shear-banding operations at lower strain rates (left) and higher strain rates (right).

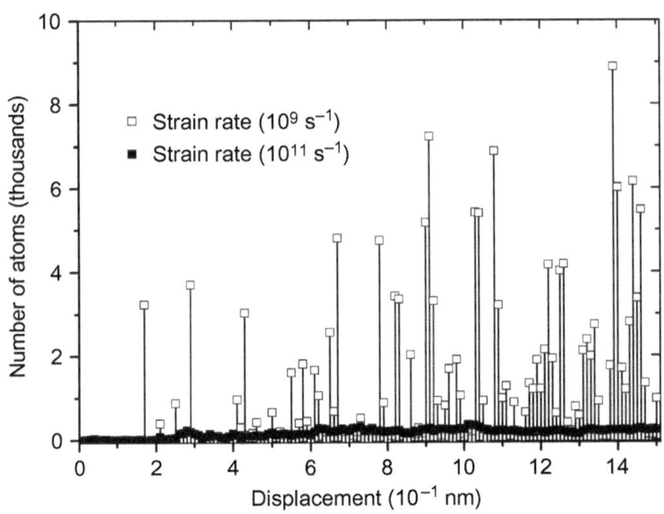

Figure 8.12 Temporal distributions of shear-banding events denoted by atoms with $D^2_{\min} > 0.15$ nm during nano-indentation at higher strain rates and lower strain rates.

Serrated plastic flow phenomena have been widely observed in BMGs under deformation-constrained loading modes such as compression [26,68,143–145] and indentation [63,95,146,147]. There is a general consensus that the shear banding in a sample should be responsible for the macroscopic flow serrations. Therefore, such serrations allow you to investigate shear-band dynamics. The flow serrations can appear as displacement bursts in the corresponding displacement–time curve. If you consider that such serrations or bursts are a result of intermittent operations of a main shear band, the velocity of the shear band can be calculated. However, it must be pointed out that the velocity calculations should first be based on the assumption regarding the mode of shear-band propagation, i.e. simultaneous or progressive in fashion. For the first mode, the shear-band velocity has been calculated to be 10^1-10^3 μm/s [143,145,148−150], which depends on the composition, ambient temperature, applied strain rate and sample size. Furthermore, the shear strain and viscosity within shear bands have been predicted [143,151]. For the second mode, the calculated velocity is about several metres per second at RT [149,152]. These velocities are many orders of magnitude smaller than one would expect, i. e. a shear-wave speed of approximately 10^2-10^3 m/s. If we take approximately 10^0 m/s as the shear-band velocity and consider that such a shear band propagates across a millimetre-sized sample, the timescale of its propagation is the order of ms. However, the experimental techniques with a temporal resolution of approximately ms have not captured the propagation process of shear bands. The underestimation of shear-band velocity maybe from the premise of a *single* shear band. In fact, Vinogradov and Khonik [140] found that an individual serration corresponds to a bundle of AE signals; a signal corresponds to a shear band operating. It is therefore reasonable to believe that the serrations are due to the operation and interaction of multiple shear bands, rather than one main shear band, especially for the ductile BMGs. Very recently, Wang and co-workers [144] have revealed that the plastic flow serrations of ductile BMGs can evolve into a self-organized critical state characterized by the power-law distribution of shear avalanches. This implies that the intermittent motion of shear bands is scale free; i.e. there is not a dominant shear band. MD simulations show that the shear-band velocity can be close to the speed of sound at the early stage of shear banding when one shear band is propagating across the sample [110]. After it penetrates across the entire sample, the shear band slips simultaneously across the entire shear plane. The first simultaneous mode was captured by Song et al. [148] using a high-speed camera (up to 5000 Hz frame).

Very recently, by developing a dynamic 'forced' shear technique of hat-shaped specimens, the shear-band propagation mode was clearly determined by Jiang and Dai [30]. They observed the longitudinal section of a deformed Vit 1 hat-shaped specimen using high resolution scanning electron microscope and found that there is a crack stopper in the forced shear zone, as shown in Figure 8.13A. It can be seen from this picture that the crack with the well-defined tip does not penetrate through the shear zone. Further AFM observation exhibits that the crack is led by a shear band (see Figure 8.13B). Obviously, the shear band has a tip that nucleates at an inhomogeneous site. This provides solid evidence for the progressive propagation of shear bands in metallic glasses. In fact, similar phenomena have been widely observed in the literature [10,11,25,44,68,106,118,139]. If the crack could be arrested, the propagating shear band would finally penetrate the entire sample to reach the opposite surface. At that time, the shear band would operate in a simultaneous fashion; this is expected to occur in some ductile systems. Ma and co-workers [110] recently captured the transition of the shear-banding mode from progressive at small strains to a simultaneous fashion at very larger strains, using MD simulations. In reality, most BMGs display very limited ductility, especially in

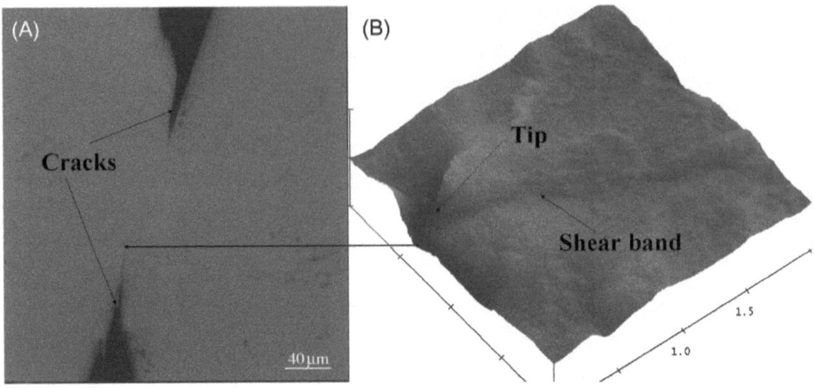

Figure 8.13 The shear zone in a deformed hat-shaped Vit 1 BMG specimen under dynamic forced shear, showing a stop crack that is led by a shear band with a well-defined tip.

tension or shearing. Therefore, the simultaneous propagation mode of a shear band occurs rarely in most cases.

8.5.2 Shear-Band Toughness

For an advancing shear band, a very basic question is how much energy the shear band can consume prior to catastrophic cracking. In other words, after the energy release Γ overcomes the critical plastic energy Γ_c, the shear band will mature to become a runaway shear crack. In this regard, the larger the Γ_c, the more significant is the toughness or ductility of BMGs. Γ_c therefore determines shear-band susceptibility that measures the intrinsic resistance of materials to the propagation of shear bands. Recent studies [153,154] have focused on the energy release Γ and further experimentally determined the Γ_c. Very recently, the shear-band toughness, initially proposed by Grady [155] for crystalline alloys, was extended to BMGs by Jiang and Dai [30]. They aim at theoretically describing the inherent susceptibility Γ_c to shear-band propagation in BMGs.

To characterize ASB susceptibility in crystalline alloys, Grady [155] introduced the concept of shear-band toughness, analogous to fracture toughness. As for BMGs, however, the shear banding exhibits some distinct features. When subjected to an external loading, some local regions, instead of the whole sample, preferentially yield via the cascade of a number of individual atomic jumps around free-volume sites [17] or STZs [18], forming local plastic regions (LPRs). The shear bands then nucleated simultaneously or successively in these LPRs. Finally, one of the nucleated shear bands dominates, propagates and causes a catastrophic fracture of the material. Furthermore, in addition to conventional energy/thermal and momentum/viscous dissipation, the free-volume dissipation should be involved in the shear-banding process in BMGs.

Based on the Grady–Kipp solution [155–157] and further considering shear-band stress softening due to both free-volume creation and temperature rise, the critical dissipation energy Γ_c can be analytically expressed (subscript '0' denotes the terms due to the sole free-volume softening) as [30]:

$$\left[1 - \Omega(\Gamma_c/\Gamma_{c0})^{1/3}(w/w_0)\right]\left[1 + \frac{(\Gamma_c/\Gamma_{c0})^{2/3}}{(w/w_0)^2}\right] = 2\frac{(\Gamma_c/\Gamma_{c0})}{(w/w_0)} \tag{8.4a}$$

with a dimensionless thermal-effect coefficient

$$\Omega = \left(\frac{2}{\alpha R}\right)\left(\frac{B}{\mathrm{Le}}\right) \tag{8.4b}$$

where B accounts for the degree of thermal softening, α is the free-volume softening coefficient, the parameter R describes the local dilatation ability [158] and the Lewis number $\mathrm{Le} = \chi/D$ measures the competition between thermal diffusivity and free-volume diffusivity. Equation (8.4) actually provides the implicit expression for Γ_c in a coupled free-volume softening and thermal-softening shear band. The concept of shear-band toughness is naturally introduced as [30]:

$$K_s = \sqrt{2G\Gamma_c} \tag{8.5}$$

which measures the internal resistance of BMG materials to propagation of shear bands. For the typical Vit 1 BMG, a shear-band toughness of $K_s \sim (2.66 - 26.58)\ \mathrm{MPa}\sqrt{\mathrm{m}}$ is calculated, if considering the critical displacement ψ_c varies from 100 nm to 10 μm [118,121,128,153,159]. This calculated value range of K_s is expected to be smaller than the mode II fracture toughness ($K_{\mathrm{IIC}} = 75\ \mathrm{MPa}\sqrt{\mathrm{m}}$) of this material measured by Flores and Dauskardt [160]. It is expected that the shear-band toughness should contribute to the fracture toughness [161], and their relationship can be linked [30].

The functional dependence of the critical dissipation energy on the shear-band width under different thermal-effect coefficients, determined by Eq. (8.4), is shown in Figure 8.14. For comparison, the case without the thermal effect, i.e. $\Omega = 0$, corresponding to the bold line, was also calculated. From this graph, it is readily seen that all these curves have two branches – left and right – and each of them intersects at a local minimum. The physical mechanism is reasonably clear. For thinner bands (moving to the left branch), the enhanced free-volume diffusion restrains the rate of free-volume softening and leads to excessive dissipation [20,162]. Wider bands (moving to the right branch) are effectively free of free-volume diffusion. However, the accelerated diffusion of momentum (inertia) into the shear-band vicinity again limits the rate of free-volume softening and also incurs additional dissipation. The shear-band thicknesses near the local minimum properly balance

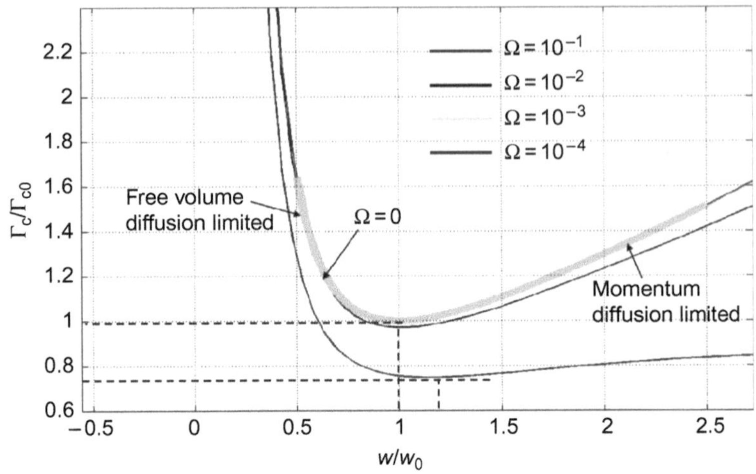

Figure 8.14 Plot of the critical energy dissipated in the shear band as a function of the shear-band thickness. All curves are provided by (8.4) in the text with different thermal-effect coefficients and are non-dimensionalized by the optimum shear-band thickness w_0 and the corresponding dissipation energy Γ_{c0} for the sole free-volume softening case.

the free-volume and momentum diffusion, providing the minimum possible shear-band dissipation. During the competition between the free-volume and momentum diffusion, the thermal softening, including its diffusion, plays a secondary role. As shown in Figure 8.14, with increasing the thermal-effect coefficient Ω from 10^{-4} to 10^{-1}, the optimum shear-band thickness increases slightly from w_0 to about $1.2w_0$; the corresponding optimum shear-band dissipation energy decreases somewhat from Γ_{c0} to $0.75\Gamma_{c0}$. This result clearly indicates that the thermal effect promotes the shear-band propagation because it decreases the critical energy barrier preventing the shear band from cracking. In particular, the insensitivity to thermal softening of the shear-band thickness indicates that the shear-band propagation in BMGs is governed by the free-volume softening.

8.5.3 Shear-Band Width

As we concluded earlier (Table 8.1), the characteristic width of shear bands is of the order of approximately 10 nm. Recently, finite STZ sizes of about 1–2 nm (not reaching 2 nm) have been identified by much research [1,3,81,84,163]. Interestingly, the '10-times-rule' in granular materials [164] seems to be roughly satisfied in metallic glasses; i.e. the shear-band thicknesses are approximately 10 times the STZ sizes, implying a similar shear-instability mechanism between the two materials. However, the quantitative relationship between the thickness of the shear band and the STZ size is still pending, and the precise underlying physics that dominates the shear-band width is not clear.

In crystalline alloys, the width of an ASB is supported by a balance of the plastic work-heat diffusion; its theoretical prediction has been successfully established by Dodd and Bai [165] and Walley [166]. In the same spirit, i.e. considering the free-volume coalescence-diffusion balance, Dai and Bai [19] derived an estimation of shear-band thickness in metallic glasses as:

$$w_0 \sim \sqrt{D \frac{\xi^*}{G^{\xi^*}}} \tag{8.6}$$

where * denotes the values within the band. Note that G^{ξ^*} is the net rate of free-volume increase and is highly dependent on the strain rate. You can reasonably assume that the thickness of an MSB is the final dimension of the perturbation with the initial wavelength, developed into the local steady-state flow in the post-instability stage. In this stage, the diffusion coefficient of the free-volume concentration is expected to obey the Stokes–Einstein equation [17,74]. Furthermore, based on the analogy between dislocation motion and the STZ operation, Eq. (8.6) can then yield [21]

$$w_0 = \frac{2\pi}{3} d \left(\xi^* \cdot \frac{1}{\gamma_c} \cdot \frac{1+\nu}{1-\nu} \right)^{1/2} \tag{8.7}$$

where d is roughly the STZ size, ξ^* is the activation free-volume concentration due to STZ-induced shear instability and γ_c is the average shear yield strain whose value is almost a universal constant of about 0.0267 for various systems [83]. This equation indicates that the correlation between the shear-band thickness and the STZ size does not follow a simple linear relation, and the influence of other parameters, such as ξ^*, γ_c and ν, should be involved. The dependence of these parameters on the shear-band width was examined. It is obvious that the shear-band thickness in metallic glasses is mainly governed by the STZ-activated free-volume concentration incurring instability and the STZ size.

Because the fundamental unit process underlying plastic flow is an STZ, the activation free volume facilitating local shear instability is actually the threshold of the free volume in activated STZs, beyond which the STZs become topologically unstable. At the atomic cluster level, the free-volume threshold must be a probability (usually Gaussian) distribution [17,18]. Egami's theory of topological fluctuations in the bonding network has predicted that if the average local transformation volume strain is larger than about 10%, the local atomic cluster site will become topologically unstable and be liquid like [167]. This average local volume strain can be analogous to the mathematical expectation of the probability distribution of a local STZ-activated free-volume concentration in plastic flow. In addition, the recent work of Lu et al. [51] has also unveiled that the dividing line of free-volume concentration is about 0.10, between a densely packed structure with a coordination number (CN) of greater than 10 and a loose one with CN $<$ 10. Therefore, we can reasonably choose $\xi^* \approx 0.10$, $d = 1.5$ nm and $\nu = 0.36$; the calculated shear-banding width is $w_0 \approx 9$ nm according to Eq. (8.7). This value agrees well with the characteristic thickness ($\sim$10 nm) of shear bands in various metallic glasses, demonstrating the intrinsic universality of local topological stability of STZs in glassy structures undergoing inhomogeneous plastic flow. The shear-band thickness is underpinned by a balance of the free-volume creation diffusion via the superposition of STZs. Thus, the local topological instability of STZs, together with their activation size and free volume, determines the width of the shear band in metallic glasses.

8.5.4 Shear-Band Spacing

During the process of loading, if one shear band is not sufficient to dissipate the applied energy, additional shear bands will form within samples. Certainly, such multiplication phenomenon of shear bands is sensitive to loading mode [26,56], sample geometry [168], material composition [169], atomic topological order [170] and so on. The operation of a single shear band will lead to stress unloading or strain relaxation in the vicinity of that shear band. Such local unloading or relaxation causes other shear bands to be excluded from that vicinity, resulting in multiple shear bands with characteristic spacing. In addition, the magnitude of shear-band extension can be represented by the shear displacement within the band. At the surface of the sample, the shear displacement behaves as a shear offset, which has been widely observed [120,121,171]. It is assumed that a critical shear displacement is required to create an MSB, and additional displacement may initiate fracture of the material along the band.

Bending is an effective method to investigate the shear-band patterns in metallic glasses [172]. Conner et al. [56,171] have carried out a series of experiments in which they bent beams made of Zr-based (Vit 106) metallic glasses of various thicknesses around mandrels of different radii. The nearly evenly spaced shear bands as well as the shear offsets at the free surfaces were observed. Furthermore, the sample thickness dependence of shear-band spacing and offset was quantitatively measured. Conner et al. [171] have performed an analysis of elastic perfectly plastic bending by treating the shear bands as mode II cracks. In their consideration, the maximum shear displacement occurs at the band end or the surface, while at the shear-band tip the displacement is zero. They derived expressions for the shear-band spacing and the shear offset, as well as their theoretical relationship. Later, Ravichandran and Molinari [173] performed a more precise analysis to capture the essential details of the shear-banding phenomena during bending. By balancing the dissipated energy as calculated from elasto-plastic beam theory and the energy dissipated along shear bands, and further introducing a failure criterion, they obtained an explicit form of shear-band spacing at failure. The shear-band offset was also derived. These predicted models for shear-band spacing offset capture the experimental observation taken from Conner et al. [56]. Based on the developed finite-deformation constitutive model of plasticity of BMGs and by implementing the model in the finite-element programme, Yang et al. [126] simulated the bending experiment performed by Conner et al. [56]. Their simulation results show that the stress relaxation indeed happens due to the presence of a shear band, indicating a linear relationship between the zone size and the plate thickness. This agrees well with the experimental observation [56]. They further investigate the sample size dependence of shear-band spacing directly by allowing for the formation of multiple shear bands. The predicted almost linear spacing–thickness relationship is also in good agreement with both that predicted by the stress relaxation analysis and the experimental measurements of Conner et al. [56].

Recently, Zhang et al. [129] developed Grady's model and explained shear-band spacing in Zr-based BMGs under dynamic loads. They calculated the variation of shear-band spacing with strain rate, normal stress and critical shear displacement. It has been found that the shear-band spacing decreases with (a) increasing strain rate, (b) decreasing shear displacement and (c) decreasing normal stress. The results explained well their observed shear-band pattern beneath a dynamic Vickers indentation. Also note that the difference with and without the thermal effect is not significant for low confinement pressure and small shear displacement. They thought this is reasonable because they did not observe obvious heating phenomenon accompanying the shear bands even under such dynamic indentation. In the model of Zhang et al. [129], the shear displacement was considered to be constant along the shear band from its tip to its end. However, the experimental observation shows a different situation [118]. The shear displacement varies between almost zero and a maximum, with a value of zero at its tip, and the maximum (up to tens of micrometres) at its end. This variation can explain the scatter values of shear displacement that different researchers reported in different experiments [113,118,128,159,160].

It is noted that the question of when the shear band occurs in one dominated mode or in multiple mode has not been answered. What is the mechanism underlying the shear-band patterns in metallic glasses? To this end, Dai and co-workers [174] performed systematic four-point bending tests on Vit 1 BMGs. By developing a theoretical model that takes into account the structural feature of BMGs, they have revealed that the shear-band propagation is controlled by the free-volume softening, and, however, the resultant momentum diffusion results in the multiplication of shear bands with a certain characteristic spacing. The shear-band propagation and nucleation (multiplication) are controlled by their respective consumed energies. Recently, Jiang and Dai [97] performed a specific loading, i.e. turn machining, to investigate the multiple shear-band behaviour. It is found that the material removed exhibits a unique lamellar chip (Figure 8.15A) due to repeated shear-band formation in the primary shear zone (PSZ). Based on the experimental observations, a coupled thermo-mechanical orthogonal cutting model, taking into account force, free volume and energy balance in the PSZ, was developed, during which the lamellar chip formation or the periodic multiple shear bands can be understood as a self-sustained limit-cycle phenomenon (Figure 8.15B): there is autonomous feedback in stress, free volume and temperature in the PSZ. More specifically,

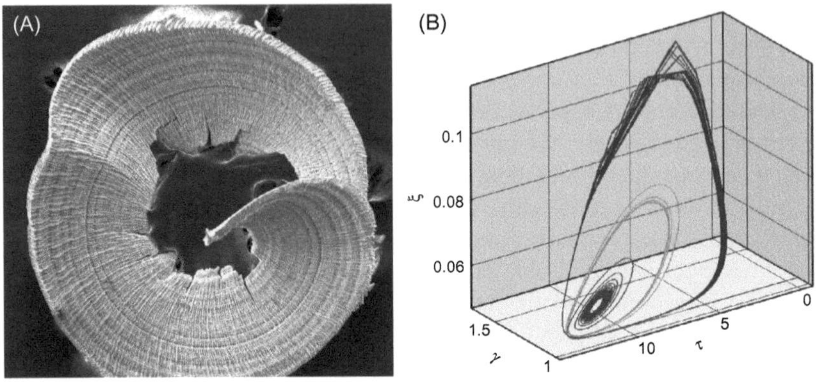

Figure 8.15 Lamellar chip morphology (A) of Vit 1 BMG and its underlying mechanism and (B) limit cycles.

such multiple shear bands occur as a result of the periodic loading–unloading cycle of material in the PSZ. It is revealed that the underlying mechanism is the symmetry breaking of free-volume flow and source, rather than thermal instability. The results, on the one hand, confirm the picture of how shear-band spacing results from stress relaxation [56,126,171,173]; on the other hand, it also provides clear experimental evidence for the physical origin of shear-banding instability in BMGs [20,27].

8.5.5 Pressure Sensitivity of Shear Banding

A lot of experimental observations [175–180] indicate that BMGs are pressure-sensitive materials. The pressure dependence of (localized) plastic deformation reflects the basic flow mechanism, which differs from that of their crystalline counterparts. Although the precise physical picture of how this dependence arises from the internal structure of BMGs remains elusive, it is plausible that it originates from atomic-scale dilatation [10,181,182]. Crystalline solids can deform at constant volume because the periodicity along slip planes provides identical atomic positions for sheared materials. However, a sheared portion of a BMG does not find such a perfect fit and thus will leave some holes. As we have known, the macroscopic flow of BMGs occurs by the cascade of STZs. As a result, STZs change into a loose configuration with a large volume, resulting in dilatation. Such dilatation induces hydrostatic stress during STZ formation, and thus the resultant macroscopic plastic flow should depend on pressure or normal stress.

Very recently, Dai and co-workers [183] derived an intrinsic theoretical correlation between the pressure-sensitivity coefficient and the dilatation factor in BMGs, taking shear-induced dilatation into consideration in STZ operations. The behaviour of STZs can be treated as an Eshelby-type inclusion problem [18]. To highlight the essential physics, let the initial spherical STZ experience a shape distortion and an accompanied bulk dilatation. It is easy to deduce that the shear strain is 2β and volume strain is 3α during such an STZ operation. The relationship between shear-induced volume strain and shear strain is assumed to be linear, $3\alpha = \ell(2\beta)$, where ℓ is the dilatation factor measuring the ratio of dilatation to shear strain. Based on these considerations, the relationship between the pressure-sensitivity coefficient m and the dilatation factor ℓ is obtained, obeying

$$m = 3\ell/[30(1-2\nu)/(7-5\nu) + 2(1+\nu)\ell^2/(1-2\nu)] \tag{8.8}$$

Now one question naturally arises: How does shear-induced dilatation affect shear-banding instability in BMGs? To answer this question, Jiang and Dai [20] recently conducted the 1D simple shear analysis for pressure-sensitive BMGs. Figure 8.16 shows the internal length and timescales versus the dilatation factor with various pressure-sensitivity indices. Note that the internal scales for shear instability decrease with an increasing dilatation factor for any fixed m. This means

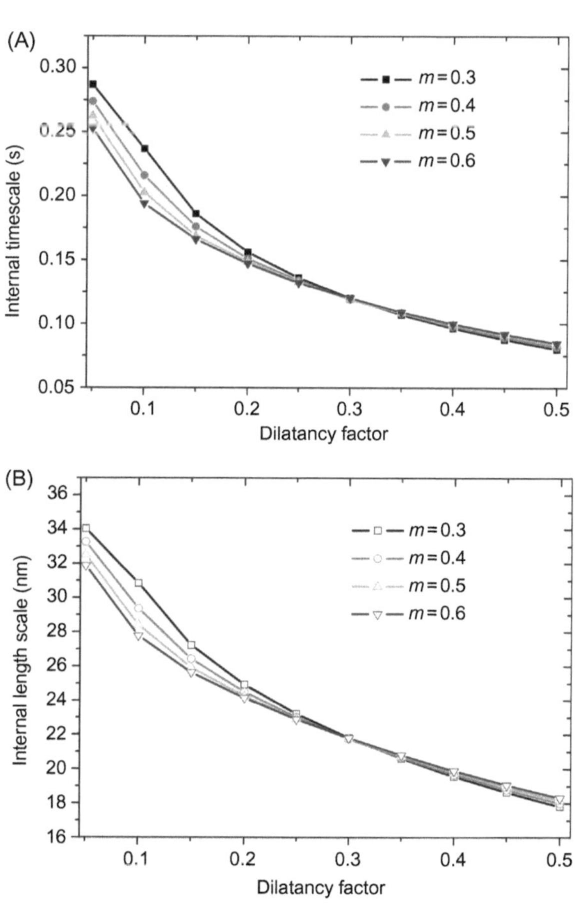

Figure 8.16 Internal (A) length scale and (B) timescale versus dilatation factor with various pressure-insensitivity indexes.

that the shear-induced dilatation strain makes the shear-banding instability easier and faster. In the fiducial interval (usually less than 0.3) of the dilatation factor, the tensile hydrostatic stress further aids shear instability according to the present results, which is consistent with the experimental observations by Flores and Dauskardt [178]. The present results show that the dilatation or the pressure sensitivity favours the origin or initiation of shear banding in metallic glasses.

The effect of pressure sensitivity on post-shear-banding behaviour has been investigated by Dubach et al. [176]. They found that with increasing pressure sensitivity, the intersection angle of two families of shear bands gradually departs from 90°. Moreover, Shi and Falk [184] have shown that the shear-band density significantly decreases with increasing pressure sensitivity, using MD simulations, in which they performed a series of uni-axial tensile tests on binary models of a glass with different degrees of structural relaxation. The enhanced pressure sensitivity or dilatation, on the one hand, facilitates the shear-banding initiation, and on the other hand, reduces the multiplication of shear bands. This is apparently paradoxical; however, it can be explained within the content of the free volume. It is well known that during structural relaxation, the free volume within samples is decreased. In such surroundings with a lower free-volume content, the STZ operations become more difficult because it requires more significant dilatation of the surrounding matrix. In fact, STZ operations occur preferentially in those regions having higher free volume as relatively less dilatation is required. Therefore, the relaxed samples with lower free-volume content hinder the formation of shear bands. However, for two samples with identical states (the same free-volume content), if one has more dilatation during deformation, i.e. creates more free volume, the shear-banding instability should initiate more easily in this sample.

8.6 Shear Bands and Global Ductility

Ductility is a mechanical property that describes the extent to which solid materials can be plastically deformed without fracture. Obviously, the ductility of BMGs is determined by the shear banding, which is the only mode for the RT plastic deformation. If the propagation of a dominant shear band can slow down or be in a stick-slip mode, then the onset of fracture might be delayed. In this case, the shear band is regarded to be ‘cold’ or stable [142]. In addition, multiplication of shear bands, i.e. distributing the plastic strain over many bands, contributes to global ductility [169,170,172]. Certainly, the most attractive way to improve the ductility is the suppression of shear bands, i.e. to achieve homogeneous deformation or necking prior to final fracture [185–187]. In this section, we discuss several approaches to this issue.

8.6.1 Loading Mode and Strain Rate

Under deformation-constrained loading conditions such as compression [26,68], rolling [188,189], bending [172] and indentation [190], obvious plastic flow can be achieved. In those cases, the shear-band propagation is geometrically constrained. The emergence of plastic flow is of great interest for understanding the flow mechanism of this type of 'brittle' material [63,190]. However, those constrained geometries are too restrictive to be generally useful for load-bearing applications. After the geometrical restriction is released, the material still undergoes catastrophic failure with very limited ductility.

As discussed in Section 8.4.1, higher strain rates can promote shear-band formation in time. For some specific composites, if the resultant shear bands can distribute within samples as homogeneously as possible, it could enhance remarkably the global ductility not only in compression [191,192] but also in tension [192,193]. However, as pointed out by Jiang et al. [68], the shear banding preferentially operates in the same regions at a higher strain rate. So the BMG under higher strain rates stands a good change of forming a dominant shear band, which tends to lead to fast catastrophic fracture [22,25,194]. In addition, it is worth noting that dynamic strain rates ($>10^3\ s^{-1}$) usually incur adiabatic heating, which also speeds up the transition of shear banding to unstable failure [20]. Therefore, although higher strain rates facilitate shear-band formation, the strain-rate dependence on ductility has a two-edged effect. The ductility is determined by the competition among the processes just mentioned: material and loading dependence.

8.6.2 Pre-treating

In conventional engineering materials, we can improve their mechanical properties by the pre-treating of samples such as introducing residual stress/strain (gradient) and severe plastic deformation. The similar pre-treating processes have also been applied to BMGs, in expectation of improving their ductility, even in tension. So far, there have been some pre-treating methods:

- Shot-peening [195]
- Pre-compression, including lateral [196], end-surface [197] and hydrostatic pressure [198]
- Surface wrap [199,200]
- Introducing a stress gradient [201]
- Releasing the boundary friction [202]
- Rolling [189].

Among them, shot-peening, pre-compression and rolling usually introduce residual stresses and/or inhomogeneous deformation, i.e. shear bands, into samples. For instance, Zhang et al. [195] carried out shot-peening of a Vit 1 BMG and found that the peened samples showed increased plasticity in bending and in compression. The enhanced ductility results from a combination of compressive residual stress, reducing the likelihood of surface cracking and more 'uniform' deformation induced by a high population of pre-existing shear bands. These pre-treatments, however, are not effective for improving the tensile ductility [189]. For the pre-compressive samples [196–198], their tensile ductility deserves further attention.

Surface wrap treatments actually allow the wrapped object, such as a Cu confinement sleeve [199] or Cu coating [200], to participate in the global deformation that should be originally taken by the sample alone. In this case, even the already fractured sample induced by a dominant shear band still undergoes an applied stress. Therefore, the observed plasticity or ductility is actually *apparent*, which is similar to that arising from tilting or bending of specimens [201]. Very recently, Scudino et al. [202] investigated the effect of boundary conditions on the compressive plasticity of the Vit 1 BMGs by using a pure Cu foil as a lubricant material between loading platens and the sample. They found that the soft metal is very effective for reducing the contact friction at the platen–specimen interface, leading to a remarkable increase in plastic deformation with respect to the conventional semi-fluid MoS_2 lubricants or no lubricant. It seems that the more homogeneously the stress state is in the sample during deformation, the more ductile the deformation of the sample is. However, as mentioned earlier, the stress inhomogeneity/gradient pre-existing in the sample favours plasticity [196,197,201]. The reason may be that the pre-existing stress gradient compensates for the stress inhomogeneity that arises during deformation. Also note that, in the present case, the Cu plates actually undertake a portion of deformation, which leads to a lower yield stress.

8.6.3 Sample Size and Machine Stiffness

In the early 1990s, Inoue et al. [203] observed that Fe-based and Co-based metallic glasses wires could exhibit good bending ductility only if the sample diameter was below a critical value. Such a size effect for ductility was studied quantitatively by Conner et al. [56,171], by bending Zr-based plates, ribbons and wires. They found that the strain to fracture increases markedly as the sample dimension decreases. Note that different sample sizes are produced in two ways. The first is samples with different sizes are cut from an identical as-cast plate or rod. Thus, the samples obtained are almost identical both chemically and structurally. In the second case, the samples with different sizes are directly cast from melt alloys and thus have distinct atomic structures due to different cooling rates. In this section, we only focus on the first case, i.e. the sole sample size effect. The structural factor will be discussed in the next section.

During MD simulations of uni-axial tension of Mg–Cu metallic glasses with scales of about 10 nm, Bailey et al. [135] have seen surprisingly that necking instability occurs before shear banding. They presumed that the disagreement with macroscopic observations comes from the local nature of a shear band compared with the 'global' nature of a necking instability. In a macroscopic sample, the effective 'wavelength' of the necking mode will be of the order of the sample size, while the characteristic size of a shear band is still of the order of tens of nanometres, which makes the latter much more likely to become unstable. Such a large difference in scale is lost in their simulations, in which the sample size is reduced to the shear-band size. Another factor that suppresses shear banding relative to necking is the necessity of breaking symmetry to choose the orientation of the shear plane. It is therefore expected that, in the rod sample down to 100 nm in diameter, necking instability is more likely to take place than the shear-banding instability. Recent experiments have confirmed this point. Actually, Jang and Greer [187] observed the necking instability in the nano-tension of a specimen with a diameter of 100 nm. Similar necking phenomena have also been observed in the tension of metallic glass plates with dimensions of the order of 100 nm [186]. Most interestingly, during the tension of Al-based metallic glass with a size less than 20 nm, even an atomic chain was formed after sample necking [204]. In particular, homogeneous deformation can also be observed in compression [185,205], bending [185], even tension [206] of metallic glasses with similar characteristic sizes, during which the samples are usually constrained to some extent. Now it can be concluded that the critical nuclei size or approximately characteristic size of a shear band (usually tens of nanometres) separates the inhomogeneous from homogeneous deformation. If the sample size is larger than this value, then shear-banding instability occurs, and the sample exhibits brittle behaviour; in the other case, the necking instability or even homogeneous flow dominates, and the sample is ductile.

As mentioned in Section 8.5.1, Masumoto and Murata [138] have noticed the effect of machine stiffness on shear banding. They found that the higher the stiffness of the machine, the shear bands progressively become more stable. Recently, Han et al. [153] investigated quantitatively this effect considering a sample-machine system. Assuming the crack-like behaviour of a shear band, they derived an instability condition for shear-band propagation, given by:

$$S = \frac{\pi E d_s^2}{4 l_s \kappa_M} > S_{cr} \tag{8.9}$$

where κ_M is the stiffness of the testing machine, S_{cr} is a critical shear-band instability index that is a material parameter related to intrinsic plasticity and E, d_s and l_s are the Young's modulus, diameter and height of the sample, respectively. Equation (8.9) indicates that the larger the S (larger d_s and/or smaller κ_M for a fixed aspect ratio of the sample) the larger the possibility of shear-banding instability. Based on a series of compressive tests with a range of controlled values of sample size and machine stiffness, $S_{cr} \sim 0.72$ and ~ 4 are determined for the aspect ratios of 2 and 1, respectively. To modify the conflict with the sample-height effect on shear banding in Eq. (8.9), Cheng et al. [142] proposed a more precise factor for shear-banding instability $l_s\,(1 + S)$. Taking the energy balance of the shear-banding propagation into consideration in the sample-machine system, Yang et al. [154] arrived at a simple geometrical relation for the size-dependence ductility of BMGs:

$$L_{ext} = l_s + \iota d_s \geq L_{int} \tag{8.10}$$

where L_{ext} is an extrinsic length scale corresponding to the elastic energy release and ι is a dimensionless parameter accounting for the effect of machine stiffness; whereas L_{int} can be viewed as the internal resistance of a BMG to unstable shear banding and is independent of sample sizes. Unstable brittle fracture induced by the shear band will occur if $L_{ext} \geq L_{int}$. If the material is fixed, i.e. $L_{int} = \text{constant}$,

L_{ext} should determine the shear-band behaviour and thus the ductility. Yang et al. [154] found that the ductility measured for the Zr-based BMGs with different simple geometries shows a very good correlation with the extrinsic length scale L_{ext}, here assuming $\iota = 1$ for simplicity. It predicts a critical intrinsic length scale of approximately 9 mm. It must be pointed out that the criteria just mentioned were obtained based on the compressive case. Extension of them to tensile cases needs to be attempted very carefully.

8.6.4 Composition and Atomic Configuration

The previous sections mainly discussed the effects of external factors on ductility of BMGs. Strictly speaking, the competition of external–internal factors determines the global ductility of metallic glasses, as indicated by Eq. (8.9) or (8.10). In this section, the intrinsic plasticity or ductility of materials is focused on. Poisson's ratio ν and equivalently the shear-bulk modulus ratio (G/K) criterion for intrinsic plasticity have been long established in crystalline alloys and are now found to be validated in metallic glasses. Lewandowski et al. [207] compiled a list of experimental data and found that there is a universal, sharp correlation for metallic glasses: they are intrinsically plastic for $G/K < 0.41 - 0.43$ or equivalently for $\nu > 0.31 - 0.32$. Recent research [169,208,209] has also demonstrated that the Poisson's ratio is a good indicator of the potential of metallic glasses to sustain plastic deformation. It is well known that elastic constants of a material are determined by its composition and microstructure. Therefore, a Poisson's ratio criterion allows us to understand the intrinsic plasticity of BMGs from the viewpoint of alloy composition and atomic configuration. For instance, through composition changes, Liu et al. [169] created some Zr-based BMGs that have relatively large ν values and display a capability to undergo multiple shear bandings and thus compressive ductility. Certainly, a specific soft–hard region structure (this will be discussed shortly) exists in their systems, which implies that the composition and structure are coupled to some degree in real cases.

When the alloy composition is fixed, different atomic configurations of BMGs can be obtained by changing the cooling history, i.e. the cooling rate. Shi and Falk [184] performed a series of uni-axial tests on binary models of metallic glasses by using MD simulations. It was revealed that the stress overshoot decreases as the cooling rate decreases. As we know, the stress overshoot is a measure of the tendency of shear banding. The larger the overshoot, the more metallic glasses are prone to shear localization. Recent experiments also show a similar tendency, although the sample size effect was included [168]. In the pure shear case, similar results have been obtained by Cheng et al. [210]. Furthermore, they examined the Poisson's ratio, configurational potential energy (CPE), shear modulus, bulk modulus and population of FI and fragmented polyhedral (FP) in the samples with different cooling rates. Table 8.2 includes these results for comparison. It can be found that the Poisson's ratio is higher in the sample with the higher cooling rate; its PEL is similar to that of a fragile liquid, with the higher CPE, the lower barrier of the megabasin and the lower effective curvature upon shearing at the basin bottom.

Table 8.2 Physical Properties of Metallic Glass Samples with High and Low Cooling Rates

Samples	High Cooling Rate	Low Cooling Rate
Stress overshoot	Low	High
Poisson's ratio	High	Low
PEL		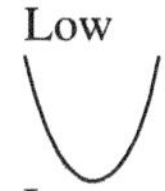
CPE	High	Low
Shear modulus	Low	High
Bulk modulus	Almost same	
Population of FI	Low	High
Population of FP	High	Low

The shear modulus G_c is then lower in this sample. With this PEL, the population of FI (solid-like clusters) is lower, whereas the concentration of FP (liquid-like clusters) is higher. A higher population of liquid-like clusters implies that there is a higher free-volume concentration within the sample. In this case, the shear localization occurs more easily or earlier, which is confirmed by Jiang and Dai [20]. In fact, the intrinsic correlation between fragility and Poisson's ratio in non-metallic glasses or metallic glasses has been reported by many researchers [211–214]. Because the bulk modulus corresponds to another dimension (hydrostatic expansion or shrinkage) in the PEL, which is correlated with valence electron density and not sensitive to configuration or cooling rates, when the composition is fixed, roughly, the lower the shear modulus of the sample, the more ductile the sample is. However, in the real case, the bulk modulus decreases with increasing cooling rate. Such a small decrease hardly affects the compressive ductility, but significantly degrades the tensile ductility, which will be discussed in detail in the following. According to the results of Cheng et al. [210], it can be concluded that the Poisson's ratio criterion for plasticity indeed has a structural origin [215].

In BMGs, the deformation mode is no longer anisotropic dislocation motion, which is sensitive to lattice types and slip planes, but rather an isotropic and universal one featured by the STZs. The STZ essentially is an SRO or MRO atomic cluster undergoing inelastic shear distortion. Assuming a Gaussian radial distribution function at SRO or MRO scale, Jiang and Dai [216] correlated the Poisson's ratio with the atomic structures of metallic glasses. The structural conditions for plasticity of BMGs can be concluded as follows:

- Atomic packing is denser.
- Such packing is more disordered.
- Atomic interaction is more anharmonic.

If the constituent atoms are fixed, i.e. the anharmonicity is identical, brittle BMG systems should have larger mean distances and less disorder. The opposite refers to plastic systems. However, the structural conditions for plasticity of BMGs, i.e. higher atomic packing density and more significant atomic dispersion, are difficult to simultaneously satisfy during the real experimental process. A higher density of atomic packing can be achieved by structural relaxation or lowering the cooling rate during glass forming, during which atomic disorder is reduced. Hence a BMG becomes stiffer or harder. This is reflected in a change in the elastic constants and a decrease in the Poisson's ratio and embrittlement [207]. However, the disordered packing of atoms implies that a BMG is soft. Often, we can obtain a soft BMG with disordered atomic packing by using higher cooling rates, during which more randomly distributed free volume is inherited from the liquid. But such BMG systems have lower density of atomic packing, which is adverse to the plasticity, especially to the tensile ductility. It is expected that a monolithic BMG may be very plastic not only in compression but also in *tension*, when a *perfect* balance among the preceding conditions is realized by careful selection of its composition, controlling liquid–glass transition and temperature and/or mechanical treatments.

8.6.5 Shear Banding in Metallic Glass Composites

The pioneering work in this aspect was performed by Johnson and co-workers [217] aiming at defence applications. They fabricated Vit 1 matrix composites reinforced with uni-axially aligned tungsten wires and used them as kinetic energy penetrators, in which the BMG matrix shows self-sharpening behaviour due to localized shear-band failure, and the tungsten wires contribute to the kinetic energy due to high density. They found that the tungsten fibre-reinforced BMG composite has approximately a 10–20% improvement in penetrator efficiency over the tungsten heavy alloy. This research arouses growing interest due to its potential military value [218–221].

On the other hand, the concept of developing composites by combining the BMG matrix with second phases has been introduced to overcome the RT brittleness. Broadly, there are two types of BMG composites. One can be called 'in situ composites', where nanocrystalline phases or polymorphic glassy phases are growing in situ within the original metallic glass matrix [222–226]. The other is 'ex situ composites', where second-phase particles, fibres or slices are added to a melt prior to casting [221,227–232]. For whichever BMG composites, it is always anticipated that more shear bands, as well as deformation or transformation of second phases in some cases, contribute to macroscopic plastic strain before fracture, i.e. ductility. Therefore, the shear-banding behaviour in these composites and its interaction with second phases are of central significance.

Usually, the second phases have three contributions to ductility. Firstly, they create stress concentrations around their sides due to the mismatch resistance of them and glassy matrix to the external stress. Large numbers of shear bands can form from these stress-concentrated sites. To this end, 'soft' elastic/plastic inhomogeneities are often introduced into a glassy matrix [223,233–235]. Secondly, they act as the obstacle to shear-band propagation across the sample. After the shear bands encounter the second phases, these bands are either arrested or bypassed. To achieve this purpose, you should carefully choose the size and volume fraction of the second phases. Matsumoto et al. [236] recently performed MD simulations of mode II deformation on a notched Zr-based BMG plate containing one nanocrystalline particle of Ni ahead of the notch base. They found that the resistance to shear banding is efficiently improved by introducing particles with sufficient size compared to the shear-band width. In addition, if you want to suppress the shear band in favour of cracking, Hofmann et al. [223] argue that the sizes of the second phases should match a material length scale related to fracture toughness, i.e. the size of a crack tip's 'plastic zone'. Schuh et al. [10] examined the effect of volume fraction of the second phase on the plastic strain to failure in compression and tension for a variety of composites. Data suggest that different second phases have different optimum volume fractions that are very sensitive to the matrix, loading mode and the respective sizes of the second phase and matrix. Thirdly, in some cases, the nanocrystalline second phase itself undergoes transformation, dislocation and twinning, which not only contributes to ductility but also produces work hardening [44,170,224,235].

Hays et al. [233] pioneered a quantitative analysis of the effects of second phase on the macroscopic plasticity. Based on the experimental observations, they concluded that the global plastic strain is controlled by the shear-band thickness and spacing. Obviously, overall plasticity or ductility can be improved by decreasing shear-band spacing by appropriately tailoring the microstructure. Very recently, Abdeljawad and Haataja [237] developed a diffuse-interface continuum model to examine the role plastic strain accumulates in ductile particles prior to shear banding. Based on a series of simulations, they suggested that the total plastic strain is a function of the particle size.

8.7 Prospects and Summary

As a unique deformation mode and a precursor of catastrophic failure, the nature of shear banding in metallic glasses has both scientific and practical significance. In this aspect, areas of interest involve (a) the geometrical configuration of shear bands such as thickness, length and spacing; (b) the structural change and temperature rise that occurs in shear bands; (c) the shear-band dynamics, including the propagating mode and velocity of shear bands; (d) the deformation accommodated by shear bands, including plastic strain and its strain rate; and (e) the shear-banding ductility relationship. Amongst them, the mechanism of shear-band formation and propagation is of central importance. In summary, we briefly provide some important conclusions:

- The shear-banding instability in metallic glasses initiates due to the coupled free-volume softening and thermal softening. Neither free-volume softening nor thermal softening can incur shear instability alone.
- During the shear-instability process, the free-volume softening plays the dominated role, whereas the thermal softening is a secondary effect. They contribute to each other. Physically, the shear band nucleates due to a cascade of stress-activated STZs or flow defects with the thermal fluctuation in the background.
- The shear-band evolution is a dissipation system, including thermal, momentum and free-volume dissipation from the shear band. The dissipated energy within the shear band is

determined by the balance between the momentum diffusion and the free-volume diffusion. The optimum dissipated energy corresponds to the shear-band thickness, about 10 nm. The momentum dissipation governs the shear-band spacing.
- The shear bands can branch, intersect, multiply, be arrested and even be suppressed, which contributes to the global plasticity of metallic glasses.

Nevertheless, there are many unresolved problems regarding shear bands in metallic glasses, which deserve further study. For example:

- *Shear-band dynamics.* The question of what modes the shear band propagates in has not been resolved well. The shear-band speed reported in the literature is confusing, from several micrometres to thousands of metres per second. How can an existing shear band disappear or stop? To answer these basic questions, it is urgent to develop proper experimental techniques with enough spatio-temporal resolution.
- *Shear-band-induced crack process.* A shear band finally evolves into a crack. It is not clear how a crack nucleates within a shear band. How to characterize the energy dissipation prior to cracking and crack nucleation needs to be studied.
- *Development of shear-band multiplication methodology.* So far, much work has focused on the compressive plasticity via the multiplication of shear bands. As for the deformation-unstrained loading case, e.g. uni-axial tension, these methods pose a great challenge. Developing shear-band multiplication methodologies suitable for the tension case is of more practical significance.
- *Shear banding in BMG derivatives.* The shear-banding process in pure BMGs has attracted much attention, in which its origin is reasonably clear, whereas the evolution process still remains unclear. However, the investigation of shear banding in BMG derivatives, including composites and porous materials, is still little understood. Current work is mainly confined to experimental observations and deals little with the quantitative description. Many questions deserve to be studied. For example, how does the introduction of a second phase (particle or void) affect the process of shear banding in the BMG matrix? If the ASB occurs in the second-phase particles, how do the two types of shear bands, i.e. the ASB and the free-volume-governed shear band, act together? How does the shear band within the matrix propagate through the second phases?

Acknowledgements

The author gratefully acknowledges useful and insightful discussions with Prof. Y. L. Bai, Prof. B. Dodd and Dr M. Q. Jiang. This work is supported by the NSFC (Grants Nos 10725211, 11132011, 11002144, 11023001 and 11021262), the National Natural Science Foundation of China – NSAF (Grant No. 10976100) and the National Basic Research Program of China (Grant No. 2009CB724401 and 2012CB937500).

References

1. Miracle, D. B. (2004). A structural model for metallic glasses. *Nature Materials* **3**, 697–702.
2. Bernal, J. D. (1960). Geometry of the structure of monatomic liquids. *Nature* **185**, 68–70.

3. Sheng, H. W., Luo, W. K., Alamgir, F. M., Bai, J. M., and Ma, E. (2006). Atomic packing and short-to-medium range order in metallic glasses. *Nature* **439**, 419–425.
4. Egami, T. (2010). Understanding the properties and structure of metallic glasses at the atomic level. *Journal of the Minerals Metals and Materials Society* **62**, 70–75.
5. Hirata, A., Guan, P. F., Fujita, T., Hirotsu, Y., Inoue, A., Yavari, A. R., Sakurai, T., and Chen, M. W. (2011). Direct observation of local atomic order in a metallic glass. *Nature Materials* **10**, 28–33.
6. Greer, A. L. (1995). Metallic glasses. *Science* **267**, 1947–1953.
7. Johnson, W. L. (1999). Bulk glass-forming metallic alloys: Science and technology. *MRS Bulletin* **24**, 42–56.
8. Inoue, A. (2000). Stabilization of metallic supercooled liquid and bulk amorphous alloys. *Acta Materialia* **48**, 297–306.
9. Wang, W. H., Dong, C., and Shek, C. H. (2004). Bulk metallic glasses. *Materials Science and Engineering Reports* **44**, 45–89.
10. Schuh, C. A., Hufnagel, T. C., and Ramamurty, U. (2007). Mechanical behavior of amorphous alloys. *Acta Materialia* **55**, 4067–4109.
11. Trexler, M. M., and Thadhani, N. N. (2010). Mechanical properties of bulk metallic glasses. *Progress in Materials Science* **55**, 759–839.
12. Ashby, M. F., and Greer, A. L. (2006). Metallic glasses as structural materials. *Scripta Materialia* **54**, 321–326.
13. Schroers, J., Kumar, G., Hodges, T. M., Chan, S., and Kyriakides, T. R. (2009). Bulk metallic glasses for biomedical applications. *Journal of the Minerals Metals and Materials Society* **61**, 21–29.
14. Wang, W. H. (2009). Bulk metallic glasses with functional physical properties. *Advanced Materials* **21**, 4524–4544.
15. Inoue, A., and Takeuchi, A. (2011). Recent development and application products of bulk glassy alloys. *Acta Materialia* **59**, 2243–2267.
16. Huang, X., Ling, Z., Liu, Z. D., Zhang, H. S., and Dai, L. H. (2012). Amorphous alloy reinforced Whipple shield structure. *International Journal of Impact Engineering* **42**, 1–10.
17. Spaepen, F. (1977). A microscopic mechanism for steady state inhomogeneous flow in metallic glasses. *Acta Metallurgica* **25**, 407–415.
18. Argon, A. S. (1979). Plastic deformation in metallic glasses. *Acta Metallurgica* **27**, 47–58.
19. Dai, L. H., and Bai, Y. L. (2008). Basic mechanical behavior and mechanics of shear banding in BMGs. *International Journal of Impact Engineering* **35**, 704–716.
20. Jiang, M. Q., and Dai, L. H. (2009). On the origin of shear banding instability in metallic glasses. *Journal of the Mechanics and Physics of Solids* **57**, 1267–1292.
21. Jiang, M. Q., Wang, W. H., and Dai, L. H. (2009). Prediction of shear band thickness in metallic glass. *Scripta Materialia* **60**, 1004–1007.
22. Jiang, M. Q., Ling, Z., Meng, J. X., and Dai, L. H. (2008). Energy dissipation in fracture of bulk metallic glasses via inherent competition between local softening and quasi-cleavage. *Philosophical Magazine* **88**, 407–426.
23. Chen, Y., Jiang, M. Q., Wei, X. J., and Dai, L. H. (2011). Failure criterion for metallic glasses. *Philosophical Magazine* **91**, 4536–4554.
24. Huang, R., Suo, Z., Prevost, J. H., and Nix, W. D. (2002). Inhomogeneous deformation in metallic glasses. *Journal of the Mechanics and Physics of Solids* **50**, 1011–1027.
25. Liu, L. F., Dai, L. H., Bai, Y. L., Wei, B. C., and Eckert, J. (2005). Initiation and propagation of shear bands in Zr-based bulk metallic glass under quasi-static and dynamic shear loadings. *Journal of Non-Crystalline Solids* **351**, 3259–3270.

26. Liu, L. F., Dai, L. H., Bai, Y. L., Wei, B. C., and Eckert, J. (2005). Behavior of multiple shear bands in Zr-based bulk metallic glass. *Materials Chemistry and Physics* **93**, 174–177.
27. Dai, L. H., Yan, M., Liu, L. F., and Bai, Y. L. (2005). Adiabatic shear banding instability in bulk metallic glasses. *Applied Physics Letters* **87**, 141916.
28. Liu, L. F., Dai, L. H., Bai, Y. L., Wei, B. C., and Eckert, J. (2006). Characterization of rate-dependent shear behavior of Zr-based bulk metallic glass using shear-punch testing. *Journal of Materials Research* **21**, 153–160.
29. Gao, Y. F., Yang, B., and Nieh, T. G. (2007). Thermomechanical instability analysis of inhomogeneous deformation in amorphous alloys. *Acta Materialia* **55**, 2319–2327.
30. Jiang, M. Q., and Dai, L. H. (2011). Shear-band toughness of bulk metallic glasses. *Acta Materialia* **59**, 4525–4537.
31. Klement, W. I., Willens, R. H., and Duwez, P. (1960). Non-crystalline structure in solidified gold–silicon alloys. *Nature* **187**, 869–870.
32. Turnbull, D. (1950). Kinetics of heterogeneous nucleation. *Journal of Chemical Physics* **18**, 198–203.
33. Turnbull, D., and Cech, R. E. (1950). Microscopic observation of the solidification of small metal droplets. *Journal of Applied Physics* **21**, 804–810.
34. Chen, H. S., and Turnbull, D. (1969). Formation, stability and structure of palladium–silicon based alloy glasses. *Acta Metallurgica* **17**, 1021–1030.
35. Drehman, A. J., Greer, A. L., and Turnbull, D. (1982). Bulk formation of a metallic glass: Pd40Ni40P20. *Applied Physics Letters* **41**, 716–717.
36. Inoue, A., Zhang, T., Nishiyama, N., Ohba, K., and Masumoto, T. (1993). Preparation of 16 mm diameter rod of amorphous Zr65Al7.5Ni10Cu17.5 alloy. *Materials Transactions, JIM* **34**, 1234–1237.
37. Peker, A., and Johnson, W. L. (1993). A highly processable metallic glass: Zr41.2Ti13.8Cu12.5Ni10.0Be22.5. *Applied Physics Letters* **63**, 2342–2344.
38. Inoue, A., Zhang, T., and Masumoto, T. (1989). Al–La–Ni amorphous alloys with a wide supercooled liquid region. *Materials Transactions, JIM* **30**, 965–972.
39. Pampillo, C. A. (1975). Flow and fracture in amorphous alloys. *Journal of Materials Science* **10**, 1194–1227.
40. Chen, H. S. (1974). Thermodynamic considerations on the formation and stability of metallic glasses. *Acta Metallurgica* **22**, 1505–1511.
41. Inoue, A., Zhang, T., and Masumoto, T. (1990). Zr–Al–Ni amorphous alloys with high glass transition temperature and significant supercooled liquid region. *Materials Transactions, JIM* **31**, 177–183.
42. Löffler, J. F. (2003). Bulk metallic glasses. *Intermetallics* **11**, 529–540.
43. Greer, A. L., and Ma, E. (2007). Bulk metallic glasses: At the cutting edge of metals research. *MRS Bulletin* **32**, 611–615.
44. Chen, M. W. (2008). Mechanical behavior of metallic glasses: Microscopic understanding of strength and ductility. *Annual Review of Materials Research* **38**, 445–469.
45. Schroers, J. (2009). Processing of bulk metallic glass. *Advanced Materials* **21**, 1–32.
46. Gene Ice, (2005). Characterizing amorphous strain. *Nature Materials* **4**, 17–18.
47. Bernal, J. D., and Mason, J. (1960). Co-ordination of randomly packed spheres. *Nature* **188**, 910–911.
48. Dong, C., Wang, Q., Qiang, J. B., Wang, Y. M., Jiang, N., Han, G., Li, Y. H., Wu, J., and Xia, J. H. (2007). From clusters to phase diagrams: Composition rules of quasicrystals and bulk metallic glasses. *Journal of Physics D: Applied Physics* **40**, R273–R291.

49. Ma, D., Stoica, A. D., and Wang, X.-L. (2009). Power-law scaling and fractal nature of medium-range order in metallic glasses. *Nature Materials* **8**, 30–34.
50. Cheng, Y. Q., and Ma, E. (2011). Atomic-level structure and structure–property relationship in metallic glasses. *Progress in Materials Science* **56**, 379–473.
51. Liu, X. J., Chen, C. L., Hui, X., Liu, T., and Lu, Z. P. (2008). Ordered clusters and free volume in a Zr–Ni metallic glass. *Applied Physics Letters* **93**, 011911.
52. Inoue, A., and Nishiyama, N. (2007). New bulk metallic glasses for applications as magnetic-sensing, chemical, and structural materials. *MRS Bulletin* **32**, 651–658.
53. Luborsky, F. E., Frischmann, P. G., and Johnson, L. A. (1980). Amorphous materials – a new class of soft magnetic alloys. *Journal of Magnetism and Magnetic Materials* **19**, 130–137.
54. Tang, M. B., Bai, H. Y., Pan, M. X., Zhao, D. Q., and Wang, W. H. (2005). Bulk metallic superconductive La60Cu20Ni10Al10 glass. *Journal of Non-Crystalline Solids* **351**, 2572–2575.
55. Eliaz, N., Fuks, D., and Eliezer, D. (1999). A new model for the diffusion behavior of hydrogen in metallic glasses. *Acta Materialia* **47**, 2981–2989.
56. Conner, R. D., Li, Y., Nix, W. D., and Johnson, W. L. (2004). Shear band spacing under bending of Zr-based metallic glass plates. *Acta Materialia* **52**, 2429–2434.
57. Grimberg, A., Baur, H., Bochsler, P., Bühler, F., Burnett, D. S., Hays, C. C., Heber, V. S., Jurewicz, A. J. G, and Wieler, R. (2006). Solar wind neon from genesis: Implications for the lunar noble gas record. *Science* **314**, 1133–1135.
58. Liu, L., Qiu, C. L., Huang, C. Y., Yu, Y., Huang, H., and Zhang, S. M. (2009). Biocompatibility of Ni-free Zr-based bulk metallic glasses. *Intermetallics* **17**, 235–240.
59. Inoue, A., and Takeuchi, A. (2010). Recent development and applications of bulk glassy alloys. *International Journal of Applied Glass Science* **1**, 273–295.
60. Taub, A. I., and Spaepen, F. (1980). The kinetics of structural relaxation of a metallic glass. *Acta Metallurgica* **28**, 1781–1788.
61. Taub, A. I. (1980). Stree-strain rate dependent of homogeneous flow in metallic glasses. *Acta Metallurgica* **28**, 633–637.
62. Wang, D., Liao, G., Pan, J., Tang, Z., Peng, P., Liu, L., and Shi, T. (2009). Superplastic micro-forming of Zr65Cu17.5Ni10Al7.5 bulk metallic glass with silicon mold using hot embossing technology. *Journal of Alloys and Compounds* **484**, 118–122.
63. Schuh, C. A., Lund, A. C., and Nieh, T. G. (2004). New regime of homogeneous flow in the deformation map of metallic glasses: Elevated temperature nanoindentation experiments and mechanistic modeling. *Acta Materialia* **52**, 5879–5891.
64. Shi, Y. F., and Falk, M. L. (2007). Stress-induced structural transformation and shear banding during simulated nanoindentation of a metallic glass. *Acta Materialia* **55**, 4317–4324.
65. Subhash, G., and Zhang, H. W. (2007). Dynamic indentation response of ZrHf-based bulk metallic glasses. *Journal of Materials Research* **22**, 478–485.
66. Jiang, S. Y., Jiang, M. Q., Dai, L. H., and Yao, Y. G. (2008). Atomistic origin of rate-dependent serrated plastic flow in metallic glasses. *Nanoscale Research Letters* **3**, 524–529.
67. Jiang, M. Q., Jiang, S. Y., Ling, Z., and Dai, L. H. (2009). Smaller Deborah number inducing more serrated plastic flow of metallic glass. *Computational Materials Science* **46**, 767–771.
68. Jiang, W. H., Fan, G. J., Liu, F. X., Wang, G. Y., Choo, H., and Liaw, P. K. (2008). Spatiotemporally inhomogeneous plastic flow of a bulk-metallic glass. *International Journal of Plasticity* **24**, 1–16.

69. Lee, S. C., Lee, C. M., Lee, J. C., Kim, H. J., Shibutani, Y., Fleury, E., and Falk, M. L. (2008). Structural disordering process of an amorphous alloy driven by the elastostatic compression at room temperature. *Applied Physics Letters* **92**, 151906.
70. Park, K. W., Lee, C. M., Lee, M. R., Fleury, E., Falk, M. L., and Lee, J. C. (2009). Paradoxical phenomena between the homogeneous and inhomogeneous deformations of metallic glasses. *Applied Physics Letters* **94**, 021907.
71. Furukawa, A., and Tanaka, H. (2009). Inhomogeneous flow and fracture of glassy materials. *Nature Materials* **8**, 601–609.
72. Falk, M. L., and Langer, J. S. (1998). Dynamics of viscoplastic deformation in amorphous solids. *Physical Review E* **57**, 7192–7205.
73. Heggen, M., Spaepen, F., and Feuerbacher, M. (2005). Creation and annihilation of free volume during homogeneous flow of a metallic glass. *Journal of Applied Physics* **97**, 033506.
74. Cohen, M. H., and Turnbull, D. (1959). Molecular transport in liquids and glasses. *Journal of Chemical Physics* **31**, 1164–1169.
75. Turnbull, D. (1961). Free-volume model of the amorphous phase: Glass transition. *Journal of Chemical Physics* **34**, 120–125.
76. Polk, D. E., and Turnbull, D. (1972). Flow of melt and glass forms of metallic alloys. *Acta Metallurgica* **20**, 493–498.
77. Argon, A. S., and Kuo, H. Y. (1979). Plastic flow in a disordered bubble raft (an analog of a metallic glass). *Materials Science and Engineering* **39**, 101–109.
78. Eshelby, J. D. (1957). The determination of the elastic field of an ellipsoidal inclusion, and related problems. *Proceedings of the Royal Society of London Series A* **241**, 376–396.
79. Eshelby, J. D. (1959). The elastic field outside an ellipsoidal inclusion. *Proceedings of the Royal Society of London Series A* **252**, 561–569.
80. Spaepen, F. (1981). Defects in amorphous metals. In "*Physics of defects*" (R. Balian, M. Kleman, and J. Poirier, Eds.), North-Holland: Amsterdam, pp. 146–162.
81. Pan, D., Inoue, A., Sakurai, T., and Chen, M. W. (2008). Experimental characterization of shear transformation zones for plastic flow of bulk metallic glasses. *Proceedings of the National Academy of Sciences* **105**, 14769–14772.
82. Schall, P., Weitz, D. A., and Spaepen, F. (2007). Structural rearrangements that govern flow in colloidal glasses. *Science* **318**, 1895–1899.
83. Johnson, W. L., and Samwer, K. (2005). A universal criterion for plastic yielding of metallic glasses with a $(T/T_g)^{2/3}$ temperature dependence. *Physical Review Letters* **95**, 195501.
84. Mayr, S. G. (2006). Activation energy of shear transformation zones: A key for understanding rheology of glasses and liquids. *Physical Review Letters* **97**, 195501.
85. Harmon, J. S., Demetriou, M. D., Johnson, W. L., and Samwer, K. (2007). Anelastic to plastic transition in metallic glass-forming liquids. *Physical Review Letters* **99**, 135502.
86. Stillinger, F. H., and Weber, T. A. (1984). Packing structures and transitions in liquids and solids. *Science* **225**, 983–989.
87. Stillinger, F. H. (1995). A topographic view of supercooled liquids and glass formation. *Science* **267**, 1935–1939.
88. Debenedetti, P. G., and Stillinger, P. H. (2001). Supercooled liquids and the glass transition. *Nature* **410**, 259–267.
89. Frenkel, J. (1926). The theory of the elastic limit and the solidity of crystal bodies. *Zeitschrift fur Physik* **37**, 572–609.

90. Yu, H. B., Wang, W. H., Bai, H. Y., Wu, Y., and Chen, M. W. (2010). Relating activation of shear transformation zones to relaxations in metallic glasses. *Physical Review B* **81**, 220201(R).
91. Steif, P. S., Spaepen, F., and Hutchinson, J. W. (1982). Strain localization in amorphous metals. *Acta Metallurgica* **30**, 447–455.
92. Vaks, V. G. (1991). Possible mechanism for formation of localized shear bands in amorphous alloys. *Physics Letters A* **159**, 174–178.
93. Wright, W. J., Hufnagel, T. C., and Nix, W. D. (2003). Free volumn coalescence and void formation in shear bands in metallic glass. *Journal of Applied Physics* **93**, 1432–1437.
94. Leamy, H. J., Chen, H. S., and Wang, T. T. (1972). Plastic flow and fracture of metallic glass. *Metallurgical Transactions* **3**, 699–708.
95. Dai, L. H., Liu, L. F., Yan, M., Wei, B. C., and Eckert, J. (2004). Serrated plastic flow in a Zr-based bulk metallic glass during nanoindentation. *Chinese Physics Letters* **21**, 1593–1595.
96. Liu, L. F., Dai, L. H., Bai, Y. L., Wei, B. C., and Yu, G. S. (2005). Strain rate-dependent compressive deformation behavior of Nd-based bulk metallic glass. *Intermetallics* **13**, 827–832.
97. Jiang, M. Q., and Dai, L. H. (2009). Formation mechanism of lamellar chip in machining bulk metallic glass. *Acta Materialia* **57**, 2730–2738.
98. Mukaia, T., Nieh, T. G., Kawamura, Y., Inoue, A., and Higashi, K. (2002). Dynamic response of a Pd40Ni40P20 bulk metallic glass in tension. *Scripta Materialia* **46**, 43–47.
99. Masumoto, T., and Maddin, R. (1971). The mechanical properties of palladium 20 a/o silicon alloy quenched from the liquid state. *Acta Metallurgica* **19**, 725–741.
100. Donovan, P., and Stobbs, W. M. (1981). The structure of shear bands in metallic glasses. *Acta Metallurgica* **29**, 1419–1436.
101. Pekarskaya, E., Kim, C. P., and Johnson, W. L. (2001). In situ transmission electron microscopy studies of shear bands in a bulk metallic glass based composite. *Journal of Materials Research* **16**, 2513–2518.
102. Li, J., Spaepen, F., and Hufnagel, T. C. (2002). Nanometre-scale defects in shear bands in a metallic glass. *Philosophical Magazine* **82**, 2623–2630.
103. Jiang, W. H., and Atzmon, M. (2006). Mechanically-assisted nanocrystallization and defects in amorphous alloys: A high-resolution transmission electron microscopy study. *Scripta Materialia* **54**, 333–336.
104. Gu, X., Livi, K. J. T., and Hufnagel, T. C. (2003). Structure of shear bands in zirconium-based metallic glasses observed by transmission electron microscopy. Materials Research Society Symposium Proceedings **754**, CC7.9.1–CC7.9.6.
105. Chen, M. W., Inoue, A., Zhang, W., and Sakurai, T. (2006). Extraordinary plasticity of ductile bulk metallic glasses. *Physical Review Letters* **96**, 245502.
106. Kim, K. B., Das, J., Lee, M. H., Li, S., Fleury, E., Zhang, Z. F., Wang, W. H., and Eckert, J. (2008). Propagation of shear bands in a CuZrAl bulk metallic glass. *Journal of Materials Research* **23**, 6–12.
107. Pampillo, C. A. (1972). Localized shear deformation in a glassy metal. *Scripta Metallurgica* **6**, 915–918.
108. Li, Q. K., and Li, M. (2006). Atomic scale characterization of shear bands in an amorphous metal. *Applied Physics Letters* **88**, 241903.
109. Li, Q. K., and Li, M. (2007). Atomistic simulations of correlations between volumetric change and shear softening in amorphous metals. *Physical Review B* **75**, 094101.

110. Cao, A. J., Cheng, Y. Q., and Ma, E. (2009). Structural processes that initiate shear localization in metallic glass. *Acta Materialia* **57**, 5146–5155.
111. Flores, K. M., Sherer, E., Bharathula, A., Chen, H., and Jean, Y. C. (2007). Sub-nanometer open volume regions in a bulk metallic glass investigated by positron annihilation. *Acta Materialia* **55**, 3403–3411.
112. Liu, C. T., Heatherly, L., Easton, D. S., Carmichael, C. A., Schneibel, J. H., Chen, C. H., Wright, J. L., Yoo, M. H., Horton, J. A., and Inoue, A. (1998). Test environments and mechanical properties of Zr-base bulk amorphous alloys. *Metallurgical and Materials Transactions A* **29A**, 1811–1820.
113. Wright, W. J., Schwarz, R. B., and Nix, W. D. (2001). Localized heating during serrated plastic flow in bulk metallic glasses. *Materials Science and Engineering A* **319–321**, 229–232.
114. Hufnagel, T. C., Jiao, T., Li, Y., Xing, L.-Q., and Ramesh, K. T. (2002). Deformation and failure of Zr57Ti5Cu20Ni8Al10 bulk metallic glass under quasi-static and dynamic compression. *Journal of Materials Research* **17**, 1441–1445.
115. Lewandowski, J. J., and Greer, A. L. (2006). Temperature rise at shear bands in metallic glasses. *Nature Materials* **5**, 15–18.
116. Miracle, D. B., Concustell, A., Zhang, Y., Yavari, A. R., and Greer, A. L. (2011). Shear bands in metallic glasses: Size effects on thermal profiles. *Acta Materialia* **59**, 2831–2840.
117. Bruck, H. A., Rosakis, A. J., and Johnson, W. L. (1996). The dynamic compressive behavior of beryllium bearing bulk metallic glasses. *Journal of Materials Research* **11**, 503–511.
118. Yang, B., Morrison, M. L., Liaw, P. K., Buchanan, R. A., and Wang, G. (2005). Dynamic evolution of nanoscale shear bands in a bulk-metallic glass. *Applied Physics Letters* **86**, 141904.
119. Yang, B., Liu, C. T., Nieh, T. G., Morrison, M. L., Liaw, P. K., and Buchanan, R. A. (2006). Localized heating and fracture criterion in bulk metallic glasses. *Journal of Materials Research* **21**, 915–922.
120. Georgarakis, K., Aljerf, M., Li, Y., LeMoulec, A., Charlot, F., Yavari, A. R., Chornokhvostenko, K., Tabachnikova, E., Evangelakis, G. A., Miracle, D. B., Greer, A. L., and Zhang, T. (2008). Shear band melting and serrated flow in metallic glasses. *Applied Physics Letters* **93**, 031907.
121. Guo, H., Wen, J., Xiao, N. M., Zhang, Z. F., and Sui, M. L. (2008). The more shearing, the thicker shear band and heat-affected zone in bulk metallic glass. *Journal of Materials Research* **23**, 2133–2138.
122. Zhang, Y., Stelmashenko, N. A., Barber, Z. H., Wang, W. H., Lewandowski, J. J., and Greer, A. L. (2007). Local temperature rises during mechanical testing of metallic glasses. *Journal of Materials Research* **22**, 419–427.
123. Spaepen, F. (2006). Must shear bands be hot? *Nature Materials* **5**, 7–8.
124. Steif, P. S., Spaepen, F., and Hutchinson, J. W. (1983). Ductile versus brittle behavior of amorphous metals. *Journal of the Mechanics and Physics of Solids* **31**, 359–388.
125. Gao, Y. F. (2006). An implicit finite element method for simulating inhomogeneous deformation and shear bands of amorphous alloys based on the free-volume model. *Modelling and Simulation in Materials Science and Engineering* **14**, 1329.
126. Yang, Q., Mota, A., and Ortiz, M. (2006). A finite-deformation constitutive model of bulk metallic glass plasticity. *Computational Mechanics* **37**, 194–204.

127. Thamburaja, P., and Ekambaram, R. (2007). Coupled thermo-mechanical modelling of bulk-metallic glasses: Theory, finite-element simulations and experimental verification. *Journal of the Mechanics and Physics of Solids* **57**, 1263–1273.
128. Zhang, H. W., Subhash, G., and Maiti, S. (2007). Local heating and viscosity drop during shear band evolution in bulk metallic glasses under quasistatic loading. *Journal of Applied Physics* **102**, 043519.
129. Zhang, H. W., Maiti, S., and Subhash, G. (2008). Evolution of shear bands in bulk metallic glasses under dynamic loading. *Journal of the Mechanics and Physics of Solids* **56**, 2171–2187.
130. Yang, Y., Ye, J. C., Lu, J., Gao, Y. F., and Liaw, P. K. (2010). Metallic glasses: Gaining plasticity for microsystems. *Journal of the Minerals Metals and Materials Society* **62**, 93–98.
131. Bai, Y. L. (1982). Thermo-plastic instability in simple shear. *Journal of the Mechanics and Physics of Solids* **30**, 195–207.
132. Anand, L., and Su, C. (2005). A theory for amorphous viscoplastic materials undergoing finite deformations, with application to metallic glasses. *Journal of the Mechanics and Physics of Solids* **53**, 1362–1396.
133. Thamburaja, P. (2011). Length scale effects on the shear localization process in metallic glasses: A theoretical and computational study. *Journal of the Mechanics and Physics of Solids* **59**, 1552–1575.
134. Jiang, F., Jiang, M. Q., Wang, H. F., Zhao, Y. L., He, L., and Sun, J. (2011). Shear transformation zone volume determining ductile–brittle transition of bulk metallic glasses. *Acta Materialia* **59**, 2057–2068.
135. Bailey, N. P., Schiøtz, J., and Jacobsen, K. W. (2006). Atomistic simulation study of the shear band deformation mechanism in Mg–Cu metallic glasses. *Physical Review B* **73**, 064108.
136. Shimizu, F., Ogata, S., and Li, J. (2007). Theory of shear banding in metallic glasses and molecular dynamics calculations. *Materials Transactions* **48**, 2923–2927.
137. Shimizu, F., Ogata, S., and Li, J. (2006). Yield point of metallic glass. *Acta Materialia* **54**, 4293–4298.
138. Masumoto, T., and Murata, T. (1976). Deformation of amorphous metals. *Materials Science and Engineering* **25**, 71–75.
139. Neuhäuser, H. (1978). Rate of shear band formation in metallic glasses. *Scripta Metallurgica* **12**, 471–474.
140. Vinogradov, A. Y., and Khonik, V. A. (2004). Kinetics of shear banding in a bulk metallic glass monitored by acoustic emission measurements. *Philosophical Magazine* **84**, 2147–2166.
141. Klaumünzer, D., Lazarev, A., Maaß, R., Dalla Torre, F. H., Vinogradov, A., and Löffler, J. F. (2011). Probing shear-band initiation in metallic glasses. *Physical Review Letters* **107**, 185502.
142. Cheng, Y. Q., Han, Z., Li, Y., and Ma, E. (2009). Cold versus hot shear banding in bulk metallic glass. *Phyical Review B* **80**, 134115.
143. Song, S. X., and Nieh, T. G. (2009). Flow serration and shear band viscosity during inhomogeneous deformation of a Zr-based bulk metallic glass. *Intermetallics* **17**, 762–767.
144. Sun, B. A., Yu, H. B., Jiao, W., Bai, H. Y., Zhao, D. Q., and Wang, W. H. (2010). Plasticity of ductile metallic glasses: A self-organized sritical state. *Physical Review Letters* **105**, 035501.

145. Maaβ, R., Klaumünzer, D., and Löffler, J. F. (2011). Propagation dynamics of individual shear bands during inhomogeneous flow in a Zr-based bulk metallic glass. *Acta Materialia* **59**, 3205–3213.
146. Ai, K., and Dai, L. H. (2007). A new modified expanding cavity model for characterizing the spherical indentation behavior of bulk metallic glass with pile-up. *Scripta Materialia* **56**, 761–764.
147. Wei, B. C., Zhang, L. C., Zhang, T. H., Xing, D. M., Das, J., and Eckert, J. (2007). Strain rate dependence of plastic flow in Ce-based bulk metallic glass during nanoindentation. *Journal of Materials Research* **22**, 258–263.
148. Song, S. X., Wang, X.-L., and Nieh, T. G. (2010). Capturing shear band propagation in a Zr-based metallic glass using a high-speed camera. *Scripta Materialia* **62**, 847–850.
149. Klaumünzer, D., Maaβ, R., Dalla Torre, F. H., and Löffler, J. F. (2010). Temperature-dependent shear band dynamics in a Zr-based bulk metallic glass. *Applied Physics Letters* **96**, 061901.
150. Liu, Z. Y., Yang, Y., and Liu, C. T. (2011). Size-affected shear-band speed in bulk metallic glasses. *Applied Physics Letters* **99**, 171904.
151. Jiang, W. H., Liu, F. X., Liaw, P. K., and Choo, H. (2007). Shear strain in a shear band of a bulk-metallic glass in compression. *Applied Physics Letters* **90**, 181903.
152. Wright, W. J., Samale, M. W., Hufnagel, T. C., LeBlanc, M. M., and Florando, J. N. (2009). Studies of shear band velocity using spatially and temporally resolved measurements of strain during quasistatic compression of a bulk metallic glass. *Acta Materialia* **57**, 4639–4648.
153. Han, Z., Wu, W. F., Li, Y., Wei, Y. J., and Gao, H. J. (2009). An instability index of shear band for plasticity in metallic glasses. *Acta Materialia* **57**, 1367–1372.
154. Yang, Y., Ye, J. C., Lu, J., Liaw, P. K., and Liu, C. T. (2010). Characteristic length scales governing plasticity/brittleness of bulk metallic glasses at ambient temperature. *Applied Physics Letters* **96**, 011905.
155. Grady, D. E. (1992). Properties of an adiabatic shear-band process zone. *Journal of the Mechanics and Physics of Solids* **40**, 1197–1215.
156. Grady, D. E. (1994). Dissipation in adiabatic shear bands. *Mechanics of Materials* **17**, 289–293.
157. Grady, D. E., and Kipp, M. E. (1987). The growth of unstable thermoplastic shear with application to steady-wave shock compression in solids. *Journal of the Mechanics and Physics of Solids* **35**, 95–118.
158. Johnson, W. L., Lu, J., and Demetriou, M. D. (2002). Deformation and flow in bulk metallic glasses and deeply undercooled glass forming liquids – a self-consistent dynamic free volume model. *Intermetallics* **10**, 1039–1046.
159. Kim, J.-J., Choi, Y., Suresh, S., and Argon, A. S. (2002). Nanocrystallization during nanoindentation of a bulk amorphous metal alloy at room temperature. *Science* **295**, 654–657.
160. Flores, K. M., and Dauskardt, R. H. (2006). Mode II fracture behavior of a Zr-based bulk metallic glass. *Journal of the Mechanics and Physics of Solids* **54**, 2418–2435.
161. Xu, J., Ramamurty, U., and Ma, E. (2010). The fracture toughness of bulk metallic glasses. *Journal of the Minerals Metals and Materials Society* **62**, 10–18.
162. Molinari, A. (1997). Collective behavior and spacing of adiabatic shear bands. *Journal of the Mechanics and Physics of Solids* **45**, 1551–1575.
163. Zink, M., Samwer, K., Johnson, W. L., and Mayr, S. G. (2006). Plastic deformation of metallic glasses: Size of shear transformation zones from molecular dynamics simulations. *Physical Review B* **73**, 172203.

164. Francois, B., Lacombe, F., and Herrmann, H. J. (2002). Finite width of shear zones. *Physical Review E* **65**, 031311.
165. Dodd, B., and Bai, Y. L. (1985). Width of adiabatic shear bands. *Materials Science and Technology* **1**, 38–40.
166. Walley, S. M. (2007). Shear localization: A historical overview. *Metallurgical and Materials Transactions A* **38A**, 2629–2654.
167. Egami, T., Poon, S. J., Zhang, Z., and Keppens, V. (2007). Glass transition in metallic glasses: A microscopic model of topological fluctuations in the bonding network. *Pysical Review B* **76**, 024203.
168. Huang, Y. J., Shen, J., and Sun, J. F. (2007). Bulk metallic glasses: Smaller is softer. *Applied Physics Letters* **90**, 081919.
169. Liu, Y. H., Wang, G., Wang, R. J., Zhao, D. Q., Pan, M. X., and Wang, W. H. (2007). Super plastic bulk metallic glasses at room temperature. *Science* **315**, 1385–1388.
170. Das, J., Tang, M. B., Kim, K. B., Theissmann, R., Baier, F., Wang, W. H., and Eckert, J. (2005). 'Work-hardenable' ductile bulk metallic glass. *Physical Review Letters* **94**, 205501.
171. Conner, R. D., Johnson, W. L., Paton, N. E., and Nix, W. D. (2003). Shear bands and cracking of metallic glass plates in bending. *Journal of Applied Physics* **94**, 904–911.
172. Schroers, J., and Johnson, W. L. (2004). Ductile bulk metallic glass. *Physical Review Letters* **93**, 255506.
173. Ravichandran, G., and Molinari, A. (2005). Analysis of shear banding in metallic glasses under bending. *Acta Materialia* **53**, 4087–4095.
174. Chen, Y., Jiang, M. Q., and Dai, L. H. (2012). Collective behavior of shear bands in metallic glasses. Unpublished work.
175. Yu, G. S., Lin, J. G., Mo, M., Wang, X. F., Wang, F. H., and Wen, C. E. (2007). Effect of relaxation on pressure sensitivity index in a Zr-based metallic glass. *Materials Science and Engineering A* **460–461**, 58–62.
176. Dubach, A., Prasad, K. E., Raghavan, R., Löffler, J. F., Michler, J., and Ramamurty, U. (2009). Free-volume dependent pressure sensitivity of Zr-based bulk metallic glass. *Journal of Materials Research* **24**, 2697–2704.
177. Zhang, Z. F., Echert, J., and Schultz, L. (2003). Difference in compressive and tensile fracture mechanisms of Zr59Cu20Al10Ni8Ti3 bulk metallic glass. *Acta Materialia* **51**, 1167–1179.
178. Flores, K. M., and Dauskardt, R. H. (2001). Mean stress effects on flow localization and failure in a bulk metallic glass. *Acta Materialia* **49**, 2527–2537.
179. Lu, J., and Ravichandran, G. (2003). Pressure dependent flow behavior of Zr41.2Ti13.8Cu12.5Ni10Be22.5 bulk metallic glass. *Journal of Materials Research* **18**, 2039–2049.
180. Tandaiya, P., Ramamurty, U., and Narasimhan, R. (2009). Mixed mode (I and II) crack tip fields in bulk metallic glasses. *Journal of the Mechanics and Physics of Solids* **57**, 1880–1897.
181. Lund, A. C., and Schuh, C. A. (2003). Yield surface of a simulated metallic glass. *Acta Materialia* **51**, 5399–5411.
182. Schuh, C. A., and Lund, A. C. (2003). Atomistic basis for the plastic yield criterion of metallic glass. *Nature Materials* **2**, 449–452.
183. Sun, L., Jiang, M. Q., and Dai, L. H. (2010). Intrinsic correlation between dilatation and pressure sensitivity of plastic flow in metallic glasses. *Scripta Materialia* **63**, 943–946.

184. Shi, Y. F., and Falk, M. L. (2005). Strain localization and percolation of stable structure in amorphous solids. *Physical Review Letters* **95**, 095502.
185. Donohue, A., Spaepen, F., Hoagland, R. G., and Misra, A. (2007). Suppression of the shear band instability during plastic flow of nanometer-scale confined metallic glasses. *Applied Physics Letters* **91**, 241905.
186. Guo, H. (2007). Tensile ductility and necking of metallic glass. *Nature Materials* **6**, 735–739.
187. Jang, D., and Greer, J. R. (2010). Transition from a strong-yet-brittle to a stronger-and-ductile state by size reduction of metallic glasses. *Nature Materials* **9**, 215–219.
188. Liu, J. W., Cao, Q. P., Chen, L. Y., Wang, X. D., and Jiang, J. Z. (2010). Shear band evolution and hardness change in cold-rolled bulk metallic glasses. *Acta Materialia* **58**, 4827–4840.
189. Cao, Q. P., Liu, J. W., Yang, K. J., Xu, F., Yao, Z. Q., Minkow, A., Fecht, H. J., Ivanisenko, J., Chen, L. Y., Wang, X. D., Qu, S. X., and Jiang, J. Z. (2010). Effect of pre-existing shear bands on the tensile mechanical properties of a bulk metallic glass. *Acta Materialia* **58**, 1276–1292.
190. Li, W. H., Wei, B. C., Zhang, T. H., Xing, D. M., Zhang, L. C., and Wang, Y. R. (2007). Study of serrated flow and plastic deformation in metallic glasses through instrumented indentation. *Intermetallics* **15**, 706–710.
191. Wakeda, M., Shibutani, Y., Ogata, S., and Park, J. (2008). Multiple shear banding in a computational amorphous alloy model. *Applied Physics A* **91**, 281–285.
192. Sergueeva, A. V., Mara, N. A., Kuntz, J. D., Lavernia, E. J., and Mukherjee, A. K. (2005). Shear band formation and ductility in bulk metallic glasses. *Philosophical Magazine* **85**, 2671–2687.
193. Yokoyama, Y., Fujita, K., Yavari, A. R., and Inoue, A. (2009). Melleable hypoeutectic Zr–Ni–Cu–Al bulk glassy alloys with tensile plastic elongation at room temperature. *Philosophical Magazine Letters* **89**, 322–334.
194. Mukaia, T., Nieh, T. G., Kawamura, Y., Inoue, A., and Higashi, K. (2002). Effect of strain rate on compressive behavior of a Pd40Ni40P20 bulk metallic glass. *Intermetallics* **10**, 1071–1077.
195. Zhang, Y., Wang, W. H., and Greer, A. L. (2006). Making metallic glasses plastic by control of residual stress. *Nature Materials* **5**, 857–860.
196. Yu, H. B., Hu, J., Xia, X. X., Sun, B. A., Li, X. X., Wang, W. H., and Bai, H. Y. (2009). Stress-induced structural inhomogeneity and plasticity of bulk metallic glasses. *Scripta Materialia* **61**, 640–643.
197. Zhang, J. L., Yu, H. B., Lu, J. X., Bai, H. Y., and Shek, C. H. (2009). Enhancing plasticity of ZrTiCuNiBe bulk metallic glass by precompression. *Applied Physics Letters* **95**, 071906.
198. Yu, P., Bai, H. Y., Zhao, J. G., Jin, C. Q., and Wang, W. H. (2007). Pressure effects on mechanical properties of bulk metallic glass. *Applied Physics Letters* **90**, 051906.
199. Yu, P., Liu, Y. H., Wang, G., Bai, H. Y., and Wang, W. H. (2007). Enhance plasticity of bulk metallic glasses by geometric confinement. *Journal of Materials Research* **22**, 2384–2388.
200. Qiu, S. B., and Yao, K. F. (2008). Novel application of the electrodeposition on bulk metallic glasses. *Applied Surface Science* **255**, 3454–3458.
201. Chen, L. Y., Ge, Q., Qu, S., Jiang, Q. K., Nie, X. P., and Jiang, J. Z. (2008). Achieving large macroscopic compressive plastic deformation and work-hardening-like behavior in

a monolithic bulk metallic glass by tailoring stress distribution. *Applied Physics Letters* **92**, 211905.

202. Scudino, S., Surreddi, K. B., Wang, G., and Eckert, J. (2010). Enhanced plastic deformation of ZrTiCuNiBe bulk metallic glass by the optimization of frictional boundary restraints. *Scripta Materialia* **62**, 750–753.
203. Inoue, A., Katsuya, A., Amiya, K., and Masumoto, T. (1995). Preparation of amorphous Fe–Si–B and Co–Si–B alloy wires by a melt extraction method and their mechanical and magnetic properties. *Materials Transactions, JIM* **36**, 802–809.
204. Luo, J. H., Wu, F. F., Huang, J. Y., Wang, J. Q., and Mao, S. X. (2010). Superelongation and atomic chain formation in nanosized metallic glass. *Physical Review Letters* **104**, 215503, 4 pages.
205. Volkert, C. A., Donohue, A., and Spaepen, F. (2008). Effect of sample size on deformation in amorphous metals. *Journal of Applied Physics* **103**, 083539.
206. Wang, Y., Li, J., Hamza, A. V., and Barbee Jr., T. W. (2007). Ductile crystalline-amorphous nanolaminates. *Proceedings of the National Academy of Sciences* **104**, 11155–11160.
207. Lewandowski, J. J., Wang, W. H., and Greer, A. L. (2005). Intrinsic plasticity or brittleness of metallic glasses. *Philosophical Magazine Letters* **85**, 77–87.
208. Gu, X. J., Poon, S. J., Shiflet, G. J., and Lewandowski, J. J. (2009). Ductile-to-brittle transition in a Ti-based bulk metallic glass. *Scripta Materialia* **60**, 1027–1030.
209. Lewandowski, J. J., Gu, X. J., Nouri, A. S., Poon, S. J., and Shiflet, G. J. (2008). Tough Fe-based bulk metallic glasses. *Applied Physics Letters* **92**, 091918.
210. Cheng, Y. Q., Cao, A. J., and Ma, E. (2009). Correlation between the elastic modulus and the intrinsic plastic behavior of metallic glasses: The roles of atomic configuration and alloy composition. *Acta Materialia* **57**, 3253–3267.
211. Novikov, V. N., and Sokolov, A. P. (2004). Poisson's ratio and the fragility of glass-forming liquids. *Nature* **431**, 961–963.
212. Novikov, V. N., and Sokolov, A. P. (2006). Correlation of fragility and Poisson's ratio: Difference between metallic and nonmetallic glass formers. *Physical Review B* **74**, 064203.
213. Jiang, M. Q., and Dai, L. H. (2007). Intrinsic correlation between fragility and bulk modulus in metallic glasses. *Physical Review B* **76**, 054204.
214. Nemilov, S. V. (2007). Structural aspect of possible interrelation between fragility (length) of glass forming melts and Poisson's ratio of glasses. *Journal of Non-Crystalline Solids* **353**, 4613–4632.
215. Zhang, L., Cheng, Y. Q., Cao, A. J., Xu, J., and Ma, E. (2009). Bulk metallic glasses with large plasticity: Composition design from the structural perspective. *Acta Materialia* **57**, 1154–1164.
216. Jiang, M. Q., and Dai, L. H. (2010). Short-range-order effects on intrinsic plasticity of metallic glasses. *Philosophical Magazine Letters* **90**, 269–277.
217. Conner, R. D., Dandliker, R. B., Scruggs, V., and Johnson, W. L. (2000). Dynamic deformation behavior of tungsten-fiber/metallic glass matrix composites. *International Journal of Impact Engineering* **24**, 435–444.
218. Lee, S. B., Lee, K., Lee, S. K., and Lee, S. (2009). Compressive and tensile properties of STS304-continuous fiber-reinforced Zr-based amorphous alloy matrix composite fabricated by liquid pressing process. *Metallurgical and Materials Transactions A* **40A**, 3159–3170.
219. Zhang, H. F., Li, H., Wang, A. M., Fu, H. M., Ding, B. Z., and Hu, Z. Q. (2009). Synthesis and characteristics of 80 vol.% tungsten (W) fibre/Zr based metallic glass composite. *Intermetallics* **17**, 1070–1077.

220. Liu, J. M., Zhang, H. F., Yuan, X. G., Fu, H. M., and Hu, Z. Q. (2011). Synthesis and properties of carbon short fiber reinforced ZrCuNiAl metallic glass matrix composite. *Materials Transactions* **52**, 412–415.
221. Zong, H. T., Ma, M. Z., Liu, L., Zhang, X. Y., Bai, B. W., Yu, P. F., Qi, L., Jing, Q., Li, G., and Liu, R. P. (2010). W(f)/Zr(41.2)Ti(13.8)Cu(12.5)Ni(10)Be(22.5) bulk metallic glass composites prepared by a new melt infiltrating method. *Journal of Alloys and Compounds* **504**, S106–S109.
222. Eckert, J., Das, J., Pauly, S., and Duhamel, C. (2007). Mechanical properties of bulk metallic glasses and composites. *Journal of Materials Research* **22**, 285–301.
223. Hofmann, D. C., Suh, J. Y., Wiest, A., Duan, G., Lind, M. L., Demetriou, M. D., and Johnson, W. L. (2008). Designing metallic glass matrix composites with high toughness and tensile ductility. *Nature* **451**, 1085–1089.
224. Pauly, S., Gorantla, S., Wang, G., Kühn, U., and Eckert, J. (2010). Transformation-mediated ductility in CuZr-based bulk metallic glasses. *Nature Materials* **9**, 473–477.
225. Liu, J. M., Zhang, H. F., Fu, H. M., Hu, Z. Q., and Yuan, X. G. (2010). In situ spherical B2 CuZr phase reinforced ZrCuNiAlNb bulk metallic glass matrix composite. *Journal of Materials Research* **25**, 1159–1163.
226. Qiao, J. W., Sun, A. C., Huang, E. W., Zhang, Y., Liaw, P. K., and Chuang, C. P. (2011). Tensile deformation micromechanisms for bulk metallic glass matrix composites: From work-hardening to softening. *Acta Materialia* **59**, 4126–4137.
227. Deng, S. T., Diao, H., Chen, Y. L., Yan, C., Zhang, H. F., Wang, A. M., and Hu, Z. Q. (2011). Metallic glass fiber-reinforced Zr-based bulk metallic glass. *Scripta Materialia* **64**, 85–88.
228. Ma, W. F., Kou, H. C., Chen, C. S., Li, J. S., Chang, H., Zhou, L., and Fu, H. Z. (2008). Compressive deformation behaviors of tungsten fiber reinforced Zr-based metallic glass composites. *Materials Science and Engineering A* **486**, 308–312.
229. Qiu, K. Q., Suo, Z. Y., Ren, Y. L., and Yu, B. (2007). Observation of shear bands formation on tungsten fiber-reinforced Zr-based bulk metallic glass matrix composite. *Journal of Materials Research* **22**, 551–554.
230. Li, H., Subhash, G., Kecskes, L. J., and Dowding, R. J. (2005). Mechanical behavior of tungsten preform reinforced bulk metallic glass composites. *Materials Science and Engineering A* **403**, 134–143.
231. Wang, G., Chen, D. M., Shen, J., Stachurski, Z. H., Qin, Q. H., Sun, J. F., and Zhou, B. D. (2006). Deformation behaviors of a tungsten-wire/bulk metallic glass matrix composite in a wide strain rate range. *Journal of Non-Crystalline Solids* **352**, 3872–3878.
232. Jiao, T., Kecskes, L. J., Hufnagel, T. C., and Ramesh, K. T. (2004). Deformation and failure of Zr(57)Nb(5)Al(10)Cu(15.4)Ni(12.6)/W particle composites under quasi-static and dynamic compression. *Metallurgical and Materials Transactions A* **35A**, 3439–3444.
233. Hays, C. C., Kim, C. P., and Johnson, W. L. (2000). Microstructure controlled shear band pattern formation and enhanced plasticity of bulk metallic glasses containing in situ formed ductile phase dendrite dispersions. *Physical Review Letters* **84**, 2901–2904.
234. Qiao, J. W., Wang, S., Zhang, Y., Liaw, P. K., and Chen, G. L. (2009). Large plasticity and tensile necking of Zr-based bulk-metallic–glass-matrix composites synthesized by the Bridgman solidification. *Applied Physics Letters* **94**, 151905.

235. Wu, Y., Xiao, Y. H., Chen, G. L., Liu, C. T., and Lu, Z. P. (2010). Bulk metallic glass composites with transformation-mediated work-hardening and ductility. *Advanced Materials* **22**, 2770–2773.
236. Matsumoto, N., Matsumoto, R., and Miyazaki, N. (2009). Estimation of shear-banding resistance in metallic glass containing nano-crystalline particles. *Journal of Non-Crystalline Solids* **355**, 29–32.
237. Abdeljawad, F., and Haataja, M. (2010). Continuum modeling of bulk metallic glasses and composites. *Physical Review Letters* **105**, 125503.

9 Adiabatic Shear Bands in Polymers

Gerard M. Swallowe

Department of Physics, Loughborough University, Loughborough, Leicestershire, UK

Nomenclature

C_V specific heat at constant volume
k thermal diffusivity
R gas constant
t_d thermal diffusion time
T temperature
T_g glass transition temperature
ν activation volume for plastic flow
W_p plastic work
α coefficient of increase of yield stress with pressure
β ratio of thermal energy to plastic work
γ shear strain
$\dot{\gamma}$ shear strain rate
ΔH activation energy for plastic flow
ε strain
ρ density
τ shear stress
σ_y yield stress

9.1 Introduction

The localization of plastic deformation into narrow shear bands has been known for many years, but despite being first reported by Tresca in 1878, it was not until the 1970s that extensive investigations of the phenomenon began. Adiabatic shear is most commonly manifested in situations involving impact or high-rate deformation and has been observed in metals, granular materials, explosives, ceramics and polymers. However, by far the greatest amount of work has been carried out on investigations, both experimental and theoretical, of the phenomenon in metals. These are fully described in other chapters in this book as well as in the book by Bai and Dodd [1] and summarized in the historical overview of the topic given by Walley [2].

Adiabatic Shear Localization. DOI: http://dx.doi.org/10.1016/B978-0-08-097781-2.00009-5

Earlier chapters in this book and also Refs [1,2] and the book by Wright [3] give a comprehensive overview of the theory of adiabatic shear band (ASB) formation. As such only the basic principles will be reviewed here.

If the time of plastic deformation of a material is a lot less than the thermal diffusion time t_d, which is approximately given by $t_d = L^2/k$, where L is the dimension of the sample and k the thermal diffusivity, it can be deduced that the deformation will be adiabatic and that the plastic work converted to heat in the deforming region will not escape. Assuming the strain is ~1, an estimation of the deformation time t_d is $1/\dot{\gamma}$, where $\dot{\gamma}$ is the strain rate. Thus the transition between an isothermal and an adiabatic process should occur when

$$\dot{\gamma} = \frac{k}{L^2} \tag{9.1}$$

The thermal diffusivity k of polymers is typically of the order of 1×10^{-7} m^2/s, and for a sample of typical dimensions 1 cm, the prediction is that the transition from isothermal to adiabatic processes should occur at a strain rate of $\sim 10^{-3}$ s^{-1}. Because high-rate tests are conducted at overall rates of 10^2 and above, and even medium-rate tests are conducted at rates of 10^{-1} s^{-1} and above, it is a reasonable assumption that the process will be adiabatic. To a first approximation, the plastic work associated with the deformation can be equated to the increase in thermal energy of the deformed material giving a temperature rise of

$$\Delta T = \frac{\tau\gamma}{\rho C_V} \tag{9.2}$$

where ρ is the density, τ is the shear stress, γ is the shear strain and C_V is the specific heat. Inserting typical values of the constants and flow stresses for polymers gives values of tens of degrees temperature rise per unit strain.

It is generally accepted that the strain rate in glassy polymers can be described by a thermally activated process [4] of the form

$$\frac{2\dot{\gamma}}{\dot{\gamma}_0} = \exp\left[-\left(\frac{\Delta H - v\tau}{RT}\right)\right] \tag{9.3}$$

where $\dot{\gamma}$ is the strain rate due to the applied stress τ, ΔH is the thermal activation energy, v is the activation volume and $\dot{\gamma}_0$ is a constant. This gives the shear yield stress in terms of strain rate as

$$\frac{\tau}{T} = \frac{\Delta H}{vT} + \frac{R}{v}\ln\frac{2\dot{\gamma}}{\dot{\gamma}_0} \tag{9.4}$$

Hence, the shear stress for a fixed temperature will increase with strain rate. The activation volume is temperature dependent (increasing with T) resulting in a drop in yield stress with temperature. The drop can be quite considerable, e.g. a factor of 3 drop in yield stress for a temperature rise from 70°C to 90°C occurs in poly(methyl methacrylate) (PMMA) [4], and the flow stress of Adiprene-L100, a vulcanizate of a polyether urethane rubber, drops from ~140 to ~15 MPa as the temperature is increased from −40°C to 25°C [5].

If the flow stress of a material is a function of temperature, strain and strain rate $\tau = f(\gamma, \dot{\gamma}, T)$, then you can write

$$\mathrm{d}\tau = \left(\frac{\partial \tau}{\partial \gamma}\right)_{\dot{\gamma},T} \mathrm{d}\gamma + \left(\frac{\partial \tau}{\partial \dot{\gamma}}\right)_{\gamma,T} \mathrm{d}\dot{\gamma} + \left(\frac{\partial \tau}{\partial T}\right)_{\gamma,\dot{\gamma}} \mathrm{d}T \tag{9.5}$$

where the three terms represent strain hardening, strain-rate hardening and temperature softening, respectively. Instability is said to occur when there is a maximum in the stress–strain curve [6]; thus, assuming constant strain rate and some constitutive relationship for the material and making use of Eq. (9.2), you can derive instability criteria for the particular material. Table 9.1 shows criteria for adiabatic shear for a range of different constitutive laws. Many of these criteria are expressed in the form of a critical strain to initiate the shear bands.

In general, a material is predicted to be sensitive to adiabatic shearing if the specific heat is low, the temperature sensitivity to flow stress is high and the strain hardening is low. A major problem in using constitutive models to predict critical strains is that the parameters $\delta\tau/\delta T$, strain-rate hardening exponent m, strain-hardening exponent n and so on are very difficult to quantify and will depend on the prevailing strain state and microstructure of the material and its temperature. Examples of the required experiments and difficulties are illustrated in the data for PMMA in Figures 9.1 and 9.2. Figure 9.1 shows uni-axial compressive stress–strain data at a strain rate of 0.001 s^{-1}, which should ensure isothermal conditions. The strain-hardening behaviour depends strongly on the temperature. The constancy of temperature at this strain rate is seen in Figure 9.2B. The effect of increasing the strain rate while carrying out tests at room temperature is shown in Figure 9.2A. The shape of the curve changes considerably, and the softening is much more pronounced as the strain rate increases. At the higher rates, softening consists of material softening and thermal softening due to the continuous temperature change in the sample, which is monitored by an infrared detector (Figure 9.2B). The drop in sample temperature near the ends of the tests is attributed to heat conduction into the stress rig at long timescales. Given that these experiments were all carried out at relatively low strain rates and much more pronounced effects will occur at high strain rates, the difficulties in extracting material parameters from the available data to fit to constitutive models are very apparent.

Table 9.1 Instability Criteria Obtained with Different Laws Describing Mechanical Behaviour

	No.	Constitutive Relation	Instability Strain	References
Linear strain hardening with yield strength	1	$\tau = A(1 + a\gamma)$	$\gamma_i = \dfrac{\rho c}{\left(-\dfrac{\partial \tau}{\partial \theta}\right)} - \dfrac{1}{a}$	[7], [8]
	2	$\tau = A(1 + a\gamma)\exp(-\beta\gamma)$	$\gamma_i = \dfrac{1}{\beta_1} - \dfrac{1}{a}$ $\beta_1 = \beta + \dfrac{\left(-\dfrac{\partial \tau}{\partial \theta}\right)}{\rho c}$	[9]
Power law without yield strength	3	$\tau = A\gamma^n$	$\gamma_i = \dfrac{n\rho c}{\left(-\dfrac{\partial t}{\partial \theta}\right)}$	[10]
	4	$\tau = A\gamma^n \exp(-\beta\gamma)$	$\gamma_i = \dfrac{n}{\beta_1}$	[9]
	5	$\tau = A\gamma^n \dot{\gamma}^m \exp\left(\dfrac{W}{\theta}\right)$	$\gamma_i = \left\{\dfrac{n\rho c}{AW}\dfrac{\theta^2}{\dot{\gamma}^m}\exp\left(-\dfrac{W}{\theta}\right)\right\}^{\frac{1}{n+1}}$	[11]
	6	$\tau = A\gamma^n \dot{\gamma}^m \theta^{-\gamma}$	$\gamma_i = \left\{\dfrac{n\rho c}{Av}\dfrac{\theta^{v+1}}{\dot{\gamma}^m}\right\}^{\frac{1}{n+1}}$	
	7	$\tau = A\gamma^n(1 + b\dot{\gamma}^m)(1 - c_1\theta)$	$\gamma_i = \left\{\dfrac{n\rho c}{Ac_1}\dfrac{1}{(1 + b\dot{\gamma})^m}\right\}^{\frac{1}{n+1}}$	[12]

Power law with yield strength	8	$\tau = A(1 + a\gamma^n)\left(1 + b\ln\frac{\dot{\gamma}}{\dot{\gamma}_0}\right) \times \frac{\theta_M - \theta}{\theta_M - \theta_0}$	$\gamma_i = \frac{n\rho c(\theta_M - \theta_0)}{A(1+a)} - \frac{\gamma_n^{1-n}}{a}$	[13]
Others	10	$\tau = A + B\log\dot{\gamma}$	$\gamma_i = \frac{B\rho C}{z\lvert\partial\tau/\partial\theta\rvert}$	[14]
	11	$\dot{\varepsilon} = ct\left(\frac{\partial\dot{\varepsilon}}{\partial\varepsilon} = 0\right)$	$\dot{\varepsilon}_c = \frac{4\pi(\varepsilon - \varepsilon_y)}{L^2\tau_y^2}\cdot kc\left(\frac{\partial\tau/\partial\varepsilon}{\partial z/\partial\theta}\right)^2$	[15]
	12	$\tau = H(\dot{\varepsilon}^R)F\left(\frac{\theta}{\theta_m}\right)$		[16]
		$= \tau_0\dot{\varepsilon}^{Rn}F\left(\frac{W}{E_m}\right)$		
		$\rightarrow F\left(\frac{W}{E_m}\right) = \left(1 - \frac{\alpha W}{E_m}\right)^{1/2}$	$\dot{\varepsilon}_t^p = \left[\frac{2\rho Emn(n+1)}{\alpha\tau_0(2n+1)}\right]^{1/(n+1)}$	
		$\rightarrow F\left(\frac{W}{E_m}\right) = \left(1 - \frac{\alpha' W}{E_m}\right)$	$\dot{\varepsilon}_t^p = \left[\frac{n\rho Em}{\alpha\tau_0}\right]^{1/(n+1)}$	
	13	$\dot{\gamma}_{xy}^p = \dot{\gamma}_0\exp\left(\frac{-\Delta H}{K\theta}\right)$	$\left[\frac{1}{\tau}\left(\frac{\partial\tau r}{\partial\gamma\rho}\right) - \frac{1}{C}\left(\frac{\partial\tau}{\partial\theta}\right)\right]\frac{\dot{\gamma}^p}{m} + \frac{k\xi^2}{C} = 0$	[17]
		$\Delta H\left(\frac{\sigma_{xy}}{r}\right)$		
		$m = \frac{\partial\ln\sigma_{xy}}{\partial\ln\dot{\gamma}_{xy}^p}$		

Source: From Ref. [2] and reproduced from Ref. [6].

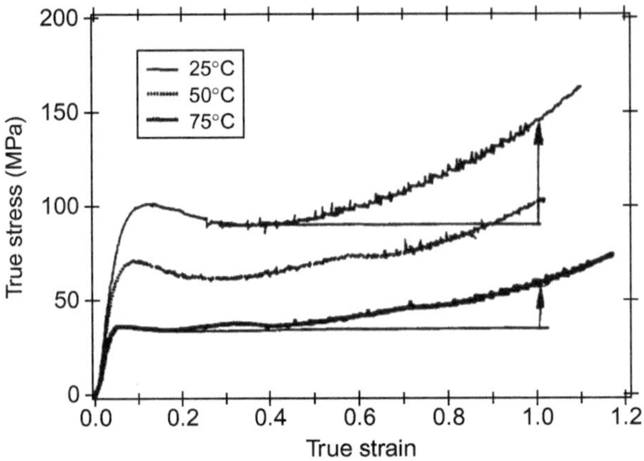

Figure 9.1 Experimental uni-axial compression stress–strain curves for PMMA at a strain rate of 0.001 s^{-1} and a range of temperatures. *Source*: From Ref. [18].

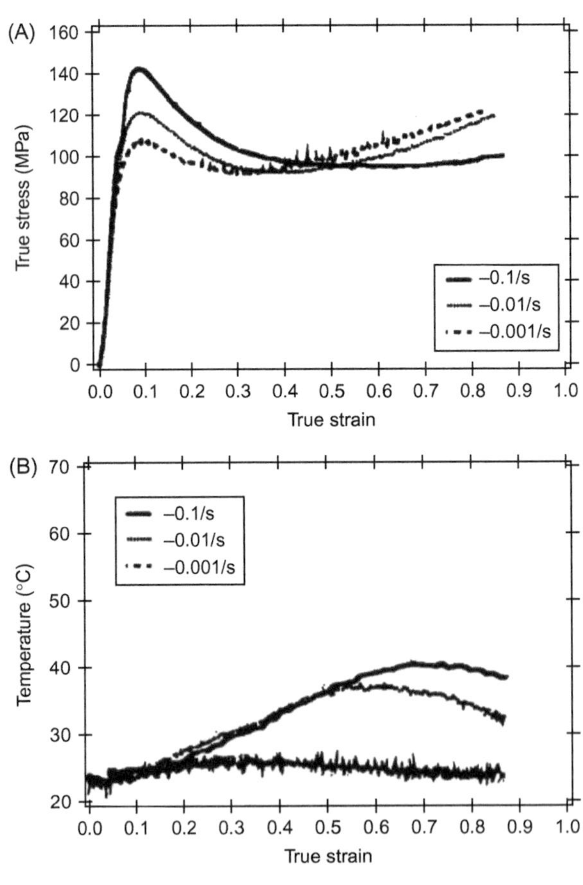

Figure 9.2 Experimental uni-axial compression stress–strain curves and temperature–strain curves for PMMA over three decades of strain rate (A) and recorded temperatures (B). *Source*: From Ref. [18].

9.2 Experimentally Measured Temperature Rises and Plastic Work Conversion

Polymers, compared to metals, have low thermal conductivities, typically 100 times less than that of metals, and high thermal softening. It is therefore expected that they should be prone to adiabatic shear banding. The first explicit recorded descriptions of adiabatic shear in polymers appear to be that of Winter [19] who impacted blocks of PMMA at room temperature with steel cylinders at speeds of up to 100 m/s and observed conical shear bands very similar to ASBs observed in titanium under similar loading conditions. Microscopic observations showed that shear was concentrated in narrow bands, which were planar regions of intense shear deformation, and it was concluded that in a similar way to metals, the localization was due to adiabatic heating causing local thermal softening that was greater than the strengthening due to strain or strain-rate hardening. Earlier, Adrianova et al. [20] reported experiments in which rectangular films of poly(ethylene terephthalate) (PET) of thickness 300 μm were extended at high rates, and a series of rapid load drops were observed in the stress time data, which were associated with the formation of narrow opaque bands in the sample. Experiments were carried out over a wide range of extension rates, but the bands were only observed when the extension rate exceeded 43 mm/min up to the maximum rates applied of 4350 mm/min (Figure 9.3). Local strain rates were not measured but must have been high. Adrianova et al. attempted to measure the temperature rises involved by dusting the samples with organic crystals containing nigrosine dye and observing where melting had occurred and also by using a copper–constantan thermocouple. Both methods produced temperature measurements of up to 140°C in the bands. It was found that the oscillations could be suppressed by carrying out experiments in cold water or by blowing cold air over the samples and that the bands became sharper if experiments were carried out in vacuum, thus suppressing heat loss. Essentially, the same experiments were repeated by Roseen, in 1974 [21], who used an AGA

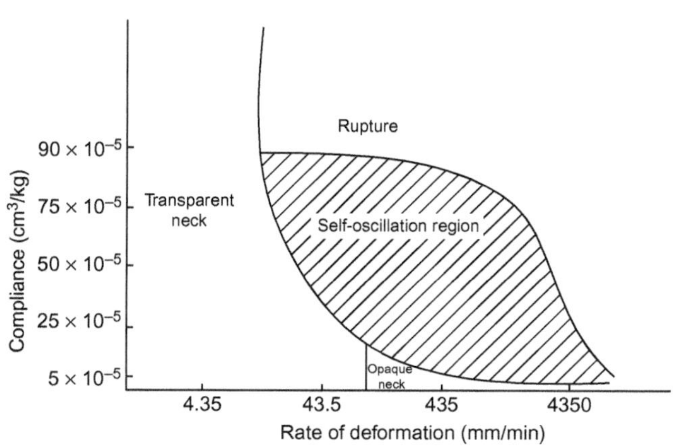

Figure 9.3 Deformation rate region in which self-oscillation (shear banding) was observed by Adrianova et al. [20].

thermovision system to monitor the temperature rises while carrying out tensile tests on PET and noted temperature of up to 100°C in the bands and also that the opacity was associated with crystallization of the polymer. Later work by Karger-Kocis et al. [22] has shown that these oscillations occur only in crystallizable polymers and that they are due to the release of elastic energy stored during a strain-hardening section of the deformation causing local cold crystallization of the polymer, each load drop giving rise to the formation of a further crystallization band. This is presumably caused by the local temperature rise being sufficient to raise the material to the cold crystallization temperature with further heating occurring due to the release of latent heat during crystallization. Figure 9.4 shows the striation bands observed in a low-rate test. These are similar to those observed by Adrianova and Roseen. Although these crystallization striations are not 'conventional' ASBs, they do demonstrate that shear localization resulting in large localized temperature increases do occur in semi-crystalline polymers. Karger-Kocis also went on to show that in high-rate tests, the type of crystallization oscillations observed at lower rates do not occur. Figure 9.5 shows the stress time records from tensile impact experiments from Karger-Kocis on PET at three different impact speeds, all close to 2 m/s. He concluded that stress oscillations observed

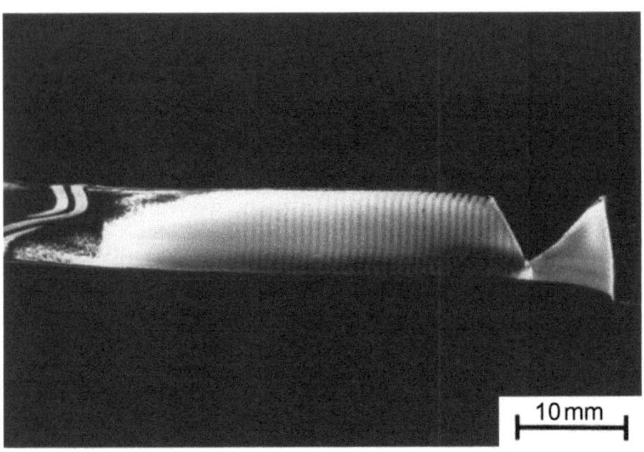

Figure 9.4 Necked specimen showing striation bands of increased crystallinity manifested in stress oscillations during tensile testing.
Source: From Ref. [22].

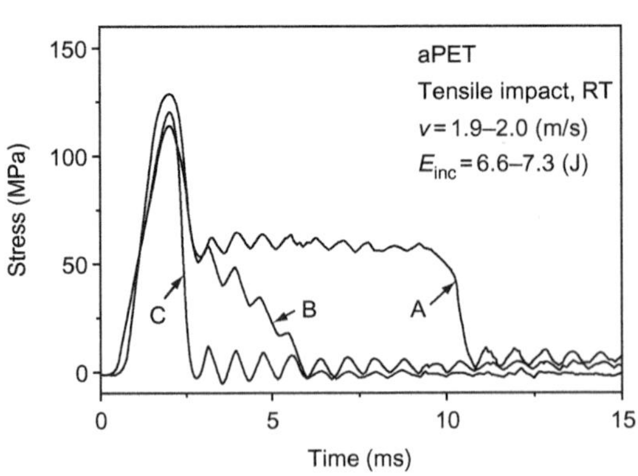

Figure 9.5 Tensile impact response of the three specimens illustrated in Figure 9.6. Specimen A necked and failed by diffuse shear and Specimen C by adiabatic shear. Specimen B showed intermediate behaviour.
Source: From Ref. [22].

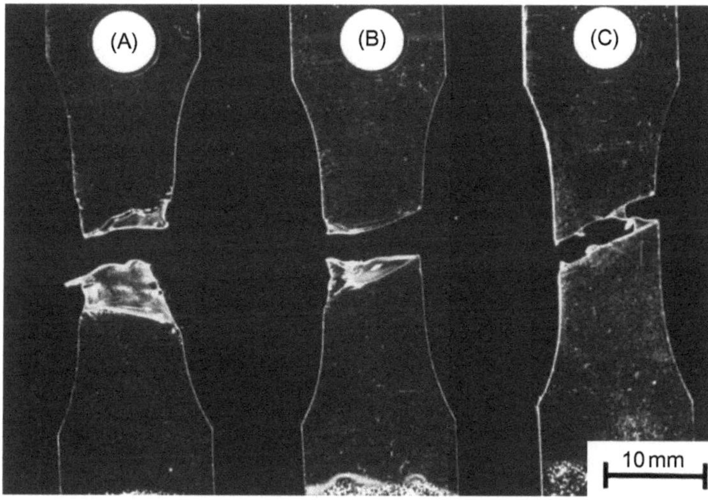

Figure 9.6 Images of specimens whose load–time plots are shown in Figure 9.5.
Source: From Ref. [22].

in the impact tests are artefacts caused by the natural frequency of the test equipment. Figure 9.6 shows the samples recovered after these three tests as illustrated in Figure 9.5. Diffuse intersecting shear banding formed a neck in sample A, which was attributed to multiple adiabatic shearing due to stress oscillations caused by ringing in the loading system. Sample C, hit at a slightly higher impact speed, failed along one dominant shear band. Sample B showed intermediate behaviour. Karger-Kocis went on to state that in tensile impact tests, both amorphous and semi-crystalline polymers show failure by adiabatic shear.

A difficulty in the use of models to predict instability strain is the assumption implicit in Eq. (9.2) that all of the plastic work is converted to heat. Kapoor and Nemat-Nasser [23] carried out a series of experiments on samples of Ti, Ta-2.5% W, 1018 steel, 6061 Al and OFHC copper in a Hopkinson-bar system at strain rates of $\sim 3000\ s^{-1}$ and monitored the temperature rise using an infrared detector. Temperature calibration was carried out by observing specimens at known temperatures over the range of 30–150°C. Initial results derived from the observed temperature rises indicated that a strain independent portion of 60–85% of the work of deformation was converted to heat in the samples. This is a surprising result because there is no obvious way that such a large fraction of the work could be stored as elastic energy in the metal. A second set of experiments was therefore carried out using an indirect temperature-measurement technique based on matching the stress–strain curves obtained in the initial experiments to those of identical samples deformed to known strains and reloaded at a temperature such that the stress and strain behaviour of the reloaded sample matches that observed in the original test. The method is described by Nemat-Nasser and Isaacs [24]. The latter method showed that a constant proportion of close to 100% of the work done on the samples was converted to heat in the sample. The error in the infrared measurement was attributed to differences between the (static) calibration of the infrared radiation (IR) system compared to its (dynamic) use in the experiments. Rosakis et al. [25] have shown from theoretical considerations that the assumption, in metals, that the fraction of plastic work converted to heat (β) is constant and close to 1 is invalid and proposed a constitutive model-based procedure which predicts that β varies with strain, strain rate, material properties and history, generally only approaching the normally assumed 0.8–1 for large strains.

Rabin and Rittel [26] carried out a theoretical investigation into the surface temperature of a cylindrical sample subjected to a step-like volumetric heat source and found that for metals, sudden loading should lead to measured temperature differences of $<0.05\%$ between surface (typically IR) and core measurements (typically embedded thermocouple). For polymers, because of their much lower thermal conductivity, temperature differences ranging from 5% to as much as 80% are predicted when steady-state conditions are achieved. However, for short time durations, such as those of an impact test, the authors showed that the differences between surface and core temperatures, even for polymeric samples, are small. Regev and Rittel [27] carried out a series of experiments in which commercial polycarbonate (PC) samples with a thermocouple embedded in their centre were deformed at a high rate in a Hopkinson-bar system while being simultaneously observed with an HgCdTe photovoltaic IR array. The agreement between core temperatures and temperatures deduced from the IR measurements was excellent showing that properly calibrated, IR systems can provide reliable temperature data and also confirming the validity of the analysis of Rabin and Rittel [26].

Work on the evaluation of β in polymers is rather rare; Adams and Farris [28] measured the heat evolved during the low-rate tensile deformation of 0.13 mm thick PC films in a specially adapted calorimeter and found that between 50% and 80% of the work of deformation was converted to heat with the remainder being retained as internal energy in the sample.

Buckley et al. [29] carried out Hopkinson-bar studies at strain rates $\sim 4500\ s^{-1}$ on thermosetting epoxy resins during which temperature data was simultaneously recorded with a time resolution of 1 μs. The proportion of the work done on the samples to appear as heat was calculated and was found to vary considerably with strain (Figures 9.7 and 9.8). It can be seen in Figure 9.8 that, for all samples, the fraction of the energy β appearing as heat rises to a maximum at a time corresponding to approximately the minimum in the stress–strain curve and thereafter falls. β is not calculated during failure. The sample in Figure 9.7B fractures at around 140 μs and shows a very rapid increase in temperature just before and during fracture, while the sample in Figure 9.7A does not fracture and does not show this rapid temperature excursion. While this seems to be a significant observation, Buckley cautions that this apparent temperature rise may be due to changes in emissivity. This result contrasts greatly with Karpoor's [23] results from metals where β is close to unity throughout. Rittel [30] using embedded thermocouples carried out similar experiments to those of Buckley on PC discs at strain rates of $5000-8000\ s^{-1}$. An example of the data is shown in Figure 9.9. The results are

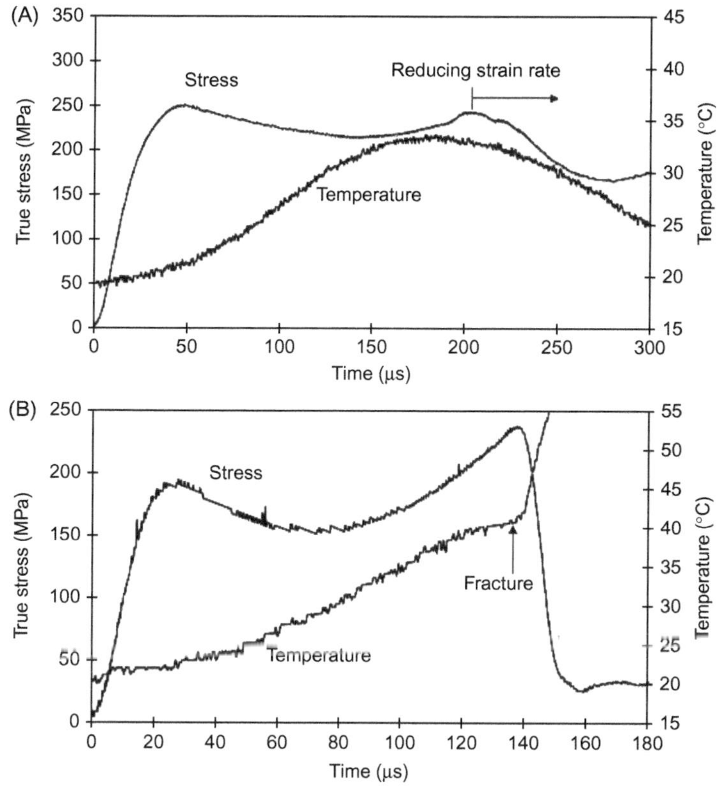

Figure 9.7 Stress and temperature time profiles for two different epoxy samples tested in the Hopkinson bar at a strain rate of 2750 s^{-1} (A) and 4500 s^{-1} (B). The minima in the stress time curves correspond to a strain of ~ 0.4.
Source: From Ref. [29].

similar to those of Buckley [29] in that almost no temperature rise is observed until the minimum in the stress–strain curve is passed. Rittel distinguishes between β_{int}, the amount of plastic work converted to thermal energy at a given strain, and β_{diff}, the rate of heat generation divided by the incremental rate of plastic working using these expressions:

$$\beta_{\text{diff}}(\varepsilon) = \frac{\rho c \dot{T}}{\mathrm{d}\dot{W}_{\text{p}}} \quad \text{and} \quad \beta_{\text{int}}(\varepsilon) = \frac{\rho c \Delta T}{\int \mathrm{d}W_{\text{p}}} \tag{9.6}$$

Results are illustrated in Figure 9.10. The peak value reached by β_{int} increases with the strain rate, i.e. it is both strain and strain-rate sensitive but is always <1. The rate of conversion of mechanical work to heat illustrated by β_{diff}, however, reaches values >1. This may seem surprising but can be explained by the concept

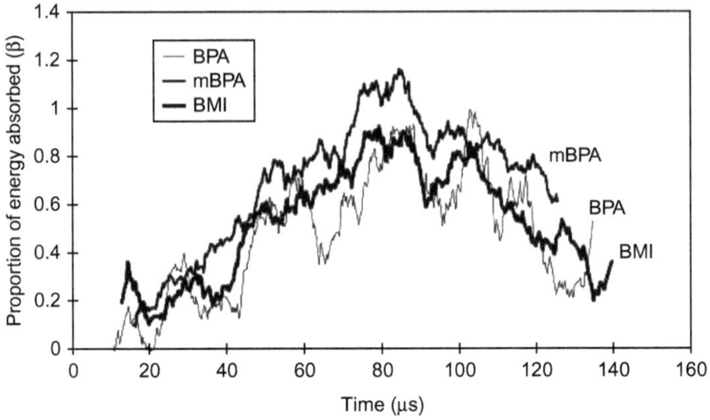

Figure 9.8 Proportion of energy of deformation appearing as heat as a function of strain for three different epoxies tested at a strain rate of 4500 s^{-1}.
Source: From Ref. [29].

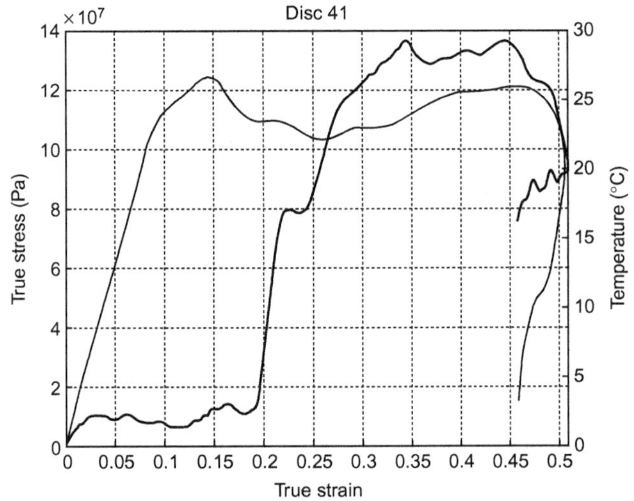

Figure 9.9 True stress strain and temperature record for a PC disc deformed at 6500 s^{-1} in a Hopkinson-bar system: continuous line, temperature; broken line, stress–strain.
Source: From Ref. [30].

that during the strain-softening regime, the stored energy in the sample decreases and transforms into heat. This latter may have consequences for the formation of ASBs.

As outlined earlier, there is now a good deal of experimental evidence regarding the fraction of plastic work converted to heat during adiabatic deformation and the

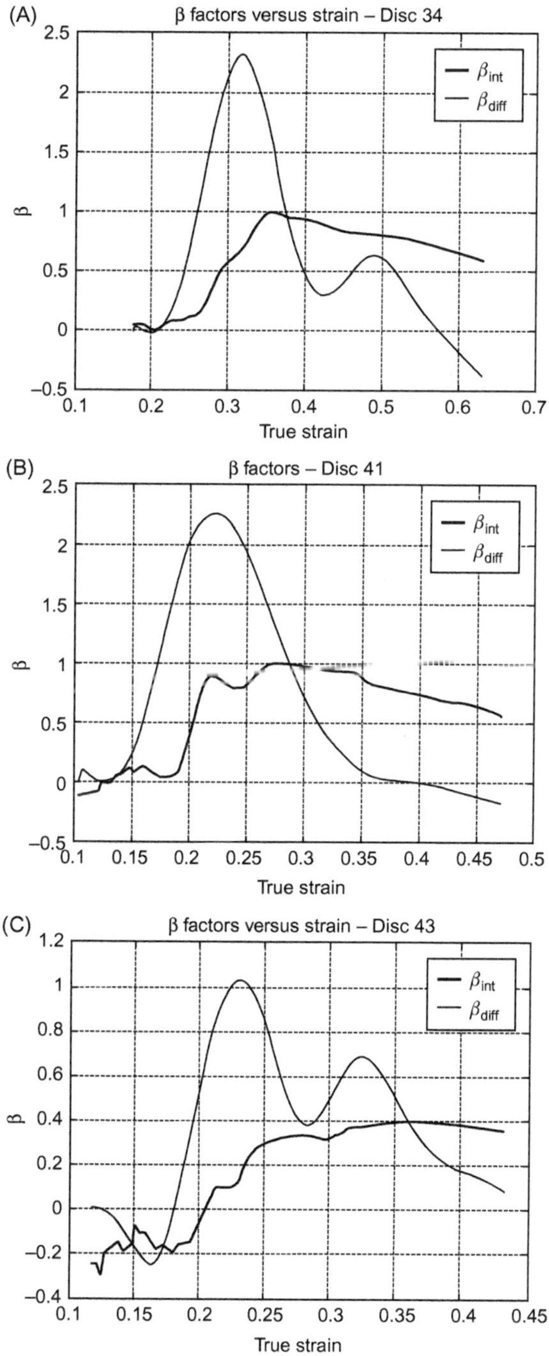

Figure 9.10 Evolution of β_{diff} and β_{int} with strain for the stress–strain and temperature data of three PC samples deformed at (A) 8500 s^{-1}, (B) 6500 s^{-1} and (C) 5000 s^{-1}. (B) is the sample in Figure 9.9. *Source*: From Ref. [30].

temperatures attained, but these are either average values over a large fraction of the sample or values obtained by one thermocouple at a specific location in the sample. There are very few measurements of the temperatures attained within a shear band. Evidence from melting (in polymers) or phase transformations (in metals) indicate that temperature increases well above the reported average values of up to 100°C. Guduru et al. [31] developed an infrared high-speed camera with sufficient temporal (500 ns) and spatial (0.1 mm) resolution to enable them to measure temperature rises in the adiabatic shear zone ahead of a running crack in martensitic steel. Images obtained showed initiation of adiabatic shear ahead of a crack when the propagating crack tip temperature rise reached 80 K. The ASB tip traversed the sample at speeds of up to 500 m/s with temperature gradients of up to 5×10^4 K/m. The temperature distribution in the band was highly non-uniform with maximum temperature regions of up to 900 K.

A polymer experiment very similar in concept to that described by Guduru et al. [20] was reported by Fuller et al. [32] in 1975. In their experiment, an infrared detector with a response time of 3 μs was used to detect radiation ahead of fast-moving cracks in glassy polymers. With crack speeds of 200–500 m/s, temperatures of up to 500 K were reported for PMMA samples and up to 600 K for polystyrene (PS) samples. These temperatures are well in excess of the glass transition and melting temperatures of these polymers and are speculated to cause decomposition in the polymer. Bjerke and Lambros [33] carried out impact experiments on PC and PMMA-grooved plates by impacting with a steel projectile on the edge of the plates at impact speeds of ~40 m/s. A linear array of 16 HgCdTe infrared detector elements of size 80 μm oriented perpendicular to the groove was used to observe the resultant shear band and crack. Surface temperature increases of up to 145 K were observed in PC and 85 K in PMMA. Examination of recovered PC samples showed clear evidence of adiabatic shearing followed by fracture. Although not as high as the temperatures reported by Fuller et al. [32], they are of the same order since Fuller's values are for the interior of the crack following shear and Bjerke's for the surface. Li and Lambros [34] used two single-element infrared detectors to observe Hopkinson-bar tests on PMMA and PC and found that β_{int} dropped from ~ 0.8 at $\varepsilon \sim 0.05$ to ~ 0.5 at $\varepsilon \sim 0.1$ and remained roughly constant thereafter. For PMMA, only modest temperature rises of about 1.5°C were observed until failure of the sample at $\varepsilon \sim 0.07$ when the temperature rapidly increased (Figure 9.11). This is very similar to the observations of Buckley et al. [29] and indicates rapid release of stored energy driving the failure process.

9.3 Adiabatic Shear Band Observations in Polymers

Direct observations of ASBs in polymers are very sparse. They are often identified by comparison to similar bands in metals subjected to the same type of deformation or by microscopic examination of recovered specimens. The latter can present difficulties because the characteristics of the deformation are not as well defined as in

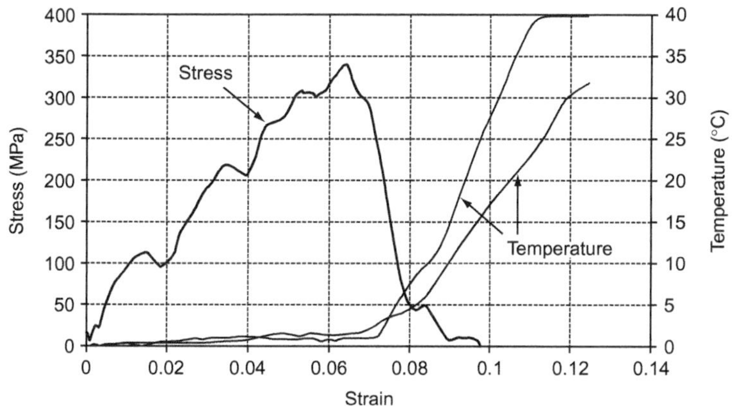

Figure 9.11 Stress–strain and temperature measurements for a Hopkinson-bar test on PMMA at a strain rate of 2000 s^{-1}. The two temperature traces are from separate infrared detectors.
Source: From Ref. [34].

metals. As mentioned earlier, Winter [9] identified adiabatic shear in PMMA by direct comparison to similar deformation in titanium. Swallowe et al. [35] carried out a series of high-rate compressive experiments on polymer discs using a falling-weight apparatus in which the samples were filmed in transmitted light along the compression axis using a high-speed camera with a 7 μs framing interval. Compression was between toughened glass anvils with a light path through the system enabling transmitted light to be used. Subsidiary experiments were carried out using a heat-sensitive film technique to record the maximum temperatures reached during compression. Experiments were carried out at room temperature on PMMA, PC, PS, nylon 6, polytetrafluoroethylene (PTFE), high-density polyethylene (HDPE), polypropylene (PP) and polyvinylchloride (PVC). The strain rates were $\sim 800\ s^{-1}$. Broadly speaking, the materials investigated fell into two categories: those which failed by shear banding or shear-initiated cracking (PMMA, PC, PS, PTFE) and those in which bulk plastic deformation was observed (PP, HDPE, nylon, PVC). Figures 9.12 and 9.13 illustrate the formation of the shear bands in PS and PC. The sequence of shear and cracking followed by PC, PS and PMMA is illustrated in Figure 9.14, and the strain to failure (defined as first indications of localized failure) and maximum observed temperatures are set out in Table 9.2. The heat-film technique used could not detect temperatures of <200°C. Temperatures recorded as <200°C may be as low as room temperature. Impacts on PMMA samples at 20°C and 50°C using an improved version of the same apparatus [36], which enables stress–strain data to be gathered, are illustrated in Figures 9.15 and 9.16. Differences in mechanical behaviour for modest changes in temperature are readily apparent with the 20°C sample undergoing catastrophic failure at low strain and the 50°C sample failing by adiabatic shear.

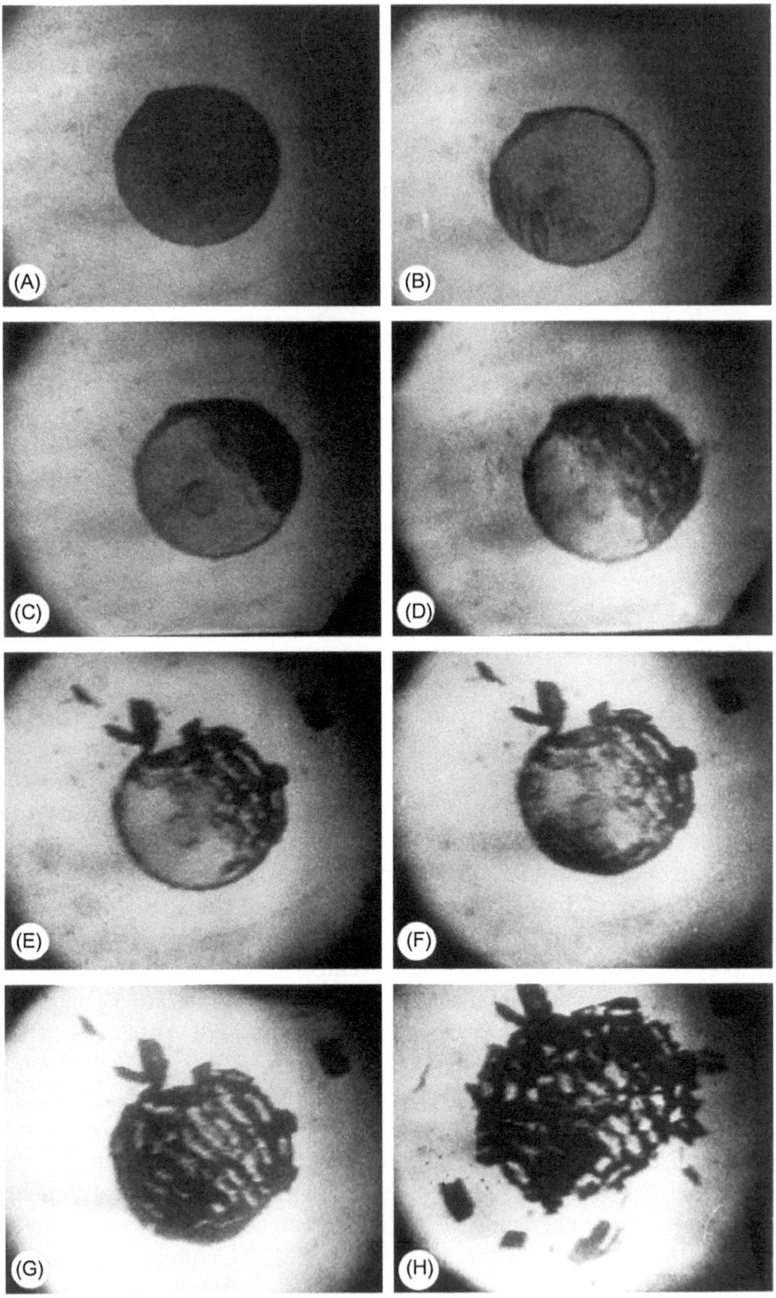

Figure 9.12 Impact on a 5 mm diameter PS disc viewed in transmitted light. Times after frame: (A) 0 μs, (B) 35 μs, (C) 42 μs, (D) 49 μs, (E) 189 μs, (F) 196 μs, (G) 203 μs and (H) 259 μs. Shear banding followed by cracking and fragmentation is evident from frame (C) onwards.
Source: From Ref. [35].

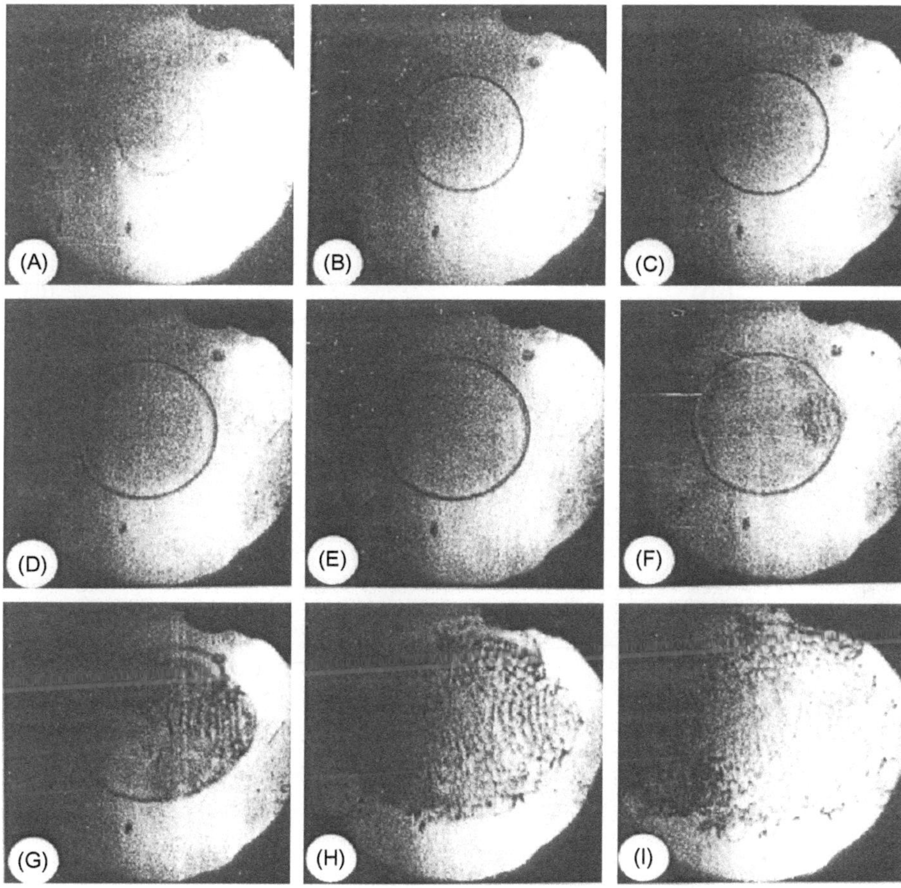

Figure 9.13 Impact on a 5 mm diameter PC disc viewed in transmitted light. Times after frame: (A) 0 μs, (B) 161 μs, (C) 188 μs, (D) 241 μs, (E) 268 μs, (F) 275 μs, (G) 281 μs and (H) 288 μs; (I) 295 μs. Shear banding commences in frame (F) and spreads rapidly across the sample.
Source: From Ref. [35].

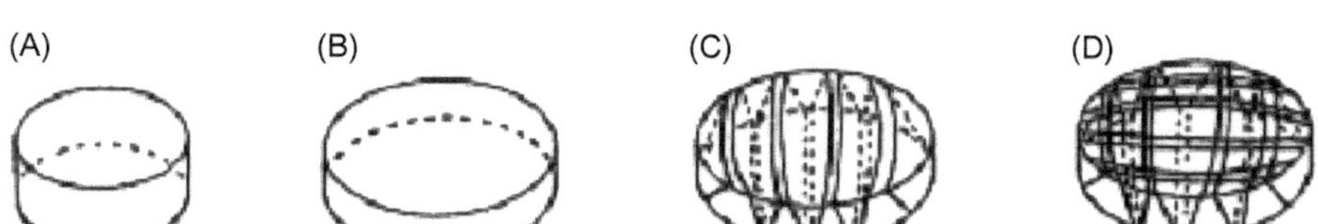

Figure 9.14 Sketches of the shear and cracking sequences reported in Ref. [35]: (A) the undeformed sample, (B) the sample at the end of uniform plastic deformation, (C) failure by parallel shear cracks and (D) continued failure by perpendicular shear cracks.

Table 9.2 Summary of Observations Reported in Swallowe et al. [35]

Polymer	Strain[a]	Max T of Plastic Deformation (°C)	Max T (°C) Shear Band or Crack	T_g (°C)
HDPE	0.9	<200		−125
PP	>2.5	230		−10
Nylon 6	1.0	400		52
PVC	0.7	450		81
PS	0.01	<200	550	100
PMMA	0.04	<200	530	105
PTFE	0.25	<200	600	117
PC	1.1	<200	700	145

[a]Strain listed is either strain at failure or maximum observed strain in the test.

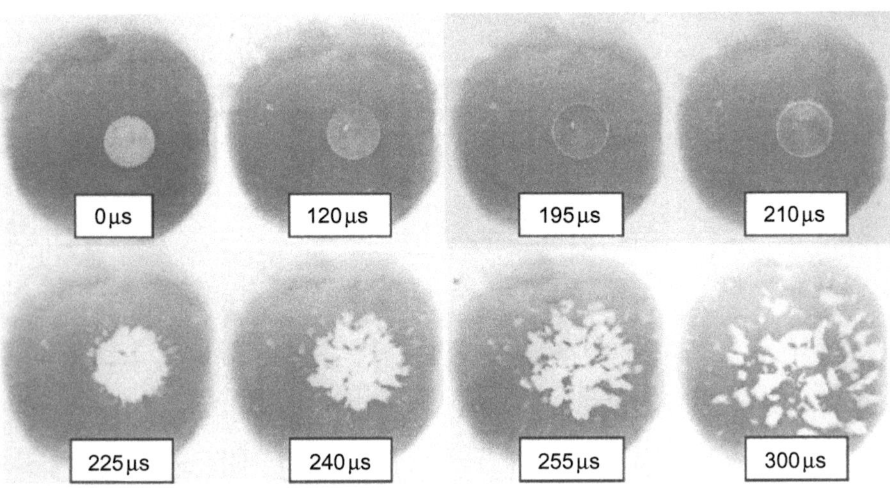

Figure 9.15 Impact on 5 mm PMMA disc at 20°C viewed in transmitted light. Times are microseconds after the first frame.
Source: From Ref. [36].

Walley et al. [37] carried out a series of experiments similar to those of Swallowe et al. [35] making use of the heat-sensitive film technique to follow temperature maxima. They observed shear banding in PMMA, PC and also in polyethersulphone (PES). Temperature rises were observed during the deformation of all of the materials, although no numerical values of the temperature rises are reported.

It is evident from Table 9.2 that those polymers exhibiting adiabatic shear (in room-temperature tests) have a glass transition temperature T_g well in excess of the test starting temperature and that they experienced relatively low bulk heating by

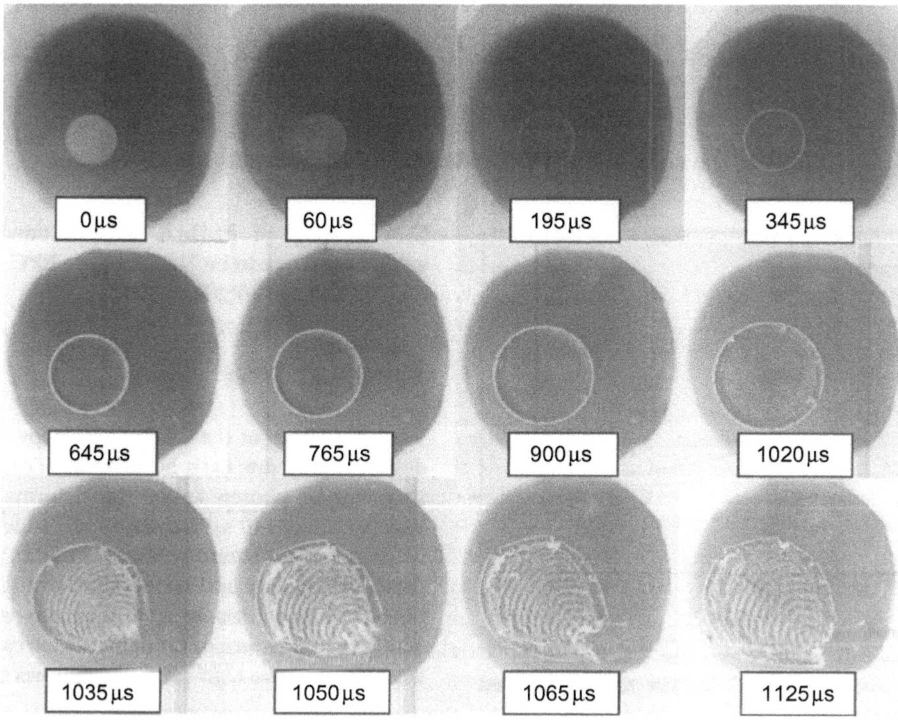

Figure 9.16 Impact on 5 mm PMMA disc at 50°C viewed in transmitted light. Times are microseconds after the first frame.
Source: From Ref. [36].

plastic deformation. As mentioned earlier, the effect of proximity to T_g on the mechanical performance is clearly illustrated by Figures 9.15 and 9.16, where brittle failure is evident in the PMMA sample tested at 20°C, whereas adiabatic shearing occurs at 50°C. At temperatures of 80°C and above, no shearing is observed, and the samples continue to deform by uniform plastic flow to high strains. Figure 9.17 shows the stress–strain response as T_g is approached.

The glass transition temperature is the most important thermal transition shown by amorphous polymers. To a first approximation, below T_g, these materials are brittle with high elastic modulus, above T_g, they have low modulus and are 'rubbery' and, at higher temperatures, they are viscous. Although quoted as a single temperature, the glass transition represents a continuous change in behaviour and typically spans ~20–30°C with the quoted T_g being at the centre of the range. Polymers have high ductility at temperatures below, but close to, T_g with post-yield softening at low strains followed by strain hardening at large strains. PC, PS, PVC and PMMA are amorphous polymers and PP, HDPE, PTFE and nylon are crystalline polymers. However, the division is not absolute. PS can come in both crystalline and amorphous form depending on the details of the molecular structure, and

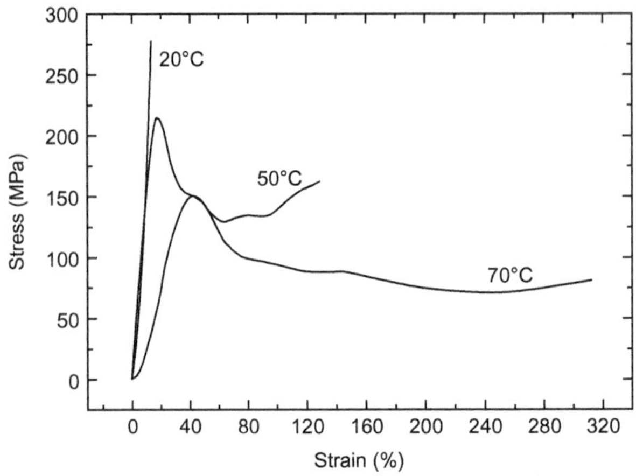

Figure 9.17 Stress–strain curves at different temperatures and a strain rate of 800 s^{-1} for the PMMA samples illustrated in Figures 9.15 and 9.16. *Source*: From Ref. [36].

crystalline polymers are actually semi-crystalline containing a large fraction (anything from 10% to 50%) of amorphous material. The amorphous/crystalline ratio will depend on many factors such as molecular structure (particularly positions and size of side groups), molecular weight and processing parameters and would need to be measured for any particular polymer sample, but all polymers will have a T_g. The glass transition is a relaxation process associated with large segmental motion of the main polymer chain and many features of the polymer structure, such as cross-linking, plasticizers and molecular structure, influence the temperature of occurrence and the magnitude of the effects passing through T_g on mechanical properties. Further relaxations (called β, γ, etc.) associated with short-range localized motions in the main chain, side chain and so on occur at lower temperatures and can significantly affect the mechanical properties. Plastic deformation in amorphous polymers is generally possible at temperatures down to the β relaxation temperature. This occurs at $\sim -100°C$ in PC and somewhere in the region of 10°C in PMMA, hence the much wider temperature range in which plastic deformation is possible in PC compared to PMMA. The crystalline portion of semi-crystalline polymers also has a large influence on the relaxation processes. The book by Ward and Sweeney [38] provides a good introduction to these processes. T_g is observed to increase with increasing strain rate. For example, the recent work by Mulliken and Boyce [39] has shown a shift of 4.7°C per decade of strain rate for PC and 11.1°C per decade of strain rate for PMMA.

Similar to other crystalline materials, polymers exhibit solid-state phase changes and therefore possible transformations from a ductile to a brittle phase must be considered. An example is the work of Rae et al. [40] who observed an abrupt ductile–brittle transition in Taylor impact experiments on PTFE, which they attributed to a pressure-induced crystalline phase change in the polymer that transforms on a millisecond timescale. Shear failures were observed at the outer annulus of the samples caused by the stress at or near the interface between the two phases. Experiments on poly(chlorotrifluoroethylene) (PCTFE), a close relative of PTFE, where one out of every four fluorine atoms along the carbon backbone is replaced by a chlorine atom, were carried out for comparison. PCTFE does not undergo a crystalline phase transition in the temperature or pressure ranges expected in the tests, and no sharp ductile–brittle transition was observed.

The vast majority of work that has been carried out related to adiabatic shearing in polymers has concentrated on the three amorphous materials: PMMA, PC and PS. This is probably because they can be formed into tough transparent windows. Ravi-Chandar and co-workers [41,42] and Rittel and co-workers [43–46] have produced a series of papers in which they explore the behaviour of impacts on PMMA and PC at high loading rates. Both teams report on experiments in which projectiles are fired at a range of speeds at the edge of notched, pre-cracked or grooved plates of PC and PMMA subjected to compressive loads and observe unexpected transitions from brittle failure at low-impact speeds to a shear-stress dominated shear-banding failure at high strain rates. The experiments were similar to those of Bjerke and Lambros [33] who recorded temperature rises of 145°C in PC and 85°C in PMMA. Post-mortem microscopic examination of their samples by Ravi-Chandar revealed a banded and heavily drawn structure suggesting a very ductile deformation mechanism (Figure 9.18). Rittel also employed electron

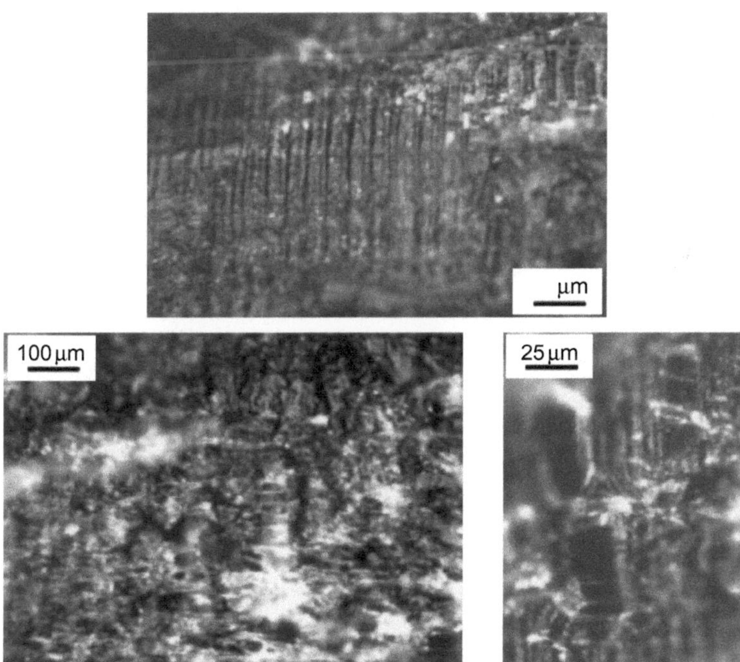

Figure 9.18 Electron micrographs of the fracture surface of PC specimens after shear fracture. The banded and very heavily drawn structure suggests a very ductile deformation mechanism.
Source: From Ref. [42].

microscopy to examine his samples and observed shear bands. Their conclusions were that provided significant (adiabatic) crack-tip heating develops, adiabatic shear failure may be observed in general mixed-mode loading. Rittel [44] studied the influence of testing temperature on the failure mode transition of high-velocity impacts on PC and PMMA plates. He noted that for PC at low temperatures (−120°C up to −40°C), fracture was exclusively by crack opening mode but that at room temperature and up to his maximum test temperature of 70°C, specimens did not fail by separation and exhibited significant crack-tip shear banding. For PMMA, he observed brittle fracture in all cases until the test temperature exceeded T_g. His conclusion was that, contrary to PC, PMMA showed only modest adiabatic crack-tip heating in these experiments. In a subsequent paper, Rittel [45] tested PMMA discs in a Hopkinson-bar apparatus using the same type of embedded thermocouple experiments as he previously reported for PC [30] and whose results are illustrated in Figure 9.9. In this test situation, he found that the bulk temperature rise in PMMA (Figure 9.19) does not occur until a much later stage in the deformation than the temperature rise observed in PC – well into the strain-softening region (compare Figure 9.9 with Figure 9.19). He also reported one of the very rare experiments where temperature rises are recorded at the tip of a shear band in a polymer during its formation. A temperature rise of 70°C was measured in a PC sample, and examination of the specimen after the experiment showed a 200 μm wide ASB ahead of the fatigue pre-crack. Figure 9.20 illustrates these shear bands and shows the characteristic elongated dimples formed by adiabatic shear in polymers. Rittel concluded that the prediction of a brittle to adiabatic shear crack failure mode was subject to the details of the thermomechanical couplings associated with the test and that the loading mode, strain levels and thermomechanical response of the material at the crack tip are all contributory factors.

To investigate the effects of a confining hydrostatic pressure on the behaviour of PMMA samples, Rittel and Brill [46] carried out a series of experiments in

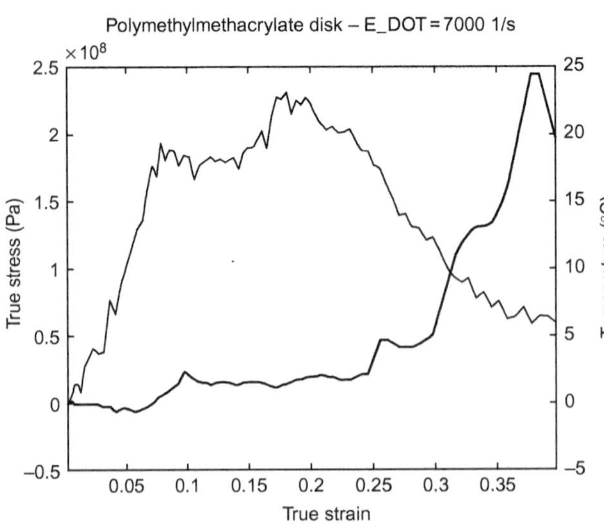

Figure 9.19 True stress–strain record and temperature rise measurement for a PMMA disc tested in compression at 7000 s^{-1}: thick line, temperature; thin line, stress–strain.
Source: From Ref. [45].

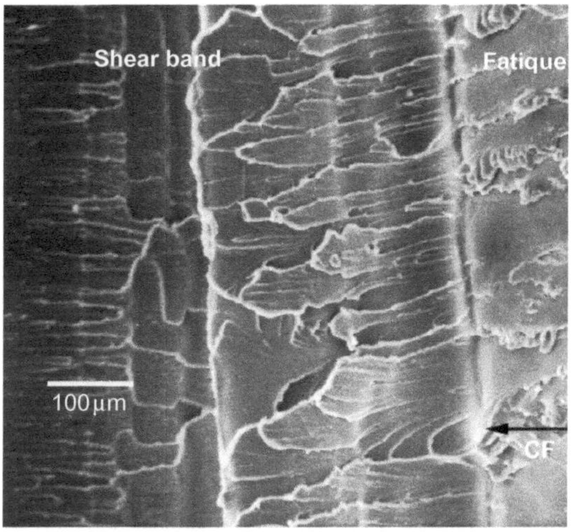

Figure 9.20 Scanning electron micrograph of a PC sample tested at −25°C. An ASB (indicated) is visible ahead of the fatigue pre-crack. CP indicates the crack propagation direction.
Source: From Ref. [45].

which cylindrical PMMA samples were enclosed in tightly fitting ductile metal sleeves of varying thickness. The enclosed samples were tested quasi-statically and using a Hopkinson bar. The metal sleeves were designed to yield in the early stages of the test, and, being made of a low strain-hardening material, they exerted an approximately constant pressure on the PMMA specimen. The range of thicknesses of the confining rings enabled a range of confining pressures to be applied with thicker rings resulting in higher pressures. The results were that at higher confining pressures (71 MPa) and at high strain rate (2000 s^{-1}), failure was by the formation of a conical plug typical of adiabatic shear. The result is reminiscent of that described by Fuller et al. [32]. The cone's envelope consisted of elongated dimples that are characteristic of adiabatic shear. Figure 9.21 illustrates the conical plug and the dimples and signs of melting on its surface. Figure 9.22 shows the shear band in cross section with multiple micro-cracks and signs of extensive flow. Rittel points out that confining pressure and strain rate have opposite effects on the brittle–ductile transition such that at low strain rates, a higher confining pressure is required to cause the transition from brittle to ductile behaviour.

It has been known for a considerable time that the application of pressure suppresses brittle failure and that ceramics and rocks undergo a brittle–ductile transition under pressure [47]. The same effect is seen in brittle polymers. The yield stress of a polymer is also well known to increase with pressure and follows a simple relationship [19]:

$$\tau = \tau_0 + \alpha p \tag{9.7}$$

where τ is the shear yield stress, τ_0 is the shear yield stress at atmospheric pressure, p is the pressure and α is the coefficient of increase in yield stress with pressure.

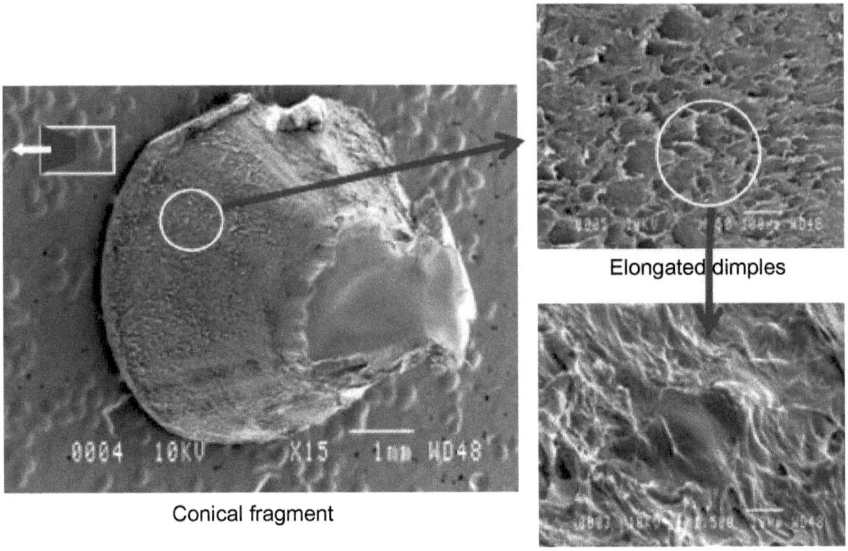

Figure 9.21 Micrographs of adiabatic shear conical cones produced in the fracture of confined PMMA cylinders at high strain rates.
Source: From Ref. [46].

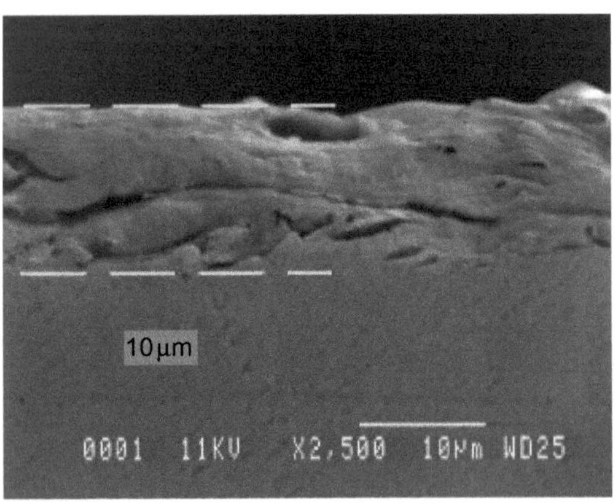

Figure 9.22 Section of ASB in PMMA showing multiple micro-cracks and extensive flow.
Source: From Ref. [46].

For the transition to occur under pressure (at a given strain rate), the pressure-dependent rate of increase in shear yield stress must therefore be less than that of the brittle fracture stress. Because the brittle–ductile transition depends critically on the proximity of test temperature to T_g, variations in T_g with pressure must also be considered. Stevens et al. [48] have carried out measurements on the increase in T_g with pressure for PS and conclude that it will increase asymptotically with

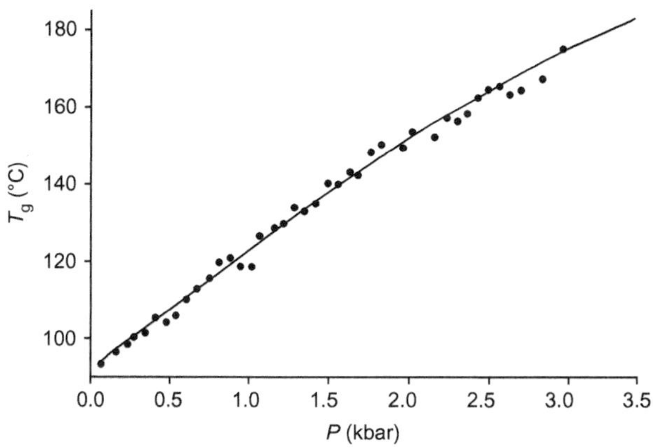

Figure 9.23 Variation in the glass transition temperature with pressure.
Source: From Ref. [48].

pressure. They produce data (Figure 9.23) showing an increase of ~80°C for an applied pressure of 3 kbar.

In a recent paper, and making use of the results of Rittel on confined cylinders, Dorogoy et al. [49] have studied the impact of steel projectiles on inclined plates of PMMA and produced a finite element (FE) model to predict the depth of penetration, probability of a ricochet, effect of projectile mass and so on. The formation of conical plugs by adiabatic shear – similar to those observed by Rittel and Brill [46] – is predicted.

Li et al. [50] carried out a series of inclined plate impact experiments using an 8 mm thick nylon 66 flyer plate to impact a 25 mm thick nylon target (nylon is a semi-crystalline polymer) at a range of angles of impact from 10° to 20° to study the attenuation of shear waves near the impact surface. Polarized optical microscopy was used to examine the microstructural changes in the sample after impact. It was found that in the centre of the samples, a layer of non-crystalline material was formed *under* the impact surface, which was interpreted as a melted layer that did not recrystallize during the unloading process. In areas closer to the periphery where the shear stress was not so great, a 2–3 μm thick layer consisting of highly deformed crystals was observed whose characteristics were typical of an ASB (Figure 9.24). Further shear, as experienced near the centre of the sample, caused melting followed by rapid solidification of this layer. Nylon 66 has a T_g of ~50°C and a melting temperature of ~250°C. The high strain rates involved in impact tests will be expected to increase T_g and therefore make the polymer susceptible to adiabatic shearing, which is normally initiated when local temperatures are below T_g. When adiabatic shear has initiated, local temperatures can rise rapidly to the melting temperature.

Elastomers are essentially cross-linked amorphous polymers and transform at a characteristic T_g from brittle to 'rubbery' behaviour. They should therefore be susceptible to adiabatic shear, provided they are deformed at high rates and at temperatures close to T_g. Elastomers present severe machining difficulties and can only be successfully machined at low temperatures. Dhokia et al. [51] reported

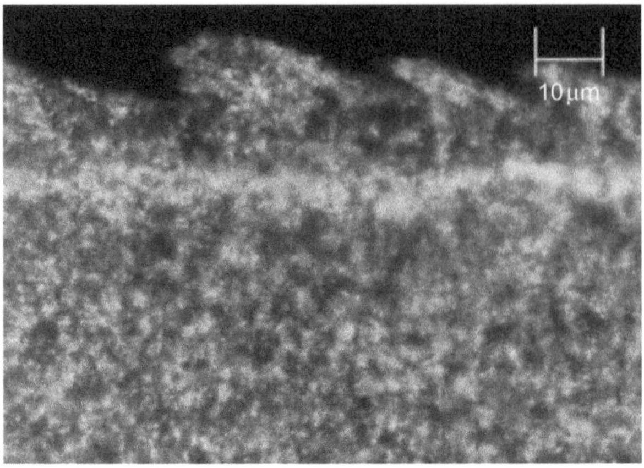

Figure 9.24 Illustration of subsurface adiabatically shear banded region in a nylon 66 sample subjected to inclined plate impact. *Source*: From Ref. [50].

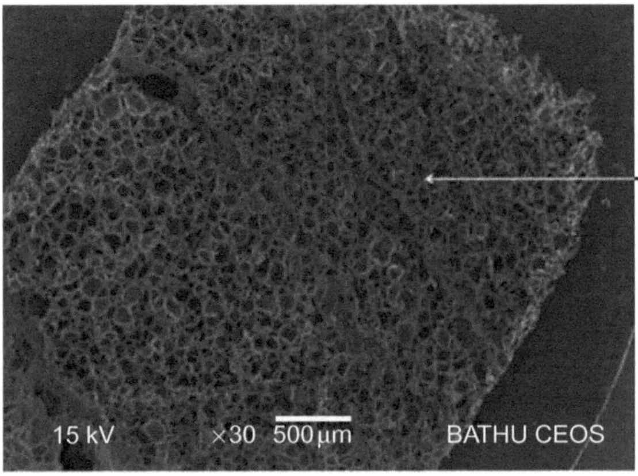

Figure 9.25 Adiabatic shear banding in a cryogenically machined EVA elastomer chip. *Source*: From Ref. [51].

experiments on the cryogenic machining of two soft elastomers: ethylene vinyl acetate (EVA) and neoprene. The T_gs were determined using dynamic mechanical-thermal analysis to be −44°C for EVA and −52°C for neoprene. Machining was done on a computer-controlled (CNC) milling machine with a cryogenic system that sprayed liquid nitrogen onto the tool tip and the workpiece and used temperature sensors to maintain the machining temperature at a desired level. At temperatures above T_g, poor quality machining occurred, which produced irregular chips and showed serrations and tearing. Tests at or just below T_g produced improved machining but led to the development of ASBs in both elastomers (Figure 9.25) with severe localized plastic deformation. Further reduction of the machining temperature eliminated or greatly reduced the shear banding and produced good quality machined surfaces.

Polymer composites are also susceptible to adiabatic shear, although the addition of the second phase can considerably alter the behaviour. Samuel et al. [52] reported on the machinability of carbon nanotube (CNT)-reinforced PC-based composites. PC composites containing 1.75% (composite A), 5% (composite B) and 15% (composite C) by weight of CNT were prepared and their material properties characterized. Machinability studies were carried out on a CNC machine, and chip morphology was used to compare the materials. ASBs were observed in plain PC samples, which were reduced by the addition of as little as 1.75% of CNTs with very few bands observed in the 5% loaded material and none in the 15% (Figure 9.26). At its highest CNT loading, the composite became brittle. The thermal conductivity of the composites increased from 0.139 W/mK for plain PC to 0.153, 0.157 and 0.236 W/mK for composites A, B and C, respectively, and the specific heat fell from 1.760 J/g°C for plain PC to 1.615 J/g°C for composite C [52]. There was a continuous reduction in strain to failure in low-rate tensile tests with CNT loading. The reduction in adiabatic shear banding was attributed to the increased thermal conductivities of the composites and the increasing importance of the CNTs in determining the machining properties.

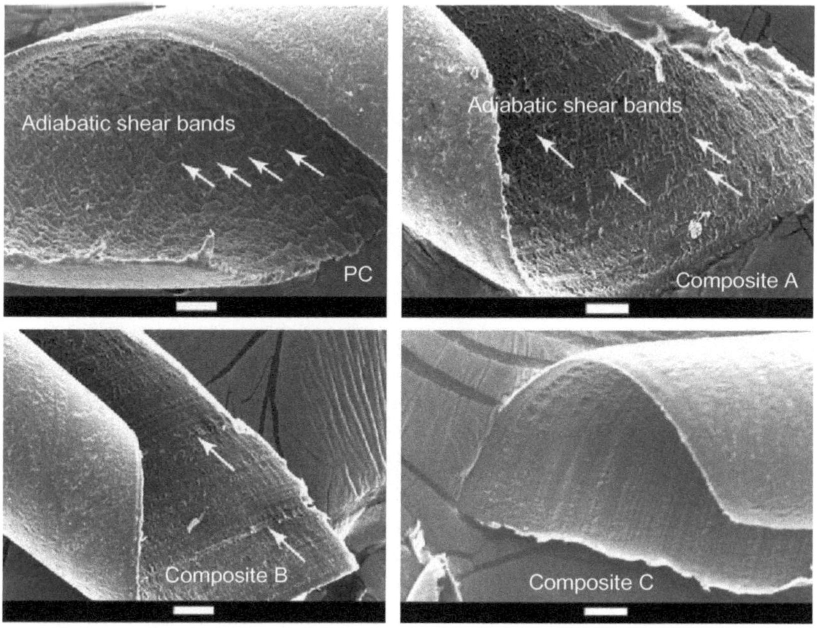

Figure 9.26 Electron micrographs of cutting chips formed while machining plain PC and three composite samples containing CNTs. The reduction in adiabatic shear banding with increasing CNT content is evident.
Source: From Ref. [52].

Other composites investigated include shock-resistant PMMA, which consists of PMMA containing small inclusions of elastomer. Segreti et al. [53] carried out experiments in which plates of the material supported by a 30 mm internal diameter tube are subjected to impact at speeds up to 83 m/s by a flat-ended steel striker of diameter 10 mm. Plugs of shock-resistant PMMA are ejected from the plate with evidence of very high local shearing at the circumferential striker-specimen contact. This is very similar to the 'plugging' mechanism caused by adiabatic shear in the impact of flat-ended projectiles on metals. Unmodified PMMA failed by fracture into many fragments. The addition of the elastomer increases the ductility of the PMMA and hence the susceptibility to adiabatic shear.

As a consequence of the Columbia shuttle disaster, investigations of impacts of polyurethane foam cylinders on structural panels have been carried out, but the failure mechanisms of the foams were not studied in the initial work. Rittel [54] carried out a series of experiments in which syntactic polyurethane foam cylinders (containing a high fraction of micro-balloons) were impacted on metal plates at speeds in the range of 235–280 m/s. He observed a failure mode in the foam that was very similar to adiabatic shear failure. The foam was a commercial variety used for packaging applications and had a density of 200 kg/m^3. Foam cylinders of 12.7 mm diameter and 55 mm length were launched against Al, Al alloy and low-carbon steel plates, resulting in severe denting or bending of the plates. Surprisingly, all of the recovered cylinders failed in the form of truncated cones (similar to failures observed in the Taylor impact test), rather than disintegrating. The sides of the cones showed plastic deformation marks and localized damage at 45° to the compression axis. The combination of shear and rapid deformation led Rittel to the conclusion that the failure mechanism was adiabatic shear failure.

9.4 Modelling of Polymer Behaviour

The mathematics, models and numerical simulations of adiabatic shear are covered elsewhere in this book, but it is worthwhile very briefly to review the use of models specifically for the description of polymer behaviour. To be useful, models should encompass the non-linear strain, strain rate and temperature dependence of the polymer's mechanical properties. Early models were essentially based on a combination of elastic springs and viscous dashpots whose parameters are adjusted to fit the experimentally observed stress–strain data for the polymers of interest. These could then be used to predict viscoelastic behaviour at different strain rates and so on. To fit observed behaviour quantitatively over a reasonable range of strain rates, it is necessary to include many springs and dashpots [4]. The problem with such models is the limited range of applicability and their non-physical nature. Models with a more physical basis are those based on the activated viscosity model of Eyring [4], which lead to Eqs (9.3) and (9.4). With the possible addition of second activation energy and activation volume, the use of the strain-rate temperature superposition and the additional terms based on the Williams–Landel–Ferry

(WLF) equation to extend the model through the glass transition region, these models can give very good descriptions of polymer behaviour. Richeton et al. [55] use this method and derive the equation

$$\sigma_y = \sigma_i(0) - mT + \frac{2kT}{v}\sinh^{-1}\left(\frac{\dot{\varepsilon}}{\dot{\varepsilon}_0}\exp\left(\frac{\Delta H}{kT}\right)\right)^{\frac{1}{n}} \tag{9.8}$$

where σ_y is the yield stress, $\sigma_i(0)$ is an athermal internal yield stress, m and n are the material parameters, $\dot{\varepsilon}$ is the strain rate, $\dot{\varepsilon}_0$ is a characteristic constant strain rate, ΔH is the activation energy, v is the activation volume, T is the temperature and k is the Boltzmann constant. Above the glass transition temperature, $\sigma_i(0)$ and m are taken to be 0, and the exponential term is modified by incorporating the WLF relationship leading to

$$\sigma_y(T \geq T_g) = \frac{2kT}{v}\sinh^{-1}\left(\frac{\dot{\varepsilon}}{\dot{\varepsilon}_0}\exp\left(\frac{\Delta H}{kT}\right)\exp - \left(\frac{\ln 10 \times c_1^g(T-T_g)}{c_2^g+T-T_g}\right)\right)^{\frac{1}{n}} \tag{9.9}$$

where c_1^g and c_2^g are the WLF parameters, and all other parameters are defined previously.

Experimental data is required to evaluate the parameters $\sigma_i(0)$, m, n, ΔH, $\dot{\varepsilon}_0$ and v. When the parameters are obtained, such models are successful in predicting the mechanical response over a wide range of temperature and strain rates, including the glass transition region and dynamic loadings [55]. They are more physically based than the spring-dashpot models, but are ‘continuum’ models rather than based on the molecular-level properties of the polymer.

An alternative approach based on group interaction modelling (GIM) is described by Porter [55] and applied to PC and PMMA by Porter and Gould [57]. This approach uses a molecular-level model to calculate the energy stored and dissipated during mechanical deformation and is based on the premise that the mechanical properties are a direct consequence of these factors. The model begins by determining the characteristic group of atoms that define the chemical structure of the polymer. Using atomistic simulations or empirically derived dimensions and main molecular energy terms, the model progresses through stages to determine all of the main characteristics of the polymer (density, thermal expansion, glass transition, elastic modulus secondary relaxations, energy dissipation in passing through transitions, Poisson’s ratio, etc.). The method is described fully in Ref. [57]. The main predictive advantage of this approach is that the independently calculated parameters that define the response do not have to be obtained by fitting to the experimental data the model is used to predict. Intuitively, these models therefore provide a much more fundamental understanding of material behaviour. In their paper on PC and PMMA, Porter and Gould [57] predict the yield stresses and strain softening of PC and PMMA up to strain rates of $10^4\ s^{-1}$ and the variation of yield stress with temperature to temperatures from well above to well below T_g (for PMMA) and find good agreement with experimental data. Energy-based failure initiation criteria have also been formulated using this approach [56].

Chowdhury et al. [58] have used a macromolecular model based on a rate-dependent statistical model of chain dynamics as described by Arruda and Boyce [59] as a constitutive model of polymer behaviour to successfully develop an FE model of the dynamic response of glassy polymers. This model predicts the formation of ASBs in high-impact situations. These results are, however, not specifically linked to the response of any particular polymer. Rapid progress is being made in the validity of the constitutive equations being used to inform FE models, and reliable quantitative data of behaviour at high rates may soon be possible.

9.5 Mechanism of Adiabatic Shear

The mechanism for adiabatic shear initiation has generally been accepted to be due to the localization of shear, for whatever reason, on a particular shear plane giving rise to an increasing temperature, which exceeds that of the surrounding material. A large fraction of the plastic work done is assumed to be converted to heat, which, because of low thermal conductivity and/or the short deformation time induced by high strain rates, leads to temperature rises being concentrated in the shear band. This will lead to an instability if the work hardening of the shearing material is exceeded by the strain softening so that deformation continues on the same plane, and homogeneous plastic deformation is then replaced by localized shear bands.

A problem with this interpretation is that several workers (see Figures 9.7, 9.9 and 9.11) do not observe any substantial increase in temperature until late in the deformation process, sometimes when the material has started to fail. This has lead Rittel et al. in a series of papers [60–63] to propose that adiabatic shear failure is more dependent on energy considerations than on the strain criteria as is usually assumed (see Table 9.1). Rittel et al. [60] report on a series of experiments on a magnesium–aluminium alloy. They carried out tests in which the samples were quasi-statically loaded to a predetermined strain and then dynamically loaded in a Hopkinson bar and found that the *total* strain required before adiabatic shear failure increased with the pre-strain. However, the dynamic mechanical energy required for failure was approximately constant irrespective of the pre-strain. In this regard, it is worth noting the results of Lee and Swallowe [64] who carried out a series of compression tests on cylinders of PMMA at fixed temperatures over a range of strain rates until the samples failed – either by cracking or by shear. Calculations showed that the tests should be adiabatic at strain rates >0.05. In the adiabatic region, the strain to failure was observed to fall with strain rate while the energy of fracture was constant (Figure 9.27).

Rittel et al. [60] also carried out dynamic–dynamic tests where cylindrical specimens were loaded using a Hopkinson bar, and the experiments stopped at predetermined strains using steel sleeves. These samples were then reloaded using

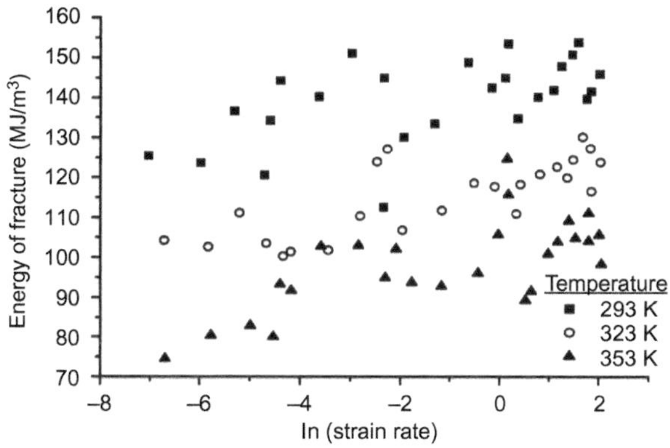

Figure 9.27 Energy of fracture versus ln strain rate for PMMA discs at a range of temperatures. The energy to fracture becomes approximately constant in the adiabatic region.
Source: From Ref. [64].

the same apparatus. In this case, all of the samples failed at approximately the same strain, but because of the differing pre-strains, the temperature increase at failure was different. On the basis of these results, Rittel proposed a criterion to identify adiabatic shear banding as being initiated at a critical energy density at which the overall dynamic mechanical energy reaches a material-dependent value. In his 2009 paper, Rittel [61] extended this work and proposed that the dynamically stored energy of cold work is the key factor in the generation of ASBs. He then went on to study the microstructure of the metallic alloys investigated and noted that dynamic recrystallization (DRX) was observed in the dynamically interrupted specimens before the formation of ASBs. He pointed out that DRX is almost always observed in adiabatic shear and was previously assumed to be a consequence of the high temperatures in the shear band. He observed DRX in interrupted test samples where the temperature rise was modest, and he proposed that DRX was driven by the energy of cold work and was a key contributor to the initiation of shear localization. In Refs [62,63], Rittel et al. further extend their DRX model by proposing that the nanograins formed by DRX soften the local stress in the gauge section and lead to material instability in the grain. This proposal is supported by a dislocation dynamics model [62] and a numerical simulation [63].

Although Rittel does point out that the DRX model is possibly only applicable to a limited subset of metals, it is the case that adiabatic shear localization is widely observed in amorphous polymers, where DRX cannot occur, and in many other materials. Because adiabatic shearing is a 'universal' phenomenon, and the currently accepted explanation is not material specific, it is desirable that any modification should have more universal applicability. In this regard it is worth noting that the observation that the temperature rises normally associated with adiabatic shearing do not occur until late in the deformation process has been reported in a range of polymers: PC [30], epoxies [29], PMMA [34] as well as in metals. The common factor in all of the polymer data is the high fraction of plastic work that is stored in the polymer before failure commences – as is the case in the metals described by Rittel – and that the rapid temperature increases observed start after the strain-softening region and are greatest at the commencement of failure.

Fleck et al. [65] reported on a series of split-Hopkinson torsion bar experiments on PC and PMMA in which he observed fracture surfaces of failed samples and attributed failure to the coalescence and growth of craze-initiated micro-cracks at both high and low strain-rate tests. He quoted Bai's perturbation analysis of thermoplastic shear instability [66] whereby the ratio

$$B = \tau \frac{\left| \frac{\partial \tau}{\partial T} \right|}{\left(\rho C_V \frac{\partial \tau}{\partial \gamma} \right)} \tag{9.10}$$

must exceed unity for a thermoplastic instability to occur and calculated the values to be ~0.2, hence precluding adiabatic shear banding as the primary failure mechanism in his experiments. However, the rapid fall in stress observed during failure must release strain energy, which drives the fracture process and can be converted to thermal energy. Anderson et al. [67] developed a model of a strain-hardening material, complementing the work of Fleck et al. [65], in which flow localization in shear was shown to occur in materials containing imperfections in the form of micro-cracks due to the reduction in stiffness in the micro-cracked region leading to mechanical instability. In this context, it is worth noting the micro-cracks observed in an ASB by Rittel and Brill (Figure 9.22) in PMMA. However, at which stage in the failure process these cracks formed cannot be ascertained.

It is clear that much work remains to be done on elucidating the precise mechanisms involved in the adiabatic shear phenomenon in polymers.

9.6 Conclusion

The adiabatic shear phenomenon is observed to occur in a wide range of polymers and polymer composites subjected to medium to high strain rates. It is most evident, or at least has been most frequently studied, in amorphous glassy polymers where it is observed in tests carried out at temperatures close to, but below, T_g. High temperatures, up to and above the melting temperature, have been observed during adiabatic shear in polymers. However, high time-resolution temperature measurements have shown that substantial temperature rises only occur after the strain-softening stage of deformation and during fracture with the energy of plastic work in the early stages of deformation being stored in the material. It is speculated that this stored energy is released during adiabatic shear, but there is not a consensus on the details of the mechanism that initiates adiabatic shear.

References

1. Bai, Y. and Dodd, B. (1992). "Adiabatic Shear Localisation: Occurrence, Theories and Applications", Pergamon Press, Oxford.
2. Walley, S. M. (2007). "Shear localisation: a historical overview", *Metallurgical and Materials Transactions* **38A**, 2007–2629.
3. Wright, T. W. (2002). "The Physics and Mathematics of Adiabatic Shear Bands", Cambridge University Press, Cambridge, UK.
4. Ward, I. M. (1983). "Mechanical Properties of Solid Polymers", 2nd ed., Wiley, Chichester.
5. Grey III, G. T., Blumenthal, C. P., Trujillo, C. P. and Carpenter II, R. W. (1997). "Influence of temperature and strain rate on the mechanical behaviour of Adiprene L-100", *Journal de Physique IV* **C3**, 523–528.
6. Bai, Y. and Dodd, B. (1992). "Adiabatic Shear Localisation: Occurrence, Theories and Applications", Chapter 6, Pergamon Press, Oxford.
7. Bai, Y. (1981). In "Shock Waves and High-Strain-Rate Phenomena in Metals" (M. A. Mayers, and L. E. Muir, Eds.), pp. 277–284. Plenum Press, New York.
8. Staker, M. R. (1981). "The Relation between Adiabatic Shear Instability Strain and Material Properties", *Acta Metall* **29**, 683–689.
9. Olson, G. B., Mescall, J. F. and Azrin, M. (1981). In "Shock Waves and High-Strain-Rate Phenomena in Metals" (M. A. Mayers, and L. E. Muir, Eds.), pp. 221–247. Plenum Press, New York.
10. Culver, R. S. (1973). In "Metallurgical Effects at High Rates of Strain" (R. W. Rohde, B. M. Butcher, J. R. Holland, and C. H. Karnes, Eds.), pp. 519–530. Plenum Press, New York.
11. Vihn, T., Afzali, M. and Roche, A (1979). in "Proceedings of the 3rd International Conference on the Mechanical Behaviour of Materials (ICM3)" eds. Miller K. J. and Smith R. F., Cambridge U.K, 633–642.
12. Burns, T. J. and Trucano, T. G. (1982). "Instability in Simple Shear Deformations of Strain Softening Materials", *Mechanics of Materials* **1**, 313–324.
13. Lindholm, U. S. and Johnson, G. R. (1983). In "Material Behaviour under High Stress and Ultra High Loading Rates" (J. Mescall, and V. Weiss, Eds.), pp. 61–79. Plenum Press, New York.
14. Pomey, J. (1966). *Annal CIRP* **13**, 93–109.
15. Recht, R. F. (1964). "Catastrophic Thermoplastic Shear", *J. Appl. Mechanics* **31**, 189–193.
16. Curran, D. R. (1979). "Computational Model for Armor Penetration", SRI Report, California, USA.
17. Clifton, R. J., Duffy, J., Hartley, K. A. and Shawki, T. G. (1984). "On Critical Conditions for Shear Band Formation at High Strain Rates", *Scripta Metall* **18**, 443–448.
18. Arruda, E. M., Boyce, M. C. and Jayachandran, R. (1995). "Effects of strain rate, temperature and thermomechanical coupling on the finite strain deformation of glassy polymers", *Mechanics of Materials* **19**, 193–212.
19. Winter, R. E. (1975). "Adiabatic shear of titanium and polymethylmethacrylate", *Philosophical Magazine* **31**, 765–773.
20. Adrianova, G. P., Kechekyan, A. S. and Kargin, A. V. (1971). "Self-oscillation mechanism of necking on extension of polymers", *Journal of Polymer Science A2* **9**, 1919–1933.
21. Roseen, R. (1974). "Temperature effect at self-oscillating necking during extension of polyethylene terephthalate", *Journal of Materials Science* **9**, 929–933.

22. Karger-Kocsis, J., Benevolenski, O. I. and Moskala, E. J. (2001). "Toward understanding the stress oscillation phenomenon in polymers due to tensile impact loading", *Journal of Materials Science* **36**, 3365–3371.
23. Kapoor, R. and Nemat-Nassar, S. (1998). "Determination of temperature rise during high strain rate deformation", *Mechanics of Materials* **27**, 1–12.
24. Nemat-Nasser, S. and Isaacs, J. B. (1996). "Direct measurement of isothermal flow stress of metals at elevated temperatures and high strain rates with application to Ta and Ta–W alloys", *Acta Materialia* **45**, 907–919.
25. Rosakis, P., Rosakis, A. J., Ravichandran, G. and Hodowany, J. (2000). "A thermodynamic internal variable model for the partition of plastic work into heat and stored energy in metals", *Journal of the Mechanics and Physics of Solids* **48**, 581–607.
26. Rabin, Y. and Rittel, D. (2000). "Infrared temperature sensing of mechanically loaded specimens: thermal analysis", *Experimental Mechanics* **40**, 197–202.
27. Regev, A. and Rittel, D. (2008). "Simultaneous transient temperature sensing of impacted polymers using infrared detectors and thermocouples", *Experimental Mechanics* **48**, 675–682.
28. Adams, G. W. and Farris, R. J. (1988). "Latent energy of deformation of bisphenol A polycarbonate", *Journal of Polymer Science B: Polymer Physics* **26**, 433–445.
29. Buckley, C. P., Harding, J., Hou, J. P., Ruiz, C. and Trojanowski, A. (2001). "Deformation of thermosetting resins at impact rates of strain. Part 1: Experimental study", *Journal of the Mechanics and Physics of Solids* **49**, 1517–1538.
30. Rittel, D. (1999). "On the conversion of plastic work to heat during high strain rate deformation of glassy polymers", *Mechanics of Materials* **31**, 131–139.
31. Guduru, P. R., Ravichandran, G. and Rosakis, A. J. (2001). "Observations of transient high temperature vertical microstructures in solids during adiabatic shear banding", *Physical Review E* **64**, 036128.
32. Fuller, K. N. G., Fox, P. G. and Field, J. E. (1975). "The temperature rise at the tips of fast moving cracks in glassy polymers", *Proceedings of the Royal Society of London* **A341**, 537–557.
33. Bjerke, T. and Lambros, J. (2002). "Heating during shearing and opening dominated dynamic fracture of polymers", *Experimental Mechanics* **42**, 107–114.
34. Li, Z. and Lambros, J. (2001). "Strain rate effects on the thermomechanical behaviour of polymers", *International Journal of Solids and Structures* **38**, 3549–3562.
35. Swallowe, G. M., Field, J. E. and Horn, L. A. (1986). "Measurements of transient high temperatures during the deformation of polymers", *Journal of Materials Science* **21**, 4089–4096.
36. Lee, S. F. and Swallowe, G. M. (2004). "Direct measurement of high rate stress strain curves using instrumented falling weight and high speed photography", *The Imaging Science Journal* **52**, 193–201.
37. Walley, S. M., Field, J. E., Pope, P. H. and Safford, N. A. (1989). "A study of the rapid deformation behaviour of a range of polymers", *Philosophical Transactions of the Royal Society of London* **A328**, 1–33.
38. Ward, I. M. and Sweeney, J. (2004). "The Mechanical Properties of Solid Polymers", 2nd ed., Wiley, Chichester.
39. Mulliken, A. D. and Boyce, M. C. (2006). "Mechanics of the rate-dependent elastic–plastic deformation of glassy polymers from low to high strain rates", *International Journal of Solids and Structures* **43**, 1331–1356.

40. Rae, P. J., Brown, E. J., Clements, B. E. and Dattelbaum, D. M. (2005). "Pressure induced phase change in poly(tetrafluoroethylene) at modest impact velocities", *Journal of Applied Physics* **98**, 063521.
41. Ravi-Chandar, K. (1995). "On the failure mode transitions in polycarbonate under dynamic mixed mode loading", *International Journal of Solids and Structures* **32**, 925–938.
42. Ravi-Chandar, K., Lu, J., Yang, B. and Zhiu, Z. (2000). "Failure mode transitions in polymers under high strain rate loading", *International Journal of Fracture* **101**, 33–72.
43. Rittel, D. and Levin, R. (1998). "Mode-mixity and dynamic failure mode transitions in polycarbonate", *Mechanics of Materials* **30**, 197–216.
44. Rittel, D. (1998). "The influence of temperature on dynamic failure mode transitions", *Mechanics of Materials* **30**, 217–227.
45. Rittel, D. (2000). "Experimental investigation of transient thermoplastic effects in dynamic failure", *International Journal of Solids and Structures* **37**, 2901–2913.
46. Rittel, D. and Brill, A. (2008). "Dynamic flow and failure of confined polymethylmethacrylate", *Journal of the Mechanics and Physics of Solids* **56**, 1401–1416.
47. Mogi, K. (1966). "Pressure dependence of rock strength and transition from brittle fracture to ductile flow", *Bulletin of the Earthquake Research Institute* **44**, 215–232.
48. Stevens, J. R., Coakley, R. W., Chau, K. W. and Hunt, J. L. (1986). "The pressure variation of the glass transition in atactic polystyrene", *Journal of Chemical Physics* **84**, 1006–1014.
49. Dorogoy, A., Rittel, D. and Brill, A. (2010). "A study of inclined impact in polymethylmethacrylate plates", *International Journal of Impact Engineering* **37**, 285–294.
50. Li, T., Zhiping, T. and Cai, J. (2005). "Micro observation of shear wave attenuation mechanism in nylon-66", *Materials Letters* **61**, 1436–1438.
51. Dhokia, V. G., Newman, S. T., Crabtree, P. and Ansell, M. P. (2011). "Adiabatic band formation as a result of cryogenic machining of elastomers", *Proceedings of the Institute of Mechanical Engineers, Part B: Journal of Engineering Manufacture* **225**, 1482–1492.
52. Samuel, J., Dikshit, A., DeVor, R. E., Kapoor, S. G. and Hsia, K. J. (2009). "Effect of carbon nanotube loading on the thermomechanical properties and machinability of CNT reinforced polymer composites", *Journal of Manufacturing Science and Engineering* **131**, 031008.
53. Segreti, M., Rusinek, A. and Klepaczko, J. R. (2004). "Experimental study on puncture of PMMA at low and high velocities, effect on the failure mode", *Polymer Testing* **23**, 703–718.
54. Rittel, D. (2005). "Adiabatic shear failure of a syntactic polymeric foam", *Materials Letters* **59**, 1845–1848.
55. Richeton, J., Ahzi, S., Vecchio, K. S., Jiang, F. C. and Adharapurapu, R. R. (2005). "Influence of temperature and strain rate on the mechanical properties of three amorphous polymers: characterization and modelling of the compressive yield stress", *International Journal of Solids and Structures* **43**, 2318–2335.
56. Porter, D. (1995). "Group Interaction Modelling of Polymers", Marcel Dekker, New York, NY.
57. Porter, D. and Gould, P. J. (2009). "Predictive nonlinear constitutive relations in polymers through loss history", *International Journal of Solids and Structures* **46**, 1981–1993.

58. Chowdhury, K. A., Benzerga, A. A. and Talreja, R. (2008). "A computational framework for analysing the dynamic response of glassy polymers", *Computational Methods in Mechanics and Engineering* **197**, 4485–4502.
59. Arruda, E. M. and Boyce, M. C. (1993). "A three-dimensional constitutive model for the large stretch behaviour of rubber elastic materials", *Journal of the Mechanics and Physics of Solids* **41**, 389–412.
60. Rittel, D., Wang, Z. G. and Merzer, M. (2006). "Adiabatic shear failure and dynamic stored energy of cold work", *Physical Review Letters* **96**, 075502.
61. Rittel, D. (2009). "A different viewpoint on adiabatic shear localisation", *Journal of Physics D: Applied Physics* **42**, 214009.
62. Rittel, D. and Osovski, S. (2010). "Dynamic failure by adiabatic shear banding", *International Journal of Fracture* **162**, 177–185.
63. Dolinski, M., Rittel, D. and Dorogoy, A. (2010). "Modeling adiabatic shear failure from energy considerations", *Journal of the Mechanics and Physics of Solids* **58**, 1759–1775.
64. Lee, S. F. and Swallowe, G. M. (2006). "Quasi-static and dynamic compressive behaviour of poly(methylmethacrylate) and polystyrene at temperatures from 293 K to 363 K", *Journal of Materials Science* **42**, 6280–6289.
65. Fleck, N. A., Stronge, W. J. and Liu, J. H. (1990). "High strain-rate shear response of polycarbonate and polymethylmethacrylate", *Proceedings of the Royal Society of London* **A429**, 459–479.
66. Bai, Y. L. (1982). "Thermo-plastic instability in simple shear", *Journal of the Mechanics and Physics of Solids* **30**, 195–207.
67. Anderson, P. M., Fleck, N. A. and Johnson, K. L. (1990). "Localization of plastic deformation in shear due to microcracks", *Journal of the Mechanics and Physics of Solids* **38**, 681–699.

10 Shear Localization in Deep Geological Layers During Seismic Slip

Jean Sulem

CERMES, Laboratoire Navier, Ecole des Ponts Paris Tech, IFSTTAR, Université Paris-Est, Marne-la-Vallée, France

Nomenclature

p pressure
T temperature
n porosity
ρ density of the porous medium
ρ_f density of the saturating fluid
σ_n normal stress
τ shear stress
p pore fluid pressure
ε volumetric strain
γ shear strain
ε^p plastic volumetric strain
γ^p plastic shear strain
ω^c Cosserat rotation
R Cosserat internal length
G elastic shear modulus
G_c Cosserat elastic shear modulus
K elastic bulk modulus of the porous medium
K_s bulk modulus of the solid matrix
b Biot coefficient
H plastic hardening modulus
μ friction coefficient
β dilatancy coefficient
β_f pore fluid compressibility
λ_f pore fluid thermal expansion coefficient
β_n pore volume compressibility
β_f pore volume thermal expansion
β^* storage coefficient
Λ undrained thermo-elastic pressurization coefficient
k_f permeability

Adiabatic Shear Localization. DOI: http://dx.doi.org/10.1016/B978-0-08-097781-2.00010-1

c_{hy} hydraulic diffusivity
c_{th} thermal diffusivity
ρC specific heat per unit volume of the material
$\Delta_r H_T^0$ enthalpy change of the chemical reaction for calcite decomposition
E_a activation energy of the chemical reaction for calcite decomposition

10.1 Introduction

In the past 30 years, the study of earthquake mechanisms has emphasized the major role of rock friction. Earthquakes appear to be the result of a frictional instability and occur by sudden slippage along a pre-existing fault because the frictional resistance to slip on the fault walls decreases with increasing slip, causing an acceleration of sliding. Field observations of mature faults, i.e. faults that have experienced a large slip, show a generally broad zone of damaged rock, but nevertheless suggest that shear in individual earthquakes occurs in very narrow localized zones of a few millimetres thick or even less. Field observations reveal that the fault material inside the localized shear zone is a finely granulated gouge that has usually undergone mechano-chemical degradation (intense grain comminution, gelification, decarbonation and dehydration reactions, melting) [1]. The actual thickness of the zone over which the shear is localized is a key parameter for the evaluation of the energy budget during earthquake rupture. It is generally estimated that more than 90% of the frictional work is converted into heat [2]. Thus, it is expected that weakening mechanisms from thermal origin are of major importance. Fault zones commonly exhibit the presence of fluid interacting with the rock. Recent theoretical studies have emphasized the role of thermal pore fluid pressurization as an important cause of slip weakening [3]. The principle of slip weakening by thermal pressurization is based on the fact that pore fluids trapped inside the fault zone are put under pressure by shear heating, thus inducing a reduction of the effective mean stress and of the shearing resistance of the fault plane [4–7]. This mechanism has also been suggested for weakening in catastrophic landslides [8–15]. The presence of clay material in fault zones also affects thermal pressurization as possible collapse of the clay under thermal loading may activate fluid pressurization [7,16]. Thermo-poro-mechanical couplings due to shear heating can also be associated with chemical effects such as dehydration of minerals or decomposition of carbonates [17–20]. Such reactions induce two competing effects: a direct increase in pore pressure because they release fluid in the system and a limit in temperature increase because these reactions are endothermic so that part of the frictional heat is actually absorbed in the chemical reactions.

In this chapter, we first present the thermo-poro-mechanical background used to analyse the effect of pore fluid and temperature in the strain localization process as the result of instability of undrained adiabatic shearing of a rock layer. In particular, we analyse the possibility of a selection of a preferred wavelength of the instability mode (i.e. the one corresponding to the fastest growth), which could control the width of the localized zone. The undrained adiabatic limit is applicable as soon as the slip event is sufficiently rapid and the shear zone broad enough to effectively preclude heat or fluid transfer as is the case during an earthquake. Then we will consider a particular example of chemical couplings during seismic slip by taking into account the possibility for some parts of the fault zone to reach the threshold temperature at which mineral decomposition can be triggered.

10.2 Position of the Problem and Governing Equations

We consider a layer of saturated rock of thickness h that is sheared in plane strain (Figure 10.1) in such a way that there is no extensional strain in the x-direction and displacements u_x, and u_z of material points vary only with z (and time t). The stress state in the layer is a shear stress τ and a normal stress σ in the y-direction. The only non-zero strains, a shear strain γ and a normal strain ε, are related to the displacements by

$$\gamma = \frac{\partial u_x}{\partial z}, \quad \varepsilon = \frac{\partial u_z}{\partial z} \tag{10.1}$$

Because the normal strains in the other directions are zero, the volumetric strain is equal to ε.

The equilibrium equations during quasi-static deformation reduce to

$$\frac{\partial \tau}{\partial z} = 0, \quad \frac{\partial \sigma}{\partial z} = 0 \tag{10.2}$$

so that the stresses must be spatially uniform and vary only with time during the deformation.

10.2.1 Constitutive Equations

The rate thermo-poro elasto-plastic relationships for the rock layer are expressed as

$$\dot{\gamma} = \frac{1}{G}\dot{\tau} + \dot{\gamma}^p, \quad \dot{\varepsilon} = \frac{1}{K}(\dot{\sigma} - b\dot{p}) - \alpha_s \dot{T} + \dot{\varepsilon}^p \tag{10.3}$$

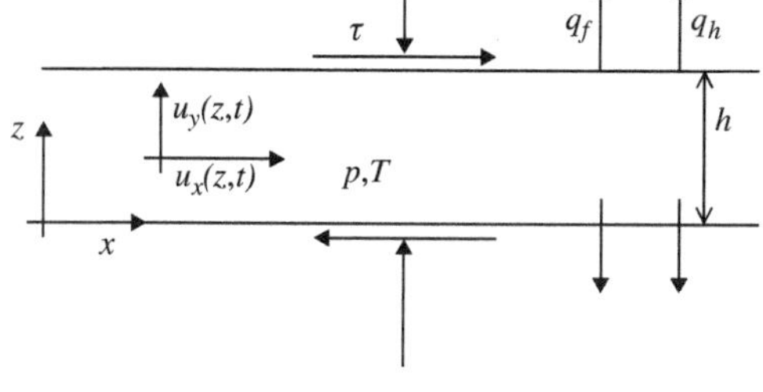

Figure 10.1 Model of a deforming shear band with heat and fluid fluxes.

where G and K are the elastic shear and bulk modulus, respectively, of the empty porous solid, p is the pore pressure, T is the temperature, b is the Biot coefficient, and α_s is the thermal dilatation coefficient of the empty porous solid. The Biot coefficient is related to the bulk modulus of the solid matrix K_s according to the following equation:

$$b = 1 - \frac{K}{K_s} \tag{10.4}$$

The rate of plastic deformation is written as [21]

$$\begin{aligned} \dot{\gamma}^p &= \frac{1}{H}(\dot{\tau} - \mu(\dot{\sigma} - \dot{p})) \\ \dot{\varepsilon}^p &= \beta\dot{\gamma}^p \end{aligned} \tag{10.5}$$

where $H(\gamma^p)$ is the plastic-hardening modulus and is related to the tangent modulus $H_{\tan}$ of the τ versus γ curve by (Figure 10.2)

$$H_{\tan} = \frac{H}{1 + H/G} \tag{10.6}$$

and is either positive or negative according to whether τ versus γ curve is rising (hardening) or falling (softening), although the falling portion may never be observed as homogeneous deformation in a drained test. In Eq. (10.5), β is the dilatancy coefficient.

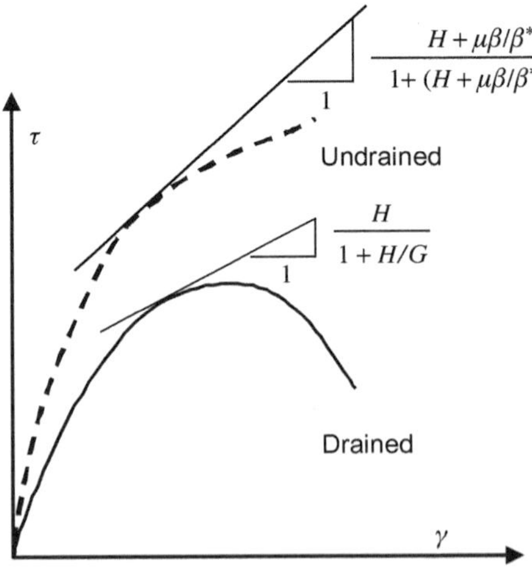

Figure 10.2 Stress–strain curve for a saturated rock in drained or undrained conditions.

10.2.2 Mass Balance Equation

Conservation of fluid mass is expressed by

$$\frac{\partial m_{\mathrm{f}}}{\partial t} = -\frac{\partial q_{\mathrm{f}}}{\partial z} \tag{10.7}$$

where m_{f} is the total fluid mass per unit volume of porous medium (in the reference state) and q_{f} is the flux of fluid. The total fluid mass per unit volume of porous medium is written as $m_{\mathrm{f}} = \rho_{\mathrm{f}} n$, where n is the pore volume fraction (Lagrangian porosity) and ρ_{f} is the density of the saturating fluid. The left-hand side of Eq. (10.7) is obtained by differentiating this product:

$$\frac{\partial m_{\mathrm{f}}}{\partial t} = n\frac{\partial \rho_{\mathrm{f}}}{\partial t} + \rho_{\mathrm{f}}\frac{\partial n}{\partial t} \tag{10.8}$$

The derivatives of the right-hand side of Eq. (10.8) are given by

$$\frac{\partial \rho_{\mathrm{f}}}{\partial t} = \rho_{\mathrm{f}}\beta_{\mathrm{f}}\frac{\partial p}{\partial t} - \rho_{\mathrm{f}}\lambda_{\mathrm{f}}\frac{\partial T}{\partial t} \tag{10.9}$$

and

$$\frac{\partial n}{\partial t} = n\beta_n\frac{\partial p}{\partial t} + n\lambda_n\frac{\partial T}{\partial t} + \frac{\partial n^p}{\partial t} \tag{10.10}$$

where $\beta_{\mathrm{f}} = \frac{1}{\rho_{\mathrm{f}}}\left(\frac{\partial \rho_{\mathrm{f}}}{\partial p}\right)_T$ is the pore fluid compressibility, $\lambda_{\mathrm{f}} = -\frac{1}{\rho_{\mathrm{f}}}\left(\frac{\partial \rho_{\mathrm{f}}}{T}\right)_{P_p}$ is the pore fluid thermal-expansion coefficient, $\beta_n = \frac{1}{n}\left(\frac{\partial n}{\partial p}\right)_T$ is the pore volume compressibility and $\lambda_n = \frac{1}{n}\left(\frac{\partial n}{\partial T}\right)_{P_p}$ is the elastic thermal-expansion coefficient of the pore volume, which is equal to the elastic thermal-expansion coefficient of the solid fraction, $\lambda_n = \alpha_{\mathrm{s}}$.

The expression of the pore volume compressibility is obtained from poro-elasticity theory and is given by [3,22]

$$\beta_n = \frac{1}{n}\left(\frac{1}{K} - \frac{1+n}{K_{\mathrm{s}}}\right) \tag{10.11}$$

In Eq. (10.10), $\frac{\partial n^p}{\partial t}$ is the rate of plastic porosity change.

Using Eqs (10.2)–(10.4), the left-hand side of Eq. (10.7) is thus evaluated as

$$\frac{\partial m_{\mathrm{f}}}{\partial t} = n\rho_{\mathrm{f}}(\beta_n + \beta_{\mathrm{f}})\frac{\partial p}{\partial t} - \rho_{\mathrm{f}} n(\lambda_{\mathrm{f}} - \lambda_n)\frac{\partial T}{\partial t} + \rho_{\mathrm{f}}\frac{\partial n^p}{\partial t} \tag{10.12}$$

The flux term in Eq. (10.7) is evaluated assuming Darcy's law for fluid flow with viscosity η_f through a material with permeability k_f:

$$q_f = -\frac{\rho_f}{\eta_f} k_f \frac{\partial p}{\partial z} \tag{10.13}$$

Substituting Eqs (10.12) and (10.13) into Eq. (10.7) gives the equation that governs the pore fluid production and diffusion:

$$\frac{\partial p}{\partial t} = c_{hy} \frac{\partial^2 p}{\partial z^2} + \Lambda \frac{\partial T}{\partial t} - \frac{1}{\beta^*} \frac{\partial n^p}{\partial t} \tag{10.14}$$

where

$$\Lambda = \frac{\lambda_f - \lambda_n}{\beta_n + \beta_f} \tag{10.15}$$

is the thermo-elastic pressurization coefficient under undrained conditions [3]. This coefficient is pressure and temperature dependent because the compressibility and the thermal-expansion coefficients of the fluid vary with pressure and temperature and also because the compressibility of the pore space of the rock can change with the effective stress [22]. In Eq. (10.14), $\beta^* = n(\beta_n + \beta_f)$ is the storage capacity of the rock and $c_{hy} = k_f/(\beta \eta_f)$ is the hydraulic diffusivity. For incompressible solid phase and fluid, $\beta^* = 1/K$.

If we assume that the solid matrix is plastically incompressible,

$$\frac{\partial n^p}{\partial t} = \frac{\partial \varepsilon^p}{\partial t} \tag{10.16}$$

10.2.3 Energy Balance Equation

Let E_F be the rate of frictional heat produced during shearing. The equation of conservation of energy is expressed as

$$\rho C \frac{\partial T}{\partial t} = E_F - \frac{\partial q_h}{\partial z} \tag{10.17}$$

where ρC is the specific heat per unit volume of the material in its reference state and q_h is the heat flux. It is assumed here that all of the plastic work is converted into heat and, thus, $E_F = \tau \dot{\gamma}^p$. The heat flux is related to the temperature gradient by Fourier's law:

$$q_h = -k_T \frac{\partial T}{\partial z} \tag{10.18}$$

where k_T is the thermal conductivity of the saturated material. Substituting Eq. (10.18) into Eq. (10.19) gives the energy conservation equation

$$\frac{\partial T}{\partial t} = c_{th}\frac{\partial^2 T}{\partial z^2} + \frac{1}{\rho C}\tau\dot{\gamma}^p \tag{10.19}$$

where $c_{th} = k_T/\rho C$ is the thermal diffusivity.

Undrained Adiabatic Limit

We assume that the drainage and the heat flux are prohibited at the boundaries of the layer ($q_f = 0$ and $q_h = 0$). This corresponds to the situation when the slip event is sufficiently rapid and the shear zone broad enough to effectively preclude heat or fluid transfer as is the case during an earthquake or a landslide. We also assume that the normal stress σ acting on the sheared layer is constant ($\dot{\sigma} = 0$). From Eqs (10.5), (10.14), (10.16) and (10.19), we obtain the following relationship between the pore pressure rate and the shear stress rate:

$$\dot{p} = \frac{\frac{\Lambda\tau}{\rho C} - \frac{\beta}{\beta^*}}{H - \mu\left(\frac{\Lambda\tau}{\rho C} - \frac{\beta}{\beta^*}\right)}\dot{\tau} \tag{10.20}$$

Substituting Eq. (10.20) into Eq. (10.3), we get

$$\dot{\gamma} = \left(\frac{1}{G} + \frac{1}{H + \mu\beta/\beta^* - \mu\Lambda\tau/\rho C}\right)\dot{\tau} \tag{10.21}$$

or

$$\dot{\tau} = \left(\frac{H + \mu\beta/\beta^* - \mu\Lambda\tau/\rho C}{1 + (H + \mu\beta/\beta^* - \mu\Lambda\tau/\rho C)/G}\right)\dot{\gamma} \tag{10.22}$$

If we neglect the thermal effect ($\Lambda = 0$), the expression for dilatant hardening effect as proposed by Rice [21] is retrieved (Figure 10.2):

$$\dot{\tau} = \left(\frac{H + \mu\beta/\beta^*}{1 + (H + \mu\beta/\beta^*)/G}\right)\dot{\gamma} \tag{10.23}$$

Equation (10.22) has to be compared with the drained case for which

$$\dot{\tau} = \left(\frac{H}{1 + H/G}\right)\dot{\gamma} \tag{10.24}$$

For dilatant material ($\beta > 0$), we identify in Eq. (10.22) a hardening effect due to dilatancy with the term $\mu\beta/\beta^*$ and a thermal-softening effect with the term $-\mu\Lambda\tau/\rho C$. When thermal softening prevails against strain hardening, plastic localization in the form of adiabatic shear banding can occur.

10.3 Stability of Analysis of Undrained Adiabatic Shearing of a Rock Layer

We consider again the shearing of the layer while drainage and heat fluxes are prevented at its boundaries. The pertinent variables are written in the form

$$\gamma = \gamma_0 + \tilde{\gamma}, \quad \varepsilon = \varepsilon_0 + \tilde{\varepsilon}, \quad \sigma = \sigma_0 + \tilde{\sigma}, \quad \tau = \tau_0 + \tilde{\tau}, \quad p = p_0 + \tilde{p}, \quad T = T_0 + \tilde{T} \tag{10.25}$$

where the quantities γ_0, τ_0 and so on represent the solution of the last section for homogeneous deformation, and $\tilde{\gamma}$, $\tilde{\tau}$ and so on are perturbation quantities. Specifically, the layer is sheared by application of a monotonically increasing shear stress $\tau_0 = \tau_0(t)$ to its boundary while the normal stress is held constant at σ_0. From the equilibrium equation (10.2), the stress field is uniform within the layer, and thus $\tilde{\sigma} = 0$ and $\tilde{\tau} = 0$. This problem is similar to the one addressed by Rice [21] with the introduction here of the thermal effect. From the conservation equations (10.14) and (10.19), and by keeping only the first-order terms, we get the same type of rate equations for the perturbation terms

$$\begin{aligned} \frac{\partial \tilde{p}}{\partial t} &= c_{\text{hy}} \frac{\partial^2 \tilde{p}}{\partial z^2} + \Lambda \frac{\partial \tilde{T}}{\partial t} - \frac{1}{\beta^*} \frac{\partial \tilde{\varepsilon}^p}{\partial t} \\ \frac{\partial \tilde{T}}{\partial t} &= c_{\text{th}} \frac{\partial^2 \tilde{T}}{\partial z^2} + \frac{1}{\rho C} \tau_0 \frac{\partial \tilde{\gamma}^p}{\partial t} \end{aligned} \tag{10.26}$$

with the following from the constitutive equation (10.5):

$$\frac{\partial \tilde{\gamma}^p}{\partial t} = \frac{\mu}{H} \frac{\partial \tilde{p}}{\partial t}, \quad \frac{\partial \tilde{\varepsilon}^p}{\partial t} = \beta \frac{\partial \tilde{\gamma}^p}{\partial t} = \beta \frac{\mu}{H} \frac{\partial \tilde{p}}{\partial t} \tag{10.27}$$

The spatial dependence of the perturbations is decomposed into Fourier modes with wavelength λ. Perturbations of the form

$$\tilde{p} = P(0)e^{st} \cos\left(\frac{2\pi z}{\lambda}\right), \quad \tilde{T} = T(0)e^{st} \cos\left(\frac{2\pi z}{\lambda}\right) \tag{10.28}$$

with $\lambda = h/n$ and n equal to an integer satisfy the zero heat and fluid flux boundary conditions at $y = 0, h$. In Eq. (10.28), s is the growth coefficient in time of the perturbation. By substituting the perturbations field (Eq. (10.28)) into Eq. (10.26), we obtain a homogeneous algebraic system for the coefficients $P(0)$ and $T(0)$. Thus, a non-zero solution is possible only if the determinant of the coefficients vanishes:

$$\det\begin{pmatrix} s\left(1+\dfrac{\beta\mu}{\beta^* H}\right)+c_{hy}\left(\dfrac{2\pi}{\lambda}\right)^2 & -\Lambda s \\ \dfrac{\mu}{\rho C H}\tau_0 s & -\left(c_{th}\left(\dfrac{2\pi}{\lambda}\right)^2+s\right) \end{pmatrix}=0 \tag{10.29}$$

The preceding condition yields a quadratic equation for the growth coefficient s:

$$\left(\frac{\Lambda\mu}{\rho C}\tau_0-\frac{\beta\mu}{\beta^*}-H\right)\lambda^4 s^2-\left((c_{th}+c_{hy})H+\frac{\beta\mu}{\beta^*}\right)\lambda^2 s-c_{th}c_{hy}H=0 \tag{10.30}$$

If a solution of Eq. (10.30) has a positive real part, then the corresponding perturbation grows exponentially in time.

If we do not consider the thermal effect, i.e. $\Lambda = 0$, the condition for stability is simply $H > 0$. Thus, the result of Rice [21] is retrieved that undrained shearing is stable only in those circumstances for which the underlying drained deformation would be stable. This problem has also been examined by Vardoulakis [23,24] who proposed a regularization of the mathematically ill-posed problem in the softening regime by resorting to a second-grade extension of plasticity theory.

If we now take into account the thermal effect ($\Lambda > 0$), we observe that if $H > 0$, the stability condition is

$$\frac{\Lambda\mu}{\rho C}\tau_0 < \frac{\beta\mu}{\beta^*}+H \tag{10.31}$$

Consequently, the condition of stability of undrained adiabatic shearing is

$$H > 0 \quad \text{and} \quad \frac{\Lambda\mu}{\rho C}\tau_0 < \frac{\beta\mu}{\beta^*}+H \tag{10.32}$$

and the system is always unstable in the softening regime.

Obviously, this condition is more restrictive than the one for undrained shearing. This result demonstrates the destabilizing effect of thermal fluid pressurization: undrained adiabatic shearing of a material with positive strain hardening is stable only when the thermal pressurization is not too high.

10.4 Wavelength Selection and Thickness of the Shear Band

The stability analysis performed as described earlier for a classical Cauchy continuum does not reveal any wavelength selection of the instability mode because the change of sign from negative to positive values of the growth coefficient s occurs for the same value of the hardening modulus independent of the considered particular wavelength λ. Consequently, the width of the localized zone is actually indeterminate. Moreover, we observe that at the onset of the instability, i.e. for $\frac{\Lambda\mu}{\rho C}\tau_0 = \frac{\beta\mu}{\beta^*} + H$, there is a sharp transition from infinitely stable to infinitely unstable behaviour, which indicates that the solution of the considered initial-value problem does not exist and consequently that the corresponding problem is mathematically ill-posed. The origin of this undesirable situation can be traced back to the fact that conventional constitutive models do not contain material parameters with dimension of length, so that the shear-band thickness (i.e. the extent of the plastically softening region) is undetermined. Rudnicki [25] proposed an ad hoc regularization by introducing a zone of finite thickness corresponding to a layer that has already weakened due to past faulting. In doing so, inhomogeneous deformation is prohibited over scales shorter than some prescribed length. Vardoulakis [24,26,27] showed that by introducing viscous effects combined with second-gradient dependency of the friction in strain may alleviate the ill-posedness of the problem. More generally, it appears necessary to resort to continuum models with microstructure to describe correctly localization phenomena. These generalized continua usually contain additional kinematical degrees of freedom (Cosserat continuum) and/or higher deformation gradients (higher grade continuum). Cosserat continua and higher grade continua belong to a general class of constitutive models, which account for the materials microstructure. The contemporary formulation of these models is based on the work of Mindlin [28] and Germain [29,30]. Rotation gradients and higher velocity gradients introduce a material length scale into the problem, which, as already mentioned, is necessary for the correct modelling of localization phenomena. This idea was widely publicized by the paper of Mühlhaus and Vardoulakis [31] who have resorted to concepts from Cosserat continuum mechanics to determine the thickness of the shear band. Thus, by allowing both particle displacements and particle rotations, they accounted for the angularities of the grains and enriched the classical continuum with additional kinematic and static fields. For dry sand, they obtained the shear-band thickness of about $16d_{50}$, where d_{50} is the mean grain diameter. Considering the non-cohesive character of the cataclastic fault gouge inside the slip zone, we follow the same idea and allow for additional rotational degrees of freedom for the fault material within the frame of the Cosserat continuum theory.

10.4.1 Constitutive Equations

The incremental constitutive equations for the considered 2D Cosserat continuum are derived from the Mühlhaus–Vardoulakis plasticity model. The main feature of the model is that a 2D flow theory of plasticity for granular media with Cosserat microstructure can be derived by keeping the same definitions for the yield surface and the plastic potential as in the classical theory and by generalizing appropriately the stress and strain invariants involved in these definitions.

In a 2D Cosserat continuum, each material point has two translational degrees of freedom (v_1, v_2) and one rotational degree of freedom ($\dot{\omega}^c$). The four components of the rate of the non-symmetric deformation tensor are given by

$$\begin{aligned} \dot{\varepsilon}_{11} &= \partial v_1/\partial x_1, \quad & \dot{\varepsilon}_{12} &= \partial v_1/\partial x_2 + \dot{\omega}^c \\ \dot{\varepsilon}_{22} &= \partial v_2/\partial x_2, \quad & \dot{\varepsilon}_{21} &= \partial v_2/\partial x_1 + \dot{\omega}^c \end{aligned} \tag{10.33}$$

and the two components of the 'curvature' of the deformation rate (gradient of the Cosserat rotation rate) are given by

$$\dot{\kappa}_1 = \partial\dot{\omega}^c/\partial x_1, \quad \dot{\kappa}_2 = \partial\dot{\omega}^c/\partial x_2 \tag{10.34}$$

Besides the four components of the non-symmetric stress tensor σ_{ij}, there are two couple stresses, m_k ($k = 1,2$). The convention used here for the stress tensor follows Vardoulakis and Sulem [32], i.e. the first index denotes the direction of the force and the second index, the face with which a stress component is associated (Figure 10.3).

As in classical small-strain plasticity theory, the deformation is decomposed into elastic and plastic parts:

$$\dot{\varepsilon}_{ij} = \dot{\varepsilon}^{\mathrm{e}}_{ij} + \dot{\varepsilon}^{\mathrm{p}}_{ij}, \quad \dot{\kappa}_k = \dot{\kappa}^{\mathrm{e}}_k + \dot{\kappa}^{\mathrm{p}}_k \quad (k = 1, 2) \tag{10.35}$$

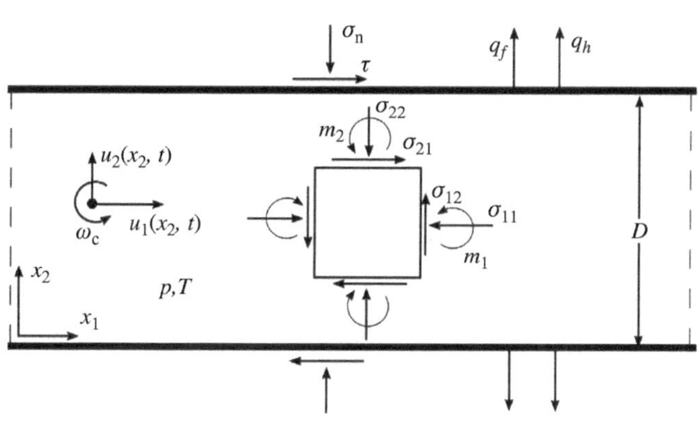

Figure 10.3 Cosserat stresses and couple stresses.

Moreover, the stress and the deformation are decomposed into spherical and deviatoric parts:

$$\begin{aligned} \sigma_{ij} &= s_{ij} + \sigma_{kk}\delta_{ij}/2 \\ \dot{\varepsilon}_{ij} &= \dot{e}_{ij} + \dot{\varepsilon}_{kk}\delta_{ij}/2 \end{aligned} \tag{10.36}$$

where δ_{ij} is the Kronecker delta.

The generalized stress and strain invariants are utilized

$$\begin{aligned} &\sigma = \sigma_{ii}/2, \quad \tau = \sqrt{h_1 s_{ij} s_{ij} + h_2 s_{ij} s_{ji} + h_3 m_k m_k / R^2} \\ &\dot{\varepsilon}^p = \dot{\varepsilon}^p_{kk}, \quad \dot{\gamma}^p = \sqrt{g_1 \dot{e}^p_{ij} \dot{e}^p_{ij} + g_2 \dot{e}^p_{ij} \dot{e}^p_{ji} + g_3 R^2 \dot{\kappa}^p_k \dot{\kappa}^p_k} \end{aligned} \tag{10.37}$$

where $\{h_i\} = \{3/4, -1/4, 1\}$ and $\{g_i\} = \{3/2, 1/2, 1\}$ for the so-called static Cosserat model, and R is the internal length of the material [32].

As in the previous section, we assume a Coulomb yield surface and plastic potential

$$F = \tau - \mu(\sigma - p), \quad Q = \tau - \beta(\sigma - p) \tag{10.38}$$

where the mobilized friction and dilatancy coefficients μ and β, respectively, are only functions of the accumulated plastic strain γ^p.

The strain-rate thermo-poro elasto-plastic relationships are formally unchanged and are expressed by Eq. (10.3). The rate of plastic deformation is given by Eq. (10.5).

For the considered 2D Cosserat continuum and the configuration of Figure 10.3, the local equilibrium equations are

$$\begin{aligned} &\frac{\partial \sigma_{12}}{\partial x_2} = 0 \\ &\frac{\partial \sigma_{22}}{\partial x_2} = 0 \\ &\frac{\partial m}{\partial x_2} + \sigma_{21} - \sigma_{12} = 0 \end{aligned} \tag{10.39}$$

with $m = m_2$ and $m_1 \equiv 0$. It is assumed that no couple stress is imposed at the boundary. Prior to localization, the state of stress and strain is uniform and considering the couple-free boundary condition, the couple stress is identically zero in the sheared layer. Therefore, the rock layer behaves as a classical Cauchy continuum prior to localization.

10.4.2 Stability of Analysis of Undrained Adiabatic Shearing for a Cosserat Layer

The detailed analysis of the stability analysis of a Cosserat layer under shear is presented in Ref. [33]. The main steps are recalled in the following. The small perturbations considered herein are defined as follows:

$$
\begin{aligned}
u_1(x_2,t) &= u_1^0(x_2,t) + \tilde{u}_1(x_2,t)\\
u_2(x_2,t) &= u_2^0(x_2,t) + \tilde{u}_2(x_2,t)\\
\omega_c(x_2,t) &= \omega_c^0(x_2,t) + \tilde{\omega}_c(x_2,t)\\
T(x_2,t) &= T^0(x_2,t) + \tilde{T}(x_2,t)\\
p(x_2,t) &= p^0(x_2,t) + \tilde{p}(x_2,t)
\end{aligned}
\tag{10.40}
$$

Dimensionless quantities are defined as

$$
\begin{aligned}
&x = \frac{x_1}{R}, \quad z = \frac{x_2}{R}, \quad \overline{u}_i = \frac{u_i}{R}, \quad i = 1,2\\
&\overline{p} = \frac{p}{\sigma_n}, \quad \overline{\sigma}_{ij} = \frac{\sigma_{ij}}{\sigma_n}, \quad \overline{t} = \frac{c_{\text{th}}}{R^2}t, \quad \overline{T} = \frac{\Lambda}{\sigma_n}T
\end{aligned}
\tag{10.41}
$$

To simplify the notations, in the following and in the rest of the text, the line over the symbols for the dimensionless variables is omitted: $t \equiv \overline{t}, \; p \equiv \overline{p}$, etc.

The dimensionless form of the fluid mass balance equation (10.14) and the energy conservation (10.19) is

$$
\begin{aligned}
&\frac{\partial p}{\partial t} - \frac{\partial T}{\partial t} - \eta \frac{\partial^2 p}{\partial z^2} + \delta \frac{\partial \gamma^p}{\partial t} = 0\\
&\frac{\partial T}{\partial t} - \frac{\partial^2 T}{\partial z^2} - \alpha \frac{\partial \gamma^p}{\partial t} = 0
\end{aligned}
\tag{10.42}
$$

where the dimensionless parameters α, δ and η are given by

$$
\alpha = \frac{\Lambda \tau_0}{\rho C \sigma_n}, \quad \delta = \frac{\beta}{\beta^* \sigma_n}, \quad \eta = \frac{c_{\text{hy}}}{c_{\text{th}}}
\tag{10.43}
$$

and

$$\frac{\partial \gamma^p}{\partial t} = \frac{1}{2\overline{H}}\left(\frac{\partial \sigma_{12}}{\partial t} + \frac{\partial \sigma_{21}}{\partial t} + \mu\left(\frac{\partial \sigma'_{11}}{\partial t} + \frac{\partial \sigma'_{22}}{\partial t}\right)\right)$$
$$\overline{H} = \frac{H}{\sigma_n} = h\frac{\sigma'}{\sigma_n} \tag{10.44}$$

The dimensionless form of the balance of linear momentum with inertia and micro-inertia terms is as follows:

$$\frac{\partial \sigma_{12}}{\partial z} - I\frac{\partial^2 u_1}{\partial t^2} = 0$$
$$\frac{\partial \sigma'_{22}}{\partial z} - \frac{\partial p}{\partial z} - I\frac{\partial^2 u_2}{\partial t^2} = 0 \tag{10.45}$$
$$\frac{\partial m}{\partial z} + \sigma_{21} - \sigma_{12} - \frac{1}{2}I\frac{\partial^2 \omega_c}{\partial t^2} = 0$$

where the dimensionless inertia parameter is $I = \frac{c_{\mathrm{th}}^2 \rho}{R^2 \sigma_n}$. The moment of inertia of the considered Cosserat continuum is $\rho R^2/2$ [32].

For the considered simple shear initial state of stress, the (dimensionless) incremental thermo-poro elasto-plastic constitutive equations are given by

$$\begin{Bmatrix} \Delta\sigma'_{11} \\ \Delta\sigma'_{22} \\ \Delta\sigma'_{12} \\ \Delta\sigma'_{21} \\ \Delta m \end{Bmatrix} = \begin{pmatrix} a & b & c & c & 0 \\ b & a & c & c & 0 \\ f & f & d & e & 0 \\ f & f & e & d & 0 \\ 0 & 0 & 0 & 0 & g \end{pmatrix} \begin{Bmatrix} \Delta\gamma_{11} \\ \Delta\gamma_{22} \\ \Delta\gamma_{12} \\ \Delta\gamma_{21} \\ \Delta\kappa \end{Bmatrix} - \overline{\alpha}_s \begin{Bmatrix} (a+b)/2 \\ (a+b)/2 \\ 0 \\ 0 \\ 0 \end{Bmatrix} \Delta T \tag{10.46}$$

where $\Delta\sigma'_{ij} = \Delta\sigma_{ij} - \Delta p \delta_{ij}$ is the increment of dimensionless (Terzaghi) effective stress:

$$\Delta\gamma_{11} = \frac{\partial \Delta u_1}{\partial x} = 0, \quad \Delta\gamma_{22} = \frac{\partial \Delta u_2}{\partial z},$$
$$\Delta\gamma_{12} = \frac{\partial \Delta u_1}{\partial z} + \Delta\omega_c, \quad \Delta\gamma_{21} = \frac{\partial \Delta u_2}{\partial x} - \Delta\omega_c = -\Delta\omega_c \tag{10.47}$$
$$\Delta\kappa = \frac{\partial \Delta\omega_c}{\partial z}$$

and the dimensionless incremental constitutive coefficients are given by

$$
\begin{aligned}
a &= \frac{1}{\sigma_n}\left(K + G - \frac{\beta\mu K^2}{G + H + \beta\mu K}\right), \quad b = \frac{1}{\sigma_n}\left(K - G - \frac{\beta\mu K^2}{G + H + \beta\mu K}\right) \\
c &= \frac{1}{\sigma_n}\left(-\frac{GK\beta}{G + H + \beta\mu K}\right), \quad f = \frac{1}{\sigma_n}\left(-\frac{GK\mu}{G + H + \beta\mu K}\right) \\
d &= \frac{1}{\sigma_n}\left(G + G_c - \frac{G^2}{G + H + \beta\mu K}\right), \quad e = \frac{1}{\sigma_n}\left(G - G_c - \frac{G^2}{G + H + \beta\mu K}\right) \\
g &= \frac{G}{\sigma_n} \\
\overline{\alpha}_s &= \alpha_s \frac{\sigma_n}{\Lambda}
\end{aligned}
\tag{10.48}
$$

Inserting the constitutive equations (10.46) in the governing equations (10.42) and (10.45), we get

$$
\begin{aligned}
& d\frac{\partial^2 \tilde{u}_1}{\partial z^2} + f\frac{\partial^2 \tilde{u}_2}{\partial z^2} + (d - e)\frac{\partial \tilde{\omega}_c}{\partial z} - I\frac{\partial^2 \tilde{u}_1}{\partial t^2} = 0 \\
& c\frac{\partial^2 \tilde{u}_1}{\partial z^2} + a\frac{\partial^2 \tilde{u}_2}{\partial z^2} - I\frac{\partial^2 \tilde{u}_2}{\partial t^2} - \frac{\partial \tilde{p}}{\partial z} - \overline{\alpha}_s\frac{a + b}{2}\frac{\partial \tilde{T}}{\partial z} = 0 \\
& (d - e)\frac{\partial \tilde{u}_1}{\partial z} + g\frac{\partial^2 \tilde{\omega}_c}{\partial z^2} + 2(d - e)\frac{\partial \tilde{\omega}_c}{\partial z} - \frac{1}{2}I\frac{\partial^2 \omega_c}{\partial t^2} = 0 \\
& \frac{\delta}{\overline{\overline{H}}}\left(\frac{d + e}{2} + \mu c\right)\frac{\partial^2 \tilde{u}_1}{\partial z \partial t} + \frac{\delta}{\overline{\overline{H}}}\left(f + \mu\frac{a + b}{2}\right)\frac{\partial^2 \tilde{u}_2}{\partial z \partial t} - \eta\frac{\partial^2 \tilde{p}}{\partial z^2} \\
& \qquad + \frac{\partial \tilde{p}}{\partial t} - \left(1 + \frac{\delta}{\overline{\overline{H}}}\mu\overline{\alpha}_s\frac{a + b}{2}\right)\frac{\partial \tilde{T}}{\partial t} = 0 \\
& -\frac{\alpha}{\overline{\overline{H}}}\left(\frac{d + e}{2} + \mu c\right)\frac{\partial^2 \tilde{u}_1}{\partial z \partial t} - \frac{\alpha}{\overline{\overline{H}}}\left(f + \mu\frac{a + b}{2}\right)\frac{\partial^2 \tilde{u}_2}{\partial z \partial t} - \frac{\partial^2 \tilde{T}}{\partial z^2} \\
& \qquad + \left(1 + \frac{\alpha}{\overline{\overline{H}}}\mu\overline{\alpha}_s\frac{a + b}{2}\right)\frac{\partial \tilde{T}}{\partial t} = 0
\end{aligned}
\tag{10.49}
$$

The spatial dependence of the perturbations is again decomposed into Fourier modes with wavelength λ. We consider perturbations of the form

$$\begin{aligned}
\tilde{u}_1(z,t) &= U_1\ \mathrm{e}^{st}\sin\left(\frac{z}{\lambda}\right) \\
\tilde{u}_2(z,t) &= U_2\ \mathrm{e}^{st}\sin\left(\frac{z}{\lambda}\right) \\
\tilde{\omega}_c(z,t) &= \Omega\ \mathrm{e}^{st}\cos\left(\frac{z}{\lambda}\right) \\
\tilde{p}(z,t) &= \Pi\ \mathrm{e}^{st}\cos\left(\frac{z}{\lambda}\right) \\
\tilde{T} &= \Theta\ \mathrm{e}^{st}\cos\left(\frac{z}{\lambda}\right)
\end{aligned} \tag{10.50}$$

By substituting the perturbation field (Eq. (10.50)) into the governing equations (10.49), we obtain a homogeneous algebraic system for the coefficients $\{U_1, U_2, \Omega, \Theta, \Pi\}$. Thus, a non-zero solution is possible only if the determinant of the linear system vanishes. The resulting characteristic polynomial equation is of degree 8 for the growth coefficient s. As before, if a root has a positive real part, then the corresponding perturbation grows exponentially in time.

10.4.3 Numerical Example

We refer here to a deep rock layer at a 7 km depth with the following values for the initial state of stress and thermo-mechanical parameters:

$$\begin{aligned}
&\rho = 2500\ \mathrm{kg/m^3}, \quad \sigma_n = 200\ \mathrm{MPa}, \quad p = 66\ \mathrm{MPa} \\
&\mu = 0.5, \quad \beta = 0\ (\text{at great depth}) \\
&\tau_0 = \mu(\sigma_n - p) = 67\ \mathrm{MPa} \\
&\Lambda = 0.5\ \mathrm{MPa/^\circ C}, \quad \rho C = 2.8\ \mathrm{MPa/^\circ C} \\
&G = 10{,}000\ \mathrm{MPa}, \quad K = 20{,}000\ \mathrm{MPa} \\
&G_c = 5000\ \mathrm{MPa}, \quad R = 0.01\ \mathrm{mm} \\
&c_{\mathrm{th}} = 1\ \mathrm{mm^2}/s, \quad c_{\mathrm{hy}} = 10\ \mathrm{mm^2}/s, \quad \alpha_{\mathrm{s}} = 2.5\times10^{-5}/^\circ\mathrm{C}
\end{aligned} \tag{10.51}$$

With these numerical values, it is obtained that there exists a critical value for the hardening modulus $h_{\mathrm{cr}} = 0.015$ (i.e. in the hardening regime of the stress–strain curve) such that for $h \geq h_{\mathrm{cr}}$, all of the real roots of the polynomial equation are negative. For $h < h_{\mathrm{cr}}$, positive real roots exist. It is worth mentioning that from these numerical computations, it is found that the root with the greatest positive real part is real (Figure 10.4). In other words, the instability mode with the fastest growth coefficient is not a so-called flutter instability.

For a given value of the hardening modulus $h < h_{cr}$, it is possible to plot the greatest root (i.e. corresponding to the instability mode with the fastest growth in time) as a function of the wavelength λ. This curve exhibits a maximum for $\lambda = \lambda_m$ so that a wavelength with the fastest growth in time appears (Figure 10.4). In Figure 10.5, the selected wavelength is plotted versus the hardening modulus. We observe that the selected wavelength tends to infinity for $h \to h_{cr}^-$. This selected wavelength decreases with decreasing hardening modulus and reaches a minimum (for $h = 0.007$, $\lambda_m = 186.7$) and then increases again and tends to infinity for $h \to 0$ (maximum shear stress). Mühlhaus and Vardoulakis [31] similarly obtained that at bifurcation state, the shear-band thickness for dry granular materials is infinite and decreases beyond the bifurcation state as the strain localization process evolves. With $R = 10\ \mu$m, which is an average grain size for a highly finely granulated (ultra-cataclastic) fault core, the obtained minimum selected wavelength is about 2 mm, which is compatible with field observations of localized shear zones in broader damaged fault zones.

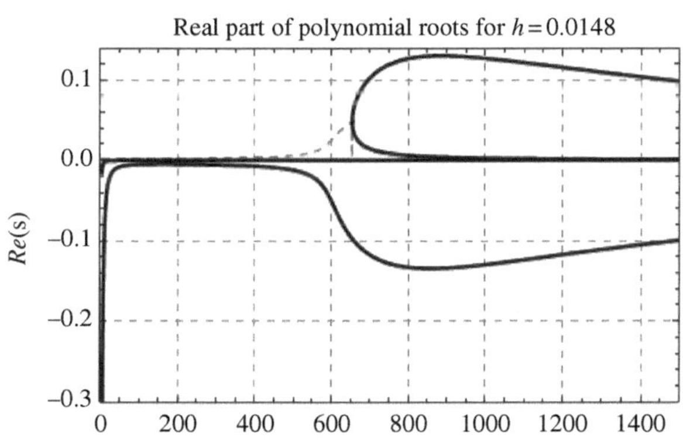

Figure 10.4 Real part of the roots of the characteristic equation (growth coefficient) for $h < h_{cr}$. The red, dashed curves represent the real part of the roots that are complex (non-zero imaginary part), while the thick, black curves depict the real roots (zero imaginary part). (For interpretation of the references to color in this figure legend, the reader is referred to the web version of this book.) *Source*: After Ref. [33].

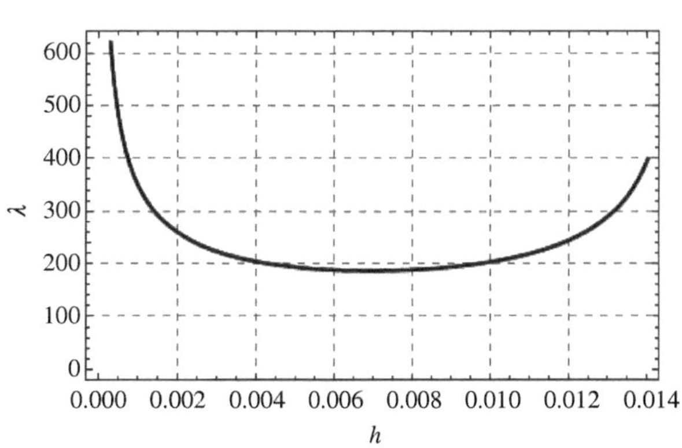

Figure 10.5 Wavelength selection.
Source: After Ref. [33].

It is important to emphasize that the wavelength selection is not obtained if Cosserat microstructure is not considered in the analysis. The critical hardening modulus for instability is unchanged ($h_{cr} = 0.015$), but the growth coefficient tends to infinity for the infinitely small wavelength limit, which indicates that the considered problem is mathematically ill-posed. Moreover, it should be mentioned that even if a Cosserat microstructure is considered, the mathematical problem remains ill-posed without the inertia terms. The main point is that the inertia terms remedy the ill-posedness of the mathematical problem, resulting in a finite growth coefficient for the instability.

10.5 Effect of Chemical Couplings During Rapid Fault Shearing

In the preceding analysis, we have emphasized the role of shear heating and fluid pressurization. The underlying assumption was that no fluid is produced or consumed during seismic slip. There is, however, growing evidence that temperature-induced decomposition of minerals may be a significant source of fluids in faulted rocks, which may enhance the generation of pore-pressure excess.

This production of fluid is in competition with the change of porosity induced by the reduction of solid volume during mineral decomposition. Obviously, the competing effects of temperature rise (due to shear heating), thermal fluid pressurization, mineral decomposition and porosity/permeability increase induce strong non-linearities in the problem of pore pressure and temperature evolution of a rapidly sheared fault zone. The kinetics of the chemical reaction of mineral decomposition is also a major factor that competes with heat- and fluid-diffusion processes. The endothermic character of the mineral decomposition reactions also plays a role in the energy balance of the system. Therefore, the effect of thermal mineral decomposition brings a very interesting chemical coupling to the problem under consideration.

As an example, in this section, we will focus on the kinetics of chemical decomposition of calcite (decarbonation) $CaCO_3 \rightarrow CaO + CO_2$, which is triggered when the temperature reaches about 700°C. Carbonates are present in every fault zone from the ductile–brittle transition (~15 km) to the subsurface, and there is evidence of CO_2 release in several active crustal faults. In the Corinth rift (Greece), e.g. chemical analysis of water springs near the seismogenic Heliki and Aegion faults revealed an anomalously high content of dissolved CO_2 compared with the regional values [34]. The surface trace of the San Andreas Fault also displays a positive anomaly of CO_2 fluxes [35], and this CO_2 comes from a shallow source, not from the mantle [36]. Moreover, there is growing evidence that CO_2 release coincides with seismic slip in crustal faults, active and/or exhumed. In the vicinity of the Nojima fault (Japan), Sato and Takahashi [37] reported that the HCO_3^- concentration of springs increased by 30 wt% immediately after the 1995 Kobe earthquake. This carbon discharge, together with other co-seismic geochemical anomalies, decreased gradually to normal values in the following 10 months. A micro-infrared analysis of exhumed pseudotachylites (i.e. friction-induced melts produced by seismic slip) from the Nojima fault revealed that shear melting destroyed the carbonates within the fault zone and released CO_2, thus providing an explanation to the co-seismic CO_2 spikes in springs [38]. The quantity of CO_2 released by friction melting during the 1995 Kobe earthquake was evaluated as $1.8-3.4 \times 10^3$ tons. As for the San Andreas Fault, the carbon isotopic signature of springs and fault rocks from Nojima is consistent with a decomposition of biogenic carbonates, not from a mantle origin [39,40]. In the Central Apennines, Italiano et al. [41] also reported enhanced fluxes of crustal CO_2 (i.e. not from the mantle) during the 1997–1998 seismic crisis of major faults and proposed that co-seismic decarbonation was responsible for the CO_2 emission. Recent studies of the Chelungpu Fault (Taiwan) responsible for the 1999 ChiChi earthquake also showed that the fault core was depleted in carbon relative to the damage zone, and the depletion was attributed to decarbonation induced by frictional heat [42,43]. In addition, recent high-velocity friction experiments on Carrara marble have shown that thermal decomposition of calcite due to frictional heating induces a pronounced fault weakening [44]. The production of co-seismic CO_2 is therefore attested by various field and experimental techniques, thus making the thermal decomposition of carbonates an important additional mechanism to be investigated among possible fault-weakening processes as recently proposed by Sulem and Famin [19].

10.5.1 Modelling

The rate of emitted CO_2 is evaluated using the kinetics of the chemical reaction of calcite thermal decomposition:

$$CaCO_3 \rightarrow CaO + CO_2 \tag{10.52}$$

The reacted fraction, α ($0 \leq \alpha \leq 1$; $\alpha = 1$ if the entire substance is reacted), is expressed by the Arrhenius equation [45]:

$$\frac{\partial \alpha}{\partial t} = f(\alpha) A \exp\left(-\frac{E_a}{RT}\right) \tag{10.53}$$

where $f(\alpha)$ is a kinetic function determined by the reaction mechanism, A is a constant (pre-exponential term of the Arrhenius law), E_a is the activation energy of the reaction and R is the gas constant (8.31447 J/K/mol). For calcite decomposition, $E_a = 319{,}000$ J/mol and $A = 2.95 \times 10^{15}\ s^{-1}$ [46].

Thus, the mass of emitted CO_2 (per unit volume) is expressed as

$$\frac{\partial m_d}{\partial t} = \chi \rho_{CaCO_3} (1 - n) A \exp\left(-\frac{E_a}{RT}\right) \tag{10.54}$$

where χ is the ratio between the molar mass M_{CO_2} of CO_2 (44 g/mol) and the molar mass M_{CaCO_3} of calcite (100 g/mol) if we assume that the total amount of calcite can be decomposed. The corresponding change of porosity due to the decomposition of the solid phase is expressed as

$$\frac{\partial n_d}{\partial t} = -\frac{1}{\rho_{CaCO_3}}\frac{\partial m_{CaCO_3}}{\partial t} - \frac{1}{\rho_{CaCO}}\frac{\partial m_{CaO}}{\partial t} = \left(\frac{1}{\rho_{CaCO_3}}\frac{M_{CaCO_3}}{M_{CO_2}} - \frac{1}{\rho_{CaCO}}\frac{M_{CaO}}{M_{CO_2}}\right)\frac{\partial m_d}{\partial t} \tag{10.55}$$

where ρ_{CaCO_3} (resp. ρ_{CaO}) and M_{CaCO_3} (resp. M_{CaO}) are the density and the molar mass of $CaCO_3$ (resp. CaO). The heat for calcite decomposition is calculated by

$$E_C = \frac{\Delta_r H_T^0}{\chi M_{CaCO_3}}\frac{\partial m_d}{\partial t} \tag{10.56}$$

where $\Delta_r H_T^0$ is the enthalpy change of the reaction (i.e. the energy consumed by the reaction), which for calcite decomposition is equal to the activation energy E_a in the isobaric mode [19].

Thus, the coupled system of production/diffusion equations (Eqs (10.14) and (10.19)) are modified as:

$$\begin{aligned} \frac{\partial p}{\partial t} &= c_{hy}\frac{\partial^2 p}{\partial z^2} + \Lambda\frac{\partial T}{\partial t} - \frac{1}{\beta^*}\frac{\partial n^p}{\partial t} + \frac{1 - \rho_f\zeta/\rho_{CaCO_3}}{\beta^*\rho_f}\frac{\partial m_d}{\partial t} \\ \frac{\partial T}{\partial t} &= c_{th}\frac{\partial^2 T}{\partial z^2} + \frac{1}{\rho C}\tau\dot{\gamma}^p - \frac{(1-n)}{\rho C}\frac{\Delta_r H_T^0}{M_{CaCO_3}}\rho_{CaCO_3} A \exp\left(-\frac{E_a}{RT}\right) \end{aligned} \tag{10.57}$$

where $\zeta = \frac{M_{CaCO_3}}{M_{CO_2}} - \frac{\rho_{CaCO_3}}{\rho_{CaCO}}\frac{M_{CaO}}{M_{CO_2}}$. With $M_{CaCO_3} = 100$ g/mol, $M_{CaO} = 56$ g/mol, $M_{CO_2} = 44$ g/mol, $\rho_{CaCO_3} = 2.71$ g/cm^3 and $\rho_{CaO} = 3.35$ g/cm^3, $\zeta = 1.24$.

10.5.2 Numerical Example

For the pore pressure and temperature range considered here for which the decarbonation of CO_2 is active ($T > 700°C$, $P_p > 70$ MPa), water and carbon dioxide are in a supercritical state. They are miscible, and their compressibility, viscosity, density and thermal-expansion coefficient are nearly equal. Thus, it is assumed in the computations that the fault zone is saturated with a unique fluid and with the thermodynamical properties of water.

The decomposition of carbonate can induce substantial change in the porosity of the rock, which affects the permeability. It is known that there is no unique relationship between porosity and permeability applicable to all porous media and that the geological evolution process of the pore space influences the permeability–porosity relationship. The commonly used cubic Carman Kozeny permeability law is assumed here to take into account the effect of porosity change due to mineral decomposition on the permeability of the rock:

$$k_f = k_{f0}\left(\frac{1-n_0}{1-n}\right)^2\left(\frac{n}{n_0}\right)^3 \tag{10.58}$$

where k_{f0} is the reference permeability corresponding to the reference porosity n_0. The assumed permeability law has a strong effect on the numerical results, and there is a need for experimental data on permeability changes resulting from the particular process of carbonate decomposition.

The initial porosity of the rock is taken to be equal to 0.03. The recent high-velocity shear experiments on Carrara marble of Han et al. [44] have shown that the friction coefficient f decreases rapidly to values as low as 0.06 due to the thermal decomposition of calcite induced by frictional heating. Here, we take $f = 0.1$.

We assume here that the initial permeability of the intact medium is $k_f = 10^{-18}\,\mathrm{m}^2$.

The computed results are presented in Figure 10.6. The evolution in time of the temperature and the pore pressure in the centre of the band is plotted in Figure 10.6A. Considering the constant slip velocity of 1 m/s, this graph can also be seen as the evolution of temperature and pore pressure with accumulated slip. The corresponding shear stress is plotted in Figure 10.6B. These results show the coupling effect of heat, which induces first a pore pressure increase. When the decomposition of the carbonate rock begins at about 700°C, the temperature increase is drastically slowed due to the energy consumed in the endothermic chemical reaction. Two competing effects act on the evolution of the pore pressure: on the one hand, the production of CO_2 induces an additional fluid mass and thus a pressurization of the pore fluid; on the other hand, the increase of porosity due to the solid decomposition induces an increase of the permeability of the medium, which limits the pressurization. The pore pressure in the centre of the band exhibits a maximum of about 145 MPa, which does not exceed the total normal stress acting (180 MPa) on the band so full liquefaction is not reached. As mentioned earlier, through the friction law, the shear stress is related to the mean effective stress inside the band (Figure 10.6B). Consequently, the shear stress decreases rapidly during initial pressurization and then increases again. Thus, the mineral decomposition of the rock can be seen as a mechanism of fault weakening in a first stage and then fault re-strengthening in a second stage.

The accumulated mass per unit area of the fault plane of emitted CO_2 is plotted on Figure 10.6C. After 1 s, the temperature reaches a quasi-constant state, and the production rate is almost constant. These results show that the mass of emitted CO_2 after 2 s is about 0.018 g/cm^2. The corresponding rate of emitted CO_2 is about

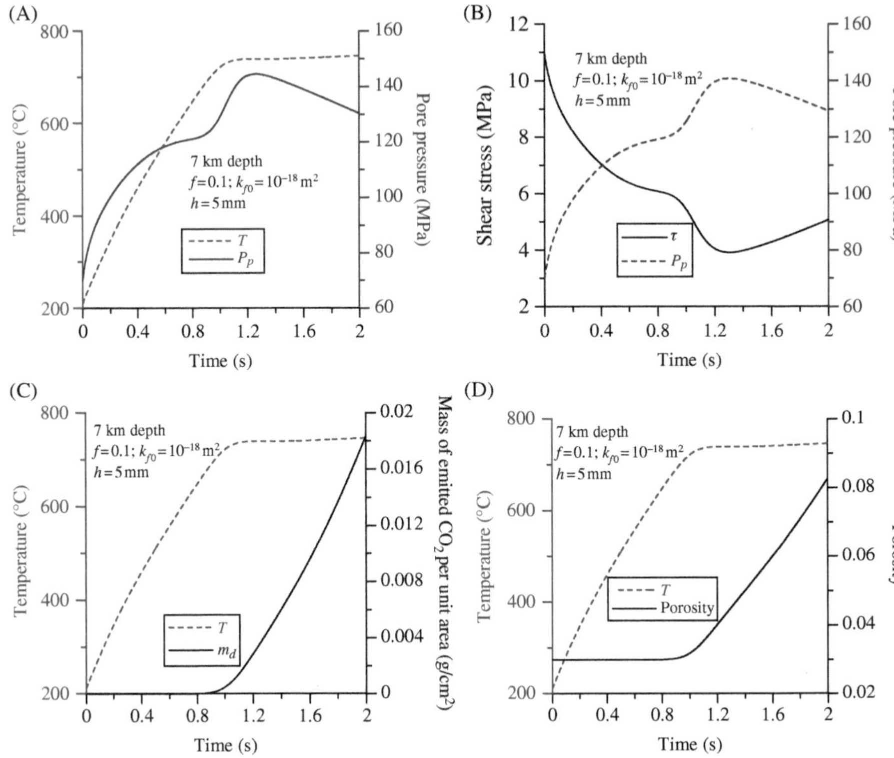

Figure 10.6 Fault at 7 km depth: (A) evolution of temperature and pore pressure in the centre of the shear band, (B) evolution of the shear stress, (C) evolution of the mass of emitted CO_2 and (D) evolution of the porosity in the centre of the band.

180 t/s/km^2 of the fault zone. The corresponding solid mass change is shown through the evolution of the porosity in the centre of the shear band (Figure 10.6D).

The temperature field around the centre of the band is shown in Figure 10.7 for $t = 2$ s with the corresponding value of the porosity. These plots show that the temperature increase is localized in the central zone of the band. The porosity is affected only in zones where the temperature exceeds 700°C, which for the computation considered corresponds to a width of about 4 mm.

The effect of mineral decomposition is also shown in Figure 10.8 where the results of the preceding computation are compared to the ones obtained without CO_2 emission. In this latter case, only thermal pressurization occurs. If no mineral decomposition occurs, the pore pressure increase is slower, and for the parameters considered, the maximum pore pressure is about 120 MPa. Consequently, the temperature increase is much stronger, and melting of the rock can eventually occur (the melting temperature is about 1600°C for calcite).

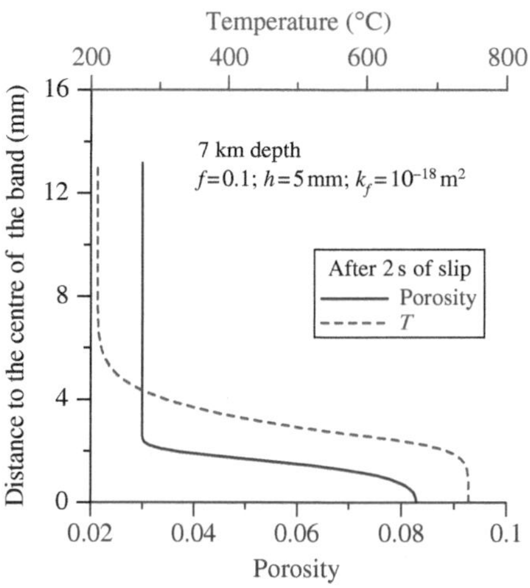

Figure 10.7 Fault at 7 km depth: temperature and porosity fields after 1.9 s.

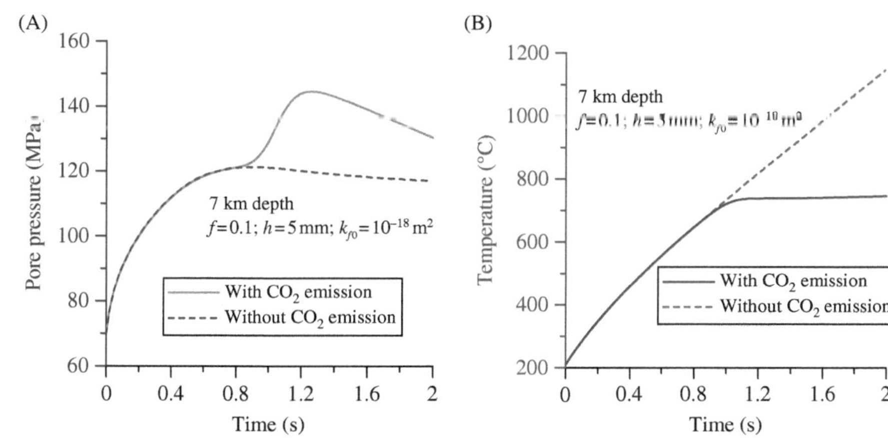

Figure 10.8 Fault at 7 km depth: effect of CO_2 emission on the evolution of temperature (A) and pore pressure (B) in the centre of the shear band.

10.5.3 Discussion

Thermal decomposition of rocks appears to be an important physical process in the phenomenon of thermal heating and pore fluid pressurization during seismic slip. The combined effects of frictional heating, temperature rise, endothermic mineral decomposition, pore pressure rise, porosity and permeability increase result in highly coupled and competing processes.

A first essential result of the preceding analysis is that the endothermic reaction of calcite decomposition limits the co-seismic temperature increase to less than ~800°C (corresponding to the initiation of the chemical reaction) within a carbonate shear band under rapid slip. Decarbonation is only one of the possible thermal reactions of mineral decomposition. Phyllosilicates are common secondary minerals in faults, and their thermal dehydration is also endothermic. Therefore, it is likely that the reaction-induced temperature limitation obtained in our model may apply qualitatively to most fault zones. In this case, mature faults with a long history of slip and a large cumulated displacement are likely to be more prone to reaction-induced temperature limitation than recent faults because of their larger content of volatile-rich secondary minerals. This would provide another explanation for the notorious absence of a positive heat flow anomaly on active crustal faults such as San Andreas [47]: a large part of the heat produced by friction would be consumed by endothermic reactions. Another consequence is that friction melting is hampered by endothermic calcite decomposition in carbonate fault rocks and probably in other faults containing a sufficiently high proportion of volatile-rich secondary minerals. This is consistent with the relative scarcity of pseudotachylytes in mature faults.

The second essential implication of the preceding analysis is that decarbonation is a source of CO_2 that significantly increases the slip-weakening effect of thermal pressurization. The pore fluid pressure exhibits an initial phase of increase due to thermal pressurization, then a sudden acceleration of pore pressure generated when solid decomposition is activated. However, the increase of permeability limits the pore pressure so that it reaches a maximum and then decreases. The numerical results reproduce this pore pressure pulse and the initial fault weakening followed by a re-strengthening of the shear stress.

10.6 Conclusion

In the first part of this chapter, we have analysed the undrained adiabatic shearing of a saturated rock layer and shown the destabilizing effect of shear heating. Instability can occur even in the hardening regime of the underlying drained stress–strain response if dilatant hardening cannot compensate for the thermal pressurization of the pore fluid. This result confirms that thermal effects play a major role in the formation and evolution of the localized zone. A remarkable result is that if we do not take into account the effect of microstructure and micro-inertia, the underlying mathematical problem is ill-posed, i.e. for a hardening modulus lower than the critical hardening modulus at instability, the growth coefficient of the instability is infinite. The complete dynamic analysis for a Cosserat continuum leads to a wavelength selection of the instability mode as the growth coefficient in time is maximum and finite for a particular wavelength. The selected wavelength is compatible with observed shear-band thickness in fault zones.

In the second part of the chapter, we analysed the effect of temperature-induced mineral decomposition as it is a major slip-weakening process. Such chemical reactions induce two competing effects: a direct increase in pore pressure because they release fluid and a limit to the temperature increase because part of the frictional heat is absorbed in the endothermic reactions. The combined effects of frictional heating, temperature rise, endothermic mineral decomposition, pore pressure rise, porosity and permeability increase result in highly coupled and competing processes.

In this chapter, we have emphasized the weakening effects of thermal pressurization and thermal decomposition of minerals. However, other important weakening mechanisms could also be included in the modelling. Flash heating and shear weakening of frictional micro-asperity contacts is a mechanism that has been first proposed for dry-metal friction and recently been applied to rock friction [3,48]. From the microscopic point of view, the contact between two rough surfaces is effective on micro-asperities with an area that is much smaller than the nominal contact area. The stresses supported by these asperities are much higher than the macroscopic applied stresses so that the local rate of heat production and consequently the local temperature rise during sliding is large and may be sufficient to activate thermal weakening of the contacts. Other mechanisms responsible for fault lubrication involve a number of physicochemical processes [1], such as the lubrication of the contact zone with the formation of silica gels [49] or the production of very fine particles with nanometric size [50], and also the formation of a molten layer along the sliding interface that also lubricates the contact zone and reduces the friction resistance

Acknowledgements

The author wishes to recall the memory of Ioannis Vardoulakis as recognition for an inspiring collaboration and a deep friendship. He gratefully acknowledges the fruitful collaboration with N. Brantut, V. Famin, A. Schubnel, I. Stefanou and E. Veveakis. He also thanks J. Rice for inspiring discussions.

References

1. Di Toro, G., Han, R., Hirose, T., De Paola, N., Nielsen, S., Mizoguchi, K., Ferri, F., Cocco, M., and Shimamoto, T. (2011). Fault lubrication during earthquakes, *Nature* **471**, 494–498.
2. Cocco, M., Spudich, P., and Tinti, E. (2006). On the mechanical work absorbed on faults during earthquake ruptures. In "Radiated Energy and the Physics of Faulting," R. Abercrombie et al. (Eds.), Geophysical Monograph Series, **170**, pp. 237–254, AGU, Washington, DC.
3. Rice, J. R. (2006). Heating and weakening of faults during earth-quake slip, *Journal of Geophysical Research* **111**, B05311.
4. Lachenbruch, A. H. (1980). Frictional heating, fluid pressure and the resistance to fault motion, *Journal of Geophysical Research* **85**, 6097–6112.
5. Mase, C. W. and Smith, L. (1985). Pore-fluid pressures and frictional heating on a fault surface, *Pure and Applied Geophysics* **122**, 583–607.

6. Sulem, J., Vardoulakis, I., Ouffroukh, H., and Perdikatsis, V. (2005). Thermo-poro-mechanical properties of the Aigion fault clayey gouge – application to the analysis of shear heating and fluid pressurization, *Soils and Foundations* **45**(2), 97–108.
7. Sulem, J., Lazar, P., and Vardoulakis, I. (2007). Thermo-poro-mechanical properties of clayey gouge and application to rapid fault shearing, *International Journal of Numerical and Analytical Methods in Geomechanics* **31**(3), 523–540.
8. Habib, P. (1967). Sur un mode de glissement des massifs rocheux, *Comptes Rendus à l'Académie des Sciences, Paris* **264**, 151–153.
9. Habib, P. (1976). Production of gaseous pore pressure during rock slides, *Rock Mechanics* **7**, 193–197.
10. Anderson, D. L. (1980). An earthquake induced heat mechanism to explain the loss of strength of large rock and earth slides. In "Engineering for Protection from Natural Disasters" P. Karasudhi et al. (Eds.), pp. 569–580, Wiley, New York, NY.
11. Voight, B. and Faust, C. (1982). Frictional heat and strength loss in some rapid landslides, *Géotechnique* **32**, 43–54.
12. Vardoulakis, I. (2002). Dynamic thermo-poro-mechanical analysis of catastrophic landslides, *Géotechnique* **52**(3), 157–171.
13. Goren, L. and Aharonov, E. (2007). Long runout landslides: the role of frictional heating and hydraulic diffusivity, *Geophysical Research Letters* **34**, L07301.
14. Veveakis, E., Vardoulakis, I., and Di Toro, G. (2007). Thermoporomechanics of creeping landslides: the 1963 Vaiont slide, northern Italy, *Journal of Geophysical Research* **112**, F03026.
15. Pinyol, N. and Alonso, E. (2010). Criteria for rapid sliding. II. Thermo-hydro-mechanical and scale effects in Vaiont case, *Engineering Geology* **114**, 211–227.
16. Sulem, J., Vardoulakis, I., Ouffroukh, H., Boulon, M., and Hans, J. (2004). Experimental characterization of the thermo-poro-mechanical properties of the Aegion fault gouge, *Comptes Rendus Geosciences* **336**(4–5), 455–466.
17. Brantut, N., Schubnel, A., Corvisier, J., and Sarout, J. (2010). Thermochemical pressurization of faults during coseismic slip, *Journal of Geophysical Research* **115**, B05314.
18. Brantut, N., Sulem, J., and Schubnel, A. (2011). Effect of dehydration reactions on earthquake nucleation: stable sliding, slow transients and unstable slip, *Journal of Geophysical Research* **116**, B05304.
19. Sulem, J. and Famin, V. (2009). Thermal decomposition of carbonates in fault zones: slip-weakening and temperature-limiting effects, *Journal of Geophysical Research* **114**, B03309.
20. Veveakis, E., Alevizos, S., and Vardoulakis, I. (2010). Chemical reaction capping of thermal instabilities during shear of frictional faults, *Journal of Mechanics and Physics of Solids* **58**, 1175–1194.
21. Rice, J. R. (1975). On the stability of dilatant hardening for saturated rock masses, *Journal of Geophysical Research* **80**(11), 1531–1536.
22. Ghabezloo, S. and Sulem, J. (2009). Stress dependent thermal pressurization of a fluid-saturated rock, *Rock Mechanics and Rock Engineering* **42**, 1–24.
23. Vardoulakis, I. (1996). Deformation of water-saturated sand. I. Uniform undrained deformation and shear banding, *Géotechnique* **46**(3), 441–456.
24. Vardoulakis, I. (1996). Deformation of water-saturated sand. II. Effect of pore water flow and shear banding, *Géotechnique* **46**(3), 457–472.
25. Rudnicki, J. W. (1984). Effects of dilatant hardening on the development of concentrated shear deformation in fissured rock masses, *Journal of Geophysical Research* **89** (B11), 9259–9270.

26. Vardoulakis, I. (1985). Stability and bifurcation of undrained, plane rectilinear deformations on water-saturated granular soils, *International Journal of Numerical and Analytical Methods in Geomechanics* **9**, 339–414.
27. Vardoulakis, I. (1986). Dynamic stability of undrained simple shear on water-saturated granular soils, *International Journal of Numerical and Analytical Methods in Geomechanics* **10**, 177–190.
28. Mindlin, R. (1964). Micro-structure in linear elasticity, *Archive for Rational Mechanics and Analysis* **16**, 51–78.
29. Germain, P. (1973). La méthode des puissances virtuelle en mécanique des milieux continus. Première partie: théorie du second gradient, *Journal de Mécanique* **12**, 235–274.
30. Germain, P. (1973). The method of virtual power in continuum mechanics. Part 2: Microstructure, *SIAM Journal of Applied Mathematics* **25**, 556–575.
31. Mühlhaus, H.-B., and Vardoulakis, I. (1987). The thickness of shear bands in granular materials, *Géotechnique* **37**, 271–283.
32. Vardoulakis, I. and Sulem, J. (1995). Bifurcation Analysis in Geomechanics, Blackie Academic and Professional, Glasgow.
33. Sulem, J., Stefanou, I., and Veveakis, E. (2011). Stability analysis of undrained adiabatic shearing of a rock layer with Cosserat microstructure, *Granular Matter* **13**, 261–268.
34. Pizzino, L., Quattrochi, F., Cinti, D., and Galli, G. (2004). Fluid geochemistry along the Eliki and Aigion seismogenic segments (Gulf of Corinth, Greece), *Comptes Rendus Geosciences* **336**(4–5), 367–374.
35. Lewicki, J. L. and Brantley, S. L. (2000). CO_2 degassing along the San Andreas fault, Parkfield, California, *Geophysical Research Letters* **27**(1), 5–8.
36. Lewicki, J. L., Evans, W. C., Hilley, G. E., Sorey, M. L., Rogie, J. D., and Brantley, S. L. (2003). Shallow soil CO_2 flow along the San Andreas and Calaveras faults, California, *Journal of Geophysical Research* **108**(B4), 2187.
37. Sato, T. and Takahashi, M. (1997). Geochemical changes in anomalously discharged groundwater in Awaji Island after the 1995 Kobe earthquake, *Chikyukagaku* **31**, 89–98.
38. Famin, V., Nakashima, S., Boullier, A.-M., Fujimoto, K., and Hirono, T. (2008). Earthquake produce carbon dioxide in crustal faults, *Earth and Planetary Science Letters* **265**(3–4), 487–497.
39. Arai, T., Tsukahara, H., and Morikiyo, T. (2003). Sealing process with calcite in the Nojima active fault zone revealed from isotope analysis of calcite, *Journal of Geography* **112**(6), 915–925.
40. Lin, A., Tanaka, N., Ueda, S., and Satish-Kumar, M. (2003). Repeated coseismic infiltration of meteoric and seawater into deep fault zones: a case study of the Nojima fault zone, Japan, *Chemical Geology* **202**(1–2), 139–153.
41. Italiano, F., Martinelli, G., and Plescia, P. (2008). CO_2 degassing over seismic areas: the role of mechano-chemical production at the study case of central Apennines, *Pure and Applied Geophysics* **165**, 75–94.
42. Hirono, T., Ikehara, M., Otsuki, K., Mishima, T., Sakaguchi, M., Soh, W., Omori, M., Lin, W., Yeh, E., Tanikawa, W., and Wang, C. Y. (2006). Evidence of frictional melting from disk-shaped black material, discovered within the Taiwan Chelungpu fault system, *Geophysical Research Letters* **33**, L19311.
43. Hirono, T., Yokohama, T., Hamada, Y., Tanikawa, W., Mishima, T., Ikehara, M., Famin, V., Tanimizu, M., Lin, W., Soh, W., and Song, S.-R. (2007). A chemical kinetic approach to estimate dynamic shear stress during the 1999 Taiwan Chi-Chi earthquake, *Geophysical Research Letters* **34**, L19308.

44. Han, R., Shimamoto, T., Hirose, T., Ree, J.-H., and Ando, J. (2007). Ultralow friction of carbonate faults caused by thermal decomposition, *Science* **316**, 878–881.
45. L'vov, B. V. (2002). Mechanism and kinetics of thermal decomposition of carbonates, *Thermochimica Acta* **386**, 1–16.
46. Dollimore, D., Tong, P., and Alexander, K. S. (1996). The kinetic interpretation of the decomposition of calcium carbonate by use of relationships other than the Arrhenius equation, *Thermochimica Acta* **282/283**, 13–27.
47. Lachenbruch, A. H., and Sass, J. H. (1980). Heat flow and energetics of the San Andreas fault zone, *Journal of Geophysical Research* **85**, 6185–6223.
48. Noda, H. E., Dunham, M., and Rice, J. R. (2009). Earthquake ruptures with thermal weakening and the operation of major faults at low overall stress levels, *Journal of Geophysical Research* **14**, B07302.
49. Di Toro, G., Goldsby, D. L., and Tullis, T. E. (2004). Friction falls toward zero in quartz rock as slip velocity approaches seismic rates, *Nature* **427**, 436–439.
50. De Paola, N., Hirose, T., Mitchell, T., Di Toro, G., Togo, T., and Shimamoto, T. (2011). Fault lubrication and earthquake propagation in thermally unstable rocks, *Geology* **39**, 35–38.

Bibliography

This list is not a complete list, but we hope it will provide an overview of the progress in the studies of adiabatic shear banding to see how understanding has progressed and to foresee possible future trends.

1798, Thompson, B.C.R. "An inquiry concerning the source of the heat which is excited by friction" *Philos. Trans. R. Soc. London* **18** 278–287.

1850, Joule, J.P. "On the mechanical equivalent of heat" *Philos. Trans. R. Soc. London.* **140** 61–82.

1859, Joule, J.P. "On some thermodynamic properties of solids" *Philos. Trans. R. Soc. London* **149** 91–131.

1859, Joule, J.P. "On the thermal effects of compressing fluids" *Philos. Trans. R. Soc. London* **149** 133–136.

1878, Tresca, H. "On further applications of the flow of solids" *Proc. Inst. Mech. Eng.* **29** 301–315.

1905, Gulliver, G.H. "Some phenomena of permanent deformation in metals" *Proc. Inst. Mech. Eng.* **68** 141–147.

1907, Gulliver, G.H. "Some phenomena of permanent deformation in metals. II" *Proc. Inst. Mech. Eng.* **72** 519–524.

1909, Portevin, A.M. "Contribution to the study of the special ternary steels" *Carnegie Scholarship Mem. Iron Steel Inst.* **1** 230–364.

1911, Mason, W. "The Lüders lines on mild steel" *Proc. Phys. Soc. London* **23** 305–333.

1915, Howe, H.M. and Levy, A.G. "Are the deformation lines in manganese steel twins or slip bands?" *Trans. AIME* **51** 881–896.

1920, Johnson, F. "Some features of the behaviour of beta-brass when cold-rolled" *J. Inst. Met.* **24** 301–314.

1921, Massey, H.F. "The flow of metals during forging" *Proc. Manchester Assoc. Eng.*, 21–66.

1922, Adcock, F. "The internal mechanism of cold-work and recrystallization in cupro-nickel" *J. Inst. Met.* **27** 73–101.

1925, Nádai, A. "Beobachtungen der Gleitflächenbildung an plastischen Stoffen", in "Proc. First Int. Congress on Applied Mechanics", C.B. Biezeno and J.M. Burgers (Eds.), pp. 318–325, Delft, The Netherlands, Technische Boekhandel en Drukkerij J. Waltman Jr.

1925, Jevons, J.D. "Strain detection in mild steel by special etching" *J. Iron Steel Inst.* **111** 191–213.

1925, Farren, W.S. and Taylor, G.I. "The heat developed during plastic extension of metals" *Proc. R. Soc. London A* **107** 422–451.

1928, Polanyi, M. "Deformation, rupture, and hardening of crystals" *Trans. Faraday Soc.* **24** 72–83.

1928, Rawdon, H.S. "Strain markings in mild steel under tension" *Bureau Stand. J. Res.* **1** 467–485.

1931, Nadai, A. "The phenomenon of slip in plastic materials" *Proc. Am. Soc. Test. Mater.* **31**(2), 11–46.

1931, Nadai, A. and Wahl, A.M. "Photoelastic contact tests and observation of slip lines under plastic impressions", in "Plasticity: A Mechanics of the Plastic State of Matter", pp. 243–257, New York, NY, McGraw-Hill.

1934, Johnson, E.R. and Buechling, W.J. "A study of banding in a chromium–molybdenum steel" *Trans. Am. Soc. Met.* **22** 249–270.

1934, Taylor, G.I. and Quinney, H. "The latent energy remaining in a metal after cold working" *Proc. R. Soc. London A* **143** 307–326.

1936, Galibourg, J. "Phénomènes d'instabilité dans les métaux" *Sciences* **64** 251–275.

1937, Quinney, H. and Taylor, G.I. "The emission of the latent energy due to previous cold working when a metal is heated" *Proc. R. Soc. London A* **163** 157–181.

1939, Barrett, C.S. "Structure of iron after compression" *Trans. Metall. Soc. AIME* **135** 296–326.

1939, Watson, J.D. "A triaxial compression apparatus for the determination of the stress-deformation characteristics of soils" *Proc. Am. Soc. Test. Mater.* **39** 1046–1057.

1941, Trent, E.M. "The formation and properties of martensite on the surface of rope wire" *J. Iron Steel Inst.* **143** 401–419.

1942, Donnell, L.H. "Plastic flow as an unstable process" *J. Appl. Mech.* **9** 91–95.

1944, Zener, C. and Hollomon, J.H. "Effect of strain rate upon plastic flow of steel" *J. Appl. Phys.* **15** 22–32.

1946, Zener, C. "A theoretical criterion for the initiation of slip bands" *Phys. Rev.* **69** 128–129.

1957, Basinski, Z.S. "The instability of plastic flow of metals at very low temperatures" *Proc. R. Soc. London A* **240** 229–242.

1957, Basinski, Z.S. and Sleeswyk, A. "On the ductility of iron at 4.2 K" *Acta Metall.* **5** 176–179.

1957, Basinski, Z.S. "Ductility of iron at 4.2 K" *Acta Metall.* **5** 763–764.

1960, Basinski, Z.S. "The instability of plastic flow of metals at very low temperatures. II" *Aust. J. Phys.* **13** 354–358.

1960, Lean, J.B. "A low temperature yield instability in iron" *Aust. J. Phys.* **13** 359–375.

1960, Phillips, A. "Pointed vertices in plasticity", in "Plasticity: Proc. Second Symp. on Naval Structural Mechanics", E.H. Lee and P.S. Symonds (Eds.), pp. 202–214, Oxford, Pergamon.

1961, Polakowski, N.H. and Mostovoy, S. "Transient and destructive instability in torsion" *Trans. Am. Soc. Met.* **54** 567–579, 850

1964, Chin, G.Y., Jr., Hosford, W.F. and Backofen, W.A. "Influence of the mechanical loading system on low-temperature plastic instability" *Trans. Metall. Soc. AIME* **230** 1043–1048.

1964, Johnson, W., Baraya, G.L. and Slater, R.A.C. "On heat lines or lines of thermal discontinuity" *Int. J. Mech. Sci.* **6** 409–414.

1964, Recht, R.F. "Catastrophic thermoplastic shear" *Trans. ASME: J. Appl. Mech.* **31** 189–193.

1964, Smith, E. "The magnitude of the strain in a localized region of plastic deformation" *Philos. Mag.* **9** 879–882.

1965, Slater, R.A.C. "Velocity and thermal discontinuities encountered during the forging of steels" *Trans. Manchester Assoc. Eng.* (5), 1–43.

1966, Holzmann, M. "Stable/unstable plastic flow transition temperature in mild steel" *J. Iron Steel Inst.* **204** 739–741.

1967, Campbell, J.D. "Plastic instability in rate-dependent materials" *J. Mech. Phys. Solids* **15** 359–370.

1968, Argon, A.S., Andrews, R.D., Godrick, J.A. and Whitney, W. "Plastic deformation bands in glassy polystyrene" *J. Appl. Phys.* **39** 1899–1906.

1968, Campbell, J.D., Cooper, R.H. and Fischhof, T.J. "The dynamics of nonuniform plastic flow in low-carbon steel", in "Dislocation Dynamics", A.R. Rosenfield, G.T. Hahn, A.L. Bement Jr. and R.I. Jaffee (Eds.), pp. 723–746, New York, NY, McGraw-Hill.

1968, Hoggatt, C.R. and Recht, R.F. "Fracture behavior of tubular bombs" *J. Appl. Phys.* **39** 1856–1862.

1969, Ernst, R.H. and Spretnak, J.W. "Effect of shear strain rate on the susceptibility of selected engineering alloys to plastic instability in torsion" *Trans. Iron Steel Inst. Jpn.* **9** 361–371.

1970, Manion, S.A. and Stock, T.A.C. "Adiabatic shear bands in steel" *Int. J. Fract. Mech.* **6** 106–107.

1970, Bowden, P.B. "A criterion for inhomogeneous plastic deformation" *Philos. Mag.* **22** 455–462.

1970, Bowden, P.B. and Raha, S. "The formation of micro shear bands in polystyrene and polymethylmethacrylate" *Philos. Mag.* **22** 463–482 .

1970, Stevenson, M.G. and Oxley, P.L.B. "An experimental investigation of the influence of speed and scale on the strain rate in a zone of intense plastic deformation" *Proc. Inst. Mech. Eng.* **184** 561–576.

1970, Wonsiewicz, B.C. and Chin, G.Y. "Inhomogeneity of plastic flow in constrained deformation" *Metall. Trans.* **1** 57–61.

1970, Stock, T.A.C. and Thompson, K.R.L. "Penetration of aluminum alloys by projectiles" *Metall. Trans.* **1** 219–224.

1971, Thornton, P.A. and Heiser, F.A. "Observations on adiabatic shear zones in explosively loaded thick wall cylinders" *Metall. Trans.* **2** 1496–1499.

1971, Wingrove, A.L. "A note on the structure of adiabatic shear bands in steel" *J. Aust. Inst. Met.* **16** 67–70.

1971, Culver, R.S. "Adiabatic heating effects in dynamic deformation", in "Proc. Third Int. Conf. of the Center for High Energy Forming", paper 4.3, Denver, CO, University of Denver.

1971, Glenn, R.C. and Leslie, W.C. "The nature of 'White Streaks' in impacted steel armor plate" *Metall. Trans.* **2** 2945–2947.

1971, Kudo, H. and Tsubouchi, M. "Strain concentration due to heat generation in fast plastic deformation" *Ann. CIRP* **19** 225–230.

1972, Manion, S.A. and Wingrove, A.L. "A note on the formation of chip fragments due to adiabatic shear" *J. Aust. Inst. Met.* **17** 158–160.

1972, Black, J.T. "On microscopic plastic instabilities in metal machining chips" *Metall. Trans.* **3** 2012–2015.

1972, Lemaire, J.C. and Backofen, W.A. "Adiabatic instability in the orthogonal cutting of steel" *Metall. Trans.* **3** 477–481.

1973, Backman, M.E. and Finnegan, S.A. "The propagation of adiabatic shear", in "Metallurgical Effects at High Strain Rates", R.W. Rohde, B.M. Butcher, J.R. Holland and C.H. Karnes (Eds.), pp. 531–544, New York, NY, Plenum Press.

1973, Wingrove, A.L. "The influence of projectile geometry on adiabatic shear and target failure" *Metall. Trans.* **4** 1829–1833.

1973, Argon, A.S. "Stability of plastic deformation", in "The Inhomogeneity of Plastic Deformation", R.E. Reed-Hill (Ed.), pp. 161–190, Metals Park, OH, American Society for Metals.

1973, Rice, J.R. "The initiation and growth of shear bands", in "Plasticity and Soil Mechanics", A.C. Palmer (Ed.), pp. 263–274, Cambridge, Cambridge University Engineering Department.

1973, Sulijoadikusimo, A.U. and Dillon, O.W., Jr. "Work softening of Ti–6Al–4V due to adiabatic heating", in "Metallurgical Effects at High Strain Rates", R.W. Rohde, B.M. Butcher, J.R. Holland and C.H. Karnes (Eds.), pp. 501–518, New York, NY, Plenum Press.

1973, Wingrove, A.L. and Wulf, G.L. "Some aspects of target and projectile properties on penetration behaviour" *J. Aust. Inst. Met.* **18** 167–172.

1973, Wu, W. and Turner, A.P.L. "Shear bands in polycarbonate" *J. Polym. Sci.: Polym. Phys. Ed.* **11** 2199–2207.

1973, Pearce, D.F. and McManus, B.R. "Dynamic chip formation and its significance to machining stability" *Proc. Inst. Mech. Eng.* **187** 273–283.

1974, Bedford, A.J., Wingrove, A.L. and Thompson, K.R.L. "The phenomenon of adiabatic shear deformation" *J. Aust. Inst. Met.* **19** 61–73.

1974, Hargreaves, C.R. and Werner, L. "Adiabatic shear – an annotated bibliography", AD/A006490, National Technical Information Service.

1974, Rogers, R.C. "Adiabatic Shearing: A Review", Philadelphia, PA, US Army Research Office, Drexel University.

1974, Kramer, E.J. "The stress–strain curve of shear-banding polystyrene" *J. Macromol. Sci. B: Phys.* **10** 191–202.

1975, Winter, R.E. "Adiabatic shear of titanium and polymethylmethacrylate" *Philos. Mag.* **31** 765–773.

1975, Winter, R.E. and Hutchings, I.M. "The role of adiabatic shear in solid particle erosion" *Wear* **34** 141–148.

1975, Kramer, E.J. "The growth of shear bands in polystyrene" *J. Polym. Sci: Polym. Phys. Ed.* **13** 509–526.

1975, Moss, G.L. and Pond, R.B., Sr. "Inhomogeneous thermal changes in copper during plastic elongation" *Metall. Trans. A* **6** 1223–1235.

1975, Petukov, B.V. and Estin, Y.Z. "Criterion of thermally activated plastic deformation" *Sov. Phys. Solid State* **17** 1333–1334.

1975, Turley, D.M. "The nature of the white-etching surface layers produced during reaming of ultrahigh strength steel" *Mater. Sci. Eng.* **19** 79–86.

1975, Winter, R.E. and Field, J.E. "The role of localized plastic flow in the impact initiation of explosives" *Proc. R. Soc. London A* **343** 399–413.

1976, Bish, R.L. "An explanation of adiabatic shear" *J. Aust. Inst. Met.* **21** 114–116.

1976, Hutchings, I.M., Winter, R.E. and Field, J.E. "Solid particle erosion of metals: the removal of surface material by spherical projectiles" *Proc. R. Soc. London A* **348** 379–392.

1976, Lindholm, U.S. and Hargreaves, C.R. "Dynamic testing of high strength steels and their susceptibility to inhomogeneous shear", in "Proc. 2nd Int. Conf. on the Mechanical Behavior of Materials", pp. 1463–1467, Metals Park, OH, American Society for Metals.

1976, Rice, J.R. "The localization of plastic deformation", in "Theoretical and Applied Mechanics", W.T. Koiter (Ed.), pp. 207–220, Amsterdam, North-Holland.

1977, Oxley, P.L.B. and Hastings, W.H. "Predicting the strain rate in the zone of intense shear in which the chip is formed in machining from the dynamic flow stress properties of the work material and the cutting conditions" *Proc. R. Soc. London A* **356** 395–410.

1978, Clifton, R.J. "Adiabatic shear in material response to ultra loading rates", report no. NMAB-356, Washington, DC, National Materials Advisory Board.

1978, Woodward, R.L. "The penetration of metal targets which fail by adiabatic shear plugging" *Int. J. Mech. Sci.* **20** 599–607.

1978, Needleman, A. and Rice, J.R. "Limits to ductility set by plastic flow localization", in "Mechanics of Sheet Metal Forming", D.P. Koistinen and N.M. Wang (Eds.), pp. 237–268, New York, NY, Plenum Press.

1978, Wada, M., Nakamura, T. and Kinoshita, N. "Distribution of temperature, strain rate and strain in plastically deforming metals at high strain rates" *Philos. Mag. A* **38** 167–185.

1978, Williams II, J.D. "Localized deformation of polymers at low temperatures and high loading rates", Ph.D. Thesis, University of North Carolina State at Raleigh.

1978, Woodward, R.L. and Aghan, R.L. "The structure of a white-etching band in an explosively fractured steel" *Met. Forum* **1** 180–184.

1978, Yamamoto, H. "Conditions for shear localization in the ductile fracture of void-containing materials" *Int. J. Fract* **14** 347–365.

1979, Bhambri, A.K. "Influence of material characteristics on adiabatic shear and chip formation" *Microstruct. Sci.* **7** 255–264.

1979, Christman, T. and Shewmon, P.G. "Adiabatic shear localization and erosion of strong aluminum alloys" *Wear* **54** 145–155.

1979, Chau, C.C. and Li, J.C.M. "Some observations of coarse shear bands in polystyrene" *J. Mater. Sci.* **14** 1593–1608.

1979, Chau, C.C. and Li, J.C.M. "Intersections of coarse shear bands in polystyrene" *J. Mater. Sci.* **14** 2172–2182.

1979, Christman, T. and Shewmon, P.G. "Erosion of a strong aluminum alloy" *Wear* **52** 57–70.

1979, Costin, L.S., Crisman, E.E., Hawley, R.H. and Duffy, J. "On the localisation of plastic flow in mild steel tubes under dynamic torsional loading" *Inst. Phys. Conf. Ser.* **47** 90–100.

1979, Dao, K.C. and Shockey, D.A. "A method for measuring shear-band temperatures" *J. Appl. Phys.* **50** 8244–8246.

1979, de Morton, M.E., Woodward, R.L. and Yellup, J.M. "Defeating armour and projectiles by fracture", in "Proc. 4th Tewksbury Symposium on Fracture", G.H. Vasey and D.S. Mansell (Eds.), paper 11, Melbourne, Australia, University of Melbourne.

1979, Rogers, H.C. "Adiabatic plastic deformation" *Annu. Rev. Mater. Sci.* **9** 283–311.

1979, Stelly, M. "Metallurgical aspects of the dynamic expansion of shells" *Inst. Phys. Conf. Ser.* **47** 252–253.

1979, Woodward, R.L. "Penetration behaviour of a high-strength aluminium alloy" *Met. Technol.* **6** 106–110.

1979, Woodward, R.L. "Metallographic features associated with the penetration of titanium alloy targets" *Metall. Trans. A* **10** 569–573.

1979, Woodward, R.L. and Baldwin, N.J. "Oblique perforation of targets by small armour-piercing projectiles" *J. Mech. Eng. Sci.* **21** 85–91.

1980, Staker, M.R. "On adiabatic shear band determinations by surface observations" *Scr. Metall.* **14** 677–680.

1980, Yellup, J.M. and Woodward, R.L. "Investigation into the prevention of adiabatic shear failure in high strength armour materials" *Res. Mech.* **1** 41–57.

1980, Seaman, L., Curran, D.R., Shockey, D.A. and Erlich, D.C. "A shear band model for 4340 steel", in "Proc. Army Symp. on Solids Mechanics: Designing for Extremes. Vol. 2", pp. 68–70, Alexandria, VA, US Army Material Development and Readiness Command.

1981, Bai, Y.L. "A criterion for thermo-plastic shear instability", in "Shock Waves and High-Strain-Rate Phenomena in Metals", M.A. Meyers and L.E. Murr (Eds.), pp. 277–284, New York, NY, Plenum Press.

1981, Doraivelu, S.M., Gopinathan, V. and Venkatesh, V.C. "Formation of adiabatic shear bands during upsetting of 18-4-1 alloy steel at high strain rates", in "Shock Waves and High-Strain-Rate Phenomena in Metals", M.A. Meyers and L.E. Murr (Eds.), pp. 263–276, New York, NY, Plenum Press.

1981, Dormeval, R. and Stelly, M. "Study of adiabatic shear bands by means of dynamic compressive tests", in "Proc. 7th Int. Conf. on High Energy Rate Fabrication", T.Z. Blazynski (Ed.), pp. 10–18, Leeds, UK, University of Leeds.

1981, Moss, G.L. "Shear strains, strain rates and temperature changes in adiabatic shear bands", in "Shock Waves and High-Strain-Rate Phenomena in Metals", M.A. Meyers and L.E. Murr (Eds.), pp. 299–312, New York, NY, Plenum Press.

1981, Rogers, H.C. and Shastry, C.V. "Material factors in adiabatic shearing in steels", in "Shock Waves and High-Strain-Rate Phenomena in Metals", M.A. Meyers and L.E. Murr (Eds.), pp. 285–298, New York, NY, Plenum Press.

1981, Staker, M.R. "The relation between adiabatic shear instability strain and material properties" *Acta Metall.* **29** 683–689.

1981, Velez, J.F. and Powell, G.W. "Some metallographic observations on the spalling of AISI 1060 steel by the formation of adiabatic shear bands" *Wear* **66** 367–378.

1981, Timothy, S.P. and Hutchings, I.M. "Microstructural features associated with ballistic impact in Ti6Al4V", in "Proc. 7th Int. Conf. on High Energy Rate Fabrication", T.Z. Blazynski (Ed.), pp. 19–28, Leeds, UK, University of Leeds.

1982, Bai, Y.L. "Thermo-plastic shear instability in simple shear" *J. Mech. Phys. Solids* **30** 195–207.

1982, Merzer, A.M. "Modelling of adiabatic shear band development from small imperfections" *J. Mech. Phys. Solids* **30** 323–338.

1982, Tirupataiah, Y. and Raju, K.R. "Observation of adiabatic shear bands at very high strain rates in a HSLA steel" *Trans. Indian Inst. Met.* **35** 212–220.

1983, Me-Bar, Y. and Shechtman, D. "On the adiabatic shear of Ti–6Al–4V ballistic targets" *Mater. Sci. Eng.* **58** 181–188.

1983, Rogers, H.C. "Adiabatic shearing – General nature and material aspects", in "Material Behavior Under High Stress and Ultrahigh Loading Rates", J. Mescall and V. Weiss (Eds.), pp. 101–118, New York, NY, Plenum Press.

1983, Timothy, S.P. and Hutchings, I.M. "Scale effects in adiabatic shear banding caused by normal and oblique impact", in "Proc. 6th Int. Conf. on Erosion by Liquid and Solid Impact", J.E. Field and N.S. Corney (Eds.), paper 43, Cambridge, Cavendish Laboratory.

1983, Timothy, S.P. "Shear bands in a titanium alloy", Ph.D. Thesis, University of Cambridge, Cambridge.

1983, Hutchings, I.M. "The behavior of metals under ballistic impact at sub-ordnance velocities", in "Material Behavior Under High Stress and Ultrahigh Loading Rates", J. Mescall and V. Weiss (Eds.), pp. 161–196, New York, NY, Plenum Press.

1984, Affouard, J.L., Dormeval, R., Stelly, M. and Ansart, J.P. "Adiabatic shear bands in metals and alloys under dynamic compressive conditions" *Inst. Phys. Conf. Ser.* **70** 533–540.

1984, Clifton, R.J., Duffy, J., Hartley, K.A. and Shawki, T.G. "On critical conditions for shear band formation at high strain rates" *Scr. Metall.* **18** 443–448.

1984, Rogers, H.C. "Adiabatic strain localization during dynamic deformation", in "Deformation, Processing, and Structure", G. Krauss (Ed.), pp. 425–443, ASM, Metals Park, OH.

1984, Semiatin, S.L., Staker, M.R. and Jonas, J.J. "Plastic instability and flow localization in shear at high rates of deformation" *Acta Metall.* **9** 1347–1354.

1984, Staker, M.R. "The influence of dynamic strain aging on the temperature dependence of flow stress and the impact of this on adiabatic shear analysis" *Scr. Metall.* **18** 735–738.

1984, Timothy, S.P. and Hutchings, I.M. "Influence of adiabatic shear bands on the fatigue strength of a titanium alloy" *Fatigue Fract. Eng. Mater. Struct.* **7** 223–227.

1984, Timothy, S.P. and Hutchings, I.M. "Adiabatic shear bands in a titanium alloy resulting from ballistic impact at low and high temperatures" *Inst. Phys. Conf. Ser.* **70** 397–404.

1984, Timothy, S.P. and Hutchings, I.M. "The effect of anisotropy on adiabatic shear band trajectories in titanium alloys", in "Proc. 8th Int. Conf. on High Energy Rate Fabrication", pp. 31–40, New York, NY, American Society of Mechanical Engineers.

1984, Woodward, R.L., Baxter, B.J. and Scarlett, N.V.Y. "Mechanisms of adiabatic shear plugging failure in high strength aluminium and titanium alloys" *Inst. Phys. Conf. Ser.* **70** 525–532.

1984, Woodward, R.L. "The interrelation of failure modes observed in the penetration of metallic targets" *Int. J. Impact Eng.* **2** 121–129.

1984, Wu, F.H. and Freund, L.B. "Deformation trapping due to thermoplastic instability in one-dimensional wave propagation" *J. Mech. Phys. Solids* **32** 119–132.

1985, Burns, T.J. "Approximate linear stability analysis of a model of adiabatic shear band formation" *Q. Appl. Math.* **43** 65–84.

1985, Curran, D.R. and Seaman, L. "Computational models for nucleation, growth, and coalescence of adiabatic shear bands" *J. Phys.* **46**(C5), 395–401.

1985, Dodd, B. and Bai, Y.L. "Width of adiabatic shear bands" *Mater. Sci. Technol.* **1** 38–40.

1985, Grady, D.E. and Kipp, M.E. "The growth of inhomogeneous thermoplastic shear" *J. Phys.* **46** 291–298.

1985, Grebe, H.A., Pak, H.-R. and Meyers, M.A. "Adiabatic shear localization in titanium and Ti6Al4V alloy" *Metall. Trans. A* **16** 761–775.

1985, Fressengeas, C. and Molinari, A. "Inertia and thermal effects on the localization of plastic flow" *Acta Metall.* **33** 387–396.

1985, Leech, P.W. "Observations of adiabatic shear band formation in 7039 aluminum alloy" *Metall. Trans. A* **16** 1900–1903.

1985, Moss, G.L. and Pritchard, D.S. "Adiabatic shear band velocities" *J. Phys.* **46**(C5), 315–321.

1985, Timothy, S.P. and Hutchings, I.M. "The structure of adiabatic shear bands in a titanium alloy" *Acta Metall.* **33** 667–676.

1985, Timothy, S.P. and Hutchings, I.M. "Initiation and growth of microfractures along adiabatic shear bands in Ti–6Al–4V" *Mater. Sci. Technol.* **1** 526–530.

1985, Tirupataiah, Y. and Raju, K.R. "Tempering behaviour of adiabatic shear bands" *Indian J. Technol.* **23** 17–20.

1985, Wright, T.W. and Batra, R.C. "The initiation and growth of adiabatic shear bands" *Int. J. Plast.* **1** 205–212.

1985, Wright, T.W. and Batra, R.C. "Further results on the initiation and growth of adiabatic shear bands at high strain rates" *J. Phys.* **46**(C5), 323–330.

1986, Backman, M.E., Finnegan, S.A., Schulz, J.C. and Pringle, J.K. "Scaling rules for adiabatic shear", in "Metallurgical Applications of Shock-Wave and High-Strain-Rate Phenomena", L.E. Murr, K.P. Staudhammer and M.A. Meyers (Eds.), pp. 675–688, New York, NY, Marcel Dekker.

1986, Bai, Y.L., Cheng, C.M. and Yu, S.B. "On evolution of thermo-plastic shear band" *Acta Mech. Sin.* **2** 1–7.

1986, Burns, T.J. "Influence of effective rate sensitivity on adiabatic shear instability", in "Metallurgical Applications of Shock-Wave and High-Strain-Rate Phenomena", L.E. Murr, K.P. Staudhammer and M.A. Meyers (Eds.), pp. 741–747, New York, NY, Marcel Dekker.

1986, Curran, D.R. and Seaman, L. "Computational models for nucleation, growth, and coalescence of adiabatic shear bands", in "Shock Waves in Condensed Matter – 1985", Y.M. Gupta (Ed.), pp. 315–320, New York, NY, Plenum Press.

1986, Kuriyama, S. and Meyers, M.A. "Numerical modeling of the propagation of an adiabatic shear band" *Metall. Trans. A* **17** 443–450.

1986, Mescall, J.F. "On the relative roles of strain-hardening and thermal-softening in adiabatic shear bands", in "Metallurgical Applications of Shock-Wave and High-Strain-Rate Phenomena", L.E. Murr, K.P. Staudhammer and M.A. Meyers (Eds.), pp. 689–704, New York, NY, Marcel Dekker.

1986, Meyer, L.W. and Manwaring, S. "Critical adiabatic shear strength of low alloyed steel under compressive loading", in "Metallurgical Applications of Shock-Wave and High-Strain-Rate Phenomena", L.E. Murr, K.P. Staudhammer and M.A. Meyers (Eds.), pp. 657–674, New York, NY, Marcel Dekker.

1986, Meyers, M.A. and Pak, H.-R. "Observation of an adiabatic shear band in titanium by high-voltage transmission electron microscopy" *Acta Metall.* **34** 2493–2499.

1986, Meyers, M.A. and Kuriyama, S. "Numerical modeling of the propagation of an adiabatic shear band" *Metall. Trans. A* **17** 433–450.

1986, Mogilevskii, M.A., Sanchaa, T.O. and Shishkin, Y.D. "Critical loading parameters for the development of adiabatic shear in titanium" *J. Appl. Mech. Tech. Phys.* **27** 430–432.

1986, Shockey, D.A. "Materials aspects of the adiabatic shear phenomenon", in "Metallurgical Applications of Shock-Wave and High-Strain-Rate Phenomena", L.E. Murr, K.P. Staudhammer and M.A. Meyers (Eds.), pp. 633–656, New York, NY, Marcel Dekker.

1986, Stelly, M. and Dormeval, R. "Adiabatic shearing", in "Metallurgical Applications of Shock-Wave and High-Strain-Rate Phenomena", L.E. Murr, K.P. Staudhammer and M. A. Meyers (Eds.), pp. 607–632, New York, NY, Marcel Dekker.

1986, Timothy, S.P. "The structure of adiabatic shear bands in metals" *Ann. Isr. Phys. Soc.* **8** 388–400.

1986, Timothy, S.P. and Hutchings, I.M. "Adiabatic shear band fracture surfaces in a titanium alloy" *J. Mater. Sci. Lett.* **5** 453–454.

1986, Swallowe, G.M., Field, J.E. and Hutchinson, C.D. "Impact experiments on thin layers of polymers and intermediate explosives", in "Shock Waves in Condensed Matter – 1985", Y.M. Gupta (Ed.), pp. 891–898, New York, NY, Plenum Press.

1986, Swallowe, G.M., Field, J.E. and Horn, L.A. "Measurements of transient high temperatures during the deformation of polymers" *J. Mater. Sci.* **21** 4089–4096.

1987, Anand, L., Kim, K.H. and Shawki, T.G. "Onset of shear localization in viscoplastic solids" *J. Mech. Phys. Solids* **35** 407–429.

1987, Bai, Y.L., Cheng, Z.M. and Ding, Y.S. "Development of thermoplastic shear bands in simple shear" *Res. Mech.* **22** 313–324.

1987, Batra, R.C. "Effect of material parameters on the initiation and growth of adiabatic shear bands" *Int. J. Solids Struct.* **23** 1435–1446.

1987, Batra, R.C. "The initiation and growth of, and the interaction among, adiabatic shear bands in simple and dipolar materials" *Int. J. Plast.* **3** 75–89.

1987, Derep, J.L. "Microstructure transformation induced by adiabatic shearing in armour steel" *Acta Metall.* **35** 1245–1249.

1987, Dormeval, R. "The adiabatic shear phenomenon", in "Materials at High Strain Rates", T.Z. Blazynski (Ed.), pp. 47–70, London, Elsevier.

1987, Fressengeas, C. and Molinari, A. "Instability and localization of plastic flow in shear at high strain rates" *J. Mech. Phys. Solids* **35** 185–211.

1987, Grady, D.E. and Kipp, M.E. "The growth of unstable thermoplastic shear with application to steady-wave shock compression of solids" *J. Mech. Phys. Solids* **35** 95–118.

1987, Hartley, K.A., Duffy, J. and Hawley, R.H. "Measurement of the temperature profile during shear band formation in steels deforming at high strain rates" *J. Mech. Phys. Solids* **35** 283–301.

1987, Johnson, W. "Henri Tresca as the originator of adiabatic heat lines" *Int. J. Mech. Sci.* **29** 301–310.

1987, Kuriyama, S. and Meyers, M.A. "Numerical analysis of adiabatic shear band in an early stage of its propagation", in "Macro- and Micro-Mechanics of High Velocity Deformation and Fracture", K. Kawata and J. Shiori (Eds.), pp. 203–212, Berlin, Springer-Verlag.

1987, Makel, D.D. and Wilsdorf, H.G.F. "An investigation of unusual surface features caused by adiabatic shear during tensile separation" *Scr. Metall.* **21** 1229–1234.

1987, Stepanov, G.V. "Effect of loading rate on the instability of uniform strain and adiabatic shear in metals" *Strength Mater.* **17** 1550–1554.

1987, Timothy, S.P. "The structure of adiabatic shear bands in metals: A critical review" *Acta Metall.* **35** 301–306.

1987, Wang, L.L., Lu, W.X., Hu, S.S. and Tang, Z.P. "Study of the initiation and development of adiabatic shear bands for a titanium alloy under high strain rates", in "Macro- and Micro-Mechanics of High Velocity Deformation and Fracture", K. Kawata and J. Shioiri (Eds.), pp. 395–406, Berlin, Springer-Verlag.

1987, Wright, T.W. and Walter, J.W. "On stress collapse in adiabatic shear bands" *J. Mech. Phys. Solids* **35** 701–720.

1987, Wright, T.W. and Batra, R.C. "Adiabatic shear bands in simple and dipolar plastic materials", in "Macro- and Micro-Mechanics of High Velocity Deformation and Fracture", K. Kawata and J. Shiori (Eds.), pp. 189–202, Berlin, Springer-Verlag.

1987, Wright, T.W. and Walter, J.W. "Analytical modeling of adiabatic shear bands", in "Constitutive Models of Deformation", J. Chandra and R.P. Srivastav (Eds.), pp. 181–182, Philadelphia, PA, Society for Industrial and Applied Mathematics.

1988, Ansart, J.P. and Dormeval, R. "Adiabatic shearing in martensitic steels", in "Impact Loading and Dynamic Behaviour of Materials", C.Y. Chiem, H.D. Kunze and L.W. Meyer (Eds.), pp. 775–784, Oberursel, Germany, DGM Informationsgesellschaft Verlag.

1988, Batra, R.C. "Effects of nominal strain-rates on the initiation and growth of adiabatic shear bands in steels" *Trans. ASME: J. Appl. Mech.* **55** 229–230.

1988, Batra, R.C. and Kim, C.H. "Effect of material characteristic length on the initiation, growth and band width of adiabatic shear bands in dipolar materials" *J. Phys.* **49**(C3), 41–46.

1988, Batra, R.C. and Wright, T.W. "A comparison of solutions for adiabatic shear banding by forward-difference and Crank-Nicholson methods" *Commun. Appl. Numer. Methods* **4** 741–748.

1988, Dormeval, R. "Adiabatic shear phenomena", in "Impact Loading and Dynamic Behaviour of Materials", C.Y. Chiem, H.D. Kunze and L.W. Meyer (Eds.), pp. 43–56, Oberursel, Germany, DGM Informationsgesellschaft Verlag.

1988, Giovanola, J.H. "Adiabatic shear banding under pure shear loading. 1: Direct observation of strain-localization and energy dissipation measurements" *Mech. Mater.* **7** 59–71.

1988, Giovanola, J.H. "Adiabatic shear banding under pure shear loading. 2: Fractographic and metallographic observations" *Mech. Mater.* **7** 73–87.

1988, Giovanola, J.H. "Observation of adiabatic shear banding in simple torsion", in "Impact Loading and Dynamic Behaviour of Materials", C.Y. Chiem, H.D. Kunze and L.W. Meyer (Eds.), pp. 705–710, Oberursel, Germany, DGM Informationsgesellschaft Verlag.

1988, Kotoul, M. and Bílek, Z. "Adiabatic shear in porous media" *J. Phys.* **49**(C3), 535–540.

1988, Kwon, Y.W. and Batra, R.C. "Effect of multiple initial imperfections on the initiation and growth of adiabatic shear bands in nonpolar and dipolar materials" *Int. J. Eng. Sci.* **26** 1177–1187.

1988, Marchand, A. and Duffy, J. "An experimental study of the formation process of adiabatic shear bands in a structural steel" *J. Mech. Phys. Solids* **36** 251–283.

1988, Meunier, Y., Sangoy, L. and Pont, G. "Metallurgical aspects of adiabatic shear phenomenon in armor steels with perforation", in "Impact Loading and Dynamic Behaviour of Materials", C.Y. Chiem, H.D. Kunze and L.W. Meyer (Eds.), pp. 711–718, Oberursel, Germany, DGM Informationsgesellschaft Verlag.

1988, Pintat, T., Scolz, B., Kunze, H.D. and Vöhringer, O. "The influence of carbon content and grain size on energy consumption during adiabatic shearing" *J. Phys.* **49**(C3), 237–244.

1988, Wang, L.L., Bao, H.S. and Lu, W.X. "The dependence of adiabatic shear banding in strain rate, strain and temperature" *J. Phys.* **49**(C3), 207–214.

1988, Wang, L.L. "A criterion of thermo-viscoplastic instability for adiabatic shearing", in "Proc. Int. Symp. on Intense Dynamic Loading and its Effects", Z. Zheng and J. Ding (Eds.), pp. 787–792, Oxford, Pergamon.

1988, O'Donnell, R.G. and Woodward, R.L. "Instability during high strain rate compression testing of 2024 T351 aluminium" *J. Mater. Sci.* **23** 3578–3587.

1989, Batra, R.C. and Kwon, Y.W. "Adiabatic shear banding in a bimetallic alloy" *Acta Mech.* **77** 281–297.

1989, Batra, R.C. and Liu, D.-S. "Adiabatic shear banding in plane strain problems" *Trans. ASME: J. Appl. Mech.* **56** 527–534.

1989, Coleman, B.D. and Newman, D.C. "On adiabatic shear bands in rigid-plastic materials" *Acta Mech.* **78** 263–279.

1989, Crisman, E.E., Duffy, J. and Chi, Y.C. "A short rise time probe for temperature measurements during the formation of adiabatic shear bands", in "Micromechanical Experimental Techniques" W.M. Sharpe Jr. (Ed.), *AMD* **102**, New York, NY, American Society of Mechanical Engineers.

1989, Dodd, B. and Bai, Y.L. "Width of adiabatic shear bands formed under combined stresses" *Mater. Sci. Technol.* **5** 557–559.

1989, Duszek-Perzyna, M.K., Perzyna, P. and Stein, E. "Adiabatic shear band localization in elastic-plastic damaged solids", in "Advances in Plasticity", A.S. Khan and M. Tokuda (Eds.), pp. 439–442, Oxford, Pergamon.

1989, Field, J.E., Parry, M.A., Palmer, S.J.P. and Huntley, J.M. "Deformation and explosive properties of HMX powders and polymer bonded explosives", in "Proc. Ninth

Symposium (Int.) on Detonation", pp. 886–896, Arlington, VA, Office of the Chief of Naval Research.

1989, Fressengeas, C. "Adiabatic shear morphology at very high strain rates" *Int. J. Impact Eng.* **8** 141–157.

1989, Kobayashi, H. and Dodd, B. "A numerical analysis for the formation of adiabatic shear bands including void nucleation and growth" *Int. J. Impact Eng.* **8** 1–13.

1989, Kotoul, M. and Bílek, Z. "Effect of porosity on adiabatic shearing" *Acta Tech. CSAV* **34** 431–442.

1989, Krishna Mohan, V., Jyothi Bhasu, V.C. and Field, J.E. "Role of adiabatic shear bands in initiation of explosives by drop-weight impact", in "Proc. Ninth Symposium (Int.) on Detonation", pp. 1276–1283, Arlington, VA, Office of the Chief of Naval Research.

1989, Timothy, S.P. and Hutchings, I.M. "Formation of adiabatic shear bands by the impact of wedges and cones" *Inst. Phys. Conf. Ser.* **102** 127–134.

1989, Wright, T.W. and Walter, J.W. "Adiabatic shear bands in one dimension" *Inst. Phys. Conf. Ser.* **102** 119–126.

1989, Wright, T.W. "Instability and localization in adiabatic shear bands", in "Advances in Plasticity", A.S. Khan and M. Tokuda (Eds.), pp. 477–480, Oxford, Pergamon.

1989, Xu, Y.B., Wang, Z.G., Huang, X.L., Xing, D. and Bai, Y.L. "Microstructure in low carbon ferrite–pearlite steel" *Mater. Sci. Eng.* **A114** 81–87.

1990, Anand, L., Dillon, O., Place, T.A. and von Turkovich, B.F. "Report of the NSF workshop on localized plastic instabilities and failure criteria" *Int. J. Plast.* **6** I–IX.

1990, Bai, Y.L. "Adiabatic shear banding" *Res Mech.* **31** 133–203.

1990, Batra, R.C. and Kim, C.H. "Adiabatic shear banding in elastic-viscoplastic nonpolar and dipolar materials" *Int. J. Plast.* **6** 127–141.

1990, Batra, R.C. and Kim, C.H. "The interaction among adiabatic shear bands in simple and dipolar materials" *Int. J. Eng. Sci.* **28** 927–942.

1990, Batra, R.C. and Kim, C.H. "Effect of integration methods on the solution of an adiabatic shear banding problem" *Int. J. Numer. Methods Eng.* **29** 1639–1652.

1990, Batra, R.C. and Kim, C.H. "An adaptive mesh refinement technique for the analysis of adiabatic shear banding" *Mech. Res. Commun.* **17** 81–91.

1990, Batra, R.C. and Liu, D.S. "Adiabatic shear banding in dynamic plane strain compression of a viscoplastic material" *Int. J. Plast.* **6** 231–246.

1990, Cho, K., Chi, Y.C. and Duffy, J. "Microscopic observations of adiabatic shear bands in three different steels" *Metall. Trans. A* **21** 1161–1175.

1990, Perzyna, P. "Thermal effects in adiabatic shear band localization failure" *Mech. Teor. Stos.* **28** 171–197.

1990, Wittman, C.L., Meyers, M.A. and Pak, H-R. "Observation of an adiabatic shear band in AISI 4340 steel by high-voltage transmission electron microscopy" *Metall. Trans. A* **21** 707–716.

1990, Wright, T.W. "Approximate analysis for the formation of adiabatic shear bands" *J. Mech. Phys. Solids* **38** 515–530.

1990, Wright, T.W. "Adiabatic shear bands" *Appl. Mech. Rev.* **43**(Suppl.), S196–S200.

1990, Woodward, R.L. "Material failure at high strain rates", in "High Velocity Impact Dynamics", J.A. Zukas (Ed.), pp. 65–126, New York, NY, Wiley.

1991, Batra, R.C. and Kim, C.H. "Effect of thermal conductivity on the initiation, growth and bandwidth of adiabatic shear bands" *Int. J. Eng. Sci.* **29** 949–960.

1991, Batra, R.C. and Zhu, Z.G. "Dynamic adiabatic shear band development in a bimetallic body containing a void" *Int. J. Solids Struct.* **27** 1829–1854.

1991, Chou, P.C., Hashemi, J., Chou, A. and Rogers, H.C. "Experimentation and finite element simulation of adiabatic shear bands in controlled penetration impact" *Int. J. Impact Eng.* **11** 305–321.

1991, Duffy, F. "Experimental study of shear band formation through temperature measurements and high speed photography" *J. Phys. IV* **1**(C3), 645–652.

1991, Duszek-Perzyna, M.K. and Perzyna, P. "Adiabatic shear band localization in elastic–plastic single crystals", in "Anisotropy and Localization of Plastic Deformation", J.P. Boehler and A.S. Khan (Eds.), pp. 51–55, London, Elsevier.

1991, Grady, D.E. "Dynamics of adiabatic shear" *J. Phys. IV* **1**(C3), 653–660.

1991, Krejcí, J., Brezina, J., Buchar, J. and Rolc, S. "Adiabatic shearing in cast austenitic steel under ballistic attack" *J. Phys. IV* **1**(C3), 667–673.

1991, Luo, L.M. and Dodd, B. "Adiabatic shear under biaxial stress conditions in sheet metal forming processes" *J. Phys. IV* **1**(C3), 775–780.

1991, Makridin, V.A. and Rodionov, D.P. "Structural features of adiabatic shear in steel shells with dynamic loading" *Strength Mater.* **23** 1028–1030.

1991, Nakkalil, R. "Formation of adiabatic shear bands in eutectoid steels in high strain rate compression" *Acta Metall. Mater.* **39** 2553–2563.

1991, Perzyna, P. "Influence of thermodynamical couplings on adiabatic shear band localization failure", in "Anisotropy and Localization of Plastic Deformation", J.P. Boehler and A.S. Khan (Eds.), pp. 81–86, London, Elsevier.

1991, Xing, D., Bai, Y.L., Cheng, C.M. and Huang, X.L. "On post-instability processes in adiabatic shear in hot rolled steel" *J. Mech. Phys. Solids* **39** 1017–1042.

1991, Boehler, J.P. and Khan, A.S. (Eds.) "Anisotropy and Localization of Plastic Deformation," London, Elsevier..

1992, Bai, Y.L. and Dodd, B. "Adiabatic Shear Localization: Occurrence, Theories and Applications", Oxford, Pergamon.

1992, Bai, Y.L., Cheng, Z.M. and Ling, Z. "Localization and pattern of deformation in thermo-visco-plastic material" *J. Mech. Behav. Mater.* **4** 19–32.

1992, Beatty, J.H., Meyer, L.W., Meyers, M.A. and Nemat-Nasser, S. "Formation of controlled adiabatic shear bands in AISI 4340 high strength steel", in "Shock-Wave and High-Strain-Rate Phenomena in Materials", M.A. Meyers, L.E. Murr and K.P. Staudhammer (Eds.), pp. 645–656, New York, NY, Marcel Dekker.

1992, Cho, K.-M., Lee, S.-H., Park, I.-M. and Lee, S.-K. "Formation of adiabatic shear band in a structural steel", in "Proc. Int. Symp. on Impact Engineering", I. Maekawa (Ed.), pp. 437–442, Sendai, Japan, ISIE.

1992, Chou, P.C., Flis, W.J. and Konopatski, K.L. "Adiabatic shear band formation in explosives due to impact", in "Shock-Wave and High-Strain-Rate Phenomena in Materials", M.A. Meyers, L.E. Murr and K.P. Staudhammer (Eds.), pp. 657–669, New York, NY, Marcel Dekker.

1992, Duffy, J. and Chi, Y.C. "On the measurement of local strain and temperature during the formation of adiabatic shear bands" *Mater. Sci. Eng. A* **157** 195–210.

1992, Duszek, M.K., Perzyna, P. and Stein, E. "Adiabatic shear band localization in elastic–plastic damaged solids" *Int. J. Plast.* **8** 361–384.

1992, Gotoh, M. and Ohno, M. "An evidence of melting along adiabatic shear band in a high-speed shearing process", in "Mechanical Behaviour of Materials 6. Vol. 1", M. Jono and T. Inoue (Eds.), pp. 487–492, Oxford, Pergamon.

1992, Grady, D.E. "Properties of an adiabatic shear-band process zone" *J. Mech. Phys. Solids* **40** 1197–1216.

1992, Krejci, J., Brezina, J. and Buchar, J. "On the substructure of adiabatic shear bands" *Scr. Metall. Mater.* **27** 611–616.

1992, Meunier, Y., Roux, R. and Moureaud, J. "Survey of adiabatic shear phenomena in armor steels with perforation", in "Shock-Wave and High-Strain-Rate Phenomena in Materials", M.A. Meyers, L.E. Murr and K.P. Staudhammer (Eds.), pp. 637–644, New York, NY, Marcel Dekker.

1992, Shawki, T.G. "The phenomenon of shear localization in dynamic viscoplasticity" *Appl. Mech. Rev.* **45** S46–S61.

1992, Taluts, G.G., Eshchenko, R.N., Borychev, A.N., Muryshev, E.Y. and Sinitsyn, A.V. "The microstructure of adiabatic shear bands" *Phys. Met. Metall.* **73** 357–361.

1992, Walter, J.W. "Numerical experiments on adiabatic shear band formation in one dimension" *Int. J. Plast.* **8** 657–694.

1992, Wang, L.L. and Bao, H. "A strain-localization analysis for adiabatic shear band at different environmental temperatures", in "Mechanical Behaviour of Materials 6. Vol. 1", M. Jono and T. Inoue (Eds.), pp. 479–486, Oxford, Pergamon.

1993, Cho, K.-M., Lee, S., Nutt, S.R. and Duffy, J. "Adiabatic shear band formation during dynamic torsional deformation of an HY-100 steel" *Acta Metall. Mater.* **41** 923–932.

1993, Duszek-Perzyna, M.K. and Perzyna, P. "Adiabatic shear band localization in elastic–plastic single crystals" *Int. J. Solids Struct.* **30** 61–89.

1993, Lee, S., Cho, K.-M., Lee, C.S. and Choo, W.Y. "Microstructural study of adiabatic shear band formed by ballistic impact in an HY-100 steel" *Metall. Trans. A* **24** 2217–2224.

1993, Ning, Q. "Numerical and experimental simulation of adiabatic shear localization in tantalum and armco iron", Ph.D. Thesis, University of Waterloo, Canada.

1993, Shahan, A.R. and Taheri, A.K. "Adiabatic shear bands in titanium and titanium alloys: A critical review" *Mater. Des.* **14** 243–250.

1993, Silling, S.A. "Shear band formation and self-shaping penetrators", report no. SAND92-2692, Sandia National Laboratories.

1993, Subhash, G. and Ravichandran, G. "High strain rate behavior and localization in hafnium", in "Experimental Techniques in the Dynamics of Deformable Solids" (*AMD* **165**), ed. K.T. Ramesh, pp. 79–87, publ. New York, American Society of Mechanical Engineers.

1994, Bai, Y.L., Xue, Q., Xu, Y.B. and Shen, L.T. "Characteristics and microstructure in the evolution of shear localization in Ti–6Al–4V alloy" *Mech. Mater.* **17** 155–164.

1994, Bair, S., Qureshi, F. and Khonsari, M. "Adiabatic shear localization in a liquid lubricant under pressure" *Trans. ASME: J. Tribol.* **116** 705–709.

1994, Cepus, E., Liu, C.D. and Bassim, M.N. "The effect of microstructure on the mechanical properties and adiabatic shear band formation in a medium carbon steel" *J. Phys. IV* **4** (C8), 553–558.

1994, Chung, D.-T., Moon, S.-K. and Yoo, Y.-H. "Numerical and experimental study of the formation of adiabatic shear band" *J. Phys. IV* **4**(C8), 547–552.

1994, Deltort, B. "Experimental and numerical aspects of adiabatic shear in a 4340 steel" *J. Phys. IV* **4**(C8), 447–452.

1994, Duszek-Perzyna, M.K. and Perzyna, P. "Analysis of the influence of different effects on criteria for adiabatic shear band localization in inelastic solids", in "Material Instabilities: Theory and Application", R.C. Batra and H.M. Zbib (Eds.), pp. 59–85, New York, NY, American Society of Mechanical Engineers.

1994, Grady, D.E. "Dissipation in adiabatic shear bands" *Mech. Mater.* **17** 289–294.

1994, Hu, B.-Y., Dong, Q.-D., Han, C.-S., Wang, D.-S. and Hu, H.-B. "Microanalysis of adiabatic shear fracture in explosive-filled cylinders", in "High Pressure Science and Technology 1993", S.C. Schmidt, J.W. Shaner, G.A. Samara and M. Ross (Eds.), pp. 1132–1229, New York, NY, American Institute of Physics.

1994, Katz, Y. and Bussiba, A. "On the propensity to adiabatic shear, localized instability or microcracking onset", in "Structures Under Shock and Impact III", P.S. Bulson (Ed.), pp. 471–478, Southampton, Computational Mechanics Publications.

1994, Klepaczko, J.R. "Some results and new experimental technique in studies of adiabatic shear bands" *Arch. Mech.* **46** 201–229.

1994, Magness, L.S., Jr. "High strain rate deformation behaviors of kinetic energy penetrator materials during ballistic impact" *Mech. Mater.* **17** 147–156.

1994, Mason, J.J., Rosakis, A.J. and Ravichandran, G. "Full field measurements of the dynamic deformation field around a growing adiabatic shear band at the tip of a dynamically loaded crack or notch" *J. Mech. Phys. Solids* **42** 1679–1698.

1994, Meyer, L.W., Staskewitsch, E. and Burblies, A. "Adiabatic shear failure under biaxial dynamic compression/shear loading" *Mech. Mater.* **17** 203–214.

1994, Meyers, M.A. "Dynamic Behavior of Materials", New York, NY, Wiley.

1994, Nemat-Nasser, S., Li, Y.-F. and Isaacs, J.B. "Experimental/computational evaluation of flow stress at high strain rates with application to adiabatic shear banding" *Mech. Mater.* **17** 111–134.

1994, Perzyna, P. "Adiabatic shear band localization fracture of solids in dynamic loading processes" *J. Phys. IV* **4**(C8), 441–446.

1994, Perzyna, P. "Instability phenomena and adiabatic shear band localization in thermoplastic flow processes" *Acta Mech.* **106** 173–205.

1994, Wang, L., Dong, X., Hu, S. and Yu, J. "A macro- and microscopic study of adiabatic shearing extension of mode-II crack at dynamic loading" *J. Phys. IV* **4**(C8), 465–470.

1994, Wright, T.W. "Towards a defect invariant basis for susceptibility to adiabatic shear bands" *Mech. Mater.* **17** 215–222.

1994, Zhou, M., Rosakis, A.J. and Ravichandran, G. "Adiabatic shear band formation in asymmetrically impacted prenotched plates: An investigation of temperature signatures", in "Material Instabilities: Theory and Application", R.C. Batra and H.M. Zbib (Eds.), pp. 35–49, New York, NY, American Society of Mechanical Engineers.

1994, Zurek, A.K. "The study of adiabatic shear band instability in a pearlitic 4340 steel using a dynamic punch test" *Metall. Mater. Trans. A* **25** 2483–2489.

1995, Batra, R.C., Zhang, X. and Wright, T.W. "Critical strain ranking of 12 materials in deformations involving adiabatic shear bands" *Trans. ASME: J. Appl. Mech.* **62** 252–255.

1995, Batra, R.C. and Adulla, C. "Effect of prior quasi-static loading on the initiation and growth of dynamic adiabatic shear bands" *Arch. Mech.* **47** 485–498.

1995, Chen, L., Li, Y.-C. and Wang, L.L. "Analysis of adiabatic shear plugging of thermo-viscoplastic plates" *DYMAT J.* **2** 167–181.

1995, Duszek-Perzyna, M.K. and Perzyna, P. "Acceleration waves in analysis of adiabatic shear band localization", in "IUTAM Symp. on Nonlinear Waves in Solids", J.L. Wegner and F.R. Norwood (Eds.), pp. 128–135, New York, NY, American Society of Mechanical Engineers.

1995, Hines, J.A. and Vecchio, K.S. "Dynamic recrystallization in adiabatic shear bands in shock-loaded copper", in "Metallurgical and Materials Applications of Shock-Wave and High-Strain-Rate Phenomena", L.E. Murr, K.P. Staudhammer and M.A. Meyers (Eds.), pp. 421–428, Amsterdam, Elsevier.

1995, Irick, E., Heinol, C., Clayton, T., Hashemi, J., Cardenas Garcia, J.F. and Sadhneni, R. "Numerical and experimental investigation of adiabatic shear bands in metals under low velocity impact conditions" *J. Mater. Eng. Perform.* **4** 709–716.

1995, Lee, C.-G., Lee, Y.-J. and Lee, S. "Observation of adiabatic shear bands formed by ballistic impact in aluminum–lithium alloys" *Scr. Metall. Mater.* **32** 821–826.

1995, White, R.N. and Pickup, I.M. "The effect of the constitutive behaviour of a range of steels on adiabatic shear propensity", in "Proc. 9th DYMAT Technical Conference: Material and Structural Modelling in Collision Research", J. Najar (Ed.), paper 12, Arcueil, France, DYMAT.

1995, Wright, T.W. "Scaling laws for adiabatic shear bands" *Int. J. Solids Struct.* **32** 2745–2750.

1995, Zhu, H.T., Zbib, H.M. and Aifantis, E.C. "On the role of strain gradients in adiabatic shear banding" *Acta Mech.* **111** 111–124.

1995, Walley, S.M., Xing, D. and Field, J.E. "Mechanical properties of three transparent polymers in compression at a very high rate of strain", in "Impact and Dynamic Fracture of Polymers and Composites", J.G. Williams and A. Pavan (Eds.), pp. 289–303, London, Mechanical Engineering Publications.

1995, Woodward, R.L., Alkemade, S.J. and Magness, L.S. "Strain localization and fracture in dynamic dumbbell compression of tungsten alloys", in "Proc. 2nd Int. Conf. on Tungsten and Refractory Metals", A. Bose and R.J. Dowding (Eds.), pp. 431–438, Princeton, NJ, Metal Powders Industries Federation.

1995, Xue, Q., Shen, L.T. and Bai, Y.L. "A modified split Hopkinson torsional bar in studying shear localization" *Meas. Sci. Technol.* **6** 1557–1565.

1995, Xue, Q., Shen, L.T. and Bai, Y.L. "Elimination of loading reverberation in the split Hopkinson torsional bar" *Rev. Sci. Instrum.* **66** 5298–5304.

1996, Batra, R.C., Adulla, C. and Wright, T.W. "Effect of defect shape and size on the initiation of adiabatic shear bands" *Acta Mech.* **116** 239–243.

1996, Duszek-Perzyna, M.K. and Perzyna, P. "Adiabatic shear localization of inelastic single crystals in symmetric double-slip process" *Arch. Appl. Mech.* **66** 369–384.

1996, Klepaczko, J.R. and Rezaig, B. "A numerical study of adiabatic shear banding in mild steel by dislocation mechanics based constitutive relations" *Mech. Mater.* **24** 125–139.

1996, Lee, C.-G., Park, W.-J., Lee, S.-H. and Shin, K.-S. "Microstructure of adiabatic shear band formed by ballistic impact in a Weldalite 049 alloy (in Korean)" *J. Korean Inst. Met. Mater.* **39** 1108–1115.

1996, Li, Q., Ma, V. and Lai, Z. "Strain evaluation model of adiabatic shear band produced by orthogonal cutting in high strength low alloy steel" *Mater. Sci. Technol.* **12** 351–354.

1996, Wright, T.W. and Walter, J.W. "The asymptotic structure of an adiabatic shear band in antiplane motion" *J. Mech. Phys. Solids* **44** 77–97.

1996, Wright, T.W. and Ockendon, H. "A scaling law for the effect of inertia on the formation of adiabatic shear bands" *Int. J. Plast.* **12** 927–934.

1996, Xu, Y.B., Bai, Y.L., Xue, Q. and Shen, L.T. "Formation, microstructure and development of the localized shear deformation in low carbon steels" *Acta Mater.* **44** 1917–1926.

1996, Yang, Y., Xinming, Z., Zhenghua, L. and Qingyun, L. "Adiabatic shear band on the titanium side in the Ti/mild steel explosive cladding interface" *Acta Mater.* **44** 561–565.

1996, Zhang, B.F., Liu, Y.J., Shen, W.C., Wang, Y.F., Tang, X.Y. and Wang, X.D. "A study of the behavior of adiabatic shear bands in impact wear" *Wear* **198** 287–292.

1997, Chen, J. and Davies, M.C.D. "Analysis of energy absorption of adiabatic shear plugging on thermoviscoplastic targets" *Int. J. Eng. Sci.* **35** 365–373.

1997, Dobromyslov, A.V., Kozlov, E.A. and Taluts, N.I. "Formation and features of adiabatic shear bands in Zr–Nb alloys in spherical stress waves" *J. Phys. IV* **7**(C3), 963–967.

1997, Duszek-Perzyna, M.K., Korbell, K. and Perzyna, P. "Adiabatic shear band localization in single crystals under dynamic loading processes" *Arch. Mech.* **49** 1069–1090.

1997, Estrin, Y. and Molinari, A. "Constitutive modeling of high strain rate deformation: Application to adiabatic shear banding", in "Advances in Fracture Research", B.L. Karihaloo, Y.-W. Mai, M.I. Ripley and R.O. Ritchie (Eds.), pp. 2755–2769, Oxford, Pergamon.

1997, Glema, A., Kakol, W. and Lodygowski, T. "Numerical modelling of adiabatic shear band formation in a twisting test" *Eng. Trans.* **45** 419–431.

1997, Han, X.P., Zhang, T.H. and Zhao, Z.H. "Experimental investigation of adiabatic shear bands formation in TNT explosives under impact" *J. Energ. Mater.* **15** 185–191.

1997, Hines, J.A. and Vecchio, K.S. "Recrystallization kinetics within adiabatic shear bands" *Acta Mater.* **45** 635–649.

1997, Lee, W.S. and Lin, C.F. "Adiabatic shear fracture of titanium alloy subjected to high strain rate and high temperature loadings" *J. Phys. IV* **7**(C3), 855–860.

1997, Mason, C.R., Worswick, M.J. and Gallagher, P.J. "Adiabatic shear in Remco iron and quenched and temperated 4340 steel" *J. Phys. IV* **7**(C3), 827–832.

1997, Mason, C.R. "Prediction of adiabatic shear in 4340 steel, remco and armco irons", M.Eng. Thesis, University of Carleton, Canada.

1997, Mazeau, C., Beylat, L., Longère, P. and Louvigné, P.V. "On the quantitative evaluation of adiabatic shear banding sensitivity of various titanium alloys" *J. Phys. IV* **7**(C3), 429–434.

1997, Mercier, S. and Molinari, A. "Propagation of adiabatic shear bands" *J. Phys. IV* **7**(C3), 803–808.

1997, Minot, J.L. and Fressengeas, C. "On the critical nominal strain in adiabatic shear banding" *J. Phys. IV* **7**(C3), 785–790.

1997, Molinari, A. "Collective behavior and spacing of adiabatic shear bands" *J. Mech. Phys. Solids* **45** 1551–1576.

1997, Nguyen, C.H. "Analysis of some thermal instability criteria in the adiabatic shear banding process" *J. Phys. IV* **7**(C3), 849–853.

1997, Wright, T.W. and Ravichandran, G. "Canonical aspects of adiabatic shear bands" *Int. J. Plast.* **13** 309–325.

1997, Zhang, B., Shen, W., Liu, Y., Tang, X. and Wang, Y. "Microstructures of surface white layer and internal white adiabatic shear band" *Wear* **211** 164–168.

1998, Batra, R.C. and Wilson, N.M. "Adiabatic shear bands in plane strain deformations of a WHA" *Int. J. Plast.* **14** 43–60.

1998, Batra, R.C. and Stevens, J.B. "Adiabatic shear bands in axisymmetric impact and penetration problems" *Comput. Methods Appl. Mech. Eng.* **151** 325–342.

1998, Chen, L., Li, Y.-C. and Wang, L.-L. "A computational investigation of adiabatic shear plugging based on thermo-viscoplastic instability" *Acta Mech.* **126** 127–138.

1998, Dinzart, F. and Molinari, A. "Structure of adiabatic shear bands in thermo-viscoplastic materials" *Eur. J. Mech. A. Solids* **17** 923–938.

1998, Hines, J.A., Vecchio, K.S. and Ahzi, S. "A model for microstructure evolution in adiabatic shear bands" *Metall. Mater. Trans. A* **29** 191–203.

1998, Kim, D.S., Nemat-Nasser, S., Isaacs, J.B. and Lischer, D. "Adiabatic shearband in WHA in high strain-rate compression" *Mech. Mater.* **28** 227–236.

1998, Kim, D.-K., Lee, S. and Baek, W.H. "Microstructural study of adiabatic shear bands formed by high-speed impact in a tungsten heavy alloy penetrator" *Mater. Sci. Eng. A* **249** 197–205.

1998, Kozlov, E.A., Elkin, V.M., Litvinov, B.V., Dobromyslov, A.V., Taluts, N.I. and Kazantseva, N.V. "Formation features and structure of the adiabatic shear bands arising in zirconium in spherical stress waves" *Dokl. Phys.* **43** 309–312.

1998, Lee, C.G., Park, W.J., Lee, S. and Shin, K.S. "Microstructural development of adiabatic shear bands formed by ballistic impact in a WELDALITE 049 alloy" *Metall. Mater. Trans. A* **29** 477–483.

1998, Liao, S.-C. and Duffy, J. "Adiabatic shear bands in a Ti–6Al–4V titanium alloy" *J. Mech. Phys. Solids* **46** 2201–2231.

1998, Minnaar, K. and Zhou, M. "An analysis of the dynamic shear failure resistance of structural metals" *J. Mech. Phys. Solids* **46** 2155–2170.

1998, Nerzak, T. "A material model describing initiation and propagation of adiabatic shear bands" *Z. Angew. Math. Mech.* **78** S643–S644.

1998, Nesterenko, V.F., Meyers, M.A. and Wright, T.W. "Self-organization in the initiation of adiabatic shear bands" *Acta Mater.* **46** 327–340.

1998, Nesterenko, V.F., Xue, Q., Meyers, M.A. and Wright, T.W. "Dependence of characteristic spacings in a system of adiabatic shear bands on material property", in "Plasticity 99: Constitutive and Damage Modeling of Inelastic Deformation and Phase Transformation", A.S. Khan (Ed.), pp. 507–510, Fulton, MD, Neat Press.

1998, Perzyna, P. and Korbel, K. "Analysis of the influence of various effects on criteria for adiabatic shear band localization in single crystals" *Acta Mech.* **129** 31–62.

1998, Roessig, K.M. and Mason, J.J. "Adiabatic shear localization in the impact of edge-notched specimens" *Exp. Mech.* **38** 196–203.

1998, Stevens, J.B. and Batra, R.C. "Adiabatic shear bands in the Taylor impact test for a WHA rod" *Int. J. Plast.* **14** 841–854.

1998, Zhang, B., Shen, W., Liu, Y. and Zhang, R. "Adiabatic shear bands in impact wear" *J. Mater. Sci. Lett.* **17** 765–767.

1998, Zhang, B., Shen, W., Liu, Y. and Zhang, R. "Some factors affecting adiabatic shear banding in impact wear" *Wear* **214** 259–263.

1999, Bassim, M.N. and Bassim, N.D. "Factors affecting occurrence of adiabatic shear bands at high strain rates", in "Proc. 8th Int. Conf. on the Mechanical Behaviour of Materials", F. Ellyin and J.W. Provan (Eds.), pp. 508–511.

1999, Chichili, D.R. and Ramesh, K.T. "Recovery experiments for adiabatic shear localization: A novel experimental technique" *Trans. ASME: J. Appl. Mech.* **66** 10–20.

1999, Feng, H. and Bassim, M.N. "Finite element modeling of the formation of adiabatic shear bands in AISI 4340 steel" *Mater. Sci. Eng. A* **266** 255–260.

1999, Fleck, N.A. "Adiabatic shear instability: Theory", in "Mechanical Properties and Testing of Polymers", G.M. Swallowe (Ed.), pp. 15–19, Dordrecht, The Netherlands, Kluwer.

1999, Li, Q.M. and Jones, N. "Shear and adiabatic shear failures in an impulsively loaded fully clamped beam" *Int. J. Impact Eng.* **22** 589–607.

1999, Roessig, K.M. and Mason, J.J. "Adiabatic shear localization in the dynamic punch test. 1: Experimental investigation" *Int. J. Plast.* **15** 241–262.

1999, Roessig, K.M. and Mason, J.J. "Adiabatic shear localization in the dynamic punch test. 2: Numerical simulations" *Int. J. Plast.* **15** 263–283.

1999, Walley, S.M. "Adiabatic shear instability: Observations and experimental techniques", in "Mechanical Properties and Testing of Polymers", G.M. Swallowe (Ed.), pp. 10–14, Dordrecht, The Netherlands, Kluwer.

1999, Wei, Z., Hu, S., Yu, J. and Li, Y. "Adiabatic shear localization of pre-twisted tungsten heavy alloy under dynamic compression/shear loading", in "Impact Response of Materials and Structures", V.P.W. Shim, S. Tanimura and C.T. Lim (Eds.), pp. 178–183, Oxford, Oxford University Press.

2000, Chen, Z.H., Tang, C.Y. and Lee, T.C. "Numerical simulation of non-adiabatic shear banding in metal forming process using finite element method" *Key Eng. Mater.* **177** 553–558.

2000, Clos, R., Schreppel, U. and Veit, P. "Experimental investigation of adiabatic shear band formation in steels" *J. Phys. IV* **10**(PR 9), 257–262.

2000, Dai, L.H., Ling, Z. and Bai, Y.L. "Strain gradient effect on initiation of adiabatic shear localization in metal matrix composites" *Key Eng. Mater.* **177** 401–406.

2000, Dobromyslov, A.V., Taluts, N.I., Kazantseva, N.V. and Kozlov, E.A. "Formation of adiabatic shear bands and instability of plastic flow in zirconium and Zr–Nb alloys in spherical stress waves" *Scr. Mater.* **42** 61–71.

2000, Faure, L., Klepaczko, J.R. and Sutter, G. "Experimental study of adiabatic shear band propagation in impacted plates" *J. Phys. IV* **10**(PR 9), 221–226.

2000, Lebouvier, A.-S., Lipinski, P. and Molinari, A. "Numerical study of the propagation of an adiabatic shear band" *J. Phys. IV* **10**(PR 9), 403–408.

2000, Molinari, A. "Adiabatic shear banding as an example of viscoplastic flow instability", in "Continuum Thermomechanics", G.A. Maugin, R. Drouot and F. Sidoroff (Eds.), pp. 313–330, Dordrecht, The Netherlands, Kluwer.

2000, Ranc, N., Pina, V. and Hervé, P. "Optical measurements of phase transition and temperature in adiabatic shear bands in titanium alloys" *J. Phys. IV* **10**(PR 9), 347–352.

2000, Tóth, L.S., Hildenbrand, A. and Molinari, A. "Dynamic recrystallization in adiabatic shear bands" *J. Phys. IV* **10**(PR 9), 365–370.

2000, Wei, Z., Yu, J., Hu, S. and Li, Y. "Influence of microstructure on adiabatic shear localization of pre-twisted tungsten heavy alloys" *Int. J. Impact Eng.* **24** 747–758.

2000, Wei, Z., Li, Y., Li, J. and Hu, S. "Formation mechanism of adiabatic shear band in tungsten heavy alloys (in Chinese)" *Acta Metall. Sin.* **36** 1263–1268.

2000, Xue, Q., Nesterenko, V.F. and Meyers, M.A. "Self-organization of adiabatic shear bands in titanium, Ti–6Al–4V and stainless steel", in "Shock Compression of Condensed Matter – 1999", M.D. Furnish, L.C. Chhabildas and R.S. Hixson (Eds.), pp. 431–434, Melville, NY, American Institute of Physics.

2000, Yu, J., Dong, X. and Zhang, J. "A study of adiabatic shear plugging in Ti6Al4V alloy" *Key Eng. Mater.* **177** 387–392.

2001, Bassim, M.N. "Study of the formation of adiabatic shear bands in steels" *J. Mater. Process. Technol.* **119** 234–236.

2001, Guduru, P.R., Ravichandran, G. and Rosakis, A.J. "Observations of transient high temperature vortical microstructures in solids during adiabatic shear banding" *Phys. Rev. E* **64** 036128.

2001, Li, J.-R., Yu, J.-L. and Wei, Z.-G. "Influence of specimen geometry on adiabatic shear sensitivity of tungsten heavy alloys: A numerical simulation", in "Proc. 4th Int. Symp. on Impact Engineering", A. Chiba, S. Tanimura and K. Hokamoto (Eds.), pp. 535–540, Amsterdam, Elsevier.

2001, Li, S., Liu, W.-K., Qian, D., Guduru, P.R. and Rosakis, A.J. "Dynamic shear band propagation and microstructure of adiabatic shear band" *Comput. Methods Appl. Mech. Eng.* **191** 73–92.

2001, Mason, C. and Worswick, M.J. "Adiabatic shear in annealed and shock-hardened iron and in quenched and tempered 4340 steel" *Int. J. Fract.* **111** 29–51.

2001, Meyers, M.A. "Plasticity: Adiabatic shear localization", in "Encyclopedia of Materials: Science and Technology", K.H.J. Buschow, R.W. Cahn, M.C. Flemings,

B. Illschner, E.J. Kramer and S. Mahajan (Eds.), pp. 7093–7103, Amsterdam, Elsevier.

2001, Pérez-Prado, M.T., Hines, J.A. and Vecchio, K.S. "Microstructural evolution in adiabatic shear bands in tantalum and Ta–W alloys" *Acta Mater.* **49** 2905–2917.

2001, Wei, Z., Yu, J., Li, J., Li, Y. and Hu, S. "Influence of stress condition on adiabatic shear localization of tungsten heavy alloys" *Int. J. Impact Eng.* **26** 843–852.

2001, Wright, T.W. "Adiabatic shear bands: Modeling and scaling laws", in "Fundamental Issues and Applications of Shock-Wave and High-Strain-Rate Phenomena", K.P. Staudhammer, L.E. Murr and M.A. Meyers (Eds.), pp. 457–466, New York, NY, Elsevier.

2002, Bäker, M., Rösler, J. and Siemers, C. "A finite element model of high speed metal cutting with adiabatic shearing" *Comput. Struct.* **80** 495–513.

2002, Bonnet, A.S., Lipinski, P. and Molinari, A. "Propagation of adiabatic shear bands: Dimensional analysis", in "Behaviour of Materials at High Strain Rates: Numerical Modelling", F.G. Benitez (Ed.), pp. 13–23, Saint Louis, France, DYMAT.

2002, Bonnet-Lebouvier, A.-S., Molinari, A. and Lipinski, P. "Analysis of the dynamic propagation of adiabatic shear bands" *Int. J. Solids Struct.* **39** 4249–4269.

2002, Burns, T.J. and Davies, M.A. "On repeated adiabatic shear band formation during high speed machining" *Int. J. Plast.* **18** 487–506.

2002, Meyers, M.A., Xue, Q. and Nesterenko, V.F. "Evolution in the patterning of adiabatic shear bands", in "Shock Compression of Condensed Matter – 2001", M.D. Furnish, N.N. Thadhani and Y. Horie (Eds.), pp. 567–570, Melville, NY, American Institute of Physics.

2002, Meyers, M.A., Perez-Prado, M.T., Xue, Q., Xu, Y. and McNelley, T.R. "Microstructural evolution in adiabatic shear localization in stainless steel", in "Shock Compression of Condensed Matter – 2001", M.D. Furnish, N.N. Thadhani and Y. Horie (Eds.), pp. 571–574, Melville, NY, American Institute of Physics.

2002, Molinari, A., Musquar, C. and Sutter, G. "Adiabatic shear banding in high speed machining of Ti6Al4V: Experiments and modeling" *Int. J. Plast.* **18** 443–459.

2002, Murr, L.E., Trillo, E.A., Pappu, S. and Kennedy, C. "Adiabatic shear bands and examples of their role in severe plastic deformation" *J. Mater. Sci.* **37** 3337–3360.

2002, Wright, T.W. "The Physics and Mathematics of Adiabatic Shear Bands", Cambridge, Cambridge University Press.

2003, Couque, H. "A hydrodynamic hat specimen to investigate pressure and strain rate dependence on adiabatic shear band formation" *J. Phys. IV* **110** 423–428.

2003, Duan, Z., Cheng, G., Li, S. and Huang, D. "Microstructures and adiabatic shear bands formed in steel plate under high rate impact (in Chinese)" *Acta Metall. Sin.* **39** 486–491.

2003, Duan, Z.Q., Li, S.X. and Huang, D.W. "Microstructures and adiabatic shear bands formed by ballistic impact in steels and tungsten alloy" *Fatigue Fract. Eng. Mater. Struct.* **26** 1119–1126.

2003, Li, J.R., Yu, J.L. and Wei, Z.G. "Influence of specimen geometry on adiabatic shear instability of tungsten heavy alloys" *Int. J. Impact Eng.* **28** 303–314.

2003, Li, Q., Xu, Y. and Bassim, M.N. "Dynamic mechanical properties in relation to adiabatic shear band formation in titanium alloy Ti17" *Mater. Sci. Eng. A* **358** 128–133.

2003, Longère, P., Dragon, A., Trumel, H., de Rességuier, T., Deprince, X. and Petitpas, E. "High strain rate model involving damage by adiabatic shear banding and related anisotropy" *J. Phys. IV* **110** 317–322.

2003, Longère, P., Dragon, A., Trumel, H., de Rességuier, T., Deprince, X. and Petitpas, E. "Nonlinear impact dynamics including adiabatic shear banding effects", in "Proc.

Fifth Int. Symp. on High Dynamic Pressures, Vol. 1", pp. 311–320, Paris, France, Commissariat à l'Energie Atomique.

2003, Longère, P., Dragon, A., Trumel, H., de Resseguier, T., Deprince, X. and Petitpas, E. "Modelling adiabatic shear banding via damage mechanics approach" *Arch. Mech.* **55** 3–38.

2003, Meyers, M.A., Xu, Y.B., Xue, Q., Pérez-Prado, M.T. and McNelley, T.R. "Microstructural evolution in adiabatic shear localization in stainless steel" *Acta Mater.* **51** 1307–1325.

2003, Chwalik, P., Klepaczko, J.R. and Rusinek, A. "Impact shear-numerical analyses of ASB evolution and failure of Ti6Al4V alloy" *J. Phys. IV* **110** 257–262.

2003, Radford, D.D., Walley, S.M., Church, P. and Field, J.E. "Dynamic upsetting and failure of metal cylinders: Experiments and analysis" *J. Phys. IV* **110** 263–268.

2004, Batra, R.C. and Zhang, G.M. "Analysis of adiabatic shear bands in elasto-thermo-viscoplastic materials by modified smoothed-particle hydrodynamics (MSPH) method" *J. Comput. Phys.* **201** 172–190.

2004, Charalambakis, N. and Baxevanis, T. "Adiabatic shearing of non-homogeneous thermoviscoplastic materials" *Int. J. Plast.* **20** 899–914.

2004, Chichili, D.R., Ramesh, K.T. and Hemker, K.J. "Adiabatic shear localization in α-titanium: Experiments, modeling and microstructural evolution" *J. Mech. Phys. Solids* **52** 1889–1910.

2004, Dai, L.H., Liu, L.F. and Bai, Y.L. "Effect of particle size on the formation of adiabatic shear band in particle-reinforced MMCs" *Mater. Lett.* **58** 1773–1776.

2004, Dai, L.H., Liu, L.F. and Bai, Y.L. "Formation of adiabatic shear band in metal matrix composites" *Int. J. Solids Struct.* **41** 5979–5993.

2004, Daridon, L., Oussouaddi, O. and Ahzi, S. "Influence of the material constitutive models on the adiabatic shear band spacing: MTS, power law and Johnson–Cook models" *Int. J. Solids Struct.* **41** 3109–3124.

2004, Grujicic, M., Zhao, C.L., DeRosset, W.S. and Helfritch, D. "Adiabatic shear instability based mechanism for particles/substrate bonding in the cold-gas dynamic-spray process" *Mater. Des.* **25** 681–688.

2004, Li, Q., Zmudzki, P., Alameeri, S. and Bassim, M.N. "Morphology of adiabatic shear bands in cylindrical specimens of AISI 4340 steel impacted by Hopkinson pressure bar" *Mater. Sci. Technol.* **20** 676–678.

2004, Longère, P., Dragon, A., Trumel, H., de Rességuier, T. and Deprince, X. "3D finite strain modelling and numerical simulation of propagating adiabatic shear bands", in "Crashworthiness and Constitutive Relationships for Engineering Materials", A. Rusinek (Ed.), pp. 57–62, Saint Louis, France, DYMAT.

2004, Wei, Q., Kecskes, L., Jiao, T., Hartwig, K.T., Ramesh, K.T. and Ma, E. "Adiabatic shear banding in ultrafine-grained iron processed by severe plastic deformation" *Acta Mater.* **52** 1859–1869.

2004, Fermen-Coker, M. "Implementation of Schoenfeld–Wright failure criterion into a three-dimensional adiabatic shear band model in CTH", report no. ARL-TR 3284, White Sands Missile Range, NM, Army Research Laboratory.

2005, Batra, R.C. and Lear, M.H. "Adiabatic shear banding in plane strain tensile deformations of 11 thermoelastoviscoplastic materials with finite thermal wave speed" *Int. J. Plast.* **21** 1521–1545.

2005, Chen, X.W., Li, Q.M. and Fan, S.C. "Initiation of adiabatic shear failure in a clamped circular plate struck by a blunt projectile" *Int. J. Impact Eng.* **31** 877–893.

2005, Dai, L.H., Yan, M., Liu, L.F. and Bai, Y.L. "Adiabatic shear banding instability in bulk metallic glasses" *Appl. Phys. Lett.* **87** 141916.

2005, Hershberger, J., Ajayi, O.O., Zhang, J., Yoon, H. and Fenske, G.R. "Evidence of scuffing initiation by adiabatic shear instability" *Wear* **258** 1471–1478.

2005, Lesuer, D.R., Syn, C.K. and Sherby, O.D. "Severe plastic deformation through adiabatic shear banding in Fe–C steels" *Mater. Sci. Eng. A* **410** 222–225.

2005, Li, J., Li, M., Guo, X. and Li, S. "Formation of adiabatic shear bands in the fatigued copper single crystal deformed at high strain rate" *Acta Metall. Sin.* **41** 161–166.

2005, Lim, J. and Huh, H. "Experiments on formation of the adiabatic shear band in sheet metal" *Met. Mater. Int.* **11** 11–18.

2005, Longère, P., Dragon, A., Trumel, H. and Deprince, X. "Adiabatic shear banding induced degradation in a thermoelastic/viscoplastic material under dynamic loading" *Int. J. Impact Eng.* **32** 285–320.

2005, Ohbuchi, Y. and Obikawa, T. "Adiabatic shear in chip formation with negative rake angle" *Int. J. Mech. Sci.* **47** 1377–1392.

2005, Rittel, D. "Adiabatic shear failure of a syntactic polymeric foam" *Mater. Lett.* **59** 1845–1848.

2005, Syn, C.K., Lesuer, D.R. and Sherby, O.D. "Microstructure in adiabatic shear bands in a pearlitic ultrahigh carbon steel" *Mater. Sci. Technol.* **21** 317–324.

2005, Yang, Y., Wang, B.F., Hu, B., Hu, K. and Li, Z.G. "The collective behavior and spacing of adiabatic shear bands in the explosive cladding plate interface" *Mater. Sci. Eng. A* **398** 291–296.

2005, Duan, C.Z. and Wang, M.J. "Characteristics of adiabatic shear bands in the orthogonal cutting of $30CrNi_3MoV$ steel" *J. Mater. Process. Technol.* **168** 102–106.

2005, Fermen-Coker, M. "Numerical simulation of adiabatic shear bands in Ti6Al4V alloy due to fragment impact", report no. ARL-RP 91, White Sands Missile Range, NM, Army Research Laboratory.

2006, Batra, R.C. and Love, B.M. "Multiscale analysis of adiabatic shear bands in tungsten heavy alloy particulate composites" *Int. J. Multiscale Comput. Eng.* **4** 95–114.

2006, Batra, R.C. and Love, B.M. "Consideration of microstructural effects in the analysis of adiabatic shear bands in a tungsten heavy alloy" *Int. J. Plast.* **22** 1858–1878.

2006, Hwang, B., Lee, S., Kim, Y.C., Kim, N.J. and Shin, D.H. "Microstructural development of adiabatic shear bands in ultrafine-grained low-carbon steels fabricated by equal channel angular pressing" *Mater. Sci. Eng. A* **441** 308–320.

2006, Longère, P., Dragon, A., Trumel, H. and Deprince, X. "Modelling and simulation of dynamic anisotropic degradation of structures under adiabatic shear banding" *J. Phys. IV* **134** 139–145.

2006, Odeshi, A.G., Bassim, M.N. and Al-Ameeri, S. "Effect of heat treatment on adiabatic shear bands in a high-strength low alloy steel" *Mater. Sci. Eng. A* **419** 69–75.

2006, Rittel, D., Wang, Z.G. and Merzer, M. "Adiabatic shear failure and dynamic stored energy of cold work" *Phys. Rev. Lett.* **96** 075502.

2006, Rittel, D., Wang, Z.G. and Merzer, M. "Some experiments on adiabatic shear failure" *J. Phys. IV* **134** 835–838.

2006, Sunwoo, A.J., Becker, R., Goto, D.M., Orzechowski, T.J., Springer, H.K., Syn, C.K. and Zhou, J. "Adiabatic shear band formation in explosively driven Fe–Ni–Co alloy cylinders" *Scr. Mater.* **55** 247–250.

2006, Xue, Q. and Gray, G.T., III "Development of adiabatic shear bands in annealed 316L stainless steel. 1: Correlation between evolving microstructure and mechanical behavior" *Metall. Mater. Trans. A* **37** 2435–2446.

2006, Xue, Q. and Gray, G.T., III "Development of adiabatic shear bands in annealed 316L stainless steel. 2: TEM studies of the evolution of microstructure during deformation localization" *Metall. Mater. Trans. A* **37** 2447–2458.

2006, Yang, Y. and Wang, B.F. "Dynamic recrystallization in adiabatic shear band in alpha-titanium" *Mater. Lett.* **60** 2198–2202.

2006, Yang, Y. and Wang, B.F. "Microstructure evolution in adiabatic shear band in alpha-titanium" *J. Mater. Sci.* **41** 7387–7392.

2006, Yang, Y., Wang, B.F., Xiong, J., Zeng, Y., Chen, Z.P. and Yang, C.Y. "Adiabatic shear bands on the titanium side in the titanium/mild steel explosive cladding interface: Experiments, numerical simulation, and microstructure evolution" *Metall. Mater. Trans. A* **37** 3131–3137.

2006, Zhou, F., Wright, T.W. and Ramesh, K.T. "A numerical methodology for investigating the formation of adiabatic shear bands" *J. Mech. Phys. Solids* **54** 904–926.

2006, Zhou, F., Wright, T.W. and Ramesh, K.T. "The formation of multiple adiabatic shear bands" *J. Mech. Phys. Solids* **54** 1376–1400.

2006, Duan, C.Z., Wang, M.J., Pang, J.Z. and Li, G.H. "A calculational model of shear strain and strain rate within shear band in a serrated chip during high speed machining" *J. Mater. Process. Technol.* **178** 274–277.

2006, Xu, Y.B., Bai, Y.L. and Meyers, M.A. "Deformation, phase transformation and recrystallization in the shear bands induced by high-strain rate loading in titanium and its alloys" *J. Mater. Sci. Technol.* **22** 737–746.

2007, Hanina, E., Rittel, D. and Rosenberg, Z. "Pressure sensitivity of adiabatic shear banding in metals" *Appl. Phys. Lett.* **90** 021915.

2007, Li, S.H., Wang, F.C., Tab, C.W., Chen, Z.Y. and Sun, Z. "Microstructural evolution in the adiabatic shear bands of predeformation TA2 (in Chinese)" *Rare Met. Mater. Eng.* **36** 1527–1529.

2007, Lins, J.F.C., Sandim, H.R.Z., Kestenbach, H., Raabe, D. and Vecchio, K.S. "A microstructural investigation of adiabatic shear bands in an interstitial free steel" *Mater. Sci. Eng. A* **457** 205–218.

2007, Martinez, F., Murr, L.E., Ramirez, A., Lopez, N.I. and Gaytan, S.M. "Dynamic deformation and adiabatic shear microstructures associated with ballistic plug formation and fracture in Ti6Al4V targets" *Mater. Sci. Eng. A* **454** 581–589.

2007, Medyanik, S.N., Liu, W.K. and Li, S. "On criteria for dynamic adiabatic shear band propagation" *J. Mech. Phys. Solids* **55** 1439–1461.

2007, Teng, X., Wierzbicki, T. and Couque, H. "On the transition from adiabatic shear banding to fracture" *Mech. Mater.* **39** 107–125.

2007, Wang, B.F., Yang, Y., Chen, Z.P. and Zeng, Y. "Adiabatic shear bands in alpha-titanium under external explosive loading" *J. Mater. Sci.* **42** 8101–8105.

2007, Wang, B.F. and Yang, Y. "Distribution and dynamic propagation of adiabatic shear bands on the titanium side in the titanium/mold steel explosive cladding plate" *Mater. Sci. Eng. A* **452** 273–277.

2007, Wang, X.B. "Peak and average temperatures in adiabatic shear band for thermo-visco-plastic metal materials" *Key Eng. Mater.* **345** 133–136.

2007, Zeb, M.A. "Adiabatic shear bands as predictors of strain rate in high speed machining", in "Proc. 2007 SEM Ann. Conf. and Exposition on Experimental and Applied Mechanics", paper 64, Bethel, CT, Society for Experimental Mechanics.

2008, Cheng, X.W., Sun, K. and Wang, F.C. "Microstructure evolution of adiabatic shear band in Ti–15Mo–3Al–2.7Nb–0.2Si alloy (in Chinese)" *Rare Met. Mater. Eng.* **37** 1756–1759.

2008, Dai, L.H. and Bai, Y.L. "Basic mechanical behaviors and mechanics of shear banding in BMGs" *Int. J. Impact Eng.* **35** 704–716.

2008, Lee, W.-S., Lin, C.-F., Chen, T.-H. and Hwang, H.-H. "Correlation of dynamic impact properties with adiabatic shear banding behaviour in Ti–15Mo–5Zr–3Al alloy" *Mater. Sci. Eng. A* **475** 172–184.

2008, Lee, W.-S., Liu, C.-Y. and Chen, T.-H. "Adiabatic shearing behavior of different steels under extreme high shear loading" *J. Nucl. Mater.* **374** 313–319.

2008, Liu, J., Li, S., Fan, A. and Sun, H. "Effect of fibrous orientation on dynamic mechanical properties and susceptibility to adiabatic shear band of tungsten alloy fabricated through hot-hydrostatic extrusion" *Mater. Sci. Eng. A* **487** 235–242.

2008, Liu, X.Q., Tan, C.W., Zhang, J., Yu, H.G., Chen, Z.Y., Wang, F.C. and Cai, H.N. "Influence of stress state on adiabatic shear sensitivity of Ti6Al4V (in Chinese)" *Rare Met. Mater. Eng.* **37** 1522–1525.

2008, Liu, J., Li, S., Zhou, X., Zhang, Z., Zheng, H. and Wang, Y. "Adiabatic shear banding in a tungsten heavy alloy processed by hot-hydrostatic extrusion and hot torsion" *Scr. Mater.* **59** 1271–1274.

2008, Odeshi, A.G. and Bassim, M.N. "Evolution of adiabatic shear bands in a dual-phase steel at very high strain rates" *Mater. Sci. Eng. A* **488** 235–240.

2008, Ranc, N., Taravella, L., Pina, V. and Herve, P. "Temperature field measurement in titanium alloy during high strain rate loading: Adiabatic shear bands phenomenon" *Mech. Mater.* **40** 255–270.

2008, Rittel, D., Landau, P. and Venkert, A. "Dynamic recrystallization as a potential cause for adiabatic shear failure" *Phys. Rev. Lett.* **101** 165501.

2008, Rittel, D. and Wang, Z.G. "Thermomechanical aspects of adiabatic shear failure of AM50 and Ti6Al4V alloys" *Mech. Mater.* **40** 629–635.

2008, Rittel, D., Wang, Z.G. and Dorogoy, A. "Geometrical imperfection and adiabatic shear banding" *Int. J. Impact Eng.* **35** 1280–1292.

2008, Sun, K., Cheng, X.W., Wang, F.C., Miao, P. and Zhao, S.Z. "Study on sensitivity of adiabatic shear of TC6 alloy under different strain rate and morphology (in Chinese)" *Rare Met. Mater. Eng.* **37** 1856–1860.

2008, Tan, C.W., Liu, X.Q., Chen, Z.Y., Ma, H.L., Wang, F.C. and Cai, H.N. "Study on the relationship between adiabatic shear susceptivity and critical fracture velocity for Ti6Al4V alloy (in Chinese)" *Rare Met. Mater. Eng.* **37** 1400–1402.

2008, Wang, B.F. and Yang, Y. "Microstructure evolution in adiabatic shear band in fine-grain-sized Ti–3Al–5Mo–4.5V alloy" *Mater. Sci. Eng. A* **473** 306–311.

2008, Wang, B.F. "Adiabatic shear band in a Ti–3Al–5Mo–4.5V titanium alloy" *J. Mater. Sci.* **43** 1576–1582.

2008, Wu, G.H., Zhu, D.Z., Chen, G.Q., Jiang, L.T. and Zhang, Q. "Adiabatic shear failure of high reinforcement content aluminum matrix composites" *J. Mater. Sci.* **43** 4483–4486.

2009, Bassim, M.N. and Odeshi, A.G. "Microstructural model for shear strain localization and occurrence of adiabatic shear band at high strain rates", in "Proc. 9th Int. Conf. on the Mechanical and Physical Behaviour of Materials under Dynamic Loading (DYMAT 2009)", pp. 935–940, Les Ulis, France, EDP Sciences.

2009, Dawson, E. "Meteorite impact shatter cones: Adiabatic shear bands?", in "Proc. 9th Int. Conf. on the Mechanical and Physical Behaviour of Materials under Dynamic Loading (DYMAT 2009)", pp. 1471–1477, Les Ulis, France, EDP Sciences.

2009, Duan, C.Z., Cai, Y.J., Wang, M.J. and Li, G.H. "Microstructural study of adiabatic shear bands formed in serrated chips during high-speed machining of hardened steel" *J. Mater. Sci.* **44** 897–902.

2009, Ji, W., Fan, Y.F., Chen, J. and Qiao, G.L. "Characteristics of the adiabatic shear bands in Mg−10Gd−2Y−0.5Zr alloy (in Chinese)" *Rare Met. Mater. Eng.* **38** 599−602.

2009, Lee, W.S., Liu, C.Y. and Chen, T.H. "Adiabatic shear localization of different steels at high strain rates", in "Proc. 9th Int. Conf. on the Mechanical and Physical Behaviour of Materials under Dynamic Loading (DYMAT 2009)", pp. 1117−1123, Les Ulis, France, EDP Sciences.

2009, Liu, X., Tan, C., Zhang, J., Hu, Y., Ma, H., Wang, F. and Cai, H. "Influence of microstructure and strain rate on adiabatic shearing behavior in Ti6Al4V alloys" *Mater. Sci. Eng. A* **501** 30−36.

2009, Liu, X.Q., Tan, C.W., Zhang, J., Wang, F.C. and Cai, H.N.A. "Correlation of adiabatic shearing behavior with fracture in Ti6Al4V alloys with different microstructures" *Int. J. Impact Eng.* **36** 1143−1149.

2009, Longère, P., Dragon, A. and Deprince, X. "Numerical study of impact penetration shearing employing finite strain viscoplasticity model incorporating adiabatic shear banding" *Trans. ASME: J. Eng. Mater. Technol.* **131** 011105.

2009, Lovinger, Z., Rikanati, A., Rittel, D. and Rosenberg, Z. "Investigation of adiabatic shear bands in thick-walled cylinders collapsed by electromagnetic driving forces", in "Shock Compression of Condensed Matter − 2009", M.L. Elert, W.T. Buttler, M.D. Furnish, W.W. Anderson and W.G. Proud (Eds.), pp. 739−742, Melville, NY, American Institute of Physics.

2009, Murr, L.E., Ramirez, A.C., Gaytan, S.M., Lopez, M.I., Martinez, E.Y., Hernandez, D. H. and Martinez, E. "Microstructure evolution associated with adiabatic shear bands and shear band failure in ballistic plug formation in Ti6Al4V targets" *Mater. Sci. Eng. A* **516** 205−216.

2009, Rittel, D. "A new perspective on adiabatic shear failure", in "Proc. 9th Int. Conf. on the Mechanical and Physical Behaviour of Materials under Dynamic Loading (DYMAT 2009)", pp. 955−961, Les Ulis, France, EDP Sciences.

2009, Rittel, D. and Wang, Z.G. "Thermomechanical aspects of adiabatic shear failure of AM50 and Ti6Al4V alloys", in "Proc. 9th Bienniel Conf. on Engineering Systems Design and Analysis", pp. 529−550, New York, NY, American Society of Mechanical Engineers.

2009, Rittel, D. "A different viewpoint on adiabatic shear localization" *J. Phys. D: Appl. Phys.* **42** 214009.

2009, Sun, K., Wang, F.C., Cheng, X.W., Miao, P. and Zhao, S.Z. "Fine structure of adiabatic shear band of TC6 alloy under different evolvement stages (in Chinese)" *Rare Met. Mater. Eng.* **38** 233−236.

2009, Sun, K., Wang, F.C., Cheng, X.W., Miao, P. and Zhao, S.Z. "Formation mechanics of adiabatic shear bands for the different microstructures of TC6 alloy (in Chinese)" *Rare Met. Mater. Eng.* **38** 34−37.

2009, Wang, L., Miao, P., Li, S.K., Sun, K., Cheng, X.W. and Wang, F.C. "Evolution process of adiabatic shear localization in C18 titanium alloy and 30CrMnSiNiMoV steel", in "Proc. 9th Int. Conf. on the Mechanical and Physical Behaviour of Materials under Dynamic Loading (DYMAT 2009)", pp. 1053−1059, Les Ulis, France, EDP Sciences.

2009, Yang, H., Xu, Y., Seki, Y., Nesterenko, V.F. and Meyers, M.A. "Analysis and characterization by electron backscatter diffraction of microstructural evolution in the adiabatic shear bands in Fe−Cr−Ni alloys" *J. Mater. Res.* **24** 2617−2627.

2009, Yazdani, F., Bassim, M.N. and Odeshi, A.G. "The formation of adiabatic shear bands in copper during torsion at high strain rates" *Procedia Eng.* **1** 225−228.

2009, Li, G., Wang, M. and Duan, C. "Adiabatic shear critical conditions in high-speed cutting" *J. Mater. Process. Technol.* **209** 1362–1367.

2009, Ryttberg, K., Wedel, M.K., Dahlman, P. and Nyborg, L. "Microstructural evolution during fracture induced by high strain rate deformation of 100Cr6 steel" *J. Mater. Process. Technol.* **209** 3325–3334.

2010, Fan, J.L., Gong, X., Huang, B.Y., Song, M., Liu, T., Qi, M.G., Tian, J.M. and Li, S.K. "Dynamic failure and adiabatic shear bands in fine-grain 93W–4.9Ni–2.1Fe alloy with Y_2O_3 addition under lower high strain rate compression" *Mech. Mater.* **42** 24–30.

2010, Gong, X., Fan, J.L., Tian, J.M., Huang, B.Y. and Li, S.K. "Formation and characteristics of adiabatic shear band in fine-grained 93W–4.9Ni–2.1Fe alloy under dynamic compression (in Chinese)" *Rare Met. Mater. Eng.* **39** 1390–1394.

2010, Lee, W.S., Chen, T.H., Lin, C.F. and Lu, G.T. "Adiabatic shearing localisation in high strain rate deformation of aluminium–scandium alloy" *Mater. Trans.* **51** 1216–1221.

2010, Lee, W.S., Chen, T.H. and Lu, G.T. "Adiabatic shearing behaviour of unweldable aluminium–scandium alloy under extremely high shear loading" *Mater. Sci. Technol.* **26** 720–725.

2010, Li, J.C., Chen, X.W. and Chen, G. "Numerical simulations on adiabatic shear behaviour of 921A steel pure shear hat-shaped specimens", in "Structures Under Shock and Impact XI", N. Jones, C.A. Brebbia and U. Mander (Eds.), pp. 323–334, Southampton, UK, WIT Press.

2010, Li, J.C., Li, S.K., Fan, Q.B. and Pei, C.H. "Effects of forging processing on adiabatic shearing sensitivity of Ti6Al4VMo alloy (in Chinese)" *Rare Met. Mater. Eng.* **39** 60–63.

2010, Li, D.H., Yang, Y., Xu, T., Zheng, H.G., Zhu, Q.S. and Zhang, Q.M. "Observation of the microstructure in the adiabatic shear band of 7075 aluminum alloy" *Mater. Sci. Eng. A* **527** 3529–3535.

2010, Love, B.M. and Batra, R.C. "Effect of particulate/matrix debonding on the formation of adiabatic shear bands" *Int. J. Mech. Sci.* **52** 386–397.

2010, McVeigh, C. and Liu, W.K. "Multiresolution continuum modeling of micro-void assisted dynamic adiabatic shear band propagation" *J. Mech. Phys. Solids* **58** 187–205.

2010, Pushkov, V., Yurlov, A., Bolshakov, A., Podurets, A., Kalmanov, A. and Koshatova, E. "Study of adiabatic localized shear in metals by split Hopkinson pressure bar method" *EPJ Web Conferences* **10** 00029.

2010, Rosenberg, Z., Ashuach, Y. and Kreif, R. "The effect of specimen dimensions on the propensity to adiabatic shear failure in Kolsky bar experiments" *Revista Matéria* **15** 283–290.

2010, Wang, L., Li, S.-K., Li, Y.-K., Hu, X.-F. and Wang, F.-C. "Influence of adiabatic shear sensitivities of steel projectiles on their ballistic performances", in "Proc. 25th International Symposium on Ballistics", pp. 1253–1259, Beijing, China, International Ballistics Society.

2010, Zhen, L., Zou, D.L., Xu, C.Y. and Shao, W.Z. "Microstructure evolution of adiabatic shear bands in AM30B magnesium alloy under ballistic impact" *Mater. Sci. Eng. A* **527** 5728–5733.

2010, Zhou, X., Li, S., Liu, J., Wang, Y. and Wang, X. "Self-sharpening behavior during ballistic impact of tungsten heavy alloy penetrators processed by hot-hydrostatic extrusion and hot torsion" *Mater. Sci. Eng. A* **527** 4881–4886.

2011, Atapek, S.H. and Karagoz, S. "Ballistic impact behaviour of a tempered bainitic steel against 7.62 mm armour piercing projectile" *Defence Sci. J.* **61** 81–87.

2011, Boakye-Yiadom, S. and Bassim, M.N. "Effect of prior heat treatment on the dynamic impact behavior of 4340 steel and formation of adiabatic shear bands" *Mater. Sci. Engng A* **528** 8700–8708.

2011, Demir, T. and Ubeyli, M. "A micro-damage investigation on a low alloy steel tested using a 7.62 mm AP projectile" *Materiali Tehnologije* **45** 115–120.

2011, Dhokia, V.G., Newman, S.T., Crabtree, P. and Ansell, M.P. "Adiabatic shear band formation as a result of cryogenic CNC machining of elastomers" *Proc. Inst. Mech. Engrs Part B: J. Engng Manuf.* **225** 1482–1492.

2011, Duan, C., Yu, H., Li, H. and Wang, M. "Some metallurgical observations of adiabatic shear phenomenon during high speed machining AISI 1045 hardened steel" *Adv. Sci. Letts* **4** 2493–2497.

2011, Duan, C.Z., Zhang, L.C., Yu, H.Y. and Wang, M.J. "An investigation into the microstructure of adiabatic shear banding in AISI 1045 hardened steel due to high speed machining" *Adv. Mater. Res.* **154** 321–324.

2011, Erice, B., Gálvez, F., Cendón, D.A., Sánchez-Gálvez, V. and Børvik, T. "An experimental and numerical study of ballistic impacts on a turbine casing material at varying temperatures" *Trans. ASME: J. Appl. Mech.* **78** 051019.

2011, Grady, D.E. "Adiabatic shear localization in brittle solids" *Int. J. Impact Eng.* **38** 661–667.

2011, Guo, W., Liu, J., Yang, J. and Li, S. "Effect of initial temperature on dynamic recrystallization of tungsten and matrix within adiabatic shear band of tungsten heavy alloy" *Mater. Sci. Engng A* **528** 6248–6252.

2011, Huo, D.M., Li, S.K., Fan, Q.B. and Wang, F.C. "Effects of electric pulse heat treatment on microstructures and dynamic deformation behaviors of Ti6441 alloys" *Mater. Sci. Engng A* **530** 161–167.

2011, Karpat, Y. "Temperature-dependent flow softening of titanium alloy Ti6Al4V: An investigation using finite element simulation of machining" *J. Mater. Process. Technol.* **211** 737–749.

2011, Li, N., Wang, Y.D., Peng, R.L., Sun, X., Liaw, P.K., Wu, G.L., Wang, L. and Cai, H.N. "Localized amorphism after high strain rate deformation in TWIP steel" *Acta Mater.* **59** 6369–6377.

2011, Liu, J.X., Li, S.K., Zhou, X.Q., Wang, Y.C. and Yang, J. "Dynamic recrystallization in the shear bands of tungsten heavy alloy processed by hot hydrostatic extrusion and hot torsion (in Chinese)" *Rare Metal Mater. Engng* **40** 957–960.

2011, Lovinger, Z., Rikanati, A., Rosenberg, Z. and Rittel, D. "Electromagnetic collapse of thick-walled cylinders to investigate spontaneous shear localization" *Int. J. Impact Engng* **38** 918–929.

2011, Mondal, C., Mishra, B., Jena, P.K., Kumar, K.S. and Bhat, T.B. "Effect of heat treatment on the behavior of an AA7055 aluminum alloy during ballistic impact" *Int. J. Impact Engng* **38** 745–754.

2011, Oussouaddi, O., Daridon, L., Ahzi, S. and Chrysochoos, A. "Influence of dissipated energy on shear band spacing in HY-100 steel" *Trans. ASME: J. Engng Mater. Technol.* **133** 021002.

2011, Peirs, J., Verleysen, P., Tirry, W., Rabet, L., Schryvers, D. and Degrieck, J. "Dynamic shear localization in Ti6Al4V" *Procedia Engng* **10** 2343–2347.

2011, Perez-Bergquist, S.J., Gray III, G.T., Cerreta, E.K., Trujillo, C.P. and Perez-Bergquist, A. "The dynamic and quasistatic mechanical response of three aluminum armor alloys: 5059, 5083, and 7039" *Mater. Sci. Engng A* **528** 8733–8741.

2011, Schuster, B.E., Ligda, J.P., Pan, Z.L. and Wei, Q. "Nanocrystalline refractory metals for extreme condition applications" *JOM* **63**(12), 27–31.

2011, Wang, H., To, S., Chan, C.Y., Cheung, C.F. and Lee, W.B. "Dynamic modelling of shear band formation and tool-tip vibration in ultra-precision diamond turning" *Int. J. Mach. Tools Manuf.* **51** 512–519.

2011, Wang, X.B. "Adiabatic shear sensitivity of ductile metal based on gradient-dependent Johnson-Cook model" *Trans. Nonferrous Metals Soc. China* **21** 1355–1361.

2011, Wu, Z., Wu, G.H., Kang, P.C., Gou, Q., Chen, G.Q. and Jiang, L.T. "High temperature fracture behavior of tungsten-fiber-reinforced copper matrix composites under dynamic compression" *Mater. Design* **32** 5022–5026.

2011, Xu, Y., Sun, K. and Yang, Y.A. "Refining mechanisms of grains in the adiabatic shear band of TC18 titanium alloy (in Chinese)" *Rare Metal Mater. Engng* **40** 1454–1457.

2011, Yang, D., An, Y., Cizek, P. and Hodgson, P. "Development of adiabatic shear band in cold-rolled titanium" *Mater. Sci. Engng A* **528** 3990–3997.

2011, Yang, Y., Jiang, F., Zhou, B.M., Li, X.M., Zheng, H.G. and Zhang, Q.M. "Microstructural characterization and evolution mechanism of adiabatic shear band in a near beta-titanium alloy" *Mater. Sci. Engng A* **528** 2787–2794.

2011, Yang, Y., Li, X.M., Tong, X.L., Zhang, Q.M. and Xu, C.Y. "Effects of microstructure on the adiabatic shearing behaviors of titanium alloy" *Mater. Sci. Engng A* **528** 3130–3133.

2011, Yang, Y., Zheng, H.G., Zhao, Z.D., Zhang, Q., Zhang, Q.M., Jiang, F. and Li, X.M. "Effect of phase composition on self-organization of shear bands in Ti-1300 titanium alloy" *Mater. Sci. Engng A* **528** 7506–7513.

2011, You, Z.P., Mi, X.J., Hui, S.X., Ye, W.J., Yu, Y. and Wang, B. "Microstructure evolution in adiabatic shear band in Ti-5Mo-5V-2Cr-3Al alloy (in Chinese)" *Rare Metal Mater. Engng* **40** 1184–1187.

2011, Yuan, F.P. and Wu, X.L. "Evolution of adiabatic shear band in ultra-fine-grained iron under dynamic shear loading" *Mater. Sci. Forum* **667** 761–765.

2011, Zhang, J., Tan, C.W., Ren, Y., Yu, X.D., Ma, H.L., Wang, F.C. and Cai, H.N. "Adiabatic shear fracture in Ti6Al4V alloy" *Trans. Nonferrous Metals Soc. China* **21** 2396–2401.

2011, Zhang, W.G., Ma, M.Z., Song, A.J., Liang, S.X., Hao, Q.H., Tan, C.L., Jing, Q. and Liu, R.P. "Temperature rise and flow of zirconium-based bulk metallic glasses under high shearing stress" *Science China: Phys. Mech. Astron.* **54** 1972–1976.

2011, Zhu, D.Z., Chen, W.P., Li, Y.Y. and Wu, G.H. "Adiabatic shearing failure mechanism of aluminum matrix composites (in Chinese)" *Rare Metal Mater. Engng* **40** 56–59.

2011, Zou, D.L., Zhen, L., Xu, C.Y. and Shao, W.Z. "Characterization of adiabatic shear bands in AM60B magnesium alloy under ballistic impact" *Mater. Charact.* **62** 496–502.

2012, Alkhader, M. and Bodelot, L. "Large strain mechanical behavior of HSLA-100 steel over a wide range of strain rates" *Trans. ASME: J. Appl. Mech.* **134** 011005.

2012, Duan, C.Z. and Zhang, L.C. "Adiabatic shear banding in AISI 1045 steel during high speed machining: Mechanisms of microstructural evolution" *Mater. Sci. Engng A* **532** 111–119.

2012, Ghomi, H.M. and Odeshi, A.G. "The effects of microstructure, strain rates and geometry on dynamic impact response of a carbon-manganese steel" *Mater. Sci. Engng A* **532** 308–315.

2012, Liu, J.X., Yang, J., Zhou, J.Y., Li, S.K. and Guo, W.Q. "Adiabatic shear banding of hot-extruded tungsten heavy alloy under cryogenic temperature" *Rare Metals* **31** 17–21.

2012, Mishra, B., Jena, P.K., Ramakrishna, B., Madhu, V., Bhat, T.B. and Gupta, N.K. "Effect of tempering temperature, plate thickness and presence of holes on ballistic impact behavior and ASB formation of a high strength steel" *Int. J. Impact Eng* **44** 17–28.

2012, Osovski, S., Rittel, D., Landau, P. and Venkert, A. "Microstructural effects on adiabatic shear band formation" *Scripta Mater.* **66** 9–12.

2012, Padilla, H.A., Lambros, J., Beaudoin, A.J. and Robertson, I.M. "Relating inhomogeneous deformation to local texture in zirconium through grain-scale digital image correlation strain mapping experiments" *Int. J. Solids Structures* **49** 18–31.

2012, Sabih, A. and Nemes, J.A. "Experimental and finite element simulation study of the adiabatic shear band phenomenon in cold heading process" *J. Mater. Process. Technol.* **212** 1089–1105.

2012, Tang, L., Chen, Z.Y., Zhan, C.K., Yang, X.Y., Liu, C.M. and Cai, H.N. "Microstructural evolution in adiabatic shear bands of copper at high strain rates: Electron backscatter diffraction characterization" *Mater. Charact.* **64** 21–26.

2012, Wan, Z.P., Zhu, Y.E., Liu, H.W. and Tang, Y. "Microstructure evolution of adiabatic shear bands and mechanisms of saw-tooth chip formation in machining Ti6Al4V" *Mater. Sci. Engng A* **531** 155–163.

2012, Wang, L.L., Chen, D., Yang, L., Zhou, F. and Dong, X. "Nonlinear wave propagations in solids and the correlated dynamic behavior of materials: An overview of the related research works by WLL group in China" *Rev. Adv. Mater. Sci.* **30** 27–59.

2012, Yang, Y., Jiang, F., Yang, M., Li, X.M., Zheng, H.G. and Zhang, Q.M. "Electron backscatter diffraction analysis of strain distribution in adiabatic shear band and its nearby area in Ti-3Al-5Mo-4.5V alloy" *Mater. Sci. Technol.* **28** 165–170.

2012, Zhang, Q.H., Li, B.L., Chen, X., Yin, J.M., Nie, Z.R. and Zuo, T.Y. "Characteristic microstructure and microstructure evolution in Al-Cu-Mn alloy under projectile impact" *Mater. Sci. Engng A* **531** 12–17.

FSC
www.fsc.org
MIX
Papier | Fördert gute Waldnutzung
FSC® C083411

Kolibri 360 GmbH
Ferdinand-Jühlke-Straße 7
99095 Erfurt, Deutschland
produktsicherheit@kolibri360.de